Human-centred Systems

Springer
Berlin
Heidelberg
New York
Barcelona
Budapest
Hong Kong
London
Milan
Paris
Tokyo

Karamjit S. Gill (Ed.)

Human Machine Symbiosis

The Foundations of Human-centred Systems Design

Springer

Karamjit S. Gill, BA (Hons), MA, MSc, Dphil
SEAKE Centre, Department of Library and
Information Studies, University of Brighton
Falmer, Brighton, BN1 9PH, UK

ISBN-13:978-3-540-76024-5

British Library Cataloguing in Publication Data
Human machine symbiosis : the foundations of human centred systems design
1. System design 2. Artificial intelligence
I. Gill, Karamjit S.
004 .2
ISBN-13:978-3-540-76024-5

Library of Congress Cataloging-in-Publication Data
Gill, Karamjit S.
Human machine symbiosis : the foundations of human centred systems design / Karamjit S. Gill.
p. cm.
Includes bibliographical references.
ISBN-13:978-3-540-76024-5 e-ISBN-13:978-1-4471-3247-9
DOI: 10.1007/978-1-4471-3247-9
1. Man-machine systems. 2. System design. 3. Human engineering.
I. Title.
TA166.G55 1996
004.2'1'019--dc20 96-4806

Typeset from author's disc by ReadyText, Bath

34 / 3830-543210 Prin

Preface

There is now a serious discussion taking place about the moment at which human beings will be surpassed and replaced by the machine. On the one hand we are designing machines which embed more and more human intelligence, but at the same time we are in danger of becoming more and more like machines. In these circumstances, we all need to consider:

- What can we do?
- What should we do?
- What are the alternatives of doing it?

This book is about the human-centred alternative of designing systems and technologies. This alternative is rooted in the European tradition of human-centredness which emphasises the symbiosis of human capabilities and machine capacity. The human-centred tradition celebrates the diversity of human skill and ingenuity and provides an alternative to the 'mechanistic' paradigm of 'one best way', the 'sameness of science' and the 'dream of the exact language'.

This alternative vision has its origin in the founding European human-centred movements of the 1970s. These include the British movement of Socially Useful Technology, the Scandinavian movement of Democratic Participation, and the German movement of Humanisation of Work and Technology. The present volume brings together various strands of human-centred systems philosophy which span the conceptual richness and cultural diversity of the human-centred movements. The core ideas of human-centredness include human–machine symbiosis, the tacit dimension of knowledge, the system as a tool rather than a machine, dialogue, participation, social shaping and usability. These ideas have become central to various design methodologies, ranging from the socio-technique, social ergonomics, user-centred design, user-involved design, and more recently, computer-supported cooperative working, informated work environments, human–computer relations, multimedia information systems and cognitive technology.

With the increasing integration of information technology into worklife, social systems and organisations, we are witnessing a growing mismatch of technological and social systems, brittleness of human–computer relations, vulnerability of social systems and instability of cultural systems.

Technology and systems rooted in the 'causal' view of science tend to exclude 'human purpose' thereby widening the mismatch between technology and society; the 'computational model' of cognitive science decouples 'intelligence from things uniquely human'; professionals bound by technical rationality face a crisis of confidence in coping with the 'complexity, instability, uncertainty, uniqueness and value conflicts'; the limit of the 'rule'-based system and the notion of 'separation' of the object from the subject symbolise the impoverishment of techno-centric approaches.

At the same time there is now an increasing recognition of the limitations of technology and the potential of human-centred systems. The human-centred tradition transcends the limitations of the techno-centric systems through concepts such as social responsibility, sustainability, social cohesion, ethics, equality, creativity, human dignity and the environment. Mike Cooley's book, *Architect or Bee?* (1987), Howard Rosenbrock's book, *Machines with a Purpose* (1990), Weizenbaum's book, *Computer Power and Human Reason* (1976), Karamjit Gill's book, *AI For Society* (1986), and Hubert Dreyfus' book on *What Computers Can't Do* (1972), have provided resource material for students and researchers. Karamjit Gill's report on *European Research in Human Centredness*, published by RISS (NTT DATA, Tokyo, 1990), has become an essential reference text for undergraduate and postgraduate students and researchers involved in interdisciplinary studies in human-centred systems design.

The present volume adds to the growing literature that approaches technology and systems design from a human-centred perspective such as:

- *Designing Human Centred Technology*, (Rosenbrock, 1990).
- *Work-oriented Design of Computer Artifacts*, (Ehn, 1988).
- *Design at Work*, (Greenbaum and Kyng, 1991).
- *Understanding Computers and Cognition*, (Winograd and Flores, 1986).
- *Plans and Situated Actions*, (Suchman, 1987).
- *Usability*, (Alder & Winograd, 1992).
- *The Reflective Practitioner*, (Schön, 1983).
- *Knowledge, Skill and Artificial Intelligence*, (Göranzon and Josefson, 1988).
- *Understanding the Artificial*, (Negrotti, 1990).

This volume is unique in the human-centred systems literature in that it spans the diversities of human-centredness to provide a comprehensive text and reference book on human-centred traditions, approaches, methodologies and practices.

The openness of human-centredness to the ideas of diversity and human purpose makes this book accessible to students, researchers and practitioners from a variety of disciplines such as information and communication systems, computer science, social sciences, humanities, science and technology policy, design studies, education, engineering, management science and business studies. The aim is to provide theoretical and methodological underpinnings for the design and applications of technological systems from a human-centred perspective, and to provide human-centred frameworks and models for designing and shaping information, communication and multimedia technologies. The book should also be of interest to researchers and practitioners who are concerned with wider social, cultural, philosophical and ethical issues of information society. Contributions in this volume emphasise multidisciplinary perspectives and reflect the diversity of fields and experiences of the contributors.

Authors in this volume have collaborated over the years in postgraduate education and doctoral programmes in human-centred systems, as well as in European projects, international research programmes and debates on technology and society. They have initiated and undertaken European and international projects, including the formation of the European Community Programme on Anthropocentric Systems, the establishment of the European inter-university postgraduate studies and doctoral programme in human-centred information systems, and the Europe–Japan collaboration in human-centred systems. Collaboration also includes the setting up of the international journal of *AI & Society* and its sister book series on human-centred systems, both promoting and stimulating debates in human-centred systems and information society.

Mike Cooley and Howard Rosenbrock have pioneered the British tradition of human-centredness and have influenced the shaping of European traditions. Lauge Rasmussen and Siv Friis have been the key researchers in bridging the Scandinavian and other European research traditions in human-centredness. Lars Qvortrup, Eunice McCarthy and Karamjit Gill extend the human-centred debate from its roots in manufacturing to the wider worklife, social and cultural spheres. Satinder Gill and Felix Schmid provide a bridge between theories of knowledge, cognitive science and learning theories. Francesco Garibaldo and David Smith situate human-centredness in the domain of industrial and service sectors in the emerging information society. The text can be read as two sections, Theory and Praxis of Human Centred Systems, and the Methodology of HCS design. The first section includes chapters on foundation principles, philosophy and praxis of human-centred systems. The second section presents methodologies of human-centred design, applications and implications of human-centred systems. Both the sections complement each other and provide a reference text for interdisciplinary studies and research into human-centred systems.

The Foundation Principles

In the opening chapter – *The Foundations of Human-centred Systems* – Gill introduces core concepts, methodologies, approaches, traditions and challenges of human-centredness. He argues that human-centred approaches such as dialogue, social action, participation, user-involved design, social shaping and valorisation of diversity provide a methodological framework for shaping information and communication networks which produce, reproduce and sustain a variety of skill and knowledge bases. The challenge is to formulate a research agenda for technological environments which recognise the dynamic relationship of local specificity and global diversity.

In his chapter, *On Human–Machine Symbiosis*, Cooley argues that the relationship between tacit knowledge, common sense and skill has been essential in the European traditions of design practised by the craftsman, originating during the Renaissance period. With the increasing use of information based technology in working life, there is a danger that technology will usurp those skills currently held by people in the name of efficiency and scientific progress. This in turn will affect the way we think of working practice and educational needs.

The subject of chapter 3 – *Rosenbrock's Account of Causality and Purpose* – a compilation of Rosenbrock's work, deals with two alternative views of the world, the 'causal' view and the 'purposive' view. The causal view is rooted in the scientific thought and belief that 'man is a machine', and that he can therefore contribute nothing which cannot be contributed just as well by the machine. In the alternative purposive view of science, technology becomes subordinate to human purpose. Human purposes can never be incorporated in technology in the way that they exist in people.

In *The Social Construction of Human-centredness*, Qvortrup presents historical roots of human-centredness. He argues that human-centredness is not an invention of our current times, but is deeply rooted in the European cultural tradition of 15th century Italian renaissance. Historically, there has always been two different concepts of 'human-centredness'. A dogmatic one, where organisational systems and technologies were developed according to an unambiguous idea of the nature of the human being, and an ambiguous one, which gave room to different interpretations, and to playing tricks.

In her chapter, *Culture, Mind and Technology: Making a Difference*, McCarthy provides an insight into theories and methods of communication and culture, complexity, chaos and organisational change, and network cultures. She notes an emerging trend towards the humanisation of technology and the impetus for change arising from the vulnerability of complex technological systems and dysfunctioning of the human–computer interfaces and increasing risk and vulnerability.

Methodological Issues

In *Human-centred Methods of Social and Technical Design*, Rasmussen introduces the human-centred approach as an individual and collective learning process. In essence, human-centred design is fundamentally a process of instrumental reflection and utopian processes confronting and reuniting the view points and practices of different design professions and practitioners towards a more sustainable and dynamic balance.

In *Information Systems Design: a User-Involved Perspective*, Friis sees users and designers of information as belonging to two cultures, and believes that dialogue between cultures needs to be part of systems design, in the sense of Freire's approach to pedagogy. She emphasises that the Scandinavian experimental approach of information systems design allows users to take responsibility in the design process, and regards users as both the problem owners and the problem solvers.

In *Designing for Knowledge Transfer*, Satinder Gill emphasises that design is about the communication of knowledge. She argues that the non person-centred perspective is inadequate as an explanation for knowledge and skill and that a person-centred perspective, rooted in practice, as an alternative framework provides a better explanation of the nature of knowledge and skill.

In *Designing Practice-Based Learning Environments*, Schmid emphasises that practice-based learning is about finding a symbiosis between practical (tacit) knowledge and general knowledge. The emphasis is on the creation of a balanced person in the sense of the renaissance period which signifies the mid-point between the historical shift from gnosis to cognis.

In *Workplace Innovations: the Making of a Human-centred Industrial Culture*, Garibaldo argues that sustainable innovation requires a dynamic equilibrium between flexibility and cooperation of group and team working on the one hand and the collective power of self-regulation on the other hand. The challenge is how can this group commitment and personal trust fit into Western concepts of democratic participation based on the individual.

Lastly, in *Meanwhile, Out in the Real World: Developing a Commercial Human-centred Software Application*, Smith raises some of the challenges, problems and compromises involved in bringing a design concept out of quasi-academic research. He emphasises that it is important not to regard devices such as computer systems in isolation. Human factors are critically important in determining the extent to which a technical innovation can be transferred into routine practice.

This book is a contribution to the evolution of human-centred systems tradition in Europe which has taken place over the last 25 years. This evolution has included research contributions of the European Research Network in Human-Centred Systems, the work of the Japanese research group in human-centred systems supported by the RISS (NTT Data, Tokyo), research into Anthropocentric Systems supported by the FAST (European Community), and research into human-centred systems and social innovation undertaken by the SEAKE Centre, Department of Library and Information Studies at the University of Brighton.

I owe special gratitude to Mike Cooley, Massimo Negrotti, Howard Rosenbrock and David Smith for supporting the work of The SEAKE Centre; Yuji Masuda, Fumihiko Satofuka and Yoshihiro Sato for their support for SEAKE-Centre–Japan research links; Satinder Gill, Bob Muller and Jeremy Potter for their cajoling and inspiration to bring out this book; the SEAKE Centre researchers over the past few years, Tania Funston, Masao Hijikata, Takao Nuki, Sean Smith, John Gøtze, and Jim Thorpe for their contribution to the human-centred movement at Brighton. I give special thanks to John Watson, Beverley Ford, David Anderson, Christiane Notarmarco of Springer-Verlag, London for their support of the *AI & Society* journal, its sister book series in human-centred systems, and the publication of this book. Above all I thank the contributors for their on-going support for the human-centred tradition.

Brighton
May, 1996

Karamjit S. Gill

Contents

Contributors

Mike Cooley

Mike Cooley studied engineering in Germany, Switzerland and England and has a PhD in computer-aided design. He has held director-level design and technical management posts in the public and private sectors and is a director of a number of high-tech companies and of the EU sponsored Technology Exchange. He is Chairman of TIA and a consultant to a number of governments and international companies. He was one of the initiators of the seminal Alternative Strategic Plan at Lucas Aerospace and is an international authority on the design and introduction of socially useful, sustainable products and processes. He is helping to implement job-creation programmes based on such products in a number of countries and also undertakes a wide range of unpaid voluntary work.

He has published over 120 scientific papers and is author or joint author of 15 books in English and German on technology and its consequences. His work has been translated into over 20 languages from Finnish to Japanese. He has been a Visiting Professor at universities throughout Europe, Australia, the USA and Japan. The Japanese edition of his book *Architect or Bee?* (Chatto '87), appeared in 1990 and there are German, Irish and Swedish editions.

Mike Cooley is a member of the Joint EU/Japanese Commission on New Technologies. He was the initiator of the successful EEC ESPRIT Project 1217 to design and build a human-centred advanced manufacturing system. He was the first to use the term 'human-centred. He was Chairman of the EEC/FAST Expert Committee which produced the report: *European Competitiveness in the 21st Century: The Integration of Work Culture and Technology*. He directed the EEC/DELTA 'Artisan Project' – the design of a multimedia learner workstation for industrial environments. His software work includes tools for 'Learning Earning Organisations' and utilises his 'Curiosity Generators'.

Mike Cooley is President of the International Research Institute in Human Centred Systems. He writes for publications world-wide, broadcasts extensively and has made several films on high technology and its implications. His awards and distinctions include the Keys of the City of Osaka, the Freedom of the cities of Dublin and Detroit and the $50,000 Right Livelihood Award (the Alternative Nobel Prize) which he donated to Socially Useful Production.

Siv Friis

Siv Friis has a PhD and is an associate professor in the Department of Informatics, Lund University, Sweden. Her work concerns both research and design. Her focus is on approaches to work organisational change, mainly towards computer-based information systems design. For this purpose, the PROTEVS approach was developed with participating white-collar and blue-collar workers in several empirical tests. The approach proposes that the two cultures of users and data-processing experts should establish a code between them, and work in dialogue with one another in 'local design shops. The research concerns organisational consequences of user-oriented information systems design, e.g., the integration of organisational learning and quality of work systems, and the design of supportive measures for data-processing experts to facilitate user-controlled information systems design.

She is concerned with the study of the role of information technologies in supporting the process of local and global dissemination and transfer of local praxis. The aim is to realise the theory and praxis of participatory design from industrial design areas to other areas of working life.

Francesco Garibaldo

Francesco Garibaldo is director of IRES (Instituto Riccerche Economiche e Sociali, Roma). Over the years, he has coordinated research between IRES and the Research Institute of the Brazilian CUT on processes of economic integration in the Southern Cone of Latin America in comparison with that in Europe; IRES–FIOM research on the integrated factory of the FIAT plants; CGIL research in Emilia Romagna where a scientific committee analysed the problem of the relationship between quality of work and firm efficiency in very small enterprises; IRES–FIOM research on working-class identity in new working realities. He has prepared one of the official reports of the CGIL's Programme Conference in Chianciano in 1994, and has organised a permanent FORUM in Italy and the Information Society.

He has also coordinated a study, carried out by a European research group: Brite-Euram, whose theme was the role of people in global world production and a study on the 'modernisation of European industry' conducted by a group of European researchers. He is currently organising several research–action networks on a European scale and is carrying out two research projects: on work organisation in the FIAT plants at Mirafiori and Melfi and in a sample of its supplier firms, and the economical and industrial consequences of information communication technology on Italian society.

He edits the Study Book Series of Ediesse publishers. His has published a number of reports and papers on the impact of technology in worklife. He is a member of the Italian Association of Sociology, a member of the Consiglio Nazionale dell'Economia e del Lavoro, collaborates with Sociology Department, University of Bologna, and is a member of the association for State Reform (CRS).

Karamjit S. Gill

Karamjit S. Gill is Founding Director of the SEAKE Research Centre at the University of Brighton, and is Honorary Professor of Human Centred Systems at University of Urbino, Italy. He studied mathematics and computer science and has a DPhil in Applied Sciences from University of Sussex. He is directing research into human-centred systems and technological innovations at the SEAKE Centre. His research activities are in the area of information, communication and media technologies with a particular focus on human-centred information systems design, information society and social innovation, knowledge networking for co-development and social cohesion, and socially sustainable technological environments. He has published more than 80 academic papers and book chapters, and has edited *AI For Society* (1986), and *New Visions of the Post-industrial Society* (1985).

His international research involvements include international and European projects on New Technology and Adult Literacy (EC); Computer Aided Animated Arts Theatre (CAAAT) Project; Transcultural and Transnational Knowledge Transfer (Italy); Culture, Language and Artificial Intelligence (EC/Sweden); Impact of Expert Systems in Production and Services on Qualifications and Working Life (ILO); and Human Centred CIM Systems (EC); Human Centred Systems Research in Europe (Japan), and the SUCCESS Group (Sustainable Competitive Change for European SMEs); the International Research Programme on Society, Culture and Technology (S-CAAAT); the Eurotecnet (EC) study on Emerging Patterns of Qualifications and Learning in Modern Manufacturing Industries, and a European Study on Team Working (IRES, Italy). He is the chairman of a European inter-university research network in human-centred systems, and is coordinating the development of a European postgraduate and a PhD programme in human-centred systems (ERASMUS, EU). He is the Editor of the international journal *AI & Society*, the series Editor of the *Human Centred Systems Book Series*, both published by Springer-

Verlag, London. He is the international coordinator of the International Institute on Human Centred Systems, and over the years has acted as a chair of international conferences on artificial intelligence and society and workshops on human-centred systems held in the UK, Italy, USA, Japan, Sweden, India and Switzerland, including the chair of the recent international conference on 'New Visions of the Post-Industrial Society' held at University of Brighton in July 1994.

Satinder P. Gill

Satinder P. Gill graduated from the University of Keele in 1987 in Philosophy, Politics and Economics. She spent the subsequent year as a research assistant with the Swedish Centre for Working Life, working with Bo Göranzon and Ingela Josefson. During this time she undertook a project comparing cultural traditions of design of Britain and Scandinavia. She has taken her interests in the Scandinavian and other European traditions of design much further in her PhD research at the University of Cambridge. Her research was upon 'Tacit Knowledge and Dialogue for Knowledge Transfer'. She received her PhD in 1995. Her fundamental interest is to investigate the communication (acquisition) of knowledge. She is currently a postdoctoral research fellow at the University of Lancaster where she is working on a project on aesthetic practice and the development and use of information technology within the domain of architecture. Her interests also cover the area of the communication (acquisition) of knowledge across the Internet and via the use of multimedia communications technology. She has been the Editorial Assistant on the internationally refereed Journal, *AI & Society*, since 1987.

Lauge Baungaard Rasmussen

Lauge Rasmussen is a sociologist and associate professor at the Institute of Technology and Social Sciences, Technical University of Denmark. His special fields of interests are: learning cultures, organisation theory, methods of participation and communication. He was project leader of the Danish part of the ESPRIT project: 'Human Centred CIM Systems' (1986–1989), and project leader of the Danish part of the FAST project 'Anthropocentric production systems' (1990–91).

He has been a member of the international research network CAPIRN since 1990, and is a member of the ERASMUS Network in Human Centred Systems. He is an author of several books about participative design processes, including *Crossing the Border* together with J. Martin Corbett and Felix Rauner (Springer-Verlag, 1991).

Howard Rosenbrock

Howard Rosenbrock was born in Ilford, England in 1920, and graduated in 1941 from University College London with a first class honours degree in Electrical Engineering. He served in the Royal Air Force Volunteer Reserve (Signals) from 1941–46, mainly on the India–Burma border. From 1947 he worked at the GEC Research Laboratories, taught high-school physics and then worked for the Electrical Research association where he took part in a wind-power survey.

In 1951 he was invited to join John Brown & Co., who were building a wind turbine on Costa Head in the Orkneys, and was later awarded a PhD for work done in industry in vibration and stability problems in large wind turbines.

In 1954 he was invited to join the John Brown subsidiary CJB where, in 1957, he was appointed Research Manager. His work at that time covered a wide range of process engineering problems, including development work in a process for producing heavy water, a liquid–liquid extraction process, high-pressure electrolysis, digital blending, high-temperature measurement, etc. From the early days of computing, when the only commercially-available computer in London was the 'Leo', he was concerned with the application of computers to chemical engineering problems – including numerical methods of optimisation, solution of stiff differential equations, digital simulation of distillation and a large number of other process engineering areas. A growing interest in control engineering from the late 1940s led, in 1962, to a move to Cambridge University where he joined John Coales's control group. In 1966 he was appointed to a chair in control engineering at UMIST, where he set up the Control Systems Centre. There, until his retirement in 1987, he worked on control theory and on the interactive use of computers for the design of control systems, and latterly in the social dimension of science and technology which after the mid-1970s clearly became problematical.

His publications include over 100 technical papers and about 30 papers on the philosophical basis of science and technology and its influence in the way technology develops and is applied, as well as seven books. He received his DSc for research in Control from London University in 1963. He has received a number of awards including an honorary doctorate of Salford University (1987), IEE Premium, Heaviside Premium and Control Achievement Award (all IEE), Control Systems Science and Engineering Award (IEEE), Moulton Medal (IChemE), Sir Harold Hartley Medal (Inst. Meas. and Cont.). He is also a fellow of the IEE and the IChemE, Hon Fellow of Inst. Meas. and Cont., Fellow of University College London, Fellow of the Royal Academy of Engineering, and Fellow of the Royal Society. He received the Oldenburger Medal of ASME in 1994 and the Nordic Press Control Award in 1995.

Felix Schmid

Born in Zurich, Switzerland, Felix Schmid graduated from the Eidgenoessische Technische Hochschule in 1976, with a degree in electrical engineering. He has worked as a computer systems analyst, locomotive and machine-tool designer, electronics researcher and as a lecturer in manufacturing engineering.

His research interests are in the areas of human-centred systems design, employee participation and control systems for railways. Felix directed the NATO Advanced Study Institute 'People and Computers', concerned with developing better human–machine interaction in the manufacturing industry. He is currently involved in researching the comparative performance of human- and computer-based railway scheduling.

David Smith

David Smith is a lecturer in multimedia communications at Gwent College of Higher Education, a University of Wales Associate College. Originally educated as a marine biologist, David Smith has followed a convoluted career track, including physiological research, lecturing in the Malawi polytechnic, teaching biology in UK schools, digging in archaeological sites and heading the Centre for Evaluation of IT in Education at the National Foundation for Educational Research. In 1991 he and a group of colleagues founded AVC Multimedia Ltd, developing commercial multimedia and data-management systems for healthcare. David is now a freelance writer and consultant. An associate of the SEAKE Centre for many years, David has held research fellowships at Huddersfield Polytechnic and Southampton University. He has published widely in both conventional and multimedia formats.

Lars Qvortrup

Lars Qvortrup is associate professor at the Centre for Cultural Studies, Odense University. Since 1977 Lars Qvortrup has written and/or edited approximately 20 books in Danish or English and a large number of articles in Danish and international readers and journals. Lars Qvortrup has lectured all over the world, including North- and South-America, Australia, China, India, and most European countries. He has been involved in a large number of studies and projects regarding the use of advanced communication for rural development and organisational, sociological and cultural impacts of telework and networking. His current study field is communication and cultural studies, focusing on organisational communication, telework, rural communication and theories of cultural studies. He is a member of the editorial board of the international journal *Cybernetics and Human Knowing*. He recently published the book *Mellem kedsomhed og dannelse – variationer over et tema af Pico*, ('Between boredom and bildung – variations on a theme by Pico'), Odense University Press, 1995.

Summaries

Chapter 1. The Foundations of Human-centred Systems

Karamjit S. Gill

Gill's opening chapter on the foundations of human-centred design provides a comprehensive synthesis and analysis of the core concepts, methodologies, approaches, traditions and challenges of human-centredness. The first part introduces scientific and intellectual challenges of human-centredness, followed by a discussion on human-centred concepts such as human–machine symbiosis, rule-following, tool perspective, dialogue, causality and purpose, language games, breakdowns, and usability. The sections on design methodology and design challenges include discussion on action research, vision-oriented design, future workshops, and CSCW; followed by a discussion on the evolution of design methodology from techno-centred design to cooperative design. The last part focuses on the diversity and similarities of the current European debates on anthropocentric systems. The chapter, in essence, summarises European research into human-centred systems over the last 25 years and emphasises its potential for shaping information and communication systems from a human-centred perspective.

The challenge of human-centredness is to extend the notion of human–machine symbiosis to varieties of human–human and human–machine relationships and networks of relationships. This means moving beyond the age of ergonomics and human factors, human–computer interaction, and cognitive performance of the individual user. Human-centred approaches such as dialogue for design, action research, vision-oriented design and future workshops provide a methodological framework for designing networks of communication for producing, reproducing and sustaining a variety of skill and knowledge bases. The challenge is to formulate a research agenda for information, communication and multimedia environments which recognises the dynamic relationship of local specificity and global diversity.

Chapter 2. On Human–Machine Symbiosis

Mike Cooley

The relationship between tacit knowledge, common sense and skill has been essential in the European traditions of design practiced by the craftsman. The chapter introduces the historical origin of the European tradition of design from the Renaissance period, and its relevance to the essence and spirit of human-centredness, human creativity, the notion of expertise, the idea of the tool, and the nature of human–machine collaboration. With the increasing use of information-based technology in working life, there is a danger that technology will usurp those skills currently held by people in the name of efficiency and scientific progress. This in turn will affect the way we think of working practice and educational needs. Cooley points to the limitations of the idea of training comparing it to that of apprenticeship. Training produces narrow, over-dedicated capabilities which are generally machine-, system- or program- specific. The practice of apprenticeship, however, was one of educating to build the common sense and tacit knowledge required to perform as an artisan, a craftsman. One's knowledge was valued.

In the human-centred system, there exists a symbiotic relation between the human and the machine in which the human would handle the qualitative subjective judgements and the machine the quantitative elements. It involves a radical design of the interface technologies and at a philosophical level the objective is to provide tools (in the Heidegger sense) which would support human skill and ingenuity rather than machines which would objectivise that knowledge.

Those involved in systems design will need to be competent in the design of adaptive tools which accord closely with traditions and practices of the domain area. Furthermore, they will need to be competent in the design of systems and organisations which display the characteristics of coherence, inclusiveness, malleability, engagement, ownership, responsiveness, purpose, being panoramic, and transcendence.

Chapter 3. Rosenbrock's Account of Causality and Purpose

A compilation of Howard Rosenbrock's works selected and annotated by Satinder P. Gill

The discussion on the causal and purposive myth in Rosenbrock's compilation on causality or purpose deals with the fundamental human-centred concerns about the mechanistic view of the world and the alternative view based on human potential and human purpose.

The causal view is rooted in the scientific thought and belief that 'man is a machine', and that he can, therefore, contribute nothing which cannot be contributed just as well by the machine. In the causal myth, the purpose effectively disappears and becomes incorporated in the machine. We are left with a device which follows certain causal laws but has no purpose and is not subject to moral judgement. The causal myth resides in the long tradition in science which accepts no explanation except those in terms of cause and effect. The result is that we regard everything outside ourselves as a machine, and a machine without purpose.

Rosenbrock proposes an alternative purposive view of science in which technology becomes subordinate to human purpose. Human purposes can never be incorporated in technology in the way that they exist in people. The purposive myth allows us to regard technology in a different light in which we are released from the tyranny of the causal view allowing us to suggest a better kind of technology, better matched to human needs and aspirations. In this view no human work should have machine-like triviality and aimlessness. This offends by equating people to automata: fulfilling a purpose appropriate only to machines. It also offends in another way by subordinating people to machines. The purposive view does not reject science but seeks to reap the benefits of science without its highly undesirable consequences. It seeks to gain insight into man-made systems with a view to producing technology which is not antagonistic to people or to their environment.

Chapter 4. Culture, Mind and Technology: Making a Difference

Eunice McCarthy

The chapter introduces the theme of culture, mind and technology and provides insight into theories and methods of communication and culture, complexity, chaos and organisational change, and network cultures. The discussion draws upon cultural relativity, cognitive technology, chaos theory, and network cultures.

Interdependence between culture and technology leads to a shift from the physical to the mental, resulting in a new synergy embodied in HCI, e.g. human–machine compatibility, which is connected with the limits and tolerance of fatigue, memory, vigilance, mental workload, i.e., a kind of synergy between human capacities and work tasks and demands. She notes the emerging trend towards the humanisation of technology and impetus for change arising from the vulnerability of complex technological systems and dysfunctioning of the human–computer interface and increasing risk and vulnerability.

The increasing integration of information and communication technologies in the organisation gives rise to complexities of socio-technical systems. This requires a balance between procedures and professionalism, a shift in resource management which facilitates communication and learning. This, in turn, requires a deep understanding of the relationship between experts' skills and the ability to deal with risks. The handling of complexities in organisations arising from the use of technology is also dependent upon management ideology and practices at the workplace, not solely upon the technology.

At the societal level, increasing complexity of technology affects the access to technology. Unequal access leads to unequal advantages. The ability to access and use technology dictates the choices and opportunities for people.

At a global level, the emergence of network cultures with their web-like structure, give rise to new forms of communication and participation. These provide a potential for building cultures of shared knowledge. The traditionally peripheral groups or individuals can become part of the mainstream network culture. This provides challenges for new forms of sharing resources and for involving people in innovating future possibilities.

Chapter 5. The Social Construction of Human-centredness

Lars Qvortrup

The aim of this chapter is to search for the historical roots of human-centredness. Following a brief introduction to the current situation of human-centred systems theory, the paper is divided into three basic sections. In the first section Qvortrup demonstrates that human-centredness is not an invention of our current times, but is deeply rooted in the European cultural tradition of 15th century Italian renaissance (of philosophy, painting, architecture etc.). More importantly, however, he argues that moving the human being into the centre of the world – and that is, of course, the essence of human-centredness – is closely related to secularisation. Implicitly or explicitly the human being replaces God. This then opens for an ethical dilemma which has ever since been part of human-centredness: the dilemma that human-centredness with its good intentions of representing 'ordinary humans' may install itself as a new indisputable dogma. In the second section, Qvortrup looks at the role of science and technology in the development of the new human-centred world view. Although we say that in a human-centred approach, science and technology must be adapted to the nature of the human, instead of human as being adapted to science and technology, still the very idea of human-centredness is a social construction which was heavily influenced by the development of science and technology. Particularly, the theories of optics and astronomy developed during several hundred years from Greece and from the Arab world are important influences.

Finally, the third section returns to the ethical dilemma of human-centredness by making a distinction between normativism and dogmatism. What does it imply that something – for instance technology – is supposed to be human-centred? Does it have to follow some specific commands or rules – in other words, is 'human-centredness' an unambiguous and regulatory concept? In my opinion the answer is no. If one examines the historical sources, one can clearly identify two different concepts of 'human-centredness'. A dogmatic one, where organisational systems and technologies were developed according to an unambiguous idea of the nature of the human being, and an ambiguous one which gave room to different interpretations, and to playing tricks.

Chapter 6. Human-centred Methods of Social and Technical Design

Lauge Rasmussen

This chapter introduces the Scandinavian design traditions within the broader contexts of the European traditions of human-centredness. It presents and reflects on the processes and paradigm of human-centred design. The human-centred approach is defined as an individual and collective learning process based on four interrelated perspectives: a dialectical orientation, shaping perspective, dialogue perspective and social sustainability. The discussion then relates mainly to Scandinavian research experiences, examples of human-centred design processes, introducing dialogical methods such as prototyping, metaphors, organisational plays, integrated interaction and interpretation.

The process of dialogue is seen in terms of human development in the sense of a continuous learning process towards a more sustainable dynamic balance between nature, society and individuals. It emphasises individual and collective processes of experimentation providing alternatives for shaping human development. Different kinds of learning models such as single-loop and double-loop learning are presented to emphasise human-centredness as not only a means to reach an end, but as a continuous reflection of rationality towards that end.

The last section of the chapter presents a model of technical and social design and discusses various ways of integrating these perspectives and further elaborates human-centred design approaches. Central to the discussion of the chapter is the conviction that human-centred design presupposes cultural diversity as well as a continuous exchange of viewpoints and practices reflecting different interpretations and interactions of different cultures.

Chapter 7. Information Systems Design: a User-Involved Perspective

Siv Friis

The idea of 'user control' is a part of a continuum in the movement for participation in design in Scandinavia. Friis believes that future users of computer-based systems must be in control of the design process of the development of systems. Traditional systems-development methods give little significance to involving users, and traditional prototyping methods focus on the end-product, not on the process of design. Friis seeks to demonstrate that future user–designer participation and user control can be achieved through processes of dialogue, where the purpose of design becomes significant. She sees users and designers as belonging to two cultures and believes that the idea of dialogue between cultures needs to be part of systems design. She cites Freire's approach to pedagogy as a basis for this idea. Freire's pedagogy is that if you want two cultures to really communicate and learn from one another then their members have to enter into each other's culture. This necessitates establishing a common code of communication in order that the cultures can learn to communicate in the design process.

Prototyping is defined as a method for mediating knowledge/dialogue between designers and users. It allows them to learn from each other. Participation between designers and users makes for a more democratic development of information systems processes. Friis calls her case studies 'experiments'. By 'experiment' she means an exploration.

She has developed a model for the design of information systems called the PROTEVS model. PROTEVS is an acronym for PROTotyping for an EVolutionary Systems development. PROTEVS is a model for user-controlled information systems development (UCISD). It allows users to take responsibility in the design process and regards users as both the problem owners and the problem solvers. The term 'evolutionary' is significant as it denotes a continuous process of learning. In short, the PROTEVS model is expected to enable active learning between designers and users, and user participation through a process of dialogue.

Chapter 8. Designing for Knowledge Transfer

Satinder P. Gill

The chapter focuses on the issue of knowledge transfer in design discourse, and analyses the limitations of representing knowledge for design. Design, here, is about the communication of knowledge. The representation of knowledge in a propositional form in traditional information systems design rests on the idea that knowledge is universal, non-contextual, time-independent and depersonalised. This idea can be traced back to Plato's discussion of reason over emotions, and survives in the current computer metaphor within which the mind–body distinction has become a discipline called 'cognitive science'. Design for knowledge transfer on the other hand is context-dependent and rests on the premise that knowledge exists in praxis/experience and has a personal and social dimension. It is argued that the non person- centred perspective of knowledge is inadequate as an explanation for knowledge and skill and that a person-centred perspective as an alternative framework provides a better explanation of the nature of knowledge and skill.

The discussion draws upon hermeneutics and provides an analysis of the relationship between tacit and propositional dimensions of knowledge in discourse. The hermeneutic discussion introduces the concepts of rule-following and practice, tacit knowledge, and dialogue. The methodologies of communication and design such as ethnomethodology, conversation analysis, discourse analysis, case studies of design and skill, have been drawn upon for the development of a model for knowledge transfer.

Chapter 9. Designing Practice-Based Learning Environments

Felix Schmid

The chapter provides a review and synthesis of theories of knowledge and theories of learning for designing practice-based learning environments. He situates these theories within the tradition of the engineering profession, its current practice and its historical formation. Practice-based learning emphasises the symbiosis between practical (tacit) knowledge and general knowledge. Experience-based learning contains both cognis (rational knowledge) and gnosis (spiritual realm). The notion of cognitive apprenticeship is seen as a method for effective learning of both explicit and informal knowledge and for creating a balanced person in the sense of the renaissance period which signifies the mid-point between the historical shift from gnosis to cognis.

The discussion of knowledge and learning is centred around the making of a good engineer, rooted in theory and practice. Different forms of learning are considered such as problem-solving capabilities, analytical and synthetic faculties, surface learning and deep learning, holistic and serial learning, and insightful learning. Experiential learning emphasises the notions of learning to learn, life-long learning and openness to change. It takes place in a cycle of four stages: concrete experiences, reflective observation, abstract conceptualisation, and testing of the implications in new situations. Often, experiential learning is implemented as team-based learning involving groups of peers. Effective learning requires effective teamwork which needs to take place over a long period in order to develop sustainable working relationships. Experience-based learning is exemplified by a case study of project-based learning for engineering students at Brunel university.

The discussion is situated within the broader contextual debates of industrial production, computing euphoria, organisational change and strategy, human-centred systems design, person-centred ergonomics, and the process of change in engineering education and training.

Chapter 10. Workplace Innovations: the Making of a Human-centred Industrial Culture

Francisco Garibaldo

The chapter deals with workplace innovation from the interrelated perspectives of the workplace and of the socio-technical environment. Workplace innovation occurs in the form of group work, teamwork, life-long learning in the context of increasing integration of roles and functions at the workplace. Socio-technical innovation involves the integration of human factors and technical factors, resulting in user-centred and user-involved approaches. Both the former and the latter forms of innovation lead to the dynamic adaptation to changing environments. Innovation is about creating sustainable systems and processes which require personal commitment and social trust. These in turn require processes of learning at both the social and political level.

This raises crucial issues of bottom-to-top participation, empowerment, involvement at the workplace, and a shift from seeing workers as objects to social subjects. Workplace innovations have been influenced by historical developments such as the rationalisation of the Fordist/Taylorist paradigm during the 1980s, and the Japanisation of the workplace culture leading to concepts of group working, quality circles, semi-autonomous groups, and just-in- time practices. These developments raise issues of debureaucratisation of management at the workplace and self-regulation at the group level.

Sustainable innovation requires a dynamic equilibrium between flexibility and cooperation of group-working and teamworking on the one hand, and the collective power of self-regulation on the other hand. These require personal trust and group commitment. The challenge is how can this group commitment and personal trust fit into the Western concepts of democratic participation based on the individual. This chapter discusses these dilemmas and raises issues of individual empowerment on the one hand and a social commitment to group rationalisation of Taylorism and Japanisation on the other. These issues are discussed at the micro, meso and macro levels.

Chapter 11. Meanwhile, Out in the Real World: Developing a Commercial Human-centred Software Application

David Smith

Smith's chapter presents a case study called Project MEDICA, which is based on an attempt to design, develop and bring to market, a software application embodying some of the principles of 'human-centredness'. The chapter lays out and explains some of the challenges, problems and compromises involved in bringing a design concept out of quasi-academic research. The project involved a multinational team of clinicians, academics and industrialists brought to the project a wide range of expertise - including medical education, clinical psychiatry, ethical, pharmaceuticals, sales and marketing, satellite communications, interactive videodisk production, systems analysis, artificial intelligence and socio-cultural research. Project MEDICA was seen as a prototyping stage, aiming to design and develop a specification and working demonstrator of the Diagnostic Assistant, grounded in current psychiatric practice, and capable of commercial exploitation by IT industries within the European Community.

The MEDICA approach to 'intelligent' support technologies steps away from the trend towards the automation of expert domains, and represents a major step towards the more 'anthropocentric' (human-centred) application of advanced informatics recently urged on the NICT community by the EC's FAST (Forecasting and Assessment in Science and Technology) Programme.

It is important not to regard devices such as computer systems in isolation. Human factors are critically important in determining the extent to which a technical innovation can be transferred into routine practice. This is not simply a matter of attention to the ergonomic aspects of design, though these are certainly important. It is clear that the design and development of such systems must proceed in the light of some conception of the broader social, professional and cultural contexts within which they will be embedded.

Project MEDICA, and its commercial offspring AMIGOS, provides evidence that this can be achieved in a way which is commercially viable and which nevertheless remains true to the general working principles of human-centredness.

Chapter 1

The Foundations of Human-centred Systems

Karamjit S. Gill

Introduction: the Foundation of Human-centredness

What role could and should science and technology research play in meeting new challenges of social vulnerability, environmental and ecological risk, the brittleness of economic, industrial and political orthodoxies, and an increasing dependency on technological systems? These challenges are products of a science and technology rooted in the 'mechanistic' paradigm of the 'one best way', 'sameness of science', and the 'dream of the exact language'. The human-centred tradition moderates science and technology by mitigating the mechanistic paradigm through concepts such as human purpose, diversity, participation, social responsibility, equality, ethics, creativity and ecology and the environment. It provides theoretical and methodological frameworks for the social and cultural shaping of technologies emphasising human–machine symbiosis, creativity and innovation, participatory and cooperative design, and the tacit dimension of knowledge. These issues are very much part of the human-centred debates whose origins lie in the European human-centred movements of the 1970s, in particular the British LUCAS Plan of socially useful technology, the Scandinavian tradition of participatory democracy, and the German programme on humanisation of work. These debates converged and were consolidated in the European Commission programme on anthropocentric systems (APS) in the 1980s. This European dimension has raised fundamental issues of interdisciplinary research and plurality of design traditions. It also highlighted a need to recognise the valorisation of diverse cultures, languages and practices (Gill, 1990). Human-centredness expounds an emancipatory tradition which places human needs, purpose, skill, creativity, and human potential at the centre of activities of human organisations and the design of technological systems. It has broader concerns in the areas of scientific traditions, culture and technology, Industrial cultures, technology transfer and development, globalisation, sustainability, and technology assessment.

The tradition shares a common belief in shaping and developing new technologies for the benefit of all people and all societies.

The theme of 'human-centredness' challenges the 'dream of the artificial', exemplified in the dream of the 'exact language', the 'ideal of certainty', 'rational method', and the 'symbolic man' which replace the 'natural by the artificial'. It questions whether the notion of the 'artificial' is about the celebration of the dream of the separation between the individual and the community, private and public, reason and emotion, function and the social, work and living, and technology and society? Or is it about redefining and redirecting societal issues and the social agenda in terms of the culture of 'sameness' and the science of the 'one best way'?

The human-centred debate can also be seen in terms of creating a 'secular technological culture' from the perspectives of advanced industrial societies. The question that needs to be asked is: 'human-centredness from whose perspective?' How does this tradition cater for the diverse social and cultural traditions of various societies? How does it cope with the variety of relationships between the human, nature and the machine reflected in the North–South divide, and the East–West divide?

Over the years the human-centred tradition has built upon the socio-technical approach and has shaped the human-factors approaches towards user-centred, user-involved, and cooperative design approaches. The idea of human–machine symbiosis emphasises the collaboration between human capabilities and machine (Figure 1), rather than the separation which is embedded in the dominant human factors tradition of Human Computer Interaction (HCI) (Gill, 1990), which focuses on the information processing component of human–computer relations (see Figure 2). The user-centred approach transcends the linear relationship between the human and the computer, and argues for the tripartite interrelationships between the human, the computer and the job (Figure 3). The user-centred approach (Bannon, 1990), is, however, limited by its focus on user practice and user modelling which focuses on a narrow view of human activity, and on the separation between the user and the designer of technological systems. From the human-centred perspective the crucial limitation of the user-centred approach is the 'work context gap' (Thomas and Kellog, 1990), which arises from its neglect of the social setting and culture of the workplace, in other words the organisational culture. The user-involved approach (Ehn and Kyng, 1987), fills this gap by emphasising the multiplicity of relationships between the social system, the technology and organisational systems (see Figure 4).

The shift from a user-centred approach to a user-involved approach emphasises collaboration, cooperation, and coordination at the workplace, and attends to human-centred concepts such as 'situated action' (Suchman, 1987), 'usability' (Alder and Winograd, 1992), and 'cooperative design' (Bødker, 1989). These human-centred approaches provide theoretical and methodological frameworks for designing information and communication technologies which could influence the current work in computer supported cooperative working, telematics, cognitive support systems, and multimedia systems.

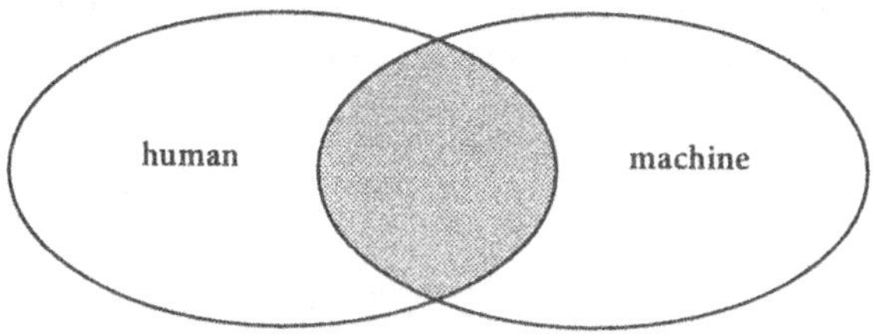

Figure 1: Human–machine symbiosis: collaboration

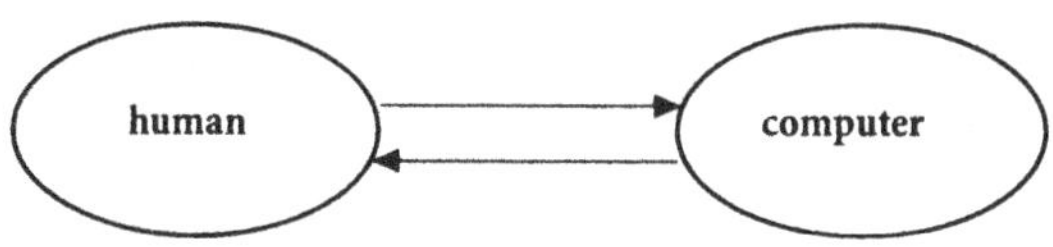

Figure 2: Human–computer interaction: separation

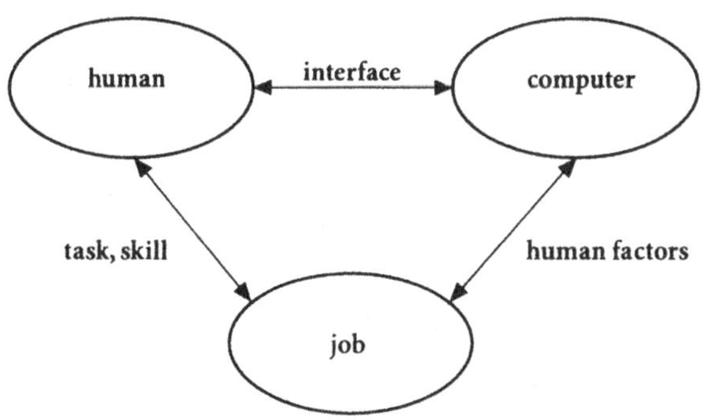

Figure 3: User-centred approach: user practice, user model

At a research level, human-centredness challenges the techno-centric focus of science and technology which ignores the diversity of human knowledge and marginalises human skill. The techno-centric focus arising from the Tayloristic and Fordist traditions has been dominated by the mechanistic paradigm of science. The human-centred movement provides an alternative focus for designing technology which enshrines human purpose and transcends beyond causality. In essence, human-centredness enshrines:

- the idea of a symbiotic relationship between the human and the machine, between cause and purpose, and between objective knowledge and the tacit dimension of human knowledge;
- a belief that it is by being proactive and not reactive, and by designing tools and not machines, that we can cope with the highly complex and synchronised systems of the microelectronics age;

- it is by adopting the concept of 'valorisation of diversity' and rejecting the notions of the 'one best way', 'one culture', and 'sameness of scientific ideas' (e.g. consistency, reliability, predictability), that we can deal effectively with societal issues of local specificity and global sustainability.

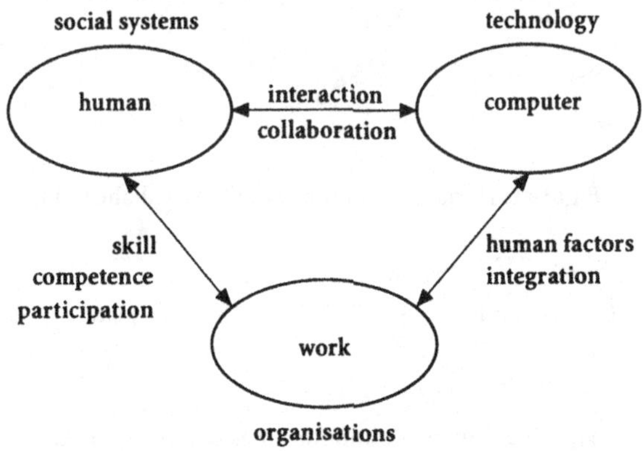

Figure 4: User-involved approach: social settings, organisational culture

This purposive view of human–machine relations provides both a theoretical and practical argument for highlighting the limitations of the human factor and information processing approaches of HCI and cognitive science, and the expert-centred approaches of expert systems (Gill, 1990). Recently, some of the innovative developments in Computer Support for Cooperative Working (CSCW), telematics, and information networking are tacitly promoting the centrality of the human-centred approach for dealing with the complexities and uncertainties of socio-technical systems such as issues of brittleness, compatibility, accountability, breakdowns, transparency, transportability, and operability. They are also highlighting the need to identify the criteria for handling uncertainties in tasks, such as disturbance control, fallibility, error responsibility and operator flexibility (Rosenbrock, 1989: 40).

The current human-centred tradition has arisen out of two complementary approaches in Britain in the 1970s, namely 'human–machine symbiosis' (UMIST tradition) and socially useful production (LUCAS Plan). The common core of these approaches is that culture-based knowledge and actions of human beings should be reflected in a dynamic way in systems instead of being subsumed by them. During the 1970s, the human-centred traditions found a receptive environment in other European nations, take, for example, the use of the concept of the 'Tool Perspective' and the projects on 'Democracy in Participation' in Scandinavia, e.g. UTOPIA Project (Ehn 1988), and the 'Humanisation of Technology and Work' (Dankbaar, 1987). During the 1980s, human-centred ideas were developed further by the ESPRIT project on human-centred CIM (Rosenbrock, 1992), the projects on Culture, Language, and Artificial Intelligence (Göranzon & Josefson, 1987), Social Action (Gill, 1986, 1990; Ennals, 1991), the 'Shaping of Technology and Work' in Germany (Corbett *et al.*, 1990), and the Culture of the Artificial (Negrotti, 1991).

These developments in human-centredness in Europe have made a major contribution to a new European tradition of 'Anthropocentric Systems' (Wobbe, 1991). Although human-centredness is essentially a European tradition, and may not map directly onto other cultures and nations, the basic ideas of valorisation of diversity, cultural and industrial renaissance, and holistic systems could provide a common basis for culturally-based, human-centred traditions. Central to these new developments are the notions of human-centredness exemplified by the foundation ideas and basic challenges of human-centredness which are presented below.

Foundation Ideas of Human-centredness

1. Valorisation of diversity is the seabed of creativity, and creative tension is at the heart of finding unity through diversity.
2. Human-centredness rejects the idea of the 'one-best-way', 'one culture', and the 'sameness' of science and technology.
3. Human-centredness is open-ended and inclusive.
4. Education in the humanistic tradition is about the transfer of cultures.
5. Human-centredness is essentially multidisciplinary, crossing academic and cultural boundaries.
6. Human-centredness is purposive in contrast to the 'techno-centric' approaches which are causal.
7. Human-centred systems design concerns itself with two fundamental questions, i.e. what could be designed?, and what should be designed? The first is about what is technically feasible, and the second is about what is socially desirable.
8. Tacit knowledge is a cornerstone of human-centred philosophy, rooted in the interdependence between the subjective and the objective, and rejects the notion of their separate existence.
9. Human-centredness regards people as both producers and consumers, as well as evaluators of knowledge, and thus rejects the notion of knowledge being the intellectual property of the chosen few.
10. Social and economic cohesion through networks of social, economic and technological systems is at the heart of the vision of the information society from the human-centred perspective.

Basic Challenges for the Human-centred Approach

1. How can we recast technology which enables a coexistence between technology and society?
2. To project the holistic ideas of human-centredness beyond the narrow confines of anthropocentricity.
3. To extend the research focus of human-centred systems from a relationship between work and technology to one of work, technology, and living.
4. To shape the debates on information society and post-industrial society from human-centred perspectives.
5. How can an orthodoxy of human-centredness be avoided, i.e. how can one ensure that human-centredness remains open and innovative?
6. How does human-centredness sustain the dialectics of human-machine symbiosis beyond its European tradition?
7. How does the emancipatory and purposive tradition of human-centredness overcome the orthodoxy of disciplines and funding bodies?
8. How does human-centredness deal with issues of vulnerability and exclusion from the combination of cognition and the virtual?
9. How does human-centredness find symbiosis among cultural rationalities?

Scientific and Intellectual Challenge

One of the most important barriers which human-centredness had to cross in the 1970s was the mechanistic view of science. This view regarded forces of production, and in particular science and technology, as ideologically neutral, and it was considered that the development of these forces was inherently positive and progressive. For it was science which, through Galilean and Darwinian revolutions, liberated humanity from the bondage of superstition. What was needed, it was argued, was to invest adequately, to plan for science and to provide a rational framework for its widespread application in the elimination of disease, poverty and toil. Science thus appeared as critical knowledge for human progress (Gill, 1986). However, by the early 1970s, as Cooley (1989) points out, there was a gradual realisation that science has embodied within it, many of the assumptions of society which has given rise to it.

This led to the growing questioning of the neutrality of science and technology, as well as to the questioning of the nature of the scientific process itself, beyond the scientific use/abuse model.

The 19th century's concern with energy gave rise to the consideration of the machine as essentially an extension of arms and hand. Bruner (1983) describes it as '... the age of mechanics and energy transfer. Not surprisingly, the machine model of mechanics, the guiding metaphor of technology, was transferred to the human domain.' With the dawn of cybernetics, information replaced energy as the central focus of the emerging technology. Interest in the notion of control and transformation of information provided the impetus for an investigation into issues such as the human ability to know, the limitations of the human information processing model, and how we deal with uncertainty and use feedback from our own action to regulate these actions. Although with cybernetics, the process of the separation of machine from the human had started, the machine was still a model of human labour.

The advent of the computer completed the process of the separation of the machine from the human. Previous ideas about the collaboration between the human and the machine shifted to ones about the collaboration between the machine and knowledge extracted from the human. This led to machine-centred production and control. Simon's 'science of the artificial' in the 1960s allowed for the separation of the subjective knowledge from the objective and provided a basis for the design of computer artefacts as machines. The science of the computer was defined in terms of artificial languages of logic, mathematics and statistics. The computer was seen to simulate the human mind. The processes of the human mind could only be conceived by an understanding of the processes of human behaviour and language. Human behaviour and language therefore could only be defined in terms of their functionality as closed systems. Just as complex computer systems are decomposed into functional units, so was human behaviour and communication. The process of design itself was therefore reduced to a problem of declarative logic (Gill, 1986).

This approach of viewing human behaviour and human communication resulted in the design of artefacts as mechanistic tools which marginalised the human subjective knowledge, and hence excluded the experiential knowledge of human practice and skills from the design process and the use of the tool. In the 1970s this view was challenged by researchers such as Dreyfus and Weizenbaum. Recently, European researchers such as Bannon (1989), Bødker (1989), Rosenbrock (1990), Ehn (1988), Floyd (1984), Brattetein & Stolterman (1995), Gill (1990, 1994), and Læssoe & Rasmussen (1989), have provided an alternative vision of participatory design to that of the rationalistic approach of Simon. The point of departure of this approach is that it emphasises human participation and dialogue throughout the design process. The design becomes action and designing for the future, a process for anticipation of possible breakdowns for the user in the future use situations. The focus is on the human use of computer artefacts rather than upon a detached reflection of design.

There was also a questioning of the mechanistic ideals of science and technology. Nobert Wiener's book, *The Human use of Human Beings* which appeared in 1950, inspired many critics. By 1965, Hubert Dreyfus was comparing artificial intelligence to medieval alchemy. He developed a critique of AI which culminated in a book entitled *What Computers Can't do* in 1972. Also in 1972, Joseph Weizenbaum produced his *On the Impact of the Computer on Society* with the challenging *subtitle How does One Insult a Machine?* By 1976 these concerns had found fuller expression in his book *Computer Power and Human Reason*. This has as its subtitle, *From Judgement to Calculation* and highlights the dangers which will surround an uncritical acceptance of computerisation. During the same period, Rosenbrock had been working on the use of computers in systems design, and his *Computer Aided Control Systems Design* appeared in 1974, to be followed in 1976 by his seminal paper *The future of Control*. In 1979, Cooley' book, *Architect or Bee?* later revised as *Architect or Bee?: Human Price of Technology* (1987), generated a pro-active international debate on human-centred systems. In 1986, a book on *Artificial Intelligence for Society*, edited by Gill (1986), became a focus for extending the human-centred debate to the social domain. In 1987, an international journal, *AI & Society*, and in 1988, a sister book series in *Human-Centred Systems*, both published by Springer-Verlag, were launched to support and disseminate future developments in human-centred systems. Since the late 1980s, these publications have provided a forum for a number of intersecting debates on society and technology; human-centred systems design; social shaping of work and technology; skill, language and artificial intelligence; culture of the artificial; technology and ethics and social and economic cohesion. Some of the basic challenges raised by technology remain rooted in the rationality of the 'rule' and 'causality', and the paradox of misalignment between the human and the technology. The notions of 'valorisation of diversity', dualism, and mediation offer stimulating possibilities of a new direction of research into human-centred information systems design.

Challenge of the Rule

The rationality of the 'rule' and the notion of 'rule following' lie at the heart of the human-centred debate. The challenge of the rule comes from the philosophy of design which accepted the separation of the hand from the brain, skill from practice, objective knowledge from the subjective. 16th century Europe witnessed the appearance of new word meaning, 'design', to describe the occupational activity of 'designing'. This proved to be a momentous event – the beginning of the separation of hand and brain, of manual from intellectual work, and of the conceptual part of work from the labour processes. It signified an historic trend that 'designing' was to be separated from 'doing'. This trend began with the separation of designing (by architects) and construction (by builders) of great churches in Europe (e.g. Italian churches).

Cooley (1987), gives an insight into the 'challenge of the rule', and into the 'separation of hand and brain'. The separation of manual from intellectual work provided the basis for further subdivisions of work, both intellectual and manual. It was the beginning of 'Taylorism'. The process of subdivision found theoretical expression in the writings of early economists such as Adam Smith. The growing factory system thrived on it. Some 19th century writers warned about human consequences: '... To subdivide a man is to assassinate him. The subdivision of labour is the assassination of a people ...' (Urquart, A., 1961). The first person to succeed in rendering visible the rules underlying various craft skills in any significant way was the great Filipo Brunelleschi, architect of the Duomo in Florence. As early as 1413, Brunelleschi had developed a systematic technique for constructing representations of buildings using true perspectives (Saalman, H., 1980). This systematic approach was further refined by Leonardo, and by the end of the 15th century, architects such as Mathias Roriczer of Regensburg were publishing generalised methods for various aspects of cathedral construction.

The consequence of the emergence of the rule and objectivity was that not only the conceptual part of the work of craftsmen (masons) was being taken away from them, but the academic hegemony over theoretical knowledge was resulting in the denigration of the skill of practitioners which was an embodiment of the fusion of intellectual and design skills.

Mitter (1986), points out that the concept of a universal rule is rooted in the deep-seated empiricist epistemology of the West, according to which knowledge can be formulated in the form of rules, and transferred without taking into account the cultural diversity. The importance of specific cultural experience is often forgotten or ignored; partly because, after great advances were made in the physical sciences during the 19th century, the notion of pure 'objectivity' came to take a firm hold on the intellectual imagination. A persistent and pervasive assertion of the social sciences in the 19th century was that scientific objectivity transcended cultural variety, and this view of 'scientific reductionism' was powerfully enforced by social Darwinists. This doctrine of 'cultural homogeneity' of human society and 'universality' of human knowledge, even now forms a great deal of scientific thinking. It is however, crucial to recognise that while there are common and universal traits that humans share, there is a great deal which is culturally specific. The cultural factor cannot be discarded in the domain that deals with human knowledge, skill and experience.

Implications of the Rule

The pervasiveness of the 'rule' is even reflected in the designing of technological systems for human use. It permeates, for example, the research into knowledge-based systems, natural language processing and human-computer interface design. In the case of knowledge-based design, important questions of contexts and background are ignored. Human knowledge which cannot be explicated in the form of rules is abandoned. This ignoring of the 'tacit' dimension of knowledge leads to the design of technological systems which map only part (may be the larger part) of the reality. This discrepancy between the reality and the technological systems results in the vulnerability and brittleness of systems. Where such systems are integrated into human organisations, human judgement becomes dependent only upon the objectivity which ultimately leads to technology-led human disasters such as the Three Mile Island, Chernobyl, and Bhopal.

The 'rule' model of systems design tends to find the solution to the 'discrepancy' problem (between the technological system and the reality) through the design of internally consistent (logically) and highly synchronised systems which aim to eliminate human subjectivity and redundancy. Thus both the design process and the use of technology becomes and remains the prisoner of the straitjacket of consistency, thus removing crucial elements of human–machine collaboration. For example, as pointed out by Gill, S. P. (1988), in the design of experts systems, knowledge acquisition is considered to be the 'quantification' of expertise. The construction of expert systems means reducing human expertise into syntax recognisable by the computer. This imposes constraints upon the user in the sense that both the dialogue and the decision-making process are expressed in terms of machine-understandable rules. The computer becomes the dominant medium of communication, and both the expert and the user become distanced from their own knowledge and skills.

The Challenge of 'Causality'

What makes the technologist design a machine that replaces the human? What makes managers adopt a uniform technology, world-wide, irrespective of the different histories and cultures, and different levels of economic and industrial concerns? Rosenbrock (1992), points out that there has been a strong tendency for technological convergence throughout the history of industrial developments. He suggest that a common explanation for adopting the same technical means to overcome economic pressures of international trade and gain competitive advantage are too simplistic. Economic calculations are notoriously arbitrary since they neglect human, social and environmental costs. This neglect of the social lies in the neglect of decision makers to place a value on human capital and the environment.

Rosenbrock's concern here is with the relative values which we place, on the one hand, on machine and mechanised calculation, and on the other hand, on human beings: human beings with shortcomings that machines and organised knowledge can help to overcome, but human beings also with a tacit knowledge, who with their unique abilities of observation and ingenuity and experience can respond to problems.

He also rejects the argument of technological determinism and points out that our current technology with computer control and communication systems is immensely flexible. It will support organisations based on an extreme centralisation, with all initiative removed from the periphery. But with equal ease it will support an organisation in which there is a very high delegation in decision making, with central coordination rather than command. Yet, in spite of these technological possibilities, the same Tayloristic regime of control and automation is applied the world over. Rosenbrock offers an explanation of the uniformity of technological practice throughout the industrialised nations. The only activity which generates an equal consistency in all its practitioners seems to be science, and it is in the causal view of science, that the ultimate source of consistency in technology lies.

Causality is a slippery concept: a cause must never be later than its affect, and the effect must be a necessary consequence of the cause, but beyond this lie endless complexities and difficulties. What is excluded, Rosenbrock argues, is human purpose. In science we are not permitted to give explanation in terms of purpose, and this embargo is enforced just as rigorously in technology. In this causal view of science, human work is turned into a machine-like character, human knowledge exists in the explicit form only, and the social part of the human becomes separated from the technical skill. There lies the dilemma of the technologist. Rosenbrock offers a glimmer of hope that causality is not something that is imposed on us but resides in our scientific view of the world. It is a presupposition which we adopt before we begin to study the world and explain it. It is equally in our gift to take an alternative 'purposive' view of science. This alternative view of science provides a foundation approach for shaping socially useful technology.

Paradox of Misalignment

Rosenbrock (1989), points out that the rationalisation of human activities leads to the creation of a work-force who are rendered more passive as technology becomes more active. The tendency is to design more and more complex technologies capable of handling a wide variety of tasks which belong to the human domain. Humans are left to handle trivial and single tasks which will be too expensive to be handled by the machine. This situation gives rise to a 'misalignment' between the human abilities and the demands of some tasks.

The process of eliminating misalignment between technology and human skill gives rise to a paradox. To use a complex machine to perform a trivial task is too expensive and hence there is a need to design a simple machine to perform the task cost-effectively. Where the task is too complex, the solution is to decompose the task into simple tasks which are manageable by the machine. In both these cases the concern is to use the machine more economically and to make full use of its abilities. There seems to be, however, no such concern shown for human skills and abilities.

Dualisms and the Challenge of Mediation

The notions on dualism and duality are concerned with dualism of the scientific method on the one hand and the interpretative method on the other hand. It is argued that the 'one best way' approach of the scientific method is impoverished and thereby inappropriate for enquiring into human systems, while the argument against the interpretative method is that it lacks rigour and is therefore not generalisable. Latour argues that the modernist project seeks purification – the separation of the objective and given natural world from a socially constructed world (Latour, 1993). Braa and Vidgen (1995: 50), point out that modernism contains a paradox insofar as it must separate the natural and social worlds while relying upon their inseparability for its success. They draw upon the work of Habermas, and the idea of the technical, practical, and emancipatory knowledge interests (Habermas, 1971), and upon Latour's argument for the middle ground between the natural and social worlds. They argue that since knowledge interests are inseparable, outcomes can only be achieved through mediation. Outcomes are not purified beginning points of the natural and the social. In the human-centred systems tradition, Cooley (1987), and Rosenbrock (1990), seek to resolve this paradox through the notion of symbiosis between the machine and the human, benefiting from the potential of the capacities of the machine and capabilities of the human.

Towards a Symbiosis of Diversity and Rationality

The idea of human–machine symbiosis has been central to the development of human-centred systems. In the age of information networks, the symbiosis is not just between the single machine and the single user, it is a matter of symbiotic relationships between the network of users and the network of machines. It is no longer a matter of interaction between the machine and the individual user; rather, it is a matter of communication between groups and between human and machine networks. It is not a matter of the interaction between a skilled worker and the machine, it is a world of collaboration between users at a variety of skill levels and the network of machines performing at a variety of functional levels.

This diversity of interactions within and between networks raises an issue of the interdependence between the objective knowledge and the tacit dimension of knowledge of the network and its culture. The 'tacit' knowledge no longer just resides in the individual artisan but resides in the community of users in the form of a social knowledge base or a network of social knowledge-bases. It is not just a question of objectification of the experiential knowledge of the individual, but it is a question of the objectification of both social and professional knowledge at a variety of levels of objectivity and ambiguity. The challenge of human-centredness is to transcend the notions of human factors, human–machine interaction, and cognitive performance, and promote notions of human–machine symbiosis, and concepts which deal with a variety of human-human and human–machine relationships and networks of relationships, both at local and global levels. (Gill, 1992, 1993, 1995).

This challenge of symbiosis is part of a bigger societal challenge of how to re-integrate technological innovations into the civil society so that technology supports new forms of work-life and living environments. This requires innovation of new forms of social and cultural interfaces which respond and cope with the changing world of work and living.

Diversity and Coherence

The idea of social shaping of information networks encompasses human-centred notions such as participation, dialogue, user-involvement, language games and future use. In this perspective of shaping the information society, the emphasis is not just on the production of knowledge but also on the reproduction and sustainability of knowledge. This, however, depends upon the capacity and capabilities of networks to transfer and diffuse this knowledge at the local and global levels, and in a variety of languages at a variety of competence levels of expression and interpretation. Any rationality of communication in the network requires a rationality of cohesion of diverse knowledge resources, and this in turns presupposes a belief in the 'valorisation' of diversity of cultures, languages and social systems. Diversity here is more than a belief in variety, it demands a commitment to a practice of diversity and its 'valorisation'.

Diversity within this broader human-centred perspective is about a deep learning experience while sustaining deep inner coherence of human values, respect, dignity of human spirit. Without a unique coherence, we cannot agree upon a coherent measurement of technological innovations, and thus cannot achieve 'valorisation' of diversity. In the human-centred tradition, this coherence can be achieved through a balance of communication within networks and a harmony of networks of relationships. This notion of diversity transcends beyond the traditional choice of alternatives and finds coherence in the innovation of choices.

Diversity Enables:

- Individual creativity while seeking social strength through harmony and dialogue.
- Innovation as a process of regeneration and social and organisational learning, dealing with diverse cultural rationalities without prejudice; this innovation process finds a balance between scientific

rationalities, logical rationalities, empathetic rationalities, and rationality of silence.
- Deep understanding of other cultures, not just at a surface level but at a deep personal, social and cultural level.
- Cultivation of balance, harmony among diverse people through a learning process for sustainability.
- Acts as a limit and boundary limit to extremes, avoids the breaking down of societies.
- Acts as a limit to otherwise unlimited one-dimensional growth.
- Acts as a true limit of the environment of democratic cultures in the deep sense, not pretence; deep understanding of oneself, community and society: faith, confidence and sincerity.
- Social and economic cohesion through cooperation, collaboration and networking.

Designing Cultural Interfaces

The idea of the symbiosis between cultural diversity and scientific rationality broadens the notion of the human–machine interface from the cognitive perspective to the cultural perspective. In this perspective, the formulation of cultural interfacing between technology and cultural systems deals with the diversities, complexities and ambiguities of the network of human and machine relationships. This formulation thereby requires not only an understanding of the cultural roots and social motivations of technological innovation, it should also be situated within a methodological framework which promotes the attainment of a practical social harmony while meeting the cultural needs of the user. Within this framework, the criteria for building technological tools is seen in terms of social and cultural utility, in other words, in terms of concrete technological activity and phenomenological functionality (e.g. communication). Here, technological innovation is seen as a learning process. It recognises that the purpose of harnessing information and communication technology is two fold:

- *Sustainability*: the production of knowledge for harmony and sustainability;
- *Innovation*: the reproduction of knowledge for creativity and learning.

Harmony in Ambiguity

In seeking a coherent methodological framework for cultural interfacing, we need to understand the role of ambiguity which is inherent in the cultural notions of diversity and plurality. Here we go back to our earlier discussion on human–machine symbiosis, a symbiosis between classical technology of production and the skill of the artisan.

According to Negrotti (1990), classical technology of production implies some entity, serving a cause (causal component) and achieving a goal (teleological component). The paradox of classical technology is that while it builds practical systems for practical purposes, it increases the distance of the practical system from the natural system upon which the practical system is modelled and built. Here, technological systems perform functions of the real systems thus raising two alternatives of communication between the human and the machine:

- automation - cognitive machine (independent of the real systems);
- the symbiosis - harmony of the object/subject - collaboration.

Even in a defined production culture, automation determines the limit of human skill and knowledge, and thereby further distances the human skill from its own work-life contexts, let alone social and cultural contexts. By seeking harmony of the object and subject in the socio-technical tradition, the symbiosis seeks collaboration between human skill and the machine capacity. However, this symbiosis only relates to production work, and excludes social and cultural skills and knowledge which are rooted in social and cultural systems. Information network technology on the other hand assumes the existence of a network of entities. Here the motivation for cultural interfacing lies in creating a network of relations between the social systems and technological systems, and its role is to:

1. reproduce the social and cultural relations of human networks, and overcome the limitations of institutions and boundaries of cultures;
2. reformulate human network systems and extend them, and become part of the extended system, and thereby increase human relations and cultural diversity.

Extending Negrotti's idea of interface based on the culture of the artificial, the idea of the cultural interface plays the role of a mediator between the cultural systems and the technological systems. The challenge of the interface is then how to harmonise 'expressive and interpretative flexibility' or 'enduring vagueness', which may arise due to the limitation of the communication languages to deal with the diversity and rationality of communication in a knowledge network.

This limitation of the communication language leads to the methodological problem: how to cope with vagueness or interpretative flexibility. The requirement for communication between the speaker and the listener, between two groups, or between two cultures, is the shared reference concept offering inter-subjectivity (*ibid.*). Ambiguity here is regarded as essential for understanding the complexity and beauty of nature and its world. Negrotti cites art and music as exemplars of the artificial which regard ambiguity as part and parcel of artistic compositions. Here the composer uses artistic tools such as language, painting, poetry, music and technology to produce the final product; the artistic composition. The composition, the final product, its harmony and completeness is defined;

it is open to different interpretations but not modifiable in content at all.

Paradox of Ambiguity

- The complexity or nature of the musical composition cannot be changed without serious consequences for the aesthetic content or its beauty.
- The ambiguity of art results from rigorous decisions and actions on the part of the artist. Its content cannot be changed but its interpretation and use allows for flexibility of interpretation. In fact, the beauty of art lies in flexible interpretation – i.e. beauty lies in the artistic ambiguity.

Whenever a human activity is objectified and formalised into a rule-based technology (objectified), we find that in eliminating ambiguity we also eliminate the possibility for multiple interpretations. For example, technical drawings or computer programs neither allow arbitrary interpretation nor a change of content. The consequences of the arbitrary interpretation is that it destroys communicative power of the technological artefact. The significance of ambiguity thus lies in enabling the diversity of interpretations while sustaining the integrity of the content. Extending this discussion to one of cultural interfacing, it is suggested that ambiguity plays a fundamental role in enabling plurality of interpretations of knowledge. If, however, there were no ambiguity in the knowledge network, there will be no need for diversity of interpretations and thereby no need for cultural interfacing. Just as the objectified technology leads to the rigidity of the human–machine interface by eliminating the tacit dimension of communication, the elimination of ambiguity makes the knowledge network nothing more than the technology network, where the cultural interface is reduced to a user-friendly interface for essentially linking machines to machines. The methodological challenge of the cultural interface is not just to design multimedia machine interfacing, but to design multimedia cultural interfacing which builds upon cultural rationality and ambiguity as resources for enriching cultural communication and knowledge transfer.

Foundation Concepts of Human-centredness

Human–Machine Symbiosis

The concept of human–machine symbiosis, which promotes the best of the combined potential of the capacities of the machine and capabilities of the human, has been central to the shaping of the emancipatory perspective of the human-centred debate in the UK since the 1970s.

Cooley (1987) and Rosenbrock (1992) challenge the notions of the separation of the tacit and the objective and the separation of cause and purpose embedded in the Western scientific tradition.

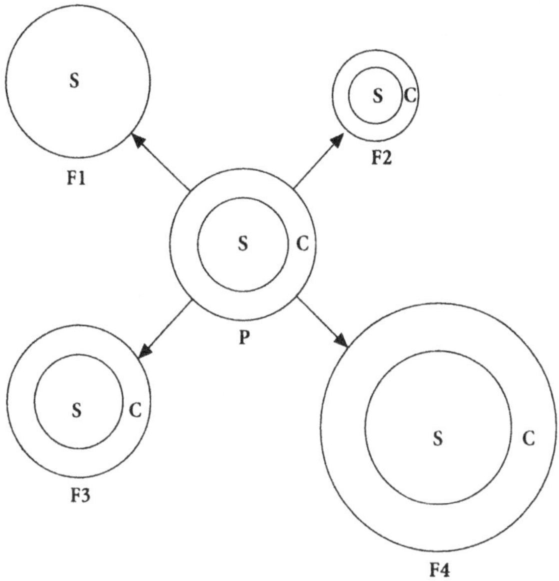

Figure 5: Four different views of the evolution of technology. In each of the diagrams S represents the 'sun' of explicit knowledge, while C represents a 'corona' of skill and tacit knowledge without which the explicit knowledge cannot be used. F1 is the Taylorist view of the future; F2 rejects much of our present technology; F3 sees the future as no different from the present. In F4 a future is envisaged in which the explicit knowledge has increased, but scope and opportunity have been given for a corresponding development of skill and tacit knowledge. *Source*: H. H. Rosenbrock (1989), *Designing Human Centred Technology*, Springer-Verlag

They propose the alternative human-centred tradition for designing socio-technological systems, rooted in the symbiotic relations between the human and the machine. Four possibilities of human–machine relations can be described as (Figure 5):

1. accept the scientific rationality of separation;
2. accept the inevitability of automation and thereby the exclusion of the human from the socio-technological loop;
3. reject the machine and all associated development consequences; and,
4. combine potentials of the human and the machine for human progress, whereby the machine expands the objective part, and the human expands the tacit dimension of knowledge, with both working in symbiosis.

Weizenbaum in his book, *Computer Power and Human Reason* (1986), highlights the dangers of uncritically accepting the inevitability of the machine, and provides two foundation criteria for designing human-centred systems: what is possible and what is desirable, both being applied together in the spirit of human–machine symbiosis. Rosenbrock points to the waste of the human potential created by automation and the machine-centred design of systems, and notes the despair which lies in rejecting the machine, and argues for adopting the purposive perspective of human–machine symbiosis. Cooley argues that this symbiosis transcends the limit of the machine intelligence (10^3) which is determined by three quantitative variables: computability, capacity, speed, and celebrates the potential of human intelligence (10^{14}) which is expanded by at least 14 variables including imagination, consciousness, will, ideology, humour, and political aspiration. The view of symbiosis emphasises that alternatives exist which reject neither human judgement, tacit knowledge, intuition nor the scientific or rule-based methods. We should rather unite them in a symbiotic totality (Figure 6).

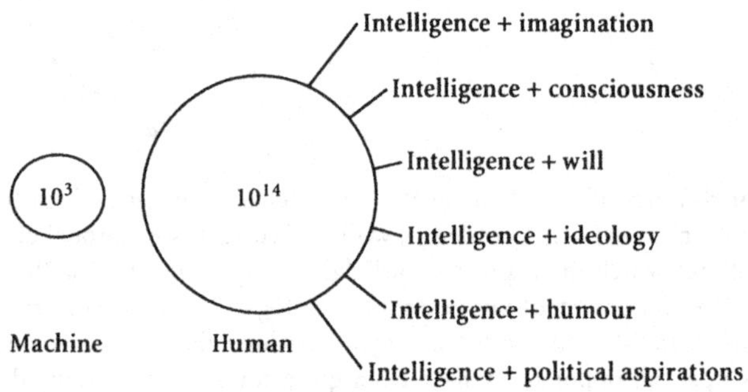

Figure 6: Comparison of units of intelligence available for total information processing. Source: M. Cooley (1987), *Architect or Bee?*, Hogarth Press

At the epistemological level, the human–machine symbiosis challenges the separation of the subject and object, and emphasises the dynamic interrelationship between the tacit dimension of knowledge (experiential knowledge and personal knowledge) and the objective knowledge. It thus responds to the increasing emphasis upon scientific or explicit knowledge at the expense of tacit knowledge, particularly in relation to the development and application of computer-based systems. This emphasis has raised two fundamental concerns within engineering, the social sciences and the information sciences for the proponents of tacit knowledge.

One is about the development of design theory and practice, and the other is about the construction of purposeful and beneficial systems. An over-emphasis on theory or scientifically-based knowledge does not adequately explain the nature of human skills. It undervalues the interrelationship between theory and practice. It also ignores the dimension of human purpose in the explanation of skills and in the design of systems (Gill, S. P., 1995). By rejecting the idea of the automated machine, and the embedded belief that expert knowledge can be completely explicated, the concept of symbiosis provides an epistemological basis for the design of participatory and cooperative information systems. At a research level, this concept has contributed to the development of multidisciplinary design, aiding cross-fertilisation between engineering, the information sciences and the social sciences.

This initially provided the foundation for the ESPRIT (1217) project on Human-centred CIM (Gill, 1990; Rosenbrock, 1989), which was used as an exemplar for the APS programme of the European Commission (Cooley, 1990). At the political level, it rejects the idea of the neutrality of science which is built into the ideology of separation between the instrument and choice. Technological inevitability is promoted by the idea of neutrality which not only takes humans out of the decision-making loop, it also promotes the elimination of the power to choose. At the societal level it provides a forum for: the 'valorisation of diversity', social and economic cohesion of societies, sustainable development, and global coexistence. It does this by promoting the acceptance of difference while making the best of the potentials of people, communities and societies. In essence it seeks symbiosis in diversity and rationality.

Causality and Purposive

Rosenbrock in his seminal book, *Machines with a Purpose* (1990), provides an insight into the purposive vision of technology as an alternative to the 'causal' vision. He argues that cause and purpose embody different conceptions of the relationship between man, nature and machine, which has direct implications for developing of human-centred systems. Causality is predominant in current science and technology. Its modern Western day usage can be traced to the 1500s. In medieval thought there was a trust in human reason which did not require observation and experiment, as is the case today. Authoritative knowledge resided in writings of the past. Between the 1500s and 1700s this gave way, with the advent of modern science, to personal (individual) recorded observation. The operation of nature was compared to the machinery of the time. Then it was the clock, and today it is the computer. Machines do not instil in us any sense of moral obligation. In addition, human behaviour, conscious or unconscious can be explained in causal terms.

In causality, a cause must never be later than its effect and the effect must be the necessary consequence of its use. It omits purpose. In science, one is not permitted to give an explanation in terms of purpose. The same rigor is applied to technology.

The translation of human beings as being machines and being purposeless allows for the Tayloristic approach to management, and a mechanistic character being applied to human work.

Purpose, however, can be considered to exist in machines or living organisms. It is simply a matter of description. A machine can be considered to be a system governed by causal laws which enable it to achieve its objectives. Likewise, in evolution, an organism has a randomness in its behaviour but, in an average sense, the future is determined by an objective which is to be attained. Evolution can be described as purposive. In a world of purpose, human purpose can take place among a multitude of purposes co-existing in the world. These can be competing and collaborative.

Human organisations with purpose allow workers to share in the overall purpose of the organisation. Machines could embody purposes which are subordinate to those of the human. The distinction is that humans care whether these purposes are fulfilled or not. Machines can thereby assist humans rather than replace them. Replacement is the natural end result of the causal description approach. It is this idea of 'replacement' that leads to automation, and thereby to the exclusion of human beings and social purpose from the shaping of technology.

Tacit Knowledge

The concept of tacit knowledge is currently the focus of reflection of two intellectual traditions, the emancipatory humanistic tradition and the Enlightenment rationalist tradition. The proponents of the emancipatory tradition (e.g. Cooley, 1987), cite Polanyi as the formulator of the modern concept, and the proponents of Enlightenment (e.g. Göranzon and Josefson, 1988), cite Wittgenstein and Didero as the two sources for a discussion of tacit knowledge. Polanyi coined the concept, 'tacit knowledge', in his work on *The Tacit Dimension* (1966), where in essence, he described it as that which we cannot express, but that which we know. Tacit knowledge is 'tacit knowing', consisting of two interdependent components, 'knowing what' and 'knowing how'. Polanyi identifies two terms of tacit knowing, the 'primal' and the 'distal', and says that we attend *from* the first to the second, thus achieving an integration of particulars to a coherent entity to which we are attending. In terms of experts in the workplace, tacit knowledge can be described as entailing:

1. *Personal knowledge* – that which we gain from our personal life experiences e.g. family culture, school, friends, i.e. social values, beliefs etc.;

2. *Experiential knowledge* – that which is specific context-based, in the workplace, e.g. work colleagues, group culture, organisational culture etc.

Dreyfus & Dreyfus (1986) propose that learners become experts in their profession and craft by a process of skill acquisition. A learner becomes an expert after going through five stages of their skill acquisition model:

- novice;
- advanced beginner;
- competent;

- proficient;
- expert.

Upon becoming an expert the learner is said to have acquired 'intuition' and 'know-how', which can be termed as 'tacit knowledge'. Cooley (1987), reasserts the attainment of tacit knowledge through *learning-by-doing*, and acquiring 'intuition' and 'know-how', and emphasises Dreyfus' assertion that analytical thinking and intuition are not two mutually conflicting ways of understanding or making judgements. Rather they are seen to be complementary factors which work together but with a growing importance centred upon intuition as the skilled performer becomes more experienced. He sees tacit knowledge in terms of common-sense knowledge, and describes the 'tacit area of knowledge' which ranges from knowledge through to wisdom and action (see Cooley's article in this volume).

Knowledge frequently applied in a domain may become wisdom, and wisdom the basis for action. It is through the symbiosis between the objective and the subjective, that we achieve a correct balance between analytical thinking and intuition. Rosenbrock (1989) deals with the fundamental issue of automation and objectivity of knowledge. He argues that as technology advances, part of the tacit knowledge of the engineer may become redundant, or may become part of the objective knowledge base of the engineering profession. In the process of losing part of the tacit knowledge, the engineer acquires new tacit knowledge. This process of technological progress can lead to the expansion of both the tacit dimension of knowledge and the objective knowledge base. The argument is that a purposeful technology contributes to human progress, rejecting automation as well as the *status quo*, thereby believing in the innovation of technology (expansion of objective knowledge) while enhancing human skill and knowledge (tacit knowledge). The essential point of Rosenbrock's argument is that human knowledge is dynamic, and it is because of this essential character of knowledge that we can shape technology and systems for human benefit.

Tacit knowledge may be expressed in the form of concepts, metaphors, examples, stories and other non-formalistic expressions. The Wittgensteinians (e.g. Johannessen, 1988; Janik, 1990) discuss the concept of 'tacit knowledge' from a hermeneutic perspective and argue that it is obtained through practice, for example, through apprenticeship. The mastery is shown in performance (practice). Tacit knowledge is in the practice, hence it is skill that can be passed down through apprenticeship. To these hermeneuticts the most basic form of regularity in human affairs results from 'rule-following' behaviour in which paradoxically 'no explicit rules are involved', i.e. rules in the sense that traffic laws are rules. The conception of tacit knowledge in the form of constitutive rules is articulated through Aristotle's concept of practical wisdom in the sense that it can be learnt by each person. In this sense practical wisdom is tacit knowledge, for it is genuine knowledge; the practically wise person 'hits the mark' but the *how* is only indirectly communicable. Janik emphasises (1990: 52) that regularity is learnt by imitation, i.e. without explicit rules, which is to say analogically rather than digitally and this enables us to innovate. This is why there can be cohabitation of innovation and tradition within the analogical tradition. Josefson (1987), discusses tacit knowledge as 'tacit knowing', practical knowledge and knowledge by familiarity. Tacit knowing is the ability to see a situation and have an intuition about the problem and how to deal with it. Practical knowledge is the performance of skill, of the tacit knowing. Knowledge by familiarity is the ability to make sound judgements by applying one's experience of previous examples at the starting point of interpreting each unique case. To understand the hermeneutic concept of tacit knowledge, human knowledge is described as consisting in three interrelated dimensions:

1. *Propositional knowledge* - scientific/theoretical knowledge.
2. *Knowledge by familiarity* - knowing when to act; it is acquired from learning within a practice by seeing or examining examples of the tradition in the work.
3. *Practical knowledge* - performance of skill; gained from practical experience.

Rule-Following

The concept of rule-following lies in the way a rule is used, i.e. a rule is meaningful only when it is applied in practice because it is only the practice which gives reality to the rule. In this sense the rule becomes 'rule-following' (Johannessen 1988), that is, it becomes embedded in the tacit dimension of knowledge gained through practice. Since the application of a rule cannot itself be determined through a rule, practice cannot be defined by rules alone. In this Wittgensteinian sense of experiential knowledge, to follow a rule is to know when to follow it, in other words, it is about making judgements about following rules (or breaking a rule, which is also to follow a rule in practice). The argument goes that since the concept of practice has a social character, to follow a rule is to practice a custom, a usage or an institution. Rules can therefore only exist as a link in social life. Kjell Johannessen challenges the very thesis of HCI and cognitive science which espouses individual interaction with the machine, when he says that an isolated individual will not in fact be capable of distinguishing between following the rule and believing that he follows it. Since rule-following activity is social, an individual cannot judge between what is a correct or incorrect way to follow a rule. This implies that information technology tools designed for individual use cannot create a user community, enable the following of rules and practising a custom.

To learn to 'follow a rule' of a practice is like learning a language and becoming a competent performer in the world of practice. Johannessen emphasises that to learn to master a language is a matter of mastering human reality in all its complexity. It is a matter of learning to adopt an attitude towards it in established ways, to reflect over it, to investigate it, to gain a foothold in it, and become familiar with it. Language and human action are intimately interwoven, and so thereby is the relationship between language, rule following, and reality. Therefore, our mastery of language must include a grasp or practical understanding of an enormously large repertoire of situations involving the use of language. In order to understand and respond, we must have situational understanding and judgmental power. The very exercise of an activity might be a constitutive part of the formation of concepts. The content of a concept can be regarded as a function of the established use of its expression. One has mastered a given concept when one is accepted as a competent performer of the series of established practices which incorporates the concept. It is our application of the rule or practice of 'rule-following' which shows how we understand something. 'The practice gives words their meaning'. The identity of a rule over time is attained through the exercise of the established set of practices which guarantees that a rule is applied in the same way from one time to another and from one person to another. The rule itself cannot give this guarantee. (Johannessen, 1988; Gill, S. P., 1995).

Janik (1990), considers the relationship between rule-following, tacit knowledge, and practice, and argues that tacit knowledge cannot be reduced to intuition in the Dreyfus's sense or to unconsciousness in the Polanyi sense, if we are to understand the concept of rule-following, to follow regulative rules and to follow constitutive rules. Human activities such as 'learning to dance' and athletics are rule-governed in the sense of being *regular*, but do not permit themselves to be expressed in the form of explicit rules; rather such *regular rules* are laid down in the form of patterns of a model or of patterns to be imitated.

Constitutive rules are exemplified by citing Aristotle's ethics (morals, goodness, excellent action, virtue) as a form of communal pursuit. This idea of community is central to Aristotle's concept of practical wisdom, it must be learnt. Here, hints and examples are more important than formal rules. In this sense practical wisdom is tacit knowledge, for it is genuine knowledge. Janik rejects the Rationalist and Structuralist argument that there must be explicit rules at the basis of rule-following. If it were necessary to have an explicit rule according to which we learn how to imitate, say, a dance teacher's steps, we would need another rule to apply that rule and so on *ad infinitum*. Regularity is learnt by imitation, i.e. without explicit rules, which is to say analogically rather than digitally. This is why we are able to invent new ways of using old models. In the tradition of a Wittgensteinian account of knowledge, the analogical nature of intellectual practices explains the cohabitation of tradition and innovation within them. In other words the nature of constitutive rules is such that they can be canonical and open-textured at the same time.

Language Games

The idea of a language game has been explored by Ehn (1988), in the context of the Scandinavian approach to participatory user-involved design and cooperative working. This is based on the view that the user is not just an object of study but is also an active agent within the design process itself. The Scandinavian research, exemplified by the UTOPIA project, sees the involvement of users in design as a means of promoting democratisation in the organisation change-process and as a key step to ensuring that the resulting computer system adequately meets the needs of the user. It is argued that users need to have the experience of being in the future use situation, in order to be able to comment on the proposed computer systems, and affect their design.

This requires the construction of a language game which not only facilitates a transcendence between the past and the present, but also facilitates learning about future use situations. In other words, it is about constructing a relationship between language, practice and design. To understand a professional language of design, or any other language game, means to master practical rules which we do not create ourselves but which are techniques and conventions of the design or language game. There are correct and incorrect way of acting, the knowledge of which only comes through socialising and familiarity with the practice. 'To use language is to participate in a language game is the Wittgensteinian notion of practice', (*ibid.*).

Just as our social games are grounded in our everyday language, so are our professional language games grounded in everyday language. We understand what counts as a game not because we have an explicit definition but because we are already familiar with other games. Professional language games can be learnt and understood because of their family resemblance with other language games which we know how to play. To play the game we learn the social rules of the game, and acquire rule-following behaviour of being able to play together, which is more important than being able to follow regulative explicit rules. Playing is interaction and cooperation. To follow the rules in a practice means to be able to act in a way that others in the game can understand. Language as a means of communication requires agreement not only in definitions but also in judgements (Wittgenstein, 1953). This activity involves inter-subjective consensus, which is more a matter of shared backgrounds and language. This consensus requires participants to transcend their own opinions. Being socially created, the rules of language games, like those of other games, can also be socially altered. There are, according to Wittgenstein, even games in which we make up and alter the rules as we go along.

In the language game of design, we use tools such as metaphors and prototypes, as reminders for our reflections on the future use of computer artefacts and their use. By using such design artefacts we bring earlier experience to mind and bend our thinking of the past and future. Ehn poses the question of how users and designers can participate in each other's language games and what means can be developed in the design to facilitate these learning processes. Where there is a gap between the language games of users and designers, mediation is a means of facilitating the closure of the gap. In the Utopia project, scenario workshops, mocks and prototyping are examples of such mediation tools.

Dialogue

In common usage, dialogue means conversation, talk, discussion. As a quality of discussion, a true dialogue does not persuade, delude or dumbfound another person. It has no pre-set goal, it has no end, it is a pure movement, movement which cannot be frozen and translated into a formula, a rule or a programme. Whether resulting in agreement or disagreement, dialogue helps to achieve clarity. In this sense, dialogue is a means of gaining insight through inner reflection. 'Intuition' places dialogue in a meaningful context, a 'presence' of mind which gives it life and depth. This concept of dialogue as a medium of transformation is well understood in the domains of theatre and drama. In theatre dialogue occurs where roles meet; it sets them in motion, leading to unexpected results. It turns the scene into a place where firm positions and sharp contours cannot be maintained. In dialogue differences can appear and be played off against one another, both within the individual and between people, and it sets all their various voices and contradictions in motion. In true dialogue, order and disorder are two sides of the same systems; together they serve renewal and growth in society. Dialogue thus requires a certain measure of scepticism: disagreement is not simply reconcilable with dialogue – it is one of the conditions. Here dialogue emphasises reciprocity and mutual dependence.

However, in our technological culture, the cognitive science view of dialogue dominates in the area of skill and knowledge transfer. This view assumes a homogeneous understanding of reality, and the world as a systems of norms and agreements. Dialogue here is merely a technical medium of transfer, a sort of monologue, separating language from the context and thereby widening the gulf between language and people's experiences. Skill in this perspective is separated from practice and knowledge from experience. This separation inculcates isolation of the individual from the group and the context. Only those who reflect on their experiences develop competence – the ability to deal with new situations similar to those they have already experienced. An unreflecting, purely habitual action does not transcend what has once been learnt. Knowledge requires inner reflection, a 'dialogue with things'. It is through reflecting on experiences, through rhythmic exchanges between participation and distance, between action and reflection, that knowledge grows. The growth of knowledge, in all its various shades and forms, is a process that requires an inner life of its own, its own breathing rate. It can never be a uniform, linear progression. Nor can it be fed with a series of technical data and technical skills. Dialogue is the concept that expresses the dynamics of knowledge (Göranzon and Florin, 1991).

Breakdowns

If 'design is where action is', then reflection is where design is. The concept of *Breakdown* is fundamental to dialogue, reflection in practice, and mediation, which are central concepts for design. When we are accustomed to a situation, we act naturally without consciously thinking about it, we are said to be in an *involved unreflected activity*. However, when we are in an unaccustomed and or unknown situation, we stop and reflect upon how to proceed and act, we are said to be involved in *detached reflection* (Bødker *et al.*, 1991). In other words, breakdowns play an essential role in our understanding of our daily practice. For example, an expert cyclist manoeuvres the cycle through familiar traffic and drives through bends and narrow lanes without any conscious reflection or obeying any rules. An experienced nurse looks after patients in her ward using her informal knowledge and intuition, and an experienced artisan uses tools without reading the manuals. It is only when the traffic is unknown or is blocked that the cyclist stops and reflects to find an alternative route. The nurse may stop and reflect on how to deal with a situation when an accident occurs in the ward or a new patient arrives with unfamiliar circumstances, and the artisan may stop and reflect when new or different tools may be required. When a breakdown occurs in the day-to-day practice and accustomed situations of the cyclist, the nurse and the artisan, it causes them to stop and reflect on how to proceed or resolve the situation. Thus *detached reflection* takes place when involved action breaks down.

Informed by the German philosopher Heidegger, Bødker *et al.* (1991), Ehn (1988), and Winograd and Flores (1986), discuss the concept of breakdown and its significance in the use of tools and other objects in practice, and in the design of tools and systems. Tools and objects are said to be *ready-to-hand* when we are in involved in unreflective activity; we are not conscious of their presence. For example, when a car mechanic repairs the car, he uses various tools without explicitly thinking about them, and the tools for his daily practice are always ready-to-hand. When involved activity breaks down, the objects we use become *present-at-hand*. For example, when the car mechanic is faced with unusual faults or has to repair an unfamiliar car or unfamiliar car parts, he may have to find innovative ways of using tools or may have to find new tools to undertake repairs. Thus the breakdown in the accustomed activity of the car mechanic makes him aware of the new faults and the new situation, and makes him reflect on possible ways of dealing with it.

Ehn (1988), uses the Heideggerian approach of human existence as *Being-in-the World* to discuss the relationship of the concept of breakdown to the design of computer artefacts. Being-in-the World is a throwness, we are always seeking an understanding of the situation we are thrown into, we act one way or the other. In our daily life, we are concerned with ready-to-hand matters as artefacts we use in our activities. As designers, we use tools and objects which are ready-to-hand. In doing so, we are involved in pre-reflective activity and not detached reflection. When we have to learn a new language or rules of communication or when artisans require more appropriate tools, we enter into an *unready-to-hand* situation from a ready-to-hand situation. This is a world of *present-at-hand* artefacts and tools. This process of change from the world of *ready-to-hand* artefacts to *present-at-hand* world is referred to as *breakdown.*

In many cooperative or participatory situations, be they cooperative design, group-working, or learning together, when breakdowns occur we enter into dialogue to reach a common understanding, use mediation to resolve misunderstandings and enter into negotiation to resolve conflicts. These breakdowns in practical situations may occur because of diverse backgrounds, differing interests, varied skills and expertise, or because of language games which participants may play to understand the process of cooperation and collaboration. Ehn (1988), elaborates Wittgenstein's concept of practice: using language is to participate in a language game (Wittgenstein, 1953) to elaborate how we follow (and sometimes break) rules. We enter into dialogue, mediation or negotiation to understand practice by understanding the way the game is played. We use our knowledge of similar games to learn about the rules of the language game of the participatory situation. To play the language game, like any other social game, we have to learn to follow rules, because rules of these games are socially constructed and are not explicitly stated. When we are familiar with the language game, we play the game naturally in a way.

It is only when the language game is not familiar and a break down in the 'rule-following' occurs, we have to learn to follow rules or learn to break rules. Similarly, in many professional domains, experts use their tacit knowledge (knowledge of experience and personal knowledge) to practice their activities, and it is when breakdowns occur in the familiar practice that experts have to stop and reflect in order to proceed further. For example a hospital consultant may need more facts and legal guidelines to diagnose patients in major emergencies arising from major accidents such as a plane crash or a rail disaster. In addition to the facts and legal guidance, the medical staff may need to negotiate the diagnostic procedures with other emergency staff and seek local mediation in the case of unfamiliar situations. These examples illustrate that the concept of breakdown is important not only for understanding the use of tools and objects, but also for understanding the nature of dialogue, mediation and negotiation in practice. Ehn emphasises the significance of the concept of breakdown for the process of design, while pointing out a contradiction arising from this concept. On the one hand, he argues, that design should not create breakdown or make obsolete the understanding of the ready-to-hand which the user has acquired in the use of existing artefacts. Moreover, new artefacts should also be ready-to-hand in already existing practice. On the other hand, it is necessary to breakdown the specific tradition and the understanding of the existing situation in order to create space for reflection, and hence to create openings for a new understanding and alternative designs. To deal with this contradiction between tradition and transcendence, he argues for a design process which makes it possible for users and designers to make use of their practical understanding in designing new situations, and ensuring that the design process also incorporates breakdowns as a means for detached reflection on existing understanding and practice. The main concern here is that the design of tools, system and artefacts is about the world of human practice which is a world in which we as designers, users and consumers are involved in pre-reflective use of these artefacts in everyday life. While it is desirable to situate tools and systems within existing traditions and practice, breakdowns are also essential elements of innovation, creativity and designing for the future. This is the essence of participatory and cooperative design within the tradition of human-centredness.

Usability

From the human-centred perspective, usability contributes to the design of the 'learning organisation', that is, an organisation which supports learning to discover and innovate. This involves learning how systems work, and learning to tackle usability challenges of managing change, inventing and adapting. Usability as a dialogue of change is concerned with the criteria for designing technological systems which support the potential of people who work with them and to understand them, to learn and make changes. Design for usability must include design for coping with novelty, design for improvisation, and design for adaptation (Alder & Winograd, 1992: 7). Usability from this perspective perceives technology as belonging to the communication dimension in the sense that a technological system acts as a cultural medium of a technological culture. It conveys design tradition and use culture. It reflects innovation and technological change, and it defines the embeddedness of technology in society. Technology as a communication product serves as a communication medium between users and designers, and between producers and consumers. Usability should focus on the users' need to learn about the tool's potential and to deal effectively with breakdowns and contingencies. Alder and Winograd point out that the traditional human factor criteria of usability may have sufficed for low level automation, but the increasing complexity of new systems, especially computer-based systems requires a new understanding of what constitutes usability and how to design for it.

The human-centred view of usability transcends the prevalent myth of usability which has been embedded in the human factors model of design and the machine-centred tradition of cognition. The human factors model sees the human as a component of the system, and therefore the job of the designer is to produce an 'interface' which ensures the most efficient fit of this component into the system. The machine-centred model of cognition considers the user primarily from a physical and mechanical point of view, focusing on lower levels of cognition such as character recognition, and the speed and quantity of information retrieval. Both these models undermine the higher cognitive potentials and capabilities of humans, and are therefore ill-suited for understanding the higher cognitive functions of complex reasoning processes and social interaction. Because of their impoverished view of human cognition, the systems usability is considered only as a marginal element in systems design, thereby giving rise to a potential for systems breakdown and redundancy. Usability as a dialogue perspective sees people as active learners, who interpret situations and adapt to their work environments, performing higher levels of cognitive functions of monitoring and changing the system.

Usability as a change agent also challenges the two traditional determinants of work design traditions. These are the deskilling myth and the hierarchical tradition of organisational power relationships. The deskilling myth undermines the human potential and quality focus of the technological age, and is already becoming irrelevant with the rise of usability requirements of multiskilling and learning at the workplace. Traditional management and worker power relationships of control are too limiting for meeting the usability requirements of cooperation, collaboration and participation arising from the new organisational innovation of distributed organisations, group-working, and networking.

The communicative perspective of usability (*ibid.*) shifts the focus of human–computer relations from the mechanisms of human–computer interface design to the design of human–computer collaboration for co-operative working, collaborative learning, and human and organisational networking. The usability challenge here is how to design for collaborative rather than individual work; how to design for informal and adaptable communication patterns; how to design technological systems which enable collaborative working and knowledge sharing across organisational, social and cultural boundaries. In essence, the usability challenge is about how to design dialogue partnerships.

The design languages perspective (Rheifrank, J. J. *et al.*, 1992), of usability shifts design from systems 'functionality' and the 'intent of communicating that functionality' to 'design as the conscious crafting of usability, through the skilful development of form and appearance elements, with the intent of providing people with the resources to perceive and construct usability themselves.' Design languages as a medium of the expression of 'unfolding meaning' of objects, are a means of learning to understand and use objects, and engage in experiences associated with objects. For example, the design language enables artisans to share design experiences with other artisans and transfer patterns of design from one domain to another. As a language for architectural design, it enables architects to design cities, expressing the linkages between geographical, environmental, social and cultural factors of the city and its surroundings. As a business language, it projects the corporate identity and its coherence and relevance to the outside world, and enables the corporation to continuously innovate new design languages to cope with rapid technological and social changes.

The human-centred perspective of usability expects designers to act as collaborators, mediators, communicators, and learners in the design process. This view of usability can be facilitated by human-centred design traditions such as democratic participation, social shaping, and emancipation, and the Scandinavian techniques such as participatory and cooperative design, partial prototyping, mock-ups, dialogues, scenario-building and consensus conferencing.

Ehn (1992) makes a cautionary note on the Scandinavian perspectives of participation and the Wittgensteinian concept of 'language games' as facilitation mediums and tools of usability. In the spirit of envisioning the future of the usability debate, he emphasises that formal democratic and participatory procedures for designing computer-based systems for democracy at work are not sufficient. Our design of language-games must also be organised in a way that makes it possible for ordinary users not only to utilise their practical skill in the design of work, but also have fun while doing so.

Tool Perspective

Building upon the craft ideal of tools, the tool perspective in design relates to the tacit dimension of user's skill. Ehn (1988), provides a philosophical insight into this concept of the tool perspective, and notes that unlike our bodies, our language, and our social institutions, all tools are designed, constructed, maintained and redesigned by humans. Tools should therefore be designed as an extension of existing practice, and their design should involve an understanding of the traditions of design and of use situations by involving designers and users in participatory and cooperative design processes. This allows users to complement their knowledge of the use situation by gaining insights into the technical design of tools, and enables designers to complement their knowledge of technical design by gaining an understanding of the use situation. These insights into other perspectives of the design process can be enabled by both designer and user participating in a 'language game', and sharing insights of the other's practices and reaching a common understanding in the design process. This communicative tool perspective emphasises the *design-by-doing* approach which supports participation through techniques such as mock-ups, partial prototyping, design workshops, and quality circles. The emphasis here is on mutual learning processes resembling the master apprentice relationship in a double sense, where designers and users learn from each other and together about the 'otherness' of their skills and experiences. In the Heideggerian sense (*ibid.*), this perspective involves designing for the future, anticipating future-use situations in the sense of users gaining *ready-to-hand* practical experience of using their future artefacts, and learning to envisage creative *breakdowns* as a means for *present-at-hand* reflections of different possible alternatives of designing for the future. In the Wittgensteinian sense, this design-by-doing tool perspective enables users to enter into dialogues with designers and learn about the possibilities and constraints of new artefacts and their future uses. In the Polanyian sense of tacit knowledge, a good tool is transparent to us; it is something that lets us have a focal awareness of the task and material we are working with. A good tool becomes an extension of our bodies. Ehn (1988: 394), warns of a fundamental risk inherent in good tools and the skilful use of them. Tools, he says, not only strengthen our instrumental abilities, they also make us extremely dependent upon them, and weaken with them. The main risk lies in our dependence on sophisticated computer tools which are designed, maintained, and redesigned for us rather than with us. The design community involved in the Scandinavian design practices and emancipatory traditions of human-centredness is becoming increasingly aware of the challenge of designing computer tools which are convivial in Illich's sense of 'allowing the user to express his meaning in action' (Illich, 1973).

Recent developments in information networking, computer-supported cooperative working, and European initiatives such as telematics argue for using information and communication tools for the transfer and exchange of information and knowledge, and for using these tools for learning and training across organisational and geographical distances. The issue and challenge is how to design information tools which enable users to construct an understanding of these tools without being constrained by the tools and their constructions. Brown & Duguid (1992: 169) propose that well designed tools must provide:

1. legitimacy of participation in the target community in which the tool is embedded;
2. peripherality: an access to the actual, authentic practices of the community;
3. fluidity of participation: tools must enable and empower learners to increase the depth of their involvement in proportion to their development of their knowledgeable skill.

This view recognises that design of tools for learning should not be distracted from the implicit social understanding and the community in which learning takes place, and cautions that a neglect of the social aspects of learning will widen the conflict between the implicit and the explicit, rather than create a synergy which purposive tools should facilitate.

Design Methodology

Design is (Ehn, 1988: 160):

- an artistic and creative process;
- an information and decision-making process;
- one of many societal planning processes;
- one of many socially-determined labour processes

Design is an Active Participation

The cooperative nature of workplace practice is such that people create, use and change information, knowledge and tasks. While traditional HCI approaches treat specific work tasks as undertaken by individuals in isolation, the participative approach looks at groups interacting in multifarious ways within complex organisational contexts. In the latter approach, design is about the active participation of the user in this creative process. Lucy Suchman in her book, *Plans and Situated Actions*, (1987), focuses on the idea that human actions are not so much guided by concrete plans as based on situations. As our circumstances change, so do our actions. Design here is seen within a broader context of *situated actions* of people at the workplace (*ibid.*: 6).

For Greenbaum and Kyng (1991), design is based upon cooperation between systems developers and those they call users. It implies that most work is cooperative, and cooperation or respect for mutual competencies is central to this approach.

Design as a Way to be Rather Than a Thing to be

Wynn (1991) observes that a 'shift in design practice also is more of a way to *be* than a thing to *do*'. What is interesting, she notes, about these things is that they are *processes* that imply a way to *be* with respect to the users, and emphasises that just as users are involved in their worlds of work in a whole way, designers also need to be involved with users in a whole way, as people. Designers should learn to *be* more sensitive, to *be* on the alert for cues to the nature of the organisation as a whole, rather than just relying on its formal description (Wynn, 1991: 63).

Design Approaches: Underlying Ideas

- users are human actors and not cut-and-dried human factors;
- work is situated within an organisational culture and all tasks are situated action; design is a situated process.

- work is fundamentally social, involving cooperation and communication;
- *mutual learning* through dialogue and language games between users and designers is central to the design process;
- *envisioning* of future work allows users to experience how emerging designs may affect the work practice;
- the designer needs to take work practice seriously.

Design-in-Use

The concept of design-in-use concerns visions of technology in use, i.e. the focus is on creating a common understanding of realistic technological visions related to specific domains and use situations. In cooperative design, users and designers meet in a series of creative and constructive workshops undertaking experiments with possible futures based on artefacts such as mock-ups and prototypes. In participatory design and cooperative design, the focus is on techniques to facilitate a common user and designer creativity, and develop visions of technology in use. The emphasis is on giving these visions a form that allows the users to apply their knowledge and experience as competent professionals. This Scandinavian design approach accomplishes these visions through simulating the future use of emerging designs. Alternative designs are embodied in mock-ups and prototypes, examples of data are produced and potential use scenarios sketched as a basis for the simulations (Grønbæk *et al.*, 1995: 26).

Dialogue for Design

One of the major challenges of participatory or cooperative design is to orchestrate and harmonise conflicts and avoid exclusion which may arise from the diversity of design traditions, variety of skills and competencies, different languages of communication, individual preconceptions and personal experiences. It is suggested that this can be achieved by creating a dialogue between different world views in order to harmonise 'collisions' and 'blend competencies' between different backgrounds. Bratteteig & Stolterman (1995), point out that the dialogical process increases individual knowledge and provides a deeper understanding of other positions by accepting a variety of competencies and individual knowledge on an equal basis, thereby creating a common knowledge-base in the group. They suggest that the Scandinavian research on user participation supports this dialogical process through techniques such as prototype presentations (Greenbaum & Kyng, 1991), future workshops (Ehn, 1988; Kensing & Masden, 1991); search conferences (Gustavsen, 1992), priority workshops (Braa, 1995), cooperative prototyping (Grønbæk, 1991), metaphorical design (Masden, 1994), and scenario-based design (Carrol, 1994). Other techniques and methods which aim to identify and describe different perspectives are soft systems methodology (Checkland & Scholes, 1990), and soft dialectics (Bratteteig & Øgrim, 1994).

Collective Resource Tradition

The Collective resource tradition consciously assumes a 'conflict perspective', and through cooperation with workers it aims to develop user participation in the design of the organisation of work and technology. While the 'socio-technique' method focused on the question, 'how do we design systems to fit people', the collective resource tradition asked, 'how do we make it possible for people to design their own systems themselves,' (Ehn, 1988, p. 270). The UTOPIA project is regarded as the best-known of the Scandinavian research projects on the collective resource tradition, especially for developing the design-by-doing approach (Bødker *et al.* 1991; Ehn, 1988). The UTOPIA project (in the Scandinavian language, UTOPIA is an acronym for Training, Technology and Products from quality of Work perspective), was launched in 1981 with aims similar to the Lucas Plan. A team consisting of Danish and Swedish researchers, social scientists and computer scientists, collaborated closely with the Graphics Union in Sweden. Their objective was to develop a demonstrator project to illustrate the compatibility of work qualities such as high craft skills, democratic decision-making, and a high level of health and safety requirements, with the production of high quality products. UTOPIA also included an education objective as part of the design of alternative systems. The design-by-doing method enables users to express their know-how in action, and involves both the researcher and the user as active contributors to problem understanding and problem solving. One of the central aspects of the UTOPIA project is the focus of the 'tool' perspective, i.e. the attempt to design computer-based tools for human work processes, in accordance with the human-centred research traditions.

Action Research

Action research is situated in the practice, and provides a way of building theory and descriptions of socio-technical systems within the practice itself. Braa & Vidgen (1995), trace its roots in the research on democratisation and organisational development, and its links with social science and socio-technical research. In particular, they identify its ties with Lewin's research on social change and social conflicts (Lewin, 1948), the Tavistock Institute's work on socio-technical theory, and Checkland's view on human activity systems and his work on soft systems (Checkland, 1991). From Checkland's perspective, action research is a cycle of continuous enquiry. His action research model is a seven-stage, cyclical process:

1. enter the problem situation;
2. establish rules;

3. declare methodology (M) and framework of ideas (F);
4. take part in the change process;
5. rethink 2, 3, and 4;
6. exit;
7. reflect and record learning in relation to the framework (F), methodology (M) and action (A).

The significance of Checkland's model is that the advanced declaration of the framework also defines the learning that will take place during the action research. Susman's (1983), approach emphasises the formation of a participatory relationship between the researcher. It consists of 5 phases:

- diagnosing;
- action planning;
- action taking;
- evaluating;
- specifying learning.

Both Checkland's and Susman's approaches emphasise learning in action research, and provide a basis for participatory action research which involved practitioners as both subjects and co-researchers. Whyte (1991) defines participatory action research as a process in which some of the people in the organisation or the community actively participate with professional researchers throughout the research process from the initial design to the final presentation of results and discussion of action implications. Braa & Vidgen (1995: 52), make an important observation on the possible conservative and non-radical nature of action research. They point out that the nature of action research itself (e.g. action planning) implies that the diversity of interest groups and their differing opinion may lead to a compromise which is based on a common denominator, rather than on the emergence of a radical solution.

Dialogue, from this point of view, may encourage a form of conservatism rather than active participation for radical alternatives and innovation. Taken together, The Lucas Plan of the 1970s and London Technology Networks of the 1980s on socially useful production (Cooley, 1987), the MEDICA project on cognitive support systems for psychiatry (see Smith's article in this volume), and the CAAAT Project on participatory learning for people with learning difficulties and the PAROSI project on technology and literacy (Gill, 1992), all exemplify the human-centred tradition of action research.

Vision-Oriented Design Process

Bratteteig & Stolterman (1995), describe the group design process as a creation of visions and new ideas, formulating specifications on the basis of a range of competencies, and mutual understanding and learning. The purpose of design is not just the design of the artefact itself, but changes in the range of possibilities for action in the social organisation which will use the artefact (Ehn, 1988). The purpose of systems design is to formulate a vision of these changes, and to describe the vision in a concrete and precise manner. The result of a design process is a set of specifications of computer systems and work processes (Anderson *et al.*, 1986). Systems design is carried out through activities like brainstorming, formulating an offer, description of system functions, architecture, modules, and work processes (Anderson *et al.* 1986: 49). Bratteteig and Stolterman see design as a process of invention in which creating visions is more essential than repair of current malfunctions (Schön, 1983; Ferguson, 1993; Gasparski, 1987). Design is about visualising possible future situations and transcending the limits of the present situation. Design activities are concerned with three levels of abstraction: at an abstract level, *vision*; at a more concrete level, the *operative image* (or *sketch*: Ferguson, 1993); and at the most concrete level, the *design specification* (or *drawing*: Ferguson). The creation of visions, operative images, and specifications are present during the whole process of design; interacting, informing and delimiting each other in a sustainable manner.

A vision-oriented design perspective emphasises design as a *problem-setting* rather than a *problem-solving* activity, and the design process is a process of 'naming and framing' (Schön, 1983). 'Vision' enables the designer to handle and navigate through the complexities of information and possibilities of the design situation, and thereby reduces these possibilities to a manageable range. The design process is seen as a continuous dynamic interplay between these three equally important levels of abstraction. The 'vision' depends upon the imagination and skills of the designers, the operative image is a result of conversations between several designers, and the specification is a translation of the operating image for constructing and realising the design. It is emphasised that in a group engaged in vision-oriented design mutual learning and the 'blending of competencies' is essential. This involves creating a common understanding of a vision and articulating this vision into a commonly agreed operating image (Peng, 1994). Shared images are created through the individual group members contributing to the articulation on the basis of their experience and competence.

Bratteteig & Stolterman (1995), emphasise that the design process requires the grasp of one's own perspective as well as the grasp of the essence and importance of perspectives of the other members of the design group. They point out that members of the group may be trained in different design traditions, and thus the point of shared understanding is not to learn a new profession, but rather to recognise the basic values of other professions, in other words, in evaluating all perspectives. A shared understanding may be based on different interpretations of the same (boundary) object (Bratteteig & Stolterman, 1995: 141). This emphasis on a variety of placements, perspectives and disciplines argues for the group design as an interdisciplinary effort, in which group activities may lead to new perspectives and new priorities, thereby offering new strength to participatory approaches of system development.

This idea of a variety of perspectives and shared images expects participants to moderate their own perspectives in favour of a shared perspective - in other words *forgetting* the individual for the benefit of the group. They argue that vision-oriented design emphasises creativity, and creative design always implies taking risks and that risk cannot be reduced in any simple fashion. Not taking risks in an innovative process, especially within the paradigm of a modifying process, implies defining the problem and its solutions in advance: safety in this context is concerned with staying well within the limits of the possible and the predictable, thereby inhibiting novel or surprising ideas (*ibid.*: 142).

This design process provides an alternative to the management perspective of risk avoidance, through increasing the control of the design process by imposing techniques such as quality assurance or by imposing restrictions of resources or by controlling the agenda. It emphasises that:

> The art of orchestrating group design is the art of balancing and managing a variety of concerns in a way appropriate to the situational conditions - several 'voices' need to be taken care of and different melodies need to be coordinated in order to create music (*ibid.*: 143).

The vision-oriented design process caters for unexpected solutions, both transcending and preserving traditional structures.

User-Controlled Information Systems Development

The idea of 'user control' is a part of a continuum in the movement for participation in design in Scandinavia. Friis (1987, 1991), believes that future users of computer-based systems must be in control of the design process of the development of systems. Traditional systems development methods give little significance to involving users, and traditional prototyping methods focus on the end product, not on the process of design. Friis seeks to demonstrate that future user-designer participation and user control can be achieved through processes of dialogue, where the purpose of design becomes significant. She sees users and designers as belonging to two cultures and believes that the idea of dialogue between cultures needs to be part of systems design. She cites Freire's approach to pedagogy (1972), as a basis for this idea. Freire's pedagogy is that if you want two cultures to really communicate and learn from one another their members have to enter into each other's culture. This necessitates the establishment of a common code of communication in order that the cultures can learn to communicate in the design process.

Prototyping (Friis, 1987), is defined as a method for mediating knowledge/dialogue between designers and users. It allows them to learn from each other. Participation between designers and users makes for a more democratic development of information systems processes. Friis's work differs from that of Ehn (1988), and action-oriented research, and is similar to the Danish project (Laessøe & Rasmussen 1989), in that she describes it as an experimental approach. She calls her case studies 'experiments'. By 'experiment' she means an exploration.

She has developed a model for the design of information systems called the PROTEVS model. PROTEVS is an acronym for PROTotyping for an EVolutionary Systems development. PROTEVS is a model for User Controlled Information Systems Development [UCISD]. It allows users to take responsibility in the design process and regards users as both the problem owners and the problem solvers. The term 'evolutionary' is significant as it denotes a continuous process of learning. Prototyping as a communication instrument between designers and users enables users to develop a better understanding of the opportunities and constraints of computer based information systems, if they are given the opportunity to both design, test and use their own prototype system. Prototyping is expected to provide users with a more understandable tool for their work. In short, the PROTEVS model is expected to enable active learning between designers and users, and user participation, through a process of dialogue.

The design model is modified with the experience of each application, i.e. case study. It is an interactive, process-oriented model which is sensitive to context and places dialogue at the centre of the design process. It provides a framework for how UCISD, and mainly the work with requirements specifications, and evaluations may be performed. As a concept, it is based on active learning possibilities for future users since they participate in the actual design. Hence, the model embodies the idea of generations of users. It is not just about the IS (information system) but about the continued learning process after the particular event. Users can later function as teachers and instructors for the next generation of users (Gill, S. P., 1995).

Future Workshops

Future workshops were originally developed as a method for citizen groups, who had limited resources, to participate in decision-making in public planning authorities (Jung & Mullert, 1987). It has developed into an established method for systems development. The thesis of the future workshops is that the future is too important to leave to the experts. Future workshops build on techniques of creativity which have been tied to the social reality of the participants. The development of social fantasy and social change become two sides of the process. Jungke (Future workshops as a method of work; *ibid.*: 77) believes in making people believe in their own experiences and in the prospect of using their fantasies mutually in a socially productive and dynamic way. The workshops involve a criticism and development of utopian ideas, joined with the aim of action-oriented plans related to the mutual situation of the participants. Future workshops involve a combination of plays and structured/goal-oriented activities. They follow a patterns of steps:

1. criticism based on mutual experiences;
2. turning criticism into positive objectives;
3. turning fantasy from something private and compensatory into a collective method of work;
4. turning utopias into products and actions by systematically working on realising the potentialities;
5. establishing a general cooperation through which a fixation on each others' different views is transgressed and thus utilising each other's strengths in a mutual endeavour.

It is proposed that the future workshops method can be turned into a method for user-researcher cooperation. To achieve this cooperation, a network of 'workshop leaders' should be established in order to facilitate the exchange of experiences among participants. The preparatory work for tutors, especially during the phase of realisation, should be strengthened by integrating researchers into the process. This approach involves two experimental processes:

1. *Extended future workshops*: a week before the workshop starts, users are presented with a 'future exhibition'. They discuss and give opinions on historical examples of utopian projects and social experiments with democratic industrial production. Users are encouraged to criticise and present proposals for improvements. A 'future exchange' is established where the participants are informed about useful contacts, meet invited experts, and then work out the necessary steps to be taken.
2. *A number of research workshops*: their aim is to develop ideas about 'how to create social conditions which make it possible to turn the utopian outlines of the future workshops into reality'. At the end of the process, a mutual public hearing and exhibition are arranged based on the concrete ideas for experiments with democratic industry.

Læssøe *et al.* (1989), discuss the application of the future workshops method, along with experimental prototyping and visionary workshops, in the design of the Electronic Sketch Pad. They comment on the role of the workshop in disclosing the daily work problems and gathering inspiration for how to solve them. They were successful with regard to the first part. The fantasy phase was filled with proposals on how to improve working conditions at the social level. But there were few concrete ideas on how to support engineers in their work with designing human-centred systems and tools. The realisation phase came to almost nothing. They also made the mistake of not involving the technical part of the project in these future workshops. However, the future workshops provided a lot of inspiration.

Kensing and Masden (1991), describe the future workshop as an envisioning method which enables participants to share the problematic situation, shed light on the situation, generate visions about the future and discuss how these visions can be realised. It presents scenarios of the situation in various settings, and uses metaphors to develop participants critique and broaden their visions of the situation. The metaphorical design helps define systems goals by focusing on issues relating to how to achieve results rather than on technical or economic problems. It encourages participants to pay attention to possible changes in the organisation and the working environment, and encourages them to express their likes and dislikes during the Critique and Fantasy phases. The approach encourages participants to contribute their knowledge to identify the 'right' problems to be solved. The focus on user-participation and enabling users to develop their ideas of more desirable systems is based on a critique of current systems. Ehn (1988), discusses the use of metaphors, such as the tool metaphor, the desk metaphor, the spatial metaphor, the model metaphor, in the design of computer artefacts. Metaphors help understanding the situation in terms of another. They may act as reminders and suggestions to the user of similar and familiar situations, and they may relate more to users' general experiences than their professional experience.

Generally, future workshops are run by one or two facilitators, with no more than 20 participants. The facilitators attempt to ensure an equal distribution of speaking time, and also ensure that all participants can follow the discussion. A future workshop is divided into three phases: the Critique phase, the Fantasy phase, and the Implementation phase. Essentially, the Critique phase is designed to draw-out specific issues about current work practice; the Fantasy phase allows participants the freedom to imagine 'what if' the workshop could be different; and the Implementation phase focuses on what resources would be needed to make realistic changes. These phases are preceded by the preparation period and followed by the follow-up period.

The Experimental Work Method

Rauner (cf. *Experimental Learning in Technical Design* [translation] 1984), emphasises that:

> The experiment as a source of learning goes a step deeper and comprises much more than a natural scientific method, partly because the rational objectives of natural science for experiments, with its rules for controlled verification/falsification of hypotheses, gradually has limited the understanding of other forms of experiments for producing knowledge.

'The experiment as a source for learning goes deeper' is a reference to anthropological-psychological studies that emphasise experimental action as fundamental for human recognition. 'We do not only examine the world out of curiosity, but primarily to fulfil our needs'. In this respect the development and use of tools become central factors not only as a means to fulfil our needs but also as a means of recognition. When we develop tools, we go through an experimental learning process constantly correcting the tool on the basis of our practical experiences for its suitability. The technical experiment is radically different from the experiments of the natural sciences because it is oriented towards utility value. Through technical experiments we not only test the validity of abstract hypotheses but also generate concrete useful knowledge.

The application of tools facilitates the experimental learning process. By using the tools we gather knowledge about the world and by applying different tools to the same object we experience the world in different ways. In this context tools are to be understood in a very broad sense. Language, for instance, is the most important tool for establishing our cognitive world. Our total development of language, our conceptual systematisation of reality can be considered as experiments confronting our sensed experience with socially transmitted concepts in a process where they transcend each other and lead to new cognitions.

In saying that the experiment comprises much 'more than and is different from the experiment of the natural sciences', it is implied that we all, mentally and action wise, experiment. The point made by Rasmussen *et al.*, is that experiments are a very important general human method for cognition. One example of experimental method is experimental prototyping: as part of the design-by-doing method carried out in the Utopia project (Ehn, 1988), and the Electronic Sketch Pad (Læssøe *et al.*, 1989).

Computer Support for Cooperative Working (CSCW)

CSCW has emerged as an area which focuses on the role of the computer in group-work and involves researchers from a range of disciplines. The common thread is the idea of group-working. The focus is predominantly upon groups working from different geographical space within a dispersed organisation or upon groups working in the same location (building). Bannon and Schmidt (1989), highlight roles and issues addressed by each of the disciplines involved in CSCW, and emphasise the multidisciplinary approach required in investigating group-work (Diaper and Sanger, 1993).

There is no cohesive point of view as to what counts as CSCW from the technological perspective. For some, e-mail does not count as part of such systems (cf. Wastell and White, 1993), for others, collaborative writing is questionable (Gilbert, 1993). Desktop conferencing is considered to be effective only if groups know each other previously (Tang *et al.*, 1993), so does collaborative writing. Some believe in process enabling systems

[PSS], based upon descriptions of tasks in specific situations (Wastell *et al.*, 1993). These support work activities by undertaking part of the work process itself. It is argued that a large part of the focus is upon office type work or office situated, and other areas of working life, such as the shop-floor in an industrial organisation, need to be considered (Bowers *et al.*, 1995). In the latter case structures are considered to be more rigid. Again, the focus is upon work within an organisation. There is a concern on the part of some (Plowman *et al.*, 1995), that researchers involved in ethnographic research feel a misplaced need to produce design recommendations from their detailed analyses of everyday working practices. This is considered to be another realm of work altogether.

There is a variance in opinion as to how far CSCW should involve a focus on information processing. Some discuss the nature of interpretation of information and how systems should provide cues to enable this process. Dourish *et al.* (1993), in addition, believe in a symbiotic system akin to the Cooley/Rosenbrock belief that the technology should not embody information that lies in the realm of action, contrary to the PSS approach.

Two general approaches exist in the development of CSCW systems. The first and most prominent approach has been to develop systems that support the exchange of information between users. The second approach is to develop systems that exploit the sharing of information to allow cooperation. Often these are both combined to allow more comprehensive co-operative systems to be developed.

Approaches

1. *Information exchange*: CSCW is concerned with the mechanisms and procedures of human–computer interfacing, for example the development of speech act systems, office procedure systems, and semi-structured message systems.
2. *Information sharing*: is concerned with the sharing of information between users, and developing mechanisms to support this sharing.

The key idea is shared information space. Direct user–user communication is normally provided by either electronic message systems or by the use of an audio or video connection, such as those used in multimedia conferencing systems. Some of the earlier CSCW methods include: textual conferencing facilities, hypertext systems for co-authoring (Fish *et al.*,1988), and real-time conferencing systems. These earlier systems traditionally address asynchronous interaction among users. Sarin and Greif (1985) outline a number of areas, e.g. crisis management, where synchronous communication is necessary. The next stage of CSCW methods include:

1. *Desktop conferencing*: the merging of real-time conferencing and workstation technology, thus allowing single-user applications to be

shared between participants in a collaboration transparent manner (Piccardi and Tissato, 1989, MMConf, cf. Crowley *et al.*, 1990);

2. *Multimedia conferencing*: combining shared screen facilities in real-time conferencing with video and audio communication services. An example is the Rapport multimedia conferencing system (cf. Ahuja *et al.*, 1989), which helps users participate in meetings without leaving their offices by providing virtual meeting rooms. The virtual room metaphor has been applied at Bellcore in the development of a system called Cruiser (Root, 1988); another example is MERMAID (Watabe *et al.*, 1990);

3. *Electronic meeting systems*: the support of face-to-face communication represents the most recent and distinct research development in CSCW.

Most such systems have developed from a class of systems known as *decision conferences*. These focus on improving decision-making by groups rather than by individuals - as earlier decision-support systems did. Decision conferences emphasise the use of structured decision processes, mainly involving statistical computer models but increasingly utilising models that embody collaborative notions (Kraemer and Kling, 1988).

The nature and form of control within message-based systems highlights three different classes of cooperative message systems: speech act systems, procedural systems, and semi-structured systems. The approach taken by more 'formal' message systems such as speech act and procedure-based systems is an attempt to capture the cooperation taking place. Semi-structured systems aim merely to provide support for cooperating users by alleviating those tasks that inhibit effective group-work. These systems do not attempt to automate or to merely represent the cooperation taking place. An alternative technological approach to support co-operation has been to focus on the sharing of information and to provide facilities for structuring this information. Most notable examples include various forms of computer conferencing and multi-user hypertext systems. Information-sharing systems assume the existence of a number of communication channels surrounding the shared information. The most striking example is of electronic meeting rooms where the exchange of information in face-to-face communication is central to the success of the system. Electronic meeting systems and the construction of purpose-built meeting rooms represent a recent trend in CSCW systems, which is likely to change many of the existing views of meetings.

CSCW Issues

CSCW is influenced by the development of previous technology. This is reflected in the use of essentially technical classifications in the discussion on CSCW systems. There, it is argued that successful CSCW will require the merging of the technical viewpoints with insights gained from other disciplines on the essentially human nature of cooperation. If cooperative working using computer systems is to succeed, it is argued, it is important to ensure that systems support the user, rather than impose rules and ways of working on them. The type of support that such systems should provide are (Brooke, 1993):

1. To allow each individual to access shared functions and information in their own preferred fashion.

2. Facilitate the exchange of information and ways of working between individuals.

3. Allow evolutionary development and dissemination of working practices and support mechanisms.

Such support requirements are understandable while basic questions about the human interface to multi-user systems remains unanswered (Gilbert, 1993). Sharples (1993), comments that the merging of practices in CSCW work in collaborative writing is haphazard since there is little in the way of a vocabulary and conceptual framework for talking about the process of collaborative writing. The writers reach a method of working mainly through tacit agreement and mutual accommodation.

Green (1991), notes that collaborative methods such as the Coordinator (Winograd, 1988; Fish *et al.*, 1988), are problematic - they can lead to premature commitment, requiring users to commit themselves to courses of action (e.g. selecting a social role or a type of message) when they may not have a rational basis for making a decision, or may prefer to leave options open. Until the 'partners' have worked together they cannot make informed choices about which of the options fits their style of working. Neuwirth *et al.* (1990), point out that roles such as co-author, and 'commenter', may alter or evolve during the task, so taking a predefined role may inhibit activities. Another example of computer support for writers is ShrEdit (Olson *et al.*, 1990). Such types of CSCW systems may well be appropriate where strategies for collaboration have been understood and agreed, but for informal or newly-formed groups the most appropriate tools may be 'blank slate' ones based on familiar media, that allow participants to impose structure to suit their needs as they arise.

Tang and Isaacs (1993), cover issues surrounding the use of video for remote collaboration, use studies, and video conferencing. On design implications of CSCW systems, Olson and Bly (1991), comment that desktop conferencing is a useful medium for distributed collaboration for a working team when participants are already familiar with each other.

Another area of research which has emerged in CSCW is workflow technology - which is designed to give order to, or record, the unfolding of work activity over time by, for example, enabling users to overview the work process or to design work processes for themselves or others, etc. It is pointed out that several systems have been developed on the basis of general theories of communication (e.g. Winograd & Flores, 1986), or abstract process models (e.g. Medina-Mora *et al.*, 1992), rather than upon empirical studies of talk at work or details of actually occurring work practice. Bowers (1994), and Orlikowski (1992), comment that such systems are insensitive to contextual details of work and interaction such that they inhibit their usability. Most CSCW research takes place in 'the office', in administrative and managerial sectors. Such work makes it easy for designers to consider adding to the functionality of office systems with the introduction of workflow applications, whereby computers are the tools/medium of work. Bowers, *et al.* (1995), look instead at production and manufacturing work, which involves special-purpose tools and materials, which may not necessarily be computational or informational in nature. They argue that design requirements for cooperative CSCW technology should:

- support awareness and mutual monitoring of cooperation and not contradict participants own methods of working (Heath & Luff, 1992);
- recognise *ad hoc* collaborative arrangements and not make additional demands on working practice (cf. Abbot & Sarin, 1994);
- consider the implications for systems requirements of understanding workflow technologies as technologies for (inter-organisational) accountability.

They (Bowers *et al.* 1995), emphasise that CSCW research needs to consider the formal (for administrative and management purposes) problems that organisations face and often impact not only on their technology policies but also the details of usage. Hence, they are concerned that CSCW is equated with informal, non-structural interaction. They want to draw attention to the multiple considerations which impinge upon the acceptability of technology in actual contexts.

Bowers (1994), discusses the introduction of a local area network for running CSCW-related applications in an organisation, and describes the kinds of problems and offers concepts to counter them:

1. Notion of boundary objects – entities which can, in some sense, be shared across different social worlds, yet have a variable significance between them. Boundary objects are seen as a solution to overcome the problem when something is perceived differently and sometimes conflictingly by users. What may been seen as useful by one person may be seen as a threat to someone else.

2. Problems and issues are socio-technical, not purely technical nor purely social. The management and use of a CSCW network will require the solution of a range of problems, be they economic, electronic, organisational or computational.

3. CSCW research has to be organisational of a new sort – that which takes account of how organisations perform – as enablers and

constrainers, and how technologies do not so much exist in an organisational context, but are part of what organisations turn out to be.

4. Understanding and managing change is a matter of sorting out what resists and what doesn't and what are the available resources which can be mobilised as forces for change.

Plowman, Rogers and Ramage (1993), give a critique of workplace studies from CSCW literature and discuss problems for the transition from field-work to systems design. The dominant research methodology is ethnography, an interpretative method. This does not impose specific research questions on the participants. It is distinct from experimental methods which are designed to investigate particular hypotheses. They argue that the ethnographic approach provides a rich material on which to ground general design recommendations, but not to make them.

Dourish *et al.* (1993), comment that with the increasing ease and power of networking technologies, shared information systems are becoming the basis of much organisational collaboration. A primary focus in the development of these systems is ease of information retrieval. They suggest that an equally important component is the problem of information interpretation. Information interpretation is guided by a context which many electronic systems do not fully acknowledge. They argue that the transition from traditional to electronic media for the management of information is not straightforward, and question the adage that 'an information system is only as good as the information it holds'. Information must also be interpreted so that an individual can decide how to use it. They investigate this process of interpretation: how and when it takes place; the resources which support it; the implications it holds for the design, deployment & evaluation of electronic shared information systems. The main issues and design implications for information systems are described as:

1. Contextual cues are important resources for information management. Electronic systems should attempt to provide them, otherwise these systems will be less usable and less flexible.

2. The design process for information systems should be sensitive to the nature of the decisions which are based on the information they supply to users, and to the factors which influence this decision-making process.

3. The practices evolving from the use of contextual cues are informal, implicit, and evolving. It is not practical to provide, within the system, actions which are based on contextual information. The contextual information is of value to the end-user in interpreting the information, rather than to the system in making inferences, even on the user's behalf. This leads to a model of shared information: the system is a focus for information and browsing, rather than for

shared action, where the system performs tasks traditionally performed by individuals. This model appreciates the value of informal contextual information in decision-making.

4. Contextual information is frequently low-level and easy for a system to collect, e.g. names, entry times etc. It is most useful for browsing. The value of this context lies earlier in the retrieval process.

Design implications for these researchers become:

1. What contextual cues should be provided in order to assist users in interpreting the information correctly?
2. How will the appropriate set of interpretative practices develop, and how can this process be seeded and guided?
3. More investigation is necessary to reveal more intricate relationships between the information which these systems carry and the context in which it is embedded.

They argue that these issues are at the heart of the design of large shared systems. Information retrieval must be suitably contextualised in order to be used. As designers of large information systems for large groups and dispersed organisations, it is important to have an understanding of the contextual factors involved which may be unique to particular organisations or environments, and to investigate ways in which systems can support the contextual interpretation of information.

The CSCW community emphasises that in essence the argument for CSCW technology is the growing need to support technical, organisational and social activity across geographical and cultural distances which is not being met by current technologies such as the phone, fax, electronic mail, and video-conferencing rooms. Digital audio and video technology allows voice and images to be computationally manipulated and transmitted over existing computer networks. This recent technology and infrastructure developments are lowering some of the barriers that have prevented the widespread adoption and use of multimedia to support remote collaboration.

From a human-centred perspective, the CSCW research is still rooted in the socio-technical domain, dealing with the workplace, predominantly in office and industrial environments. There is an urgent need to investigate the broader issues of cooperation and collaboration arising within the emerging forms of communications networks in different social and cultural contexts.

Design Ideas

Social Encounters and Social Boundaries

Eleanor Wynn (1991), argues that design is a social encounter and social boundaries are a normal feature of human societies. They serve a healthy function in providing meaningful identities and closely-woven backgrounds of practices for their members, so that there is predictability, form, and order within groups as well as across them. These boundaries pose problems in terms of our contemporary values when they rigidly support and enforce systems of social stratification. Certain activities are done within the group and others between groups. Social distance can be maintained without physical distance, in the form of stereotypes and assumptions. Real differences such as education, income, and social dialect may be taken to imply differences that are not real in intelligence, ability, responsibility, and credibility, but they are constructs of 'false consciousness'. Such a false consciousness inhibits many encounters between people of different backgrounds. It affects assumptions that we hold about others, and hence affects our interaction and observation. It therefore affects what we learn about both them and the situations they are in.

The important point to note is that everyone brings to a task a set of presuppositions which are not necessarily grounded in anything other than practice. As the problems in the field change, the presuppositions may become inadequate, and may block the development of a more inclusive paradigm. Thus there is the possibility that invisible assumptions in systems development, whether they are social or methodological in nature, stand in the way of creative innovation during a time of pervasive technological change.

In the case of observations within our own culture, there is the additional difficulty that much of the behaviour is 'transparent' to us. It looks normal and routine. It is part of our own background. The researcher or designer must reverse the field, pull the background to the foreground, and begin to see how portions of behaviour function as part of the process. In the case of workplace studies, this means seeing how those unarticulated or 'glossed' practices support the work (Wynn, 1991: 51–55).

Sociality and Design

Suchman and Trigg (1991), suggest that in designing technology for the workplace, an important assumption is that work practice is fundamentally social. It is social in the sense that any activity, whether characterised by conflicts or by cooperation, relies on a foundation of meaningful, mutually intelligible interaction. Moreover, it is the community rather than the individual, that defines what a given domain of work is and what it means to accomplish it successfully. Finally, every occasion of work, however individualistic it may appear, involves some others, either in the form of co-workers or of recipients. The basic sociality recommends that wherever we go we look for the human interactions that make up the work and define what counts as competent practice, and 'sociality is what makes analyses of human interaction so relevant to technology design' (*ibid.*: 73).

On Reflection in Practice

Bødker *et al.* provide an insight into the concept of reflective practice, and draw a distinction between *involved unreflected activity*, and *detached reflection*. They define involved unreflected activity as 'a basic way of being', acting without detached reflection, adjusting naturally to the specific circumstances, just as we would walk though the kitchen door without any thought or an experienced driver would drive the car without consciously following rules. Only when an involved action breaks down, such as the kitchen door is stuck or a road is closed for repairs, we would reflect (detached reflection) and take an appropriate action. Detached reflection is thus something secondary, taking place only when the involved action breaks down. In an unreflected activity, an experienced craftsman would use tools in a ready-to-hand way, in a natural way without reflection. When involved activity breaks down, the tools become ready-at-hand, and the craftsman may reflect on their use. The breakdown makes us aware of the possible ways to remedy the situation. From the perspective of design, it is only through use and not just by reflection, that we can get to know how the future application will work (*ibid.*: 145).

Design Challenges

From Techno-Centred Design to Cooperative Design

Terry Winograd and Fernando Flores in their book on *Understanding Computers and Cognition* (1986), point out that traditional system design approaches (techno-centric) are rooted in the rationalist tradition of Western scientific thought. The rationalist tradition not only pervades information systems design but also much of cognitive science, management science and linguistics. They comment that there exists a deep-rooted conflict in the action and reflection of practice, and note that many researchers acknowledge the phenomena that are not subject to the rationalistic style of analysis, but in their day-to-day work they proceed as though everything were, and build systems based on the rationalist tradition avoiding areas in which they break down.

With the arrival of the computer, the human-machine relationship has become synonymous with the human-computer relationship, which has been seen in terms of the relationship between psychology and computer science under the umbrella of human-computer interaction (HCI). This relationship focuses on two aspects:

1. *cognitive psychology* – a computational metaphor in psychology; and,
2. *human factors* – aiming to understand human psychology and its relevance to the design of the human-computer interface.

The 1980s saw a shift from traditional concerns of human factors to cognitive science – information processing psychology, with a heavy emphasis on cognitive ergonomics. Bannon (1990), notes that although cognitive science spans many disciplines such as psychology, sociology, linguistics, and philosophy, its central concern lies in the study of intelligence, and mechanisms, whereby it can be realised, whether in natural or artificial organisms.

The key idea is the *essential similarity* of processes that lie behind human and artificial 'reasoning'. Cognitive science perceives both brains and digital computers as physical symbol systems, thereby focusing on the study of symbolic representations of human thought processes. This view of the human mind as an information processor was challenged by the research into connectionism and neural networks in the late 1980s. The questioning of the 'information processing' thesis of cognitive science included a questioning of its emphasis on the individuals without reference to their community, their history or their culture, while expecting the subject of study to perform according to a certain 'ideal', a rational model of problem solving. The assumption was that 'problem solving' activity is a generic cognitive activity which can be measured and transferred across domains and cultures in the same sense as Taylorist skill can be measured and transferred according to the rationalist model of deviation from the quantitative norm. It assumes that the subject should play the 'problem solving' game but without participating either in defining the game, its rule, or its outcome (Bannon, 1990; Norman & Draper, 1986). The theoretical impoverishment of this cognitive approach to human–computer relations was observed by Lave (1988):

> Cognition observed in everyday practice is distributed – stretched over, not divided among – mind, body, activity and culturally organised settings (which included other actors).

From Human Factors to User-Centred

The questioning of the information-processing model shifted the focus of HCI research from human factors to the user-centred approach, focusing on user practice and user modelling, and emphasising the practicality and utility of cognitive science to the design of HCI. Allen Newell's memorable phrase 'Design is where action is' became the cornerstone of making cognitive science as the 'science' of HCI. Bannon notes that user modelling, for example Programming User Models (PUMs), (Young, Green and Simon, 1989), continues to focus on the narrow view of human activity and on the separation between the user and designer, thereby divorcing human cognition from human practice. Thomas and Kellog (1989), discuss this limitation of the HCI approach and argue for bridging the 'ecological gaps' between the abstract world of the laboratory and the real world.

They identify a number of a number of gaps between the abstract nature of HCI and the real-world needs:

- 'user gap': a gap based on individual difference and motivation;
- 'task gap': where the laboratory task may not generalise to an actual work situation;
- 'problem formulation gap': a gap between the functionality of the tool and its appropriateness for the user needs;
- 'artefact gap': where the application may not fit into other applications;
- 'the extensionality gap': a gap between the laboratory use of a tool and its general applicability;
- 'work context gap' the crucial gap (from the human-centred perspective) which they identify, and which concerns the social setting and the culture of the workplace.

It is noted that the 'bridging' metaphor requires some stepping-stones to cross the limitations and boundaries of the dominant 'representational formalism' of cognitive science, and the 'boxology' of the separation which isolates cognitive processes from actions (Velichkovsky & Zinchenko, 1982). It was recognised that if technology were to be designed as a tool for the user(s), then there has to be shift from the user-centred to user-involved design, which should act as a facilitator for collaboration, cooperation, coordination at the workplace and other group and participatory activities. This led to an increasing attention to concepts of 'situated action', and 'usability' and research developments in Computer Support for Cooperative Work (CSC), (Bannon and Schmidt, 1989).

One of the major thrusts of the shift from user-centred system design to cooperative design is illustrated by the shift from seeing the user as the human *factor* to seeing the user as the human *actor*. The human factor (HF) notion connotes the user as 'a passive, fragmented, depersonalised, unmotivated individual' while the 'human actors' notion connotes the user as 'an active and controlling one' (*ibid.*: 27). In the HF approach, the human is often regarded as another component of the machine system, that can be factored into the design equation of the overall human–machine system. It neglects human characteristics such as communication, motivation, creativity, values, goals and beliefs about life and work. The human actors approach regards the person as an autonomous agent, and emphasises the holistic nature of the person acting in a setting rather than acting as a mere information processing mechanism of technology. The HF approach is illustrated by most of the HCI research which is biased towards technology and views people as 'users' of technology, and 'naive users' at that, often blind to the reality that the user's view of technology may be very different to that of the designer's view. This approach also ignores the roles and functions of workers/users as competent practitioners; people with responsibility for tasks and involved in the development of workplace relationships. Traditionally, it is based on the separation of system functions into machine functions and user functions, and the design of the human–computer interface is then about synchronising these separate functions to achieve the maximum performance of the system.

These approaches reside in the North American developments in *human factor engineering* and the European developments in *ergonomics*, undertaken by behaviour scientists or industrial engineers, contributing to the understanding of human capabilities and limitations in workplace settings – effects of stress, psycho-motor ability, perceptual acuity, mental processing and work loads etc. (*ibid.*: 32). The 1980s saw a shift in HCI research from HF to cognitive science, promoting the design of better cognitive coupling between the human and the computer, and highly interactive interfaces which focus on better dialogues with users, and proposing slogans such as 'ease of use' and 'user friendliness' (*ibid.*: 33). The concerns of the HCI research community have recently, however, expanded to building better, more usable and more useful artefacts, rising to Allen Newell's phrase (Bannon, 1989), 'Design is where the action is'. This has led to research into user modelling, prototyping, and looking at the structure, content, and dynamics of individual user cognition at the interface.

From User-Centred to User-Involved Design

During the 1980s there emerged a further shift from user-centred to user-involved design, which was based on the view that the user is not just an object of study but is an active agent within the design process itself (Bannon, 1989). This involvement of users in design is seen not only as a means of promoting democratisation in the organisation change-process but also as a key step to ensure that the resulting computer system adequately meets the needs of the user. It is argued that users need to have experience of being in the future-use situation, in order to be able to comment on the proposed computer systems, and affect their design. Users and designers can participate in the cooperative design process through mock-ups or prototypes. This perspective takes the work as being embodied in the work process, and attempts to support workers by providing them with skill-enhancing computerised tools (Ehn and Kyng, 1987). The essence of the concept of cooperative design lies in respecting users' skills and promoting democratisation of all phases of design and working practice, and not in the tools or their use *per se*.

The challenge was to design technological systems which are situated in the current practice but also enable future users to reshape their practice and skills. One such approach is the 'human activity framework' (Bødker, 1989), which recognises that the use of an artefact is part of social activity, and the design of computer artefacts means that we design new conditions for collective activity, e.g. new divisions of labour, and alternative ways of coordination, control and communication. The future use situation is the origin for design, and we undertake design with this in mind.

Use, as a process of learning, is a prerequisite to design. Through use, new needs arise, either as a result of changing conditions of work or as a recognition of problems with the present artefacts. To design with future use activity in mind also means to start out from the present practice(s) of future users. Cooperative design is a meeting place for many different practices, for sharing experiences and learning new skills. Viewed from this perspective the design is a learning process for individuals as well as groups about each other's practices and 'languages games' of the practice. This design approach requires that designers and users must be prepared to acknowledge each other's competence and to realise that effort must be made by both parties to develop a mutually agreed vocabulary of concepts that can be shared across different groups that comprise a design project. This human-activity based approach extends the user-centred approaches to participatory approaches from the human-centred perspective.

The main influence in this new direction comes from the Scandinavian tradition of 'democracy and participation'. Two alternative conceptions of the computer, the 'tool' perspective and the 'design-by-doing' perspective has influenced a radical shift from the earlier systems design processes. The tool approach stresses the crucial role of the tacit knowledge people have as opposed to just focusing on their theoretical knowledge. From this perspective, the design must be based on participation with people who really know the work-process that is to be redesigned, and on the tacit skills that workers have acquired through long and diverse experiences of learning-by-doing, not only for democratic reasons but also for strong epistemological reasons. One of the crucial epistemological issues for systems design is what Wittgenstein called the 'language games'. Language games is the expression of the practices of both the users and the designers. The users are experienced in the language games of their work or use situations, and the designers are experienced in the language games of the design. Systems designers and users in the design teams have to learn each other's language games in order to be involved in the cooperative design process. The user-involved approach uses the concept of the 'use model' for drawing together the relation between the user interface of the computer artefact, the users' professional language and competence, and the design situation. The practices of users and designers are different. The use model is the means to bridge the gap in design, as well as in use, of the artefact under construction. Participation techniques such as mock-ups and prototyping, and design approaches such as future workshops, experimental design and consensus conferences are used to enrich the user-involved design process.

Anthropocentric Systems (APS): a New European Tradition

The human-centred debates of the 1970s, reflecting philosophies of democratic participation and emancipation, and traditions of 'collective resource', 'co-determination' and 'humanisation of work', found their consolidation in the Anthropocentric Production Systems initiatives in the 1980s. The notions of 'lean production' flexibility, and adaptability arising from an increasing emphasis on global economic competitivity are in danger of creating a new orthodoxy of 'lean' rationality.

This trend seems to sit uncomfortably in the emerging world of networks of economies, organisations, and working life. It may be time to go back to the social roots of the human-centred debates of the 1970s, and shape education and research traditions which support 'inclusion', social innovation and co-development from wider social and cultural perspectives.

This requires a rethinking of the role of information technology from being just a production tool to being a tool for both the production and reproduction of knowledge. This means shaping the tool as a technological resource for social innovation and learning in the emancipatory and participatory traditions of human-centredness. But what can we learn from the APS tradition?

APS Perspectives: a Journey of the European Dimension

A journey through the European dimension may provide an insight and answer to the above question. The journey below is a review of a number of selected APS projects and publications sponsored by the FAST (MONITOR) Programme of the European Community in the 1980s. The review aims to illustrate the commonalities and differences of APS perspectives reflecting the national and cultural traditions of European research into human-centred systems. It also reflects the diversity and similarity of European debates on anthropocentricity of the 1980s and the early 1990s.

British Perspectives

The British human-centred tradition has arisen out of two complementary approaches in Britain in the 1970s, socially useful production (Cooley, 1987), and 'human–machine symbiosis' (Rosenbrock, 1989). The socially useful production is rooted in the 'Lucas Workers Plan' in the early 1970s, and later in the London Technology Networks in the early 1980s. Rosenbrock's concept of alternative technological development rests on two basic ideas. The first of these he calls, the 'Lushi Hill effect' which basically signifies that there are many ways to reach the peak of Lushi Hill, and one cannot say which is the best, i.e. there is no one-best way. The second idea encapsulates the 'symbiosis of the tacit and the objective' which is supportive of the use of human skill and potential of skilled workers. The third approach, of 'social innovation', extended the production contexts of the first two approaches into the socio-economic context to include human organisations such as the education, health and welfare sectors (Gill, 1986, 1993; Ennals, 1991). The common core of these human-centred approaches is that culture-based knowledge and actions of human beings should be reflected in a dynamic way in systems instead of being subsumed.

Although the human-centred research tradition in Britain has been established since the 1970s, its influence on mainstream British industrial culture has been limited. Charles *et al.* (1991), point out that the academic research on new technology and work organisation in Britain has been mainly situated in the social science research tradition. It has been concerned with the critique and analysis of Taylorism, the labour process, deskilling and managerial strategies in the "unique 'Anglo-Saxon' research tradition" of empirical research.

As compared to human-centred research traditions of academia in Scandinavia and Germany as being an integral part of the mainstream industrial culture, the mainstream British research tradition remains rooted into the techno-centric paradigm, and human-centred systems research until now has remained at the periphery. Although anthropocentric concepts such as group technology, quality circles, JIT and multi-skilling have become part of British management repertoire, their practice is undercut by the British 'top-down' managerial culture which sees enterprise flexibility more in terms of numerical rather than functional flexibility. In spite of the awareness of Japanese industrial culture of quality circles and JIT, and the Scandinavian and German production cultures of 'semi-autonomous working groups' and participation, as well as a wealth of social science research into human-centred systems, British industrial culture remains 'less conducive to APS and new forms of work organisation than neo-corporatist systems'. This may partly be due to the voluntarist nature of workplace training, exclusive nature of public and private education systems, and an underlying framework of adversarial industrial relations, managerial belief in 'top-to-bottom' innovation, and the techno-centric nature of R&D policies in Britain (Charles *et al.*, 1991).

Keywords: socially useful production, 'human–machine symbiosis', purpose, social innovation, top-to-bottom' innovation, voluntarist tradition, numerical rather than functional flexibility.

Scandinavian Perspectives

The Scandinavian tradition relates to the two Scandinavian traditions, of 'collective resource' and 'action research'. In the industrial work, this tradition is concerned with human working conditions and the cooperation of users. The emergence of the Scandinavian tradition of user-oriented working life research goes back to the 1960s when the 'socio-technique' was introduced in Norway with the aim of creating new production relations and new types of organisational design which would guarantee new forms of participation, even within national politics. The ideas did not take root in Norway but found a ready market in Sweden. In Norway, the keywords were 'industrial democracy and participation'. In Sweden they became 'job satisfaction and productivity'. In Sweden, the project on 'humanisation of work' and 'science of work' became dominated by the management. In the 1970s, slogans such as 'Research for People' became a focus of attention in Denmark, while in Norway, 'Action Research' became the keyword for methodological discussions and a basis for a large number of user-oriented projects within all areas of society (Laessøe and Rasmussen, 1989).

Banke *et al.* (1991), point out that by virtue of their social position, the Scandinavian labour movements and Social Democratic parties have been able to exert more influence upon societal development than in most of the other Western industrial nations.

This has resulted in the active participation of workers in the development of their own workplace, and research on user-oriented working life. The tradition is based on the idea that employees themselves should be actively involved in designing their own equipment and organisation. The UTOPIA project (Ehn, 1988), the Danish initiatives, 'Technology and Society Initiative' in 1983, the 'Management and Cooperation of Technological Innovation' in 1987, and the 'Improved Utilisation of Advanced Production Systems' in 1990, have contributed to expanding the socio-technical development to allow for broader, more satisfying jobs, and more autonomy for the individual worker and groups of workers.

The Danish Employers Confederation's campaign, UPS, 'Renewal From Within' in 1983, though rational and elitist in its conception, accepted the participation of employees in practice, thereby supporting the central contention of APS that new technological possibilities and their shaping should be based on work culture, rather than being based solely on some ideal originating at the management level. Banke *et al.* note that considerable interest in the APS tradition in Denmark in even small companies is due to a growing interest in the methods and principles of job development projects of the 1970s, and these are seen in terms of flexibility and quality, thereby improving competitivity. This interest in APS among Danish companies, even during the recession, indicates the Danish emphasis on the collective resource tradition and participatory democracy.

Keywords: participation, user-oriented work-life, Denmark/Norway: 'collective resource', 'action research'. Sweden: 'humanisation of work', 'science of work'.

French Perspectives

Linhart (1990), points out that the debate in France on man's place in the working process in the late 1960s, provided a basis for the creation of ANACT, a national agency for the betterment of working conditions, and the revalorisation of manual work. She emphasises that Taylorism was very much developed in France in the 1960s and had an intensive and extensive impact, with a specific component, the 'fayolism', after the name of a French engineer, Foyal, a contemporary of Taylor, whose ideas on centralism, authority and hierarchy influenced French rationalism and the French organisational culture. During the 1970s, the social critique of Taylorism highlighted the importance of workers' involvement and the role of the 'tacit' dimension of knowledge. In 1982, France instituted 'Auroux laws' (after the name of the French minister of labour), which provided workers with a totally new right: 'a collective right to direct expression' (a unique right in Europe), leading to the setting up of thousands of 'collective expression groups'. The French government also set up an interdisciplinary research programme, 'Programme Mobilisateur Technologie-Emploi-Travail', to study the place of man in the workplace, taking account of his specificity, his capacity and his possibilities. During the 1980s, a new orientation towards more anthropocentrism was promoted in the name of 'after Taylorism'. This provided a way to modernisation and flexibility through new technology, and supported a new place and new role for workers, moving towards a new philosophy of social relationships at the workplace. Linhart says that one of the major handicaps for spreading the anthropocentric tradition in France is the French ideology, which has a traditional contempt for manual work, especially industrial work. Embedded in this ideology is the antagonism of two cultures, the culture of workers and the culture of managers, inhibiting an open and participatory dialogue, which is at the heart of the anthropocentric workplace culture.

Keywords: 'fayolism', French rationalism, the place of man in the workplace, collective expression groups, French ideology, antagonism of two cultures.

German Perspectives

von Bandemer *et al.* (1991), suggest that probably the most important institutional factor that has shaped the development of anthropocentric production in Germany has been work councils' 'co-determination' rights in matters of work design. 'Co-determination' provides German employees and their representatives the most far-reaching participation rights in the matters of evaluation and regulating the integration of technology at the workplace. While in other European countries, employers may volunteer information and consultation rights to workers and von Bandemer *et al.* (1991), suggest that probably the most important institutional factor that unions, in Germany they have a right to be informed and consulted. This unique aspect of worker and union participation in Germany involves them in the rationalisation processes and flexibilisation necessitated by the global market place as partners in enterprise rather than as adversaries as in the UK. This German partnership stresses two mutual objectives:

1. to humanise work; and,
2. to increase the competitiveness of enterprises.

The technological infrastructure, the central feature of the Tayloristic vision, is not considered as deterministic but rather as favourable or restrictive of APS. Since the early 1980s, influenced by the 'humanisation of work' and the Japanese enterprises of JIT and Quality Circles, the design of work and technology initiatives in Germany have aimed at enhancing the flexibility and quality of the production process by fostering the qualification of work and workers.

One of the key assumptions of this anthropocentric strategy for the use of technology is that skills and competencies of the workforce on the shop-floor is a major source of innovation, flexibility and productivity, and have become determining factors for meeting the increasing demand for quality and consumer orientation – the characteristics of the world market. von Bandemer *et al.* suggest that the management and organisational concept of Taylorism have not influenced the German industry as much as they have been dominant in other countries. One reason for this is the strong craft tradition which is upheld by the dual vocational training system which is a feature of small, medium and large scale companies in Germany. The models of 'semi-autonomous work-groups' in Germany, they note, have been influenced by the Swedish experiments on group-working and by the 'Japanese small group activities'.

Keywords: 'co-determination', rationalisation, 'humanisation of work', qualifications, enterprise competitiveness.

Indicative Problems:
It is also interesting to note that the Swedish idea of 'semi-autonomous group-work' challenged the old German craft tradition of 'Meister', because the middle and lower management feared losing 'power, privileges and even job by the decentralisation of decision and control' (*ibid.*: 35). It was no longer the 'Meister' who was guiding and controlling the work, it was the 'group leaders' who were putting more effective, and often more subtle pressure on group members.

The non-hierarchical and almost egalitarian nature of anthropocentric production systems raises another important issue of career and occupational advancement. A traditional hierarchical work system is built around the notions of progression and career paths based on the differential skills categories and competencies, which an egalitarian anthropocentric system of workers cooperating under similar conditions may not subscribe to.

Irish Perspectives

Like its historical links to Britain and the USA, the technological and organisation innovations in Ireland have also been historically tied to the industrial cultures of these countries. Because of the colonial links with Britain, the post-colonial Irish industrial culture, reflected by its hierarchical organisational and management systems, and class-based educational system, shows a close identity with the British traditions of 'corporatism' and 'social contract'. O'Siochru and Dillon (1990) note that, like many other less developed small economies, Irish industry is dominated by multinationals who may not be inclined to invest in R&D which promotes innovations for the long-term sustainability of the small country. They have also suffered from the lack of a skill-based education, experience-based training, multiskilling and life-long learning at the workplace, which are the basic determinants of anthropocentric working life. They observe, however, that recent growing awareness of the limitations of the hierarchical forms of work organisations among unions and management in Ireland reflect the wider European dimension including the developments in anthropocentric production systems. The particular Irish notion of 'social partnership' is rooted in the 'close association between national and class struggles' which encouraged trade unions to become involved voluntarily in tripartite attempts at economic and industrial development. O'Siochru & Dillon make an important observation about the nature of century-old craft–skill culture which still exists in Ireland as well as in traditional craft cultures such as Greece and Portugal. They suggest that these old craft traditions of these periphery countries which provided the backbone of production during pre-capitalist times, may still provide qualitative skills for many small enterprises to go directly to human-centred approaches, using sophisticated microelectronic technologies. The crucial point here is that the anthropocentric paradigm is not tied to the large pool of skill and the technological base of advanced industrial countries such as Germany. The developing countries may gain a comparative advantage in combining advanced information and communication technologies with their own old traditional craft–skill bases, and by creating small enterprises which bypass the technical straightjacket of the Taylorist production and organisation cultures. It is plausible to argue that countries like Ireland, which do not have embedded technological industrial cultures in the sense of industrialised countries such as Britain and France, may be in a better position to cultivate an industrial culture such as APS, one that 'valorises the worker and human skills' and complements the 'enterprise culture' of small enterprises.

Keywords: post-colonial Irish industrial culture, class-based educational system, 'corporatism', 'social contract', 'social partnership', 'leap-frog over the Fordist stage.

Glimpses of European APS Perspectives

Glimpses of the European perspectives on APS above illustrate a diversity of perspectives, reflecting the nature of industrial cultures, socio-political traditions, and historical situations. For example, the British perspective reflects the human-centred tradition of socially useful production, and the notions of 'human–machine symbiosis', purpose, and social innovation. This perspective also reflects the constraints and opportunities of innovation of the individualistic culture and voluntarist traditions, as well as the limits of the industrial culture rooted in the separation of work and learning. Perhaps it's not surprising that the 'home of human-centredness' has not been open to its practice.

The Scandinavian tradition emphasises participation and user-oriented work-life, with the Danish and Norwegian perspectives reflected in the 'collective resource' and 'action research', and the Swedish perspective defined by the 'humanisation of work', 'science of work' ideas. On the one hand the Scandinavian tradition reflects the 'inclusiveness' of the emancipatory and democratic practices, and on the other hand, it reflects the 'exclusiveness' of the rationalist tradition. It may be that this dialectics of the Scandinavian tradition attracts the insiders (of the rationalist tradition), and that the ambiguity of this dualism attracts the outsiders. This may also say something about the amenability of this tradition to both social innovation and competitive anthropocentricity.

The German perspective is shaped by the ideas of 'co-determination', the practice of rationalisation, the initiatives of 'humanisation of work', education rooted in the notion of qualifications, and working culture sustained by the enterprise competitiveness. This perspective of 'integrated rationality' deepens the mechanisms and processes of the 'productive' industrial culture, which may not be easily amenable either to the ideology of the antagonism of two cultures of industrial work, workers and managers' (such as the French ideology, and the British tradition) or to the 'pragmatic' cultures rooted in individualistic and voluntarist traditions of the Anglo-Saxon world.

The French perspective is influenced by 'fayolism', French rationalism, the notion of the place of man in the workplace, and ideas of 'collective expression groups', and the French ideology of the antagonism of two cultures, workers and managers. Just like the British tradition of the separation of managers and workers, the French ideology of antagonism does not seem to provide a hospitable work environment for the APS. Alternatively, the 'integrative rationalism' of the APS may be inappropriate to the ideology centred and individualistic cultures.

The Irish perspective is cultivated by the notions of 'social contract', and the concept of 'social partnership' at work. The developing nature of Irish industrial culture together with a network of craft skills and small-scale enterprises, may make Ireland, like Greece and Portugal, more receptive to the more emancipatory ideas of anthropocentricity. Reflecting upon the APS traditions, there emerges a glimmer of emancipation and 'inclusiveness' in its diversity, while raising issues of the 'exclusive' nature of the 'integrative' and 'collective' rationalities embedded in the notions of 'co-determination' and 'semi-autonomous group-work', as well as, the limits of 'separation' inherent in the individualistic and hierarchical cultures.

While the voluntarist and 'individualist' traditions may be amenable to the diversity of culture, in the sense of a 'network of diversities', the 'integrative' rationalities may be too restrictive to respond to the 'fuzzy' and 'ambiguous' nature of the diversity. A challenge to APS is how to become hospitable to the notions of the 'valorisation' of diversity, and how to support developmental notions of 'inclusion' in the emerging networks of the 'information society'.

Concluding Remarks: Towards a New Symbiosis?

The notion of human-centredness, defined in terms of 'human–machine' symbiosis, is rooted in the production culture of the industrial society. It seeks collaboration between the human and the machine on the basis that the machine supports the human skills in performing human tasks and decision-making processes. It argues that the production and reproduction of knowledge is a social process and is therefore embedded in the culture itself. The 'tacit' dimension of knowledge (personal knowledge plus experiential knowledge) cannot be separated from the personal and social experiences, and is therefore not amenable to objectification.

Knowledge consisting of facts and rules, 'objective' knowledge, can be amenable to formalisation in terms of the language of the machine, and is therefore computable. Thus the notion of the symbiosis between the human and the machine requires a symbiosis between the 'tacit' dimension and the objective knowledge. While the objective knowledge may be expressed in the universal language, the expression and interpretation of the tacit knowledge is dependent upon the social and cultural contexts in which it is produced and sustained. Human-centredness thus promotes the notion of diversity of cultures, languages and knowledges.

Another notion which human-centredness promotes is that of the human as both the producer and user of knowledge. This notion basically stems from the idea that the artisan (the skilled worker) is both the producer of products and the user of products. This notion thus promotes the worker as a stake holder in the whole process of production and not just as a component of the machine and as merely an instrument of the production of goods and services. The worker therefore is not just the producer and user of products and knowledge, but is also the evaluator of production and knowledge. Historically, human-centredness promotes the design of emancipatory and socially useful technology which enhances human skill, knowledge and social potential.

Basically the motivation of human-centredness is to provide an alternative model to the machine-centred model of technological innovations, and the purpose is to promote socially useful and culturally responsive technologies. The ideas and notions underpinning human-centredness are the notions of the 'tacit' dimension of knowledge, dialogue, diversity of culture, language, and knowledge.

Over the years, the human-centred philosophy has contributed to the development of socio-technical design, user-centred systems, user involved systems, and cooperative design processes, emphasising that dialogue and participation are central beliefs of shaping work-life environments. However, because of the historical roots of the human-centred debate in the production culture, the notions of skill and symbiosis have until now remained rooted in the production culture. The world of work and living are, however, changing rapidly. In this age of information networks, the symbiosis is not between the single machine and the single user, but is a matter of symbiotic relationships between the network of users and the network of machines. It is a matter of communication between groups and between the human and machine networks. It is not a matter of the interaction between the skilled worker and the machine, but of a world of collaborating users, who have a variety of skill levels, and networks of machines performing at a variety of functional and cognitive levels. The 'tacit' knowledge no longer resides in the individual artisan but resides in the community of users in the form of the social knowledge base or a network of social knowledge bases. It is not just a question of objectification of the experiential knowledge, but a question of objectification of social and professional knowledge at a variety of levels of objectivity and ambiguity.

In the technological age of networks of consumer, user and producer communities, networks of economies, and networks of communication technologies, the challenge of human-centredness is how to move beyond the traditional notion of human–machine symbiosis, and promote notions and concepts which deal with a variety of human–human and human–machine relationships and networks of relationships, both at local and global levels. We have moved beyond the age of ergonomics and human factors, human–computer interaction, machine usability and cognitive performance of the individual user. It is an age of communication networks and networks of communication producing, reproducing, and sustaining varieties of skill- and knowledge-bases. This challenge of technology design is part of a bigger societal challenge as to how to reintegrate technological innovations into the civil society so that technology supports new forms of work-life and living environments. This requires the innovation of new forms of social structures and organisational cultures which respond to and cope with the changing world of work and living.

At the intellectual level, the way forward is to cross the wall of 'causality' embedded in the scientific method, and find a harmonious relationship between 'cause' and 'purpose'. Causality is not something that is imposed on us but resides in our scientific view of the world. It is a presupposition which we adopt before we begin to study the world and explain it. It is equally in our gift to take an alternative 'purposive' view of technology (Rosenbrock, 1990).

At the epistemological level, the way forward is to rethink about the epistemological issues arising from the changing nature of the relationships between humans and machines. One way is to seek harmony between the 'tacit' knowledge and the 'objective' knowledge in a network of human–machine relationships.

At the methodological level, the way forward is move beyond the cognitive and social spaces of the individual embedded in human–machine interaction, to human–machine relationships required of the network of cognitive and social spaces. Here we can learn from the human-centred approaches, such as the social action approaches, the participatory approaches, social shaping, and the culture of the artificial.

At the level of 'cultural pragmatics', the way forward is to build on the 'valorisation of cultural rationalities', in order to support the symbiotic relationships between local identity and cultural plurality. The issue here is to formulate a research agenda for multimedia environments, which recognises the dynamic relationship of local specificity and global diversity.

References

AI & Society: Journal of human centred systems and machine intelligence, Springer-Verlag.

Abbott, K. and Sarin, S. (1994). 'Experiences with workflow management issues for the next generation' in *Proceedings of Computer Supported Working, CSCW' 94.* ACM, Chapel Hill, New York, USA.

Ahuja, S. R., Ensor, J. R. and Horn, D. N. (1988). 'The rapport multi-media conferencing system' in Alder, P. S. and Winograd, T. A. (1992), *USABILITY: turning technologies into tools*, O.U.P.

Allen R. B. (ed.). *Proceedings of conference on Office Information Systems* (COIS-88). Palo Alto, CA.

Andersen, N. E. *et al.* (1986). *Professional Systems Development.* Teknisk Forlag, Copenhagen.

Bannon L. J. 'From Cognitive Science to Cooperative Design' *in Theories and Technologies of the Knowledge Society*, Centre for Cultural Studies, Aarhus University, September 1989.

Bannon L. J., and Schmidt, K. (1989). 'CSCW: Four Characters in search of a context' in *Proceedings of First European Conference on Computer Support for Cooperative Work*: 358-372

Bannon, L. J. (1990). 'A Pilgrim's Progress: From Cognitive Science to Cooperative Design', *AI & Society*, vol. 4.4: 259–275.

Banke, P., Clematide, B. and Rasmussen L. B. (1991). *Prospects for anthropocentric systems in Denmark*. FOP 255, APS Series, MONITOR, CEC, Brussels.

Bødker, S. 'Human Activity Approach to User Interface', DIAMI PB-291, Aarhus University, September 1989.

Bødker, S. *et al.* (1991). 'Setting the Stage for Design as Action' in Greenbaum, J. and Kyng, M. (eds.) *Design for Work: cooperative design of computer systems*. Lawrence Erlbaum Associates, Hillsdale, New Jersey.

Bowers, J., Button, G. and Sharrock, W. (1995). 'Workflow from within and without: Technology and Cooperation Work on the Print Industry Shopfloor', *ESCW' 95*, Stockholm.

Bowers, J. (1994). 'The work to make a network work: studying CSCW in action' in *proceedings CSCW '94*.

Braa, K. and Vidgen, R. (1995). 'Action Case: Exploring the Middle Kingdom inn Information Systems Research Methods' in The Proceedings of the Third Decennial Conference: *Computers in Context: Joining Forces in Design*. Aarhus, Denmark, August 14–18, pp. 50–59.

Braa, K. (1995). 'Priority Workshops: Springboard for Participatory Design in Redesign Activities' in *Proceedings of the Conference on Organisational Computing Systems*. COOCS '95. ACM SIGOIS, California.

Bratteteig, T. and Øgrim, L. (1994). 'Soft Dialectics – Structured Handling of Problem Situations in System Development' in Bates, W. R. J. (ed), (1994*). proceedings of the Second European Conference on Information Systems*. NIjenrdoe University, Breukelen, April 28–29, pp. 681–690.

Bratteteig, T. and Stolterman, E. (1995). 'Design in Groups – and all that Jazz' in *Proceedings of the Computers in Context: joining forces in design*. Aarhus, Denmark, August 14–18, pp. 137–146.

Brödner, P. (1992). *The Shape of Future Technology*. Springer-Verlag.

Brook, J. (1993). 'User interfaces for CSCW' in Diaper (eds.), (1993).

Bruner J. (1983). 'The Growth of Cognitive Psychology: developmental psychology' in Miller, J. (ed.) *States of Mind*. British Broadcasting Corporation: 30–41.

Carrol, J. (1994). 'Making Use of a Design Representation', *Communication of the ACM*, vol. 37, no. 12. pp. 28–35.

Charles, T., Charles R. and Roulstone, A. (1991). 'Prospects for Anthropocentric Production Systems in Britain', FOP 252, APS Series, MONITOR, CEC, Brussels.

Checkland, P. (1991). 'From Framework through Experience to Learning: the essential nature of Action Research' in Nissen, H.E. *et al.* (eds.) *Information Systems Research: Contemporary Approaches and Emergent Traditions*. Elsevier (North Holland).

Checkland, P., Scholes, J. (1990). *Soft Systems Methodology in Action*. Wiley, Chichester.

Cooley, M.J. (1987). *Architect or Bee?* Hogarth Press,

Cooley, M.J. (1987). 'Human Centred Systems: An Urgent Problem for Systems designers', *AI & Society*, 1.1: 37–46.

Cooley, M.J. (1990). *European Competitiveness in the 21st Century: Integration of Work, Culture and Technology*. FAST: Commission of the EC.

Corbett, J.M., Rasmussen, L. B. and Rauner, F. (1991). *Crossing the Border*. Springer-Verlag, London.

Crowley, T., Milazzo, P., Baker, E., Forsdick H., and Tomlinson, R. (1990). 'MM Conference: an infrastructure for building shared multi-media applications' *in Proceedings of CSCW'90*.

Dankbaar, B. (1987), 'Social assessment of workplace technology – some experiences with the German programme 'Humanization of work'' in *Research Policy* 16 (1987), 337–352. North-Holland.

Diaper, D. and Sanger, C. (eds.), (1993). *CSCW in Practice: An introduction and case studies.* Springer-Verlag, London.

Dourish, P., Bellotti, V., Mackay, W., and Ma, C. Y. (1993). 'Information and Context: Lessons from a study of two shared information systems' in *Proceedings of COOCS'93.*

Dreyfus, H. (1972). *What Computers Can't Do: a critique of artificial reason.* Harper & Row, New York.

Dreyfus, H. and Dreyfus, S. (1986). *Mind over Machine: The power of human intuition and expertise in the era of the computer.* The Free Press, New York.

Drucker, P. (1994). *Post-Capitalist Society.* Buttersworth-Heinemann, Oxford.

Ehn, P. and Kyng, M. (1987). 'The Collective Resource Approach to Systems Design' in Bjerknes, G., Ehn P. and Kyng M. *Computers and Democracy - a Scandinavian Challenge*, pp. 17–58. Aldershot, UK: Avebury.

Ehn, P. (1988). *Work-oriented design of computer artifacts.* Arbetslivscentrum, Almqvist & Wiksell International, Stockholm, Sweden.

Ehn, P. (1992). 'Scandinavian Design: On participation and skill' in Adler, P. S. and Winograd, T. A. (1992), *USABILITY: turning technologies into tools.* O.U.P, Oxford.

Ennals, R (1991). *Artificial Intelligence and Human Institutions.* Springer-Verlag.

Ferguson, E. (1993). *Engineering and the mind's eye.* The MIT press, Cambridge, Mass.

Florin, M. and Göranzon, B. (1991). *Dialogue and Technology: Art and Knowledge.* Springer-Verlag, London.

Floyd, C. (1984). *A Systematic Look at Prototyping.* Springer-Verlag, London.

Freire, P. (1972). *Pedagogy of the Oppressed.* London: Shedd and Ward.

Friis (1987). 'User developed prototype systems' in Rasmussen, J. and Pranas, Z. (eds.) *Empirical Foundations of Information and Software Science III.* Plenum Press, New York.

Friis, S. (1991). 'User involvement in Requirements analysis and specifications: in theory and practice' in van Besselaar, P., Clement, A. and P. Jarvinen, (eds.) *Information Systems, Work and Organization Design*: proceedings of IFIP WG 9.1. Berlin, 1989. North Holland.

Fish, R. S., Kraut, R. E., Leland, M. D P. and Cohen, M. (1988). 'QUILT: A Collaborative Tool for Cooperative Writing' in Allen, R. B. (ed.) *Proceedings of the Conference on Office Information Systems (COIS-88)*, Palo Alto, CA, 23–25 March. ACM, New York, pp. 30–37.

Garsparski, W. (1987). 'Design Methodology: A personal statement' in Durbin P. T. (ed.) *Philosophy of Technology.* Kluwer Press, pp. 153-167.

Gilbert, N. (1993). 'CSCW for Real: Reflections on Experience' in Diaper, D. *et al.* (1993)

Gill, K. S. (1986). *Artificial Intelligence for Society.* Wiley & Sons, Chichester,

Gill, K. S. (1990). *A Summary of Human Centred Systems Research in Europe.* NTT DATA.

Gill, K. S. (1992). 'Human Centredness: A 21st Century Paradigm for Industrial Cultures' in Masuda, Y. (ed.) *Human Centred Systems in the Global Economy.* Springer-Verlag, London,

Gill, K. S. (1993). 'Socially Sustainable Technology: an agenda for human promotion' in Ardigò, A. (ed.), (1993). *le Nuove tecnologie per lepromozione umana.* Angeli, Milano.

Gill, K. S. (1993). 'Education and Social Innovation' in Wobbe, W. (ed.), (1993). *The Future of Industry in the Global Context.* FOP 359, FAST (EC), pp. 69–80.

Gill, K. S. (1995). 'Information Society and Co-development: designing cultural interfacing for knowledge networking' in *International Workshop on Revival of Humanity in the Information Society.* RISS, NTT DATA, Tokyo, Japan. 30 January, 1995.

Gill, K. S. (1994). Keynote Paper: 'Human Centred Shaping of Social Innovation' in Schmid, F. *et al.* (eds.) *Computer Integrated Production Systems and Organisation.* Springer-Verlag.

Gill, K. S. (1994). 'Knowledge Networking for Co-development' in the *International Conference on Globalisation and IT Networking* (Global Information Society). IFIP WG9.4, LSE, London, May 16–17, 1994.

Gill, S. P. (1988). 'On Two AI Traditions', *AI & Society*, vol. 2.4.

Gill, S. P. (1995). *Dialogue and Tacit Knowledge for Knowledge Transfer*, PhD dissertation, University of Cambridge.

Göranzon, B. and Josefson, I. (1988). *Knowledge, Skill and Artificial Intelligence*. Springer-Verlag, London.

Green, T. R. G. (1991). 'Describing information artifacts with cognitive dimensions and structure maps' in Diaper, D. and Hammond, N (eds.) *People and Computers VI*, proceedings of HCI'91. CUP.

Greenbaum, J. and Kyng, M. (eds.), (1991). *Design at Work: Cooperative Design of Computer Systems*. Lawrence Erlbaum, Hillsdale NJ.

Greenbaum, J. and Kyng, M. (1991). *Design at Work, Cooperative design of computer systems*. Lawrence Erlbaum Associates, Hillsdale, New Jersey, Hove, London. ISBN 0-8058-0612-1 (pbk).

Grønbæk, K. (1991). *Prototyping and Active User Involvement in System Development: Towards a Cooperative Prototyping Approach*, OPhD Thesis, DAIMI, University of Aarhus.

Grønbæk, K., Kyng, M. and Mogensen, P. (1995). 'Cooperative Experimental System Development – Cooperative Techniques Beyond Initial Design and Analysis' in *Proceedings of the Computers in Context: joining forces in design*. Aarhus, Denmark, August 14–18, pp 137–146.

Gustavsen, B. (1992), *Dialogue and Development*, Arbetslivscentrum & Van Gorcum, Assen/Maastricht

Habermas, J. (1971). *Toward a Rational Society*. Heinemann, London.

Heidegger, M. (1962). *Being and Time*. Harper & Row, New York.

Heath C. and Luff, P. (1990). 'Disembodied Conduct: Task Coordination in London Underground Control Rooms' in Bannon, L., Robinson, M. and Schmidt, K. (eds.) *Proceedings of the Second European Conference on Computer Supported Cooperative Working (EC-CSCW'91)*. Amsterdam, September, Kluwer, Dordrecht.

Illich, I. (1973). *Tools for conviviality*. Calder & Boyars, London.

Janik, A. (1990). 'Tacit Knowledge, Rule-following and Learning' in Göranzon, B. and Florin, M. (eds.) *Artificial Intelligence, Culture and Language: on education and work*,. Springer-Verlag: 45–56.

Johannessen, K. S. (1988). 'Rule Following and Tacit Knowledge', *AI & Society*, Springer-Verlag, vol. 2, no. 4: 287–302.

Josefson (1987). 'The Nurse as an Engineer', *AI & Society*, vol. 1.2.

Jungk, R. and Mullert, N. (1987). *Future Workshops: how to create desirable futures*. London, Institute for Social Inventions.

Kraemer, K.L. and Kling, J.L. (1988). *Computer-based systems for cooperative work and group decision making*. ACM Computing Surveys, 20 (2).

Laessøe, J. and Rasmussen, L. B. (1989). *Human-centred Methods – development of computer-aided work processes*. Technical University of Denmark, Building 301, Institute of Social Sciences, DK-2800 Lyngby, Denmark.

Laessøe J. and Rasmussen L. B. (1989). *The Electronic Sketch Pad- prototype observation and organizational context*. Technical University of Denmark, Building 301, Institute of Social Sciences, DK-2800 Lyngby, Denmark.

Latour, B. (1993). *We Have Never Been Modern*. Harvester, Wheatsheaf.

Lewin, K. (1948). *Resolving Social Conflict*. Harper, New York.

Linhart (1990)*Prospects of anthropocentric production systems in France*. FOP 253, APS series, MONITOR, CEC, Brussels.

Lave, J. (1988). *Cognition in Practice*. CUP, Cambridge.

Kensing, F. and Masden, K.H. (1991). 'Generating Visions: future workshops and metaphorical design' in Greenbaum and King (1991).

Masden, K. H. (1994). 'A Guide to Metaphorical Design' in *Communication of the ACM*, vol. 37, no. 12, pp. 57–62.

Medina-Mora, R., Winograd, T., Flores, R. and Flores, F. (1992). 'The action workflow approach to workflow management' in *Proceedings of CSCW'92*. Toronto, Canada, New York: ACM.

Mitter, P. (1986). 'Should Artificial Intelligence Take Culture into Consideration?' in *Artificial Intelligence For Society*, John Wiley & Sons, Chichester.

Morgan, G. (ed.), (1983). *Beyond Method: strategies for social research*. Sage, Newbury Park, CA.

Negrotti, M. (ed.), (1990). *Understanding The Artificial*. Springer-Verlag, 1990.

Neuwirth, C. M., Kaufer, D.S., Chandhok, R. and Morris, J.H. (1990). 'Issues in the design of computer support for co-authoring and commenting' in *CSCW '90*: 183–195.

Norman, D. A. and Draper, S. W. (eds.), (1986). *User Centred System Design – New perspectives on human computer interaction*. Lawrence Erlbaum Associates, Hillsdale, NJ.

Olson, J., Olson, G., Mack, L. and Wellner, P. (1990). 'Concurrent editing: the group's interface' in Diaper *et al.*, *HCI proceedings*. Elsevier, Amsterdam: 835–840.

Olson. M. H. and Bly, S. A. (1991). 'The Portland Experience: A report on a distributed research group', *International Journal of Man-Machine studies*, 34(2): 211–228.

O'Siochru, S. and Dillon, B. (1990). *Prospects for Anthropocentric Production Systems in Ireland*. FOP 258, APS Series, MONITOR, CEC, Brussels.

Peng, C. (1994). Exploring Communication in Collaborative Design: cooperative architectural modelling in design studies, vol. 15.1.

Piccardi, S. and Tisato, F. (1989). 'Conference Desk: an experiment and model for application sharing' (in). Technical report, systems software laboratory, Direzione Olivetti Ricerca, Milan.

Polanyi, M. (1967). *The Tacit Dimension*. Anchor Books, Doubleday & Company, New York.

Polanyi, M. (1973). *Personal knowledge*. Routledge, Kegan and Paul; London.

Plowman, L., Rogers, Y. and Ramage, M. (1995). 'What are workplace studies for?' in the proceedings of ECSCW'95.

Orlikowski, W. (1992). 'Learning from Notes: Organisational issues in groupware implementation' in *Proceedings of CSCW' 92*, Tronto, Canada. ACM, New York.

Rauner, F. (1984). 'Experimentierendes lernen in der Technischen Bildung' in Steffens, K. (Hrsg.). *Experiementalle Statik an Fachhochschulen*. Leuchtturm-Verlag.

Rheifrank, J. J. *et al.* (1992). 'Design for Usability: crafting a strategy for the design of a new generation of Xerox copiers' in Alder, P. S. and Winograd, T. A. (eds.) *USABILITY: turning technologies into tools*. OUP.

Root, R. W. (1988). 'Design of a multimedia vehicle for social browsing' in *Proceedings of the Conference on Computer Supported Cooperative Work (CSCW-85)*, Portland, September. ACM, New York.

Rosenbrock H. H. (1977). 'The Future of Control', *Automatica*, vol. 13.

Rosenbrock, H. H. (1988). 'Engineering as an Art', *AI & Society*, vol. 2.4.

Rosenbrock, H. H. (ed.), (1989). 'Designing Human Centred Technology: A Cross Disciplinary Project' in *Computer Aided Manufacture*. Springer-Verlag.

Rosenbrock H. H. (1989). 'The Technical Problem' in [Rosenbrock, H.H. (1989)].

Rosenbrock, H. H. (1990). *Machines with a Purpose*. OUP.

Rosenbrock, H. H. (1993). 'Technology and its Environment', *AI & Society*, vol. 7, no. 2, Springer-Verlag, pp. 117–126.

Saalman W. (1980). *Filippo Brunelleschi*. Zwemmer.

Sarin, S. and Greif, I. (1985). 'Computer based real time conferencing systems', *IEEE Computer*, 18(10): 33–45.

Sharples, M. (1993). 'Adding a little structure to collaborative writing' in Diaper *et al.* (eds.).

Schön, D. A. (1983). *The Reflective Practitioner*. Basic Books.

Sean O'Siochru and Brian Dillon (1990). *Prospects for Anthropocentric Production Systems in Ireland*. FOP 258, APS Series, MONITOR, CEC, Brussels.

Seely Brown, J. and Duguid, P. (1992). 'Enacting design for the workplace' in Alder, P. S. and Winograd, T.A. (eds.), (1992). USABILITY: *turning technologies into tools*. O.U.P, Oxford.

Suchman, L. A. (1987). *Plans and Situated Actions: the problem of human-machine communication*. CUP, Cambridge.

Suchman, L. and Trigg, R. H. (1991). *Understanding Practice: video as a medium for reflection and design*, pp. 65–89.

Susman, G. (1983). 'Action Research: a socio-technical perspective' in Greenbaum, J. and Kyng, M., *op. cit.*

Tang, J. C. and Isaacs, E. (1993). 'Why do users like video? Studies of multi-media supported collaboration' in *Computer Supported Cooperative Work* (CSCW). 1: 163–196. Kluwer Academic Publishers.

Thomas, J. and Kellog, W. (1989). 'Minimizing Ecological Gaps in User Interface Design', *IEEE Software*, January, pp. 78–86.

Urquart, A. (1978). *Intellectual and manual labour: A critique of epistemology*. Macmillan, London.

von Bandemer, S., Henning, J. and Hilbert, J. (1991). 'Prospects of Anthropocentric Production Systems in West Germany', *FOP 251*, APS Series, MONITOR, CEC, Brussels.

Wastell, D.G. and White, P. (1993). 'Using process technology to support cooperative work: prospects and design issues' in *Diaper et al.* (1993).

Watabe, K., Sakata, S., Fukuoka, H. and Ohmori, T. (1990). 'Distributed multi-party desktop conferencing system: Mermaid' in *CSCW '90*.

Weizenbaum, J. (1976). *Computer Power and Human Reason*. W.H. Freeman and Company, San Francisco.

Wiener, N. (1950). *The human use of human beings*. Avon Books-Discuss Edition, 1967.

Winograd, T. (1988). 'Where the action is', *Byte*, December: 256–258.

Winograd, T. and Flores, F. (1986). *Understanding Computers and Cognition*. Ablex.

Wittgenstein, L. (1953). *Philosophical investigations*. OUP.

Whyte, W. F. (ed.), (1991). *Participatory Action Research*. Sage, Newbury Park, CA.

Wobbe, W. (1991). 'Anthropocentric Production Systems: a strategic issue for Europe', FOP 245, APS Series, MONITOR, CEC, Brussels.

Wynn, E. (1991). 'Taking Practice Seriously' in Greenbaum, J. and Kyng, M., *op. cit.*

Young, R. M., Green, T. P. and Simon, T. (1989). 'Programmable User Models for predictive evaluation of user interface design' in Bice, K. and Lewis C. (eds.) *Proceedings of CHI' 89 Conference on Human Factors in Computing Systems*. ACM Press: New York.

Chapter 2

On Human-Machine Symbiosis

Mike Cooley

Introduction: Technological Change

Firstly, human-centredness asserts that we must always put people before machines, however complex or elegant the machine might be. Secondly, it marvels and delights at the ability and ingenuity of human beings. It offers an insight into the way we work, and through our work the way we relate to each other and to nature. It is intended to highlight some of the problems associated with our top-heavy political structures and their total inability to respond to creative energy from below.

The central issue of our time is our overweening faith in science and in technological change. Science is a shallow and arid soil in which to transplant the sensitive and precious roots of our humanity. Faith is indeed the correct word to use in this context. Science and technology are now leading-edges in society– in rather the same way religion was in medieval times. Furthermore, the zealots of science and technology display much of the missionary zeal of the colonial era. Those who do not understand, or more particularly, accept the dictates of science and technology are almost viewed as lost souls who must be redeemed from their appalling ignorance and, if they cannot be redeemed, are sacrificed at the stake of exclusion and unemployability. The human-centred systems movement looks sensitively at those forms of science and technology which meet our cultural, historical and societal requirements, and seeks to develop more appropriate forms of technology to meet our long-term aspirations.

Technology and Skill

There continues to be a widespread belief that automation, computerisation and the use of robotic devices frees human beings from soul-destroying routine, back-breaking tasks and leaves them free to engage in more creative work. It is further suggested that this is automatically going to lead to a shorter working-week, longer holidays and more leisure time – that it is going to result in 'an improvement in the quality of life'. It is usually added, as a sort of occupational bonus, that the masses of data we will have available to us from computers will make our decisions much more creative, scientific and logical, and that as a result we will have a more rational form of society.[1]

Some of these assumptions are questionable. We are already repeating in the field of intellectual work most of the mistakes already made in the field of skilled manual-work at an earlier historical stage when it was subjected to the use of high-capital equipment. The old separation between the manual and the intellectual is no longer meaningful for the study of the impact of technological change. Consequently, there is a need to look critically at technological change as a whole in order to provide a framework for questioning the way computers are used today.

In my view it would be a mistake to regard the computer as an isolated phenomenon. It is necessary to see it as part of a technological continuum discernible over the last 400 years or so. I see it as another means of production and as such it has to be viewed in the context of the political, ideological and cultural assumptions of the society that has given rise to it. I am not at all surprised, given the questions we have asked of science and technology, and the 'problems' we have used them to solve, that we now end up with the kind of systems we see all around us. I hold that we have been asking the wrong questions and therefore we have come up with the wrong answers. It is, however, extremely difficult for the public at large to intervene in this process since, instead of baffling them with Latin, the new religion confuses them with mathematics and scientific jargon. They are led to believe that there is something great and profound going on out there, and it is their own fault that they don't understand it. If only they had a PhD in computer science or theoretical physics they would be able to grapple with the new theological niceties. The scientific language, the symbols, the mathematics and the apparent rationality bludgeon ordinary people's common-sense. A concern that things simply are not right and could and should be otherwise is flattened into abject silence. However, those who do have the appropriate 'qualifications' are also increasingly uncertain, confused and disoriented. The discussions among physicists about the limits of their existing 'objective' techniques and the concern among computer scientists about the implications of artificial intelligence all indicate that the fortress of science and technology in its present form is beginning to show gaping cracks.

> Above all this, there is a seething unhappiness among both manual and intellectual workers because the resultant systems tend to absorb the knowledge from them, deny them the right to use their skill and judgement, and render them abject appendages to the machines and systems being developed. Those who are not directly involved in using the equipment are merely confused bystanders. I find a deep concern that individuals feel frustrated because their common sense and knowledge, and their practical experience, whether as a skilled worker, a designer, a mother, a father, a teacher or a nurse, are less and less relevant and are almost an impediment to 'progress'.[2]

Common Sense and Tacit Knowledge

'Common sense' in some respects is a serious misnomer. Indeed, it may be held to be particularly uncommon. What I mean is a sense of what is to be done and how it is to be done, held in common by those who will have had some form of apprenticeship and practical experience in the area.

This craftsman's common-sense is a vital form of knowledge which is acquired in that complex 'learning by doing' situation which we normally think of as an apprenticeship in the case of manual workers, or perhaps practice in law or medicine.

I shall also refer frequently to tacit knowledge. This knowledge is likewise acquired through practice, or 'attending to things'.

These considerations are of great importance when we consider which forms of computerised systems we should regard as acceptable.

It is said that we are now in an 'information society'. This is held to be so because we are said to have around us 'information systems'. Most of such systems I encounter could be better described as data systems. It is true that data suitably organised and acted upon may become information. Information absorbed, understood and applied to people may become knowledge. Knowledge frequently applied in a domain may become wisdom, and wisdom the basis for positive action.

All this may be conceptualised as Figure 7 in the form of a noise-to-signal ratio. There is much noise in society, but the signal is frequently dimmed. Another way of viewing it would be the objective as compared with the subjective.

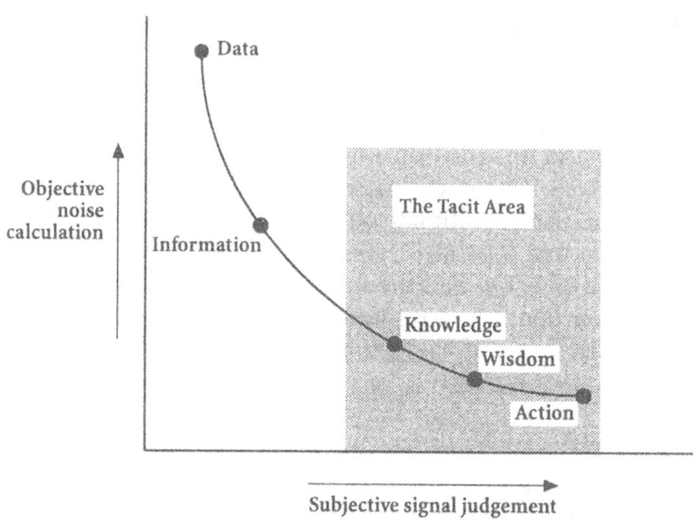

Figure 7: The Tacit Area. Source: M.Cooley (1987), Architecture or Bee?, Hogarth Press

At the data end, we may be said to have calculation; at the wisdom end, we may be said to have judgement. Throughout, I shall be questioning the desirability of basing our design philosophy on the data/information part rather than on the knowledge/wisdom part. It is at the knowledge/wisdom part of the cybernetic loop that we encounter this tacit knowledge to which I will frequently refer.

The interaction between the subjective and the objective, as indicated in Figure 8, is of particular importance when we consider the design of expert systems. In this context, I hold a skilled craftworker to be an expert just as much as I hold a medical practitioner or a lawyer to be an expert in those areas.

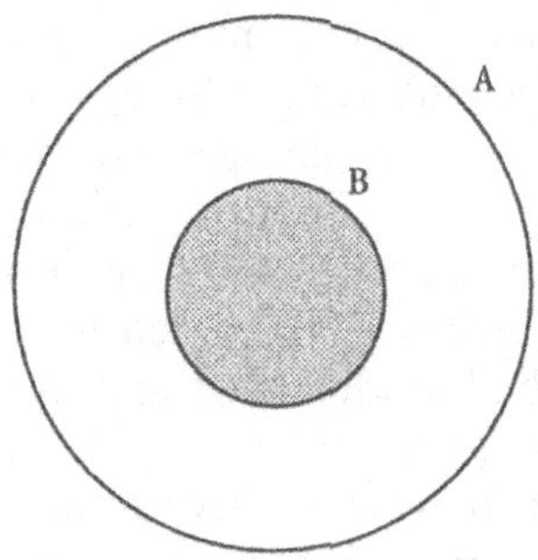

Figure 8: The limits of rule-based systems: (B) facts of the domain; (A) expertise including tacit knowledge. Source: M.Cooley (1987), Architecture or Bee? Hogarth Press

If we regard the total area of knowledge required to be an expert as that represented by (A), we will find that within it there is a core of knowledge (B) which we may refer to as the facts of the domain – the form of detailed information to be found in a textbook.

The area covered by (B) can readily be reduced to a rule-based system. The annulus (AB) may be said to represent heuristics, fuzzy reasoning, tacit knowledge and imagination. I hold that well-designed systems admit to the significance of that tacit knowledge and facilitate and enhance it. I reject the notion that the ultimate objective of an expert system should be so to expand (B) that it totally subsumes (A). It is precisely that interaction between the objective and the subjective that is so important, and it is the concentration upon the so-called objective at the expense of the subjective that is the basis of the concern expressed in respect of existing systems design.

The Acquisition of Skill

In the processes and systems that I describe in this chapter, my concern is not merely with the production, but also with the reproduction of knowledge. I frequently refer to learning by doing, for as a result of this, human beings acquire 'intuition' and 'know how' in the sense in which Dreyfus uses these. This is not in contradiction with Polyani's[3] concept of tacit knowledge; rather, it is a description of a dynamic situation in which, through skill acquisition, people are capable of integrating analysis and intuition. Dreyfus and Dreyfus[4] distinguished five stages of skill acquisition:

1. novice;
2. advanced beginner;
3. competent;
4. proficient;
5. expert.

I think learning-development situations are absolutely vital, and when someone has reached the knowledge/wisdom end of the cybernetic transformation (see Figure 7) and has become an 'expert' in the Dreyfus sense, they are able to recognise whole scenes without decomposing them into their narrow features. I do not counterpose tacit knowledge, intuition or know-how against analytical thinking, but rather believe that a holistic work situation is one which provides the correct balance between analytical thinking and intuition. Broadly stated Dreyfus views skill acquisition as discussed below.

Stage 1: Novice

At this stage, the relevant components of the situation are defined for the novice in such a way as to enable him or her to recognise them without reference to the overall situation in which they occur. That is, the novice is following 'context-free rules'. The novice lacks any coherent sense of the overall task and judges his or her performance mainly by how well the learned rules are followed. Following these rules, the novice's manner of problem solving is purely analytical and any understanding of the activities and the outcome in relation to the overall task is detached.

Stage 2: Advanced Beginner

Through practical experience in concrete situations the individual gradually learns to recognise 'situational' elements, that is, elements which cannot be defined in terms of objectively recognisable context-free features. The advanced beginner does it by perceiving a similarity to prior examples. The growing ability to incorporate situational components distinguishes the advanced beginner from the novice.

Stage 3: Competence

Through more experience the advanced beginner may reach the competent level. To perform at the competent level requires choosing an organisational plan or perspective. The method of understanding and decision making is still analytical and detached, though in a more complex manner than that of the novice and the advanced beginner.

The competent performer chooses a plan which affects behaviour much more than the advanced beginner's recognition of particular situational elements, and is more likely to feel responsible for, and be involved in, the possible outcome. The novice and the advanced beginner may consider an unfortunate outcome to be a result of inadequately specified rules or elements, while the competent performer may see it as a result of a wrong choice of perspectives.

Stage 4: Proficiency

Through more experience, the competent performer may reach the stage of proficiency. At this stage the performer has acquired an intuitive ability to use patterns without decomposing them into component features. Dreyfus calls it 'holistic similarity recognition', 'intuition' or 'know-how'. He uses them synonymously and defines them as:

> ...the understanding that effortlessly occurs upon seeing similarities with previous experiences ... intuition is the product of deep situational involvement and recognition of similarity.

Though intuitively organising and understanding a task, the proficient performer is still thinking analytically about how to perform it. The difference between the competent and the proficient performer is that the proficient performer has developed an intuitive way of understanding based on more experience while the competent performer is still forced to rely on the detached and analytical way of understanding the problem.

Stage 5: Expertise

With enough experience, the proficient performer may reach the expert level. At this level, not only situations but also associated decisions are intuitively understood. Using still more intuitive skills, the expert may also cope with uncertainties and unforeseen or critical situations. Dreyfus and Dreyfus's essential point is to assert that analytical thinking and intuition are not only two mutually conflicting ways of understanding or of making judgements. Rather, they are seen to be complementary factors which work together but with growing importance centred on intuition when the skilled performer becomes more experienced. Highly experienced people seem to be able to recognise whole scenarios without decomposing them into elements or separate features.

My criticism of the prevailing systems-design methodology and philosophy and my deep concerns about 'training' stem from the fact that in both cases they deny us that 'deep situational involvement'. Our development tends to be constrained within the novice end of the skill-acquisition spectrum.

Later, I describe those experiences, systems and machines which could reverse this approach and provide instead developmental situations to facilitate the acquisition of those attributes to be found at the expert end of the skill spectrum.

Many designers fear to discuss these concerns because they may be accused of being 'unscientific'. There is no suggestion in this line of argument that one should abandon the 'scientific method'; rather we should understand that this method is merely complementary to experience and should not override it, and that experience includes 'experience of self as a specifically and differentially existing part of the universe of reality'.

Human-Machine Interaction

When a human being uses a machine, the interaction is between two dialectical opposites. The human is slow, inconsistent, unreliable but highly creative, whereas the machine is fast, reliable but totally non-creative.[5]

Originally, it was held that these opposite characteristics – the creative and the non-creative – were complementary and would provide for a perfect symbiosis between human and machine; for example, in the field of computer-aided design. However, design methodology is not such that it can be separated into two disconnected elements which can then be combined at some particular point like a chemical compound. The process by which these two dialectical opposites are united by the designer to produce a new whole is a complex area. The sequential basis on which the elements interact is of extreme importance.

The nature of that sequential interaction, and indeed the ratio of the quantitative to the qualitative, depends on the commodity under design consideration. Even where an attempt is made to define the proportion of the work that is creative and the proportion that is non-creative, what cannot readily be stated is the point at which the creative element has to be introduced when a certain stage of the non-creative work has been completed. The process by which the designer reviews the quantitative information assembled and then makes the qualitative judgement is extremely subtle and complex. Those who seek to introduce computerised equipment into this interaction attempt to suggest that the quantitative and the qualitative can be arbitrarily divided and that the computer can handle the quantitative.

Where computer-aided design systems are installed, the operators may be subjected to work which is alienating, fragmented and of an ever increasing tempo. As the human being tries to keep pace with the rate at which the computer can handle the quantitative data in order to be able to make the qualitative value judgements, the resulting stress is enormous. The crude introduction of computers into the design activity, in keeping with the Western ethic 'the faster the better', may well result in the quality of design plummeting. Clearly, human beings cannot stand this pace of interaction for long.

Taylor's philosophy is being introduced into the field of intellectual work and in order to condition us to this subordinate role to the machine and to the control of human beings through the technology, Frederick Winslow Taylor once said:

> In my system the workman is told precisely what he is to do and how he is to do it, and any improvement he makes upon the instructions given to him is fatal to success.[6]

Why Suppress the Intellect?

The more I look at human beings, the more impressed I become with the vast bands of intelligence they can use. We often say of a job: *it's as easy as crossing a road*, yet as a technologist I am ever-impressed with people's ability to do just that. They go to the edge of the pavement and work out the velocity of the cars coming in both directions by calling up a massive memory bank which will establish whether it's a mini or a bus because the size is significant. They then work out the rate of change of the image and from this assess the velocity. They do this for vehicles in both directions in order to assess the closing velocity between them. At the same time they are working out the width of the road and their own acceleration and peak velocity. When they decide they can go, they will just fit in between the vehicles. The above computation is one of the simpler ones we do, but you should watch a skilled worker going through the diagnostic procedures of finding out what has gone wrong with an aircraft generator. There you see real intelligence at work. A human being using total information-processing capability can bring to bear synaptic connections of 10^{14}, but the most complicated robotic device with pattern-recognition capability has only about 10^{3} intelligence units.

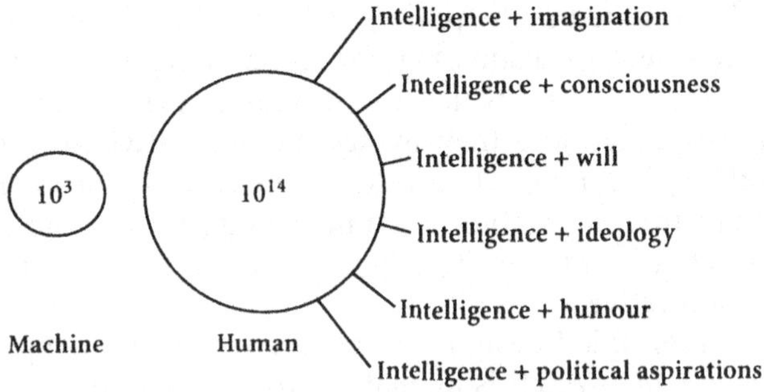

Figure 9: Comparison of units of intelligence available for total information processing. Source: M. Cooley (1987), Architect or Bee?, Hogarth Press

Why do we deliberately design equipment to enhance the 10^{3} machine intelligence and diminish the 10^{14} intellect? Human intelligence brings with it culture, political consciousness, ideology and other aspirations. In our society these are regarded as subversive – a very good reason, then, to try and suppress them or eliminate them altogether. This is the ideological assumption present all the time - see Figure 9 opposite.

As designers we don't even realise we are suppressing intellects, we are so preconditioned to doing it. That is why there is a boom in certain fields of artificial intelligence. The late Fred Margulies, former chairman of the Social Effects Committee of the International Federation of Automatic Control (IFAC), commenting on this waste of human brainpower, said:

> The waste is a twofold one, because we not only make no use of the resources available, we also let them perish and dwindle. Medicine has been aware of the phenomenon of atrophy for a long time. It denotes the shrinking of organs not in use, such as muscles in plaster. More recent research of social scientists supports the hypothesis that atrophy will also apply to mental functions and abilities.[7]

To illustrate the capabilities of human brainpower, I quote Sir William Fairbairn's definition of a millwright of 1861:

> The millwright of former days was to a great extent the sole representative of mechanical art. He was an itinerant engineer and mechanic of high reputation. He could handle the axe, the hammer and the plane with equal skill and precision; he could turn, bore or forge with the despatch of one brought up to these trades and he could set out and cut furrows of a millstone with an accuracy equal or superior to that of the miller himself. Generally, he was a fair mathematician, knew something of geometry, levelling and mensuration, and in some cases possessed a very competent knowledge of practical mathematics. He could calculate the velocities, strength and power of machines, could draw in plan and section, and could construct buildings, conduits or water courses in all forms and under all conditions required in his professional practice. He could build bridges, cut canals and perform a variety of tasks now done by civil engineers.[8]

All the intellectual work has long since been withdrawn from the millwright's function.

Too Old at 24

Just as machines are becoming more and more specialised and dedicated, so is the human being, the 'appendage' to the machine. In spite of all the talk in educational circles about wider and more generalised education, the historical tendency is towards greater specialisation in spite of all the talk about universal machines and distributed systems.

The people who interface with the machine are also required to be specialised. However, as indicated above, this is accompanied by a growing rate of knowledge-obsolescence. It was pointed out by Eugene Wigner, the internationally acclaimed physicist, when talking about the way our education system is going to meet this problem of specialisation, that it is taking longer and longer to train a physicist. 'It is taking so long to train him to deal with these problems that his is already too old to solve them.'[9] This is at 23 or 24 years of age.

Lack of Foresight

One of the founders of modern cybernetics, Norbert Wiener, once cautioned:

> Although machines are theoretically subject to human criticism, such criticism may be ineffective until long after it is relevant.[10]

It is surprising that the design community, which likes to pride itself on its ability to anticipate problems and to plan ahead, shows little sign of analysing the problems of computerisation 'until long after it is relevant'. Indeed, in this respect, the design community is displaying in its own field the same lack of social awareness which it displays when implementing technology in society at large.

Undoubtedly, most of these problems arise from the economic and social assumptions that are made when equipment of this kind is introduced. Another significant problem is the assumption that so-called scientific methods will result inevitably in better design, when in fact there are grounds for questioning whether the design process lends itself to these would-be scientific methods.[11]

Related to this is one of the unwritten assumptions of our scientific methodology - namely, that if you cannot quantify something you pretend it doesn't actually exist. The number of complex situations which lend themselves to mathematical modelling is very small indeed. We have not yet found, nor are we likely to find, a means of mathematically modelling the human mind's imagination. Perhaps one of the positive side effects of computer-aided design is that it will require us to think more fundamentally about these profound problems and to regard design as a holistic process. As Professor Lobell, the American design methodologist, has put it:

> It is true that the conscious mind cannot juggle the numbers of variables necessary for a complex design problem, but this does not mean that systematic methods are the only alternative. Design is a holistic process. It is the process of putting together complex variables whose connection is not apparent by any describable system of logic. It is precisely for that reason that the most powerful logics of the deep structures of the mind, which operate free of the limitations of space, time and causality, have traditionally been responsible for the most creative work in all of the sciences and arts. Today it has gone out of fashion to believe that these powers are in the mind.[12]

Creative Minds

It is a fact that the highly constrained and organised intellectual environment of a computerised office is remarkably at variance with the circumstances and attributes which appear to have contributed to creativity in the arts and sciences. I have heard it said that if only Beethoven had a computer available to him for generating musical combinations, the Ninth

Symphony would have been even more beautiful. But creativity is a much more subtle process. If you look historically at creative people, they have always had an open-ended, child-like curiosity. They have been highly motivated and had a sense of excitement in the work they were doing. Above all, they have possessed the ability to bring an original approach to problems. It is our ability to use our imagination that distinguishes us from animals. As Karl Marx wrote:

> A bee puts to shame many an architect in the construction of its cells. But what distinguishes the worst of architects from the best of bees is namely this: the architect will construct in his imagination that which he will ultimately erect in reality. At the end of every labour process, we get that which existed in the consciousness of the labourer at its commencement.[13]

If we continue to design systems in the matter described earlier, we will be reducing ourselves to bee-like behaviour.

It may be regarded as romantic or succumbing to mysticism to emphasise the importance of imagination and of working in a non-linear way. It is usually accepted that this type of creative approach is required in music, literature and art. It is less well recognised that this is equally important in the field of science, even in the so-called 'harder sciences' like mathematics and physics. Those who were creative recognised this themselves. Isaac Newton said:

> I seem to have been only like a boy playing on the seashore and diverting myself in now and then finding a smoother pebble or a prettier shell than ordinary, while the great ocean of truth lay all undiscovered before me.

Einstein said, 'Imagination is far more important than knowledge.' He went on to say:

> The mere formulation of a problem is far more important than its solution which may be merely a matter of mathematical or experimental skills. To raise new questions, new possibilities and to regard old problems from a new angle requires creative imagination and marks real advances in science.

On one occasion, when being pressed to say how he had arrived at the idea of relativity, he is supposed to have said:

> When I was a child of 14 I asked myself what the world would look like if I rode on a beam of light.

A beautiful conceptual basis for all his subsequent mathematical work. Central to the Western scientific methodology is the notion of predictability, repeatability and quantifiability. If something is unquantifiable we have to rarefy it away from reality, which leads to a dangerous level of abstraction, rather like a microscopic Heisenberg principle. Such techniques may be acceptable in narrow mathematical problems, but where much more complex considerations are involved, as in the field of design, they may give rise to questionable results.

The risk that such results may occur is inherent in the scientific method which must abstract common features away from concrete reality in order to achieve clarity and systemisation of thought. However, within the domain of science itself, no adverse results arise because the concepts, ideas and principles are all interrelated in a carefully structured matrix of mutually supporting definitions and interpretations of experimental observation. The trouble starts when the same method is applied to situations where the number and complexity of factors is so great that you cannot abstract without doing some damage, and without getting an erroneous result.[14] More recently, these questions have given rise to a serious political debate on the question of the neutrality of science and technology[15] and there is already a growing concern over the ideological assumptions built into our scientific methodologies.

Competence, Skill and 'Training'

The Origins of Design

Around the 16th century, there appeared in most European languages the term 'design' or its equivalent. The emergence of the word coincided with the need to describe the occupational activity of designing. That is not to suggest that designing was a new activity. Rather, it was separated out from a wider productive activity and recognised as an activity in its own right. This can be said to constitute a separation of hand and brain, of manual and intellectual work and of the conceptual part of work from the labour process. Above all, it indicated that designing was to be separated from doing. It is clearly difficult to locate a precise historical turning point at which this occurred; rather, we will view it as a historical tendency.

Up to the stage in question, a great structure such as a church would be 'built' by a master builder. We may generalise and say that the conceptual part of work would be integral to that labour process. Thereafter, however, came the concept of 'designing the church' an activity undertaken by architects, and the 'building of the church', an activity undertaken by builders. In no way did this represent a sudden historical discontinuity, but it was rather the beginning of a discernible historical tendency which has still not worked its way through many of the craft skills, so that as recently as the last century, Fairbairn was able to give his comprehensive description of the skills of a millwright quoted earlier. To this day, there are many jobs in which the conceptual part of work is still integrated with the craft basis. The significant feature of the stage in question is, however, that separating manual and intellectual work provided the basis for further subdivisions in the field of intellectual work itself – or as Braverman put it, 'Mental labour is first separated from the manual labour and then itself is subdivided rigorously according to the same rules.'[16]

Dreyfus[17] locates the root of the problem in the Greek use of logic and geometry, and the notion that all reasoning can be reduced to some kind of calculation. He suggests that the start of artificial intelligence probably began around the year 450 BC with Socrates and his concern to establish a moral standard. He asserts that Plato generalised this demand into an epistemological demand where one might hold that all knowledge could be stated in explicit definitions which anybody could apply. If one could not state one's know-how in such explicit instructions, then that know-how was not knowledge at all but mere belief. He suggests a Platonic tradition in which, for example, cooks who proceed by taste and intuition and people who work from inspiration, like poets, have no knowledge. What they do does not involve understanding and cannot be understood. More generally, what cannot be stated explicitly in precise instructions – that is, all areas of human thought which require skill, intuition or a sense of tradition – is relegated to some kind of arbitrary fumbling.[18]

Gradually, a view evolved which put the objective above the subjective, and the quantitative above the qualitative. That the two should and can interact was not accepted, in spite of a systematic effort and intellectual struggle to assert it. One important example of the attempt to do so was the work of Albrecht Dürer (1471–1528). Dürer was not only a 'Master of the Arts', but a brilliant mathematician as well, who reached the highest academic levels in Nuremberg. Dürer sought to use his abilities to develop the mathematical forms which would succeed in preserving the unity of hand and brain. Kantor[19] points out the significance of Dürer's ability to put complex mathematical techniques to practical uses, while Olschki[20] compares his mathematical achievements with those of the leading mathematicians in Italy and elsewhere at that time. Indeed, some 90 years after Dürer's death Kepler was still discussing his geometric construction techniques. Alfred Sohn Rethel points out, in speaking of Dürer, 'Instead of, however, using this knowledge in a scholarly form, he endeavoured to put it to the advantage of the craftsman. His work was dedicated 'to the young workers and all those with no one to instruct them truthfully'. What is novel in his method is that he seeks to combine the workman's practice with Euclidian Geometry.

And further: what Dürer had in mind is plain to see. The builders, metalworkers, etc., should on the one hand, be enabled to master the tasks of military and civil technology and of architecture which far exceeded their traditional training. On the other hand, the required mathematics should serve them as a means, so to speak, of preserving the unity of head and hand. They should benefit from the indispensable advantages of mathematics without becoming mathematic or brainworkers themselves. They should practise socialised thinking yet remain individual producers, and so he offered them an artisan schooling in draughtsmanship permeated through and through with mathematics – not to be confused in any way with applied mathematics.[21]

It was said that on one occasion Dürer proclaimed it would be possible to develop forms of mathematics that would be as amenable to the human spirit as human language. Thereby one could integrate into the use of the instruments of labour the conceptual parts of work, thus building on the tradition in which the profiles of complex shapes were defined and constructed with such devices as sine bars.

Holistic Design

Thus theory, itself a generalisation of practice, could have been reintegrated with practice to extend the richness of that practice and application while retaining the integration of hand and brain.

The richness of that practical tradition may be found in the sketchbook of Villard de Honnecourt, in which he introduced himself thus:

> Villard de Honnecourt greets you and begs all who will use the devices found in this book, to pray for his soul and remember him. For in this book will be found sound advice on the virtues of masonry and the uses of carpentry. You will also find strong help in drawing figures according to the lessons taught by the art of geometry.[22]

This extraordinary document by a true 13th-century cathedral builder contains subjects which might be categorised as follows:

1. mechanics;
2. practical geometry and trigonometry;
3. carpentry;
4. architectural design;
5. ornamental design;
6. figure design;
7. furniture design;
8. Subjects foreign to the special knowledge of architects and designers.

The astonishing breadth and holistic nature of the skills and knowledge are in the manuscripts for all to see. There are those who, while admitting to the extraordinary range of capabilities of craftspeople of this time, hold that it was a 'static' form of knowledge which tended to be handed unaltered from master to apprentice. In reality, these crafts and their transmission embodied dynamic processes for extending their base and adding new knowledge all the time. Some of the German manuscripts describe die Wanderjahre - a form of sabbatical in which craftspeople travel from city to city to acquire new knowledge. Villard de Honnecourt travelled extensively, and thanks to his sketchbook[23] we can trace his travels through France, Switzerland, Germany and Hungary.

He was also passionately interested in mechanical devices, and one system he designed was subsequently adapted to keep mariners' compasses horizontal and barometers vertical. He devised a variety of clock mechanisms, from which we learn 'how to make the Angel keep pointing his finger towards the sun', and he displayed extraordinary engineering skills in a range of lifting and other mechanisms to provide significant mechanical advantage. For example, he invented a screw combined with a lever with appropriate instructions – 'How to make the most powerful engine for lifting weights.'

In all this, we see brilliantly portrayed the integration of design with doing – a tradition which was still discernible when Fairbairn[24] described his millwright.

Villard was also concerned with 'automation', but in a form which freed the human being from back-breaking physical effort but retained the skilled base of work. In woodworking, he thought of a system for replacing the strenuous sawing activity – 'How to make a saw operate itself.'

He was profoundly interested in geometry as applied to drawings: 'Here begins the method of drawing as taught by the art of geometry but to understand them one must be careful to learn the particular use of each. All these devices are extracted from geometry.' He proceeds to describe 'How to measure the height of a tower', 'How to measure the width of a water course without crossing it', 'How to make two vessels so that one holds twice as much as the other'.

Many modern researchers have testified to Villard's significant grasp of geometry. Side-by-side with this we find his practical advice to stonecutters on the building elements divisions: 'How to cut an oblique voussoir', 'How to cut the springing stone of an arch', 'How to make regular pendants'. All of the latter, drawn from his own practical experience and skill, is a vivid portrayal of the integration of hand and brain.

Another 13th-century manuscript, written in the same dialect as Villard's is still preserved and can be consulted in the Bibliotheque St Genevieve in Paris. Its author likewise concerned himself with mathematical problems: 'If you want to find the area of an equilateral triangle', 'If you want to know the area of an octagon', 'If you want to find the number of houses in a circular city'.

Throughout this period, the intellectual and the manual, the theory and the practice were integral to the craft or profession. Indeed, so naturally did the two coexist that we find practical builders (architects) with university titles like Doctor Lathomorum.

The epitaph of Pierre de Montreuil, the architect who reconstructed the nave and transepts of Saint Denis, runs, 'Here lies Pierre de Montreuil, a perfect flower of good manners, in this life a Doctor of Stones.'

I have cited these sketchbooks and quoted from these manuscripts in order to demonstrate that the craft at that time embodied powerful elements of theory, scientific method and the conceptual or design base of the activity. In doing so, I am myself guilty of a serious error. I accept that a matter can only be scientific or theoretical when it is written down. I did not provide an illustration of a great church or complex structure and state that the building of such a structure must itself embody a sound theoretical basis, otherwise the structure could not have been built in the first instance.

We can also detect in the written form the basic elements of Western scientific methodology: predictability, repeatability and mathematical quantifiability. These, by definition, tend to preclude intuition, subjective judgement and tacit knowledge.

Furthermore, we begin to regard design as something that reduces or eliminates uncertainty, and since human judgement, as distinct from calculation, is itself held to constitute an uncertainty, it follows some kind of Jesuitical logic that good design is about eliminating human judgement and intuition. Furthermore, by rendering explicit the 'secrets' of craft, we prepare the basis for a rule-based system.

Rules for Design

In the two successive centuries there followed systematic attempts to describe, and thereby render visible, the rules underlying various craft skills. This applied right across the spectrum of skills of people who were artists, architects and engineers, in the Giotto tradition, from the theory of building construction through to painting and drawing. Giotto's method was not precisely optical. The receding beams of the ceiling converge to a reasonably convincing focus, but it is only approximate and does not coincide with the horizontal line as it should, according to the rules of linear perspective. 'This method is, however, systematic and rational, factors which no doubt provided a powerful stimulus for the more fully scientific rule seekers of the subsequent centuries. Priority amongst those who preceded Leonardo in searching for precise optical laws in picture making must go to the great architect and sculptor Filippo Brunelleschi.'[25]

According to Manetti, at some time before 1413 Brunelleschi constructed two drawings which showed how buildings could be represented:

> ... in what painters today call perspective, for it is part of that science which in effect sets down well and with reason the diminutions and enlargements which appear to the eyes of man from things far away and close at hand.

One of the paintings showed the octagonal baptistry (S. Giovanni) as seen from the door of the cathedral in Florence. The optical 'truth' was verified by drilling a small hole in the baptistry panel, so that the spectator could pick up the panel and press an eye to the hole on the unpainted side and, with the other hand, hold a mirror in such a way that the painted surface was visible in reflection through the hole. By these means, Brunelleschi established precisely the perpendicular axis along which his representation should be viewed.

By the use of a mirror, there was a precise matching of the visual experience and the painted representation, and this was to become Leonardo's theory of art and indeed his whole theory of knowledge.[26] He applied the same scientific methods to his architectural and other designs. One interpretation of these events is that they represented a significant turning point in the history of design and design methodology. Thereafter, there is a growing separation of theory and practice, a growing emphasis on the written 'theoretical forms of knowledge' and in my view, a growing confusion in Western society between linguistic ability and intelligence (in which the former is taken to represent the latter). Furthermore, this is accompanied by a growing denigration of tacit knowledge in which there are 'things we know but cannot tell'.[27] We may cite that most illustrious embodiment of theory and practice - Leonardo da Vinci:

> They will say that not having learning, I will not properly speak of that which I wish to elucidate. But do they not know that my subjects are to be better illustrated from experience than by yet more words? Experience, which has been the mistress of all those who wrote well and thus, as mistress, I will cite her in all cases.[28]

In spite of such assertions, the tendency to produce generalised, written-down, scientific or rule-based design systems continued to build on earlier work. In 1486, the German architect Mathias Roriczer published in Regensburg his deceptively named 'On the Ordination of Pinnacles'. In this, he set out the method of designing pinnacles from plan drawings, and in fact produced a generalised method of design for pinnacles and other parts of a cathedral. These tendencies had already elicited bitter resistance from the craftsmen-cum-designers whose work was thereby being deskilled.

The Master Masons

In 1459, master masons from cities like Strasbourg, Vienna and Salzburg met at Regensburg in order to codify their lodge statutes. Among the various decisions, they decided that nothing was to be revealed of the art of making an elevation from a plan drawing to those who were not in the guild. 'Therefore, no worker, no master, no wage earner or no journeyman will divulge to anyone who is not of our Guild and who has never worked as a mason, how to make the elevation from the plan.' Of particular note is the exclusion of those who had never [worked] as a mason. There is, as our German colleagues would put it, a Doppelnatur to this craft reaction. On the one hand, there is the negative elitist attempt to retain privileges of the profession rather as the medical profession seeks to do to this day. On the other, there is a highly positive aspect, that of attempting to retain the qualitative and the quantitative elements of work, the subjective and the objective, the creative and the non-creative, the manual and the intellectual, and the work of hand and brain, embodied in the one craft.

The pressures on the master masons were twofold. Not only was the conceptual part of the work to be taken away from them, but those workers who still embodied the intellectual and design skills were being rejected by those who sought to show that theory was above, and separate from, practice. The growing academic elite resented the fact that carpenters and builders were known as masters, for example, Magister Cementarius or Magister Lathomorum. The academics attempted to ensure that 'Magister' would be reserved for those who had completed the study of the liberal arts. Indeed, as early as the 13th century, doctors of law were moved to protest formally at these academic titles for practical people.

It would be both fascinating and illuminating to trace these tendencies through the five intervening centuries which take us up to the information society of computer-aided design and expert systems. Suffice it to say that a number of researchers, drawing on historical perspective and viewing the implications of these information-based systems, conclude that we may now be at another historical turning point where we are about to repeat, in the field of design and other forms of intellectual work, many of the mistakes made in the field of craftsmanship in the past.[29]

Separation of Theory From Practice

It is significant that J. Weizenbaum, a professor of computer science at the Massachusetts Institute of Technology, uses the subtitle 'from judgement to calculation' in his seminal work *Computer Power and Human Reason*[30] and highlights the dangers which will surround an uncritical acceptance of computerised techniques.

The spectrum of problems associated with them is already becoming manifest. They include such spectacular separation of theory and practice as to result in some of those who have been weaned on computer-aided design being unable to recognise the object that they have 'designed'. Epitomising this was the designer of an aircraft igniter who calculated the dimensions on the CAD screen and then set them out with the decimal point one place to the right (which, in an abstraction, is very much like one place to the left). He then generated the numerical control tapes with which deskilled workers on the shop floor produced an igniter ten times larger than it should have been.[31] Perhaps the most alarming aspect of this extraordinary state of affairs was that when confronted with the monstrosity, the designer saw nothing wrong with it.

Given the scale and nature of these problems and the exponential rate of technological change within which they are located, it behoves all of us to seek to demonstrate, as Dürer did, that alternatives exist which reject neither human judgement, tacit knowledge, intuition and imagination nor the scientific or rule-based method. We should rather unite them in a symbiotic totality.

Consumer Incompetence

Efforts to deskill the producers can only become operational if they are accompanied by the deskilling of the consumers. The deskilling of bakers, for example, can come about only if that awful cotton-wool stodge in plastic wrappers is regarded as bread by millions of consumers. Highly automated and factory farming techniques are only possible if the public believes that there are only two kinds of potato, 'new' and 'old', that cookers and eaters are the only forms of apple, and if it cannot distinguish the taste of free-range poultry and eggs from those produced under battery-farm conditions.

The elimination of high-level skills in carpentry and cabinet-making is possible because large sections of the public do not appreciate the difference between a tacky chipboard product and one handmade with real wood and fitted joints, or between a plastic container and (say) an inlaid needlebox.

The concern for quality should not be misunderstood as an elitist tendency. Quite ordinary working-class and rural families used to pass pieces of furniture from one generation to another which, although simple, embodied fine craftsmanship and materials. A skilled joiner recently told me with great feeling how monstrous he found it that beautiful pieces of wood which could have been hand-turned and carved were being burned on a demolition site by 'builders' who couldn't distinguish between one piece of wood and another.

Given time, more and more sections of the community will lose the capacity to appreciate craftsmanship and goods of quality. As you 'break the refractory hand of labour', you must also break the refractory will of the consumer. To do so it is of course necessary to have ranges of accomplices. These are in advertising, marketing, more generally, of the 'Waste Makers' type. The accomplices are in relation to production and consumption, and there are also partners in crime in the areas of the reproduction of knowledge. The duality of the master and apprentice, teacher and student, has now been replaced by the trainer and the trainee. In large occupational areas, we no longer have education, we have 'training'.

Apprenticeships and Training

An apprenticeship in the classical sense was not merely a process for the acquisition of technical skills. It was far more significantly the transmission of a culture, a way of understanding and respecting quality and acquiring a love of good materials. Even to this day, this cultural outlook is alive and well among craftspeople. Ken Hunt is a master engraver whose work is sought worldwide. He served his apprenticeship with Purdeys, the London Sporting Gunmakers, who arranged for him to work with Henry Kell, one of the specialist firms engraving the gun actions. This is how Ken Hunt described his work in 1987:[32]

> I think engraving creates an intensely personal relationship between the work and the craftsman. It's the most lovely feeling when everything is going right; the cutting tool is working well, the steel doesn't fight you.
>
> To me, the beauty of a cut on steel with a graver is similar to the mark made by a quill pen on paper. It flows and tapers and is far removed from the straight line drawn by a ball point. I get so involved sometimes that I lose all track of time, and I get lost in all sorts of ideas, almost fantasies, I suppose. I find myself thinking of craftsmen centuries ago who worked metal in exactly the same way as I do now. Nothing has changed, neither the medium nor the tools.
>
> It may sound strange, but occasionally I get pieces back that I might have worked on in the sixties or even earlier, and I only have to touch them to recall exactly what I was doing and thinking when I was working on them all those years ago. Perhaps it's because each job represents and absorbs a large part of your life – maybe even your soul, who knows?
>
> Michaelangelo used to claim that all he did when confronted with a block of stone was to chip away and release the sculpture which was inside it, and I feel that too.[33]

Ken does not use preliminary drawings of his intended work. Nor does he have tracing on the metal which will then be simply followed by the engraving tool. 'No, I go straight in and just do it. I've got an idea in my head as to how the finished work will look, but I don't believe in drawing it out carefully first.'

There is a tendency to regard such craft skills as being static and devoid of development. But the environment created by an apprenticeship encourages experimentation and innovation within a given tradition. Ken Hunt recalls that in his early days he would visit museums to admire and wonder at masterpieces from the past:

> I would stand and stare at a certain piece for ages wondering how it was done. Sometimes, I would even stay so long that the wardens would begin to eye me with suspicion! I was intrigued with everything to do with metal work, though especially gold inlaying. I eventually worked out my own way of keying gold to steel using a series of undercut crisscross lines which have a dovetail effect.[34]

It would be unthinkable that craftspeople like Ken Hunt would waste materials or mishandle or damage tools and equipment. All of this was integral to the totality which was embodied in a traditional apprenticeship. It was also a process by which one learned, in a very practical way, the logistics of procuring such materials, treating them and forming them in a creative process which linked hand, eye and brain in a meaningful productive process. It embodied 'design by doing' – methods of work in which the conceptual aspects of work were integrated within the overall labour process.

Apprenticeships served to develop significant skills in the field of planning and coordination, and produced quite astonishing levels of ability in the handling of materials. I marvel at St Paul's Cathedral, for even given our modern means of project management and complex techniques for handling material, we may question whether anyone would be capable of constructing it today. Even if we could, what an infinitely greater task it was in the 17th century, given the limited equipment for lifting and placing the building elements into their locations.

The kind of apprenticeships those builders had, gave them a deep sense of total machines as operating systems, epitomised by the vast knowledge of the great millwrights. It is true that with the introduction of Taylorism,[35] apprenticeships did embody the most anecdotal aspects, where considerable time was spent in making tea for others or in irrelevant activities, but that is not what we are addressing here; it is rather the great apprenticeships which produced those of the calibre described earlier.

Against this richness and competence can be counterposed 'training'. The work is very apt in the modern context. My own hierarchy of verbs in terms of competence transmission would be the following: you program a robot, you train a dog (or possibly a soldier), but for human beings you provide educational environments. Training produces narrow, overdedicated capabilities which are generally machine, system or program-specific. With the ever-increasing rate of technological change, the 'knowledge' required to cope with a particular machine or system may be obsolete in a couple of years' time. The trainee is then lost, and requires further 'training'. Much of what now passes for 'training' is nothing more than a form of social therapy. Instead of putting people on Valium you put them on a training course. It is questionable whether you produce anything more then a slightly better quality dole queue. 'Training' often hides a cruel deception.

Some companies have very competent training officers who themselves have actual knowledge of the processes involved. What I am referring to here is that new band of 'training advisers', 'training coordinators', 'training outreach workers', and 'training planners' who seem to believe that there is some separate activity called 'training' which transcends all other forms of professional knowledge. Some of the ones I have encountered seem to believe that if you've trained a Labrador to retrieve you can also train a nuclear physicist, and if you've trained somebody to make doughnuts in the catering industry, you can also train them to design a Rolls-Royce aero engine; it is, after all, just training! Because these people have no knowledge of the skills involved, they behave in a high-handed and arrogant fashion. Furthermore, because they are in a position to allocate funds, they are often able to impose their nonsense on people who could have provided a rich developmental environment. The disadvantage of using this type of 'trainer' is twofold. They don't know what they are doing and are overpaid for doing it, and, more significantly, they prevent people who do have the skills and knowledge from enjoying the experience and gaining the dignity of transmitting it to a future generation.

A Challenge For the 21st Century

This Extraordinary Millennium

> ... no longer will I follow you oblique like through the inspired form of the third person singular and the moods and hesitancies of the deponent but address myself to you, with the empirative of my vendettative, provocative and out direct ...
>
> *James Joyce*

The year 2000 marks the end of the most extraordinary millennium in human history. During it, humanity has witnessed the decline of feudalism, the growth of capitalism and weakening of religion as the leading-edge in European society. We have facilitated the emergence of Cartesian Science, the concentration of populations into modern cities and the development of 'earth-shrinking' transport systems. Above all there has been the growth of industrial society. We have allowed the great story-telling traditions to all but wither away. We have devised the means of flying, declared Jackson Pollock a great artist, bounced on the moon and killed 17 million people in just one war.

Through our science we have become the first generation of the only species to apparently have it within its power to destroy itself and life on the planet as we know it. We have become far too smart scientifically to survive much longer without wisdom.

We should reflect upon the beauty and upon the devastation we have wrought on our own two-edged way to the 21st century. The delinquent genius of our species has produced the beauty of Venice and also the hideousness of Chernobyl; the playful linguistic delights of Shakespeare and the ruthlessness of British imperialism; the musical treasures of Mozart and the stench of Bergen Belsen, the caring medical potential of Rontgen's X-rays and the horrific devastation of Hiroshima.

The last century of this millennium has been characterised by a convulsed and exponential rate of technological change in which our speed of communication has increased by 10^7, our speed of travel by 10^3 and of data-handling by 10^6. Over the same period our depletion of energy resources has increased by 10^4 and weapon power by 10^7.

We have seen the polarisation of wealth and activity. In developed countries there are computer programmes to help people diet, whilst out of the 122 million babies born in developing countries in 1982, 11 million died before their first birthday and a further 5 million died before their fifth birthday. In some countries it is more today!

At the end of the millennium we appear to stand as the masters of nature. We scrabble millions of tons of material around each year and in doing so we shift the equivalent of three times the sediment moved each year by the world's rivers. We mine and burn billions of tons of coal each year so venting the waste which includes carbon dioxide the principal contributor to the greenhouse effect.

Our agriculture, now worked on an industrial scale, causes the erosion of 25 million tons of soil each year which is 0.7% of the total arable land formed over a period of several thousand years. We put down 30kg of fertilizer per person per year to increase the crop yield, so polluting the very water we drink. In many parts of the world we have turned the soil into a craving junkie, incapable of producing without its next fix. If we continue in the present manner, we will reduce by 50% all the species of flora and fauna in less than two centuries. In fact, this is likely to be a matter of decades taking into account the greenhouse effect. This will constitute a terrible reduction in bio-diversity but is also being accompanied by a reduction in diversity amongst ourselves.

Among the global issues confronting industrial society today, two are particularly pressing. They are:

1. resource depletion; and,
2. environmental changes brought about by human activity.

Stimulus

> Symbols in turn determine the kind of stories we tell and the stories we tell determine the kind of history we make and remake.
>
> *Mary Robinson*

The year 2000 could, and should, provide a powerful stimulus to examine where, as an industrial society, we are going. To do so at the macro level, we will require the perspective of a historian, the imagination of a poet, the analytical ability of a scientist, and the wisdom of a Chief Seattle (an American Indian). We shall have to be capable of thinking holistically, working in multidisciplinary groups, coping with change and developing systems and products which are sustainable and caring of nature and humanity. Our current educational systems are fundamentally inappropriate and woefully inadequate to address this historical task.

Thus an EEC/FAST report[36] is quite unambiguous about the changes to educational activities that will be necessary. It states:

> The tendency for education to concentrate on narrow specialist areas is counterproductive and must give way to holistic forms. The concern should be education rather than training. Above all, education should be the transmission of a culture which values proactive, sensitive and creative human beings.

In relation to manufacturing and industry it seems self-evident that developing the skill and competence necessary in the 21st century will require nothing short of a 'cultural and industrial renaissance'.

For this to come about we will require citizens possessed not just of knowledge but also of wisdom. It will require the courage and the dignity to ask simple questions of profound significance. Why is it that if we grow our own lettuce and repair our own car the gross national product (GNP) goes down whereas if they have a pile-up on the motorway and in the carnage scores of people are killed and piles of cars are destroyed, the GNP goes up? How come we design products to fall apart after five years? What is the deranged mentality of an expert in artificial intelligence who can say:

> Human beings will have to accept their true place in the evolutionary hierarchy namely animals, human beings and intelligent machines?

These are key issues as we approach the 21st century and our educational system should be preparing people to discuss them in an informed, creative and imaginative fashion.

The Industrial Future

> If you can look into the seeds of time and say which grains will grow and which will not ...
>
> *William Shakespeare*

Educationalists, industrialists, trade unionists and politicians should be urged to enter into a creative dialogue with the community at large to decide what the future might be like. Such a debate would encourage human beings to perceive themselves in their dual role in society - as producers and also as consumers and to recognise that they are the subjects of history rather than the objects through which it wends its painful way. The urgency of the debate is highlighted by the crisis now facing industrial societies worldwide.

There is a growing recognition that the future cannot merely be an extrapolation of the past. We cannot assume it is possible to have an ever-increasing rate of production and consumption. The mass production of throw-away products based on energy, capital and chemically-intensive forms of production, whether in manufacturing or in agriculture, is no longer possible nor acceptable. The ecological damage we are doing is now making this clear to growing numbers of concerned citizens. Furthermore, these intensive forms of production are also giving rise to political and social tensions and are contributing significantly to growing structural unemployment worldwide. In the EEC countries there are now 16.2 million people out of work and it is predicted that this will increase to some 20 million by 1996. This is in spite of the fact that there are dramatic demographic shifts taking place in Europe. In addition, 100000 small towns and villages will cease to be economically viable.

In addition to these questions there is concern that the so-called 'lucky ones' who retain their jobs are increasingly involved in using forms of new technology which result in processes which are not only intensive and stressful for people but which frequently reduce them to mere machine appendages. The EEC report highlighted these issues and advocated Anthropocentric Systems as a more long-term sustainable alternative. The report suggested:

- That industrial society would have to move from an economy of scale to an economy of scope and on to an economy of networking.
- The need for a society of proactive, creative, involved citizens at all levels.
- That society would gradually become one of 'continuous innovation', where the capacity to design and build prototypes and to have skill-based, short batch manufacturing capabilities would be of growing importance.
- That whilst during the era of Fordism a dominant mono-industrial culture based on Taylorism gave the United States important competitive advantages this would no longer be so significant and is now beginning to be counterproductive.
- That linguistic, cultural and geographical diversity which had been perceived by some to be a weakness in the European situation should in future be perceived as an advantage and a source of innovation. Technological and educational support systems should enhance that diversity rather than diminish it.

A Tool Rather Than a Machine!

> When the fact fails him, he questions his senses;
> When the fact fails me, I approve my senses.
>
> *Robert Graves*

In the human-centred system, there exists a symbiotic relation between the human and the machine in which the human being would handle the qualitative subjective judgements and the machine the quantitative elements. It involves a radical redesign of the interface technologies and at a philosophical level the objective is to provide tools (in the Heidegger sense) which would support human skill and ingenuity rather than machines which would objectivise that knowledge.

The focus of human-centredness is to design forms of technology which reverse one of the main processes of technology which is to render systems active and human beings passive. It means providing powerful analogical systems in which it is possible to programme the devices in a manner which accords with the traditional ways of working but which enhances those by providing very modern software and hardware tools. In the context of high-level intellectual work (e.g. designing) it is necessary to challenge the concept of menu-driven systems which frequently reduce the designer to being like a child with a LEGO set. The child can make a pleasing pattern of predetermined elements but cannot change those elements. This raises very important questions for education at every level. It further requires that the skill and competence at every level of the organisation should be changed and expanded, thereby changing the worker's perception of him or herself. For example, on the shop-floor, those who functioned and thought of themselves as 'machine operators' can metamorphose into 'cell managers'. They can take an overview of the functioning of the cell and acquire additional competencies in the field of planning, costing and systems maintenance. They can use powerful support tools such as workstations with adaptive interfaces. Such interfaces should acknowledge and celebrate traditional craft skills. It is suggested that those able to make best use of such multimedia systems are those who emerge from a quasi-apprenticeship system and therefore start from a high 'competence platform'. The whole process is one in which the operator 'builds on the familiar to create the new'.

Part of the competence and some of the skills required to work in such environments will be the following:

1. The ability to absorb new knowledge and transform it.
2. The ability to draw conclusions about the unknown from the known.
3. The ability to take initiatives.
4. The capacity to make decisions.
5. The ability to work as a team.
6. The ability to adopt a systematic, analytical approach.
7. The ability to plan independently.
8. The ability to take on responsibility.

Those involved in systems design will need to be competent in the design of adaptive tools which accord closely with traditions and practices of the domain area. Furthermore, they will need to be competent in the design of systems and organisations which display the following characteristics:

- **Coherence:** the embedded meanings, if not immediately evident, at least must not be cloaked or obscure. A related concept here is 'transparence' which means rendering what is going on and what is possible as being highly visible.
- **Inclusiveness:** the system should be inviting and tend to 'invite you in' and make you feel part of a community of activities with which you are familiar and on friendly terms.
- **Malleability:** a possibility to 'mould' the situation to suit, to pick-and-mix and sculpt the environment to suit one's own instrumental needs, aesthetic tastes and craft traditions.

- **Engagement:** a sense that one is being invited to participate in the process and which creates a sense of empathy.
- **Ownership:** a feeling that you have created and thereby 'own' parts of the system. A sense of belonging and even companionship as traditional craftsmen may feel with a favourite machine tool.
- **Responsiveness:** a general sense that you can get the system to respond to your requirements and your individual needs and ways of doing things. A system which makes visible its own rules and then encourages one to learn them and to change them at will.
- **Purpose:** purpose is meant in the sense in which Rosenbrock describes it. The system is capable of responding to the purpose one has in mind and then encouraging one to go beyond it.
- **Panoramic:** most current systems tend to encourage the user to converge on narrower activities. With good embedded systems it should also provide windows or apertures through which one can take a wider or more panoramic view. This encourages the acquisition of 'boundary knowledge' and allows the user to act more effectively and competently by locating what he or she is doing in the understanding of a wider context.
- **Transcendence:** when operating the system, the user should be encouraged, enticed and even provoked to transcend the immediate task requirements. The possibility of acquiring boundary knowledge and a macro level vision of the process as a whole should be self evident.

Hard-nosed industrialists and their compliant foot-soldiers - industrial engineers - have tended to regard the type of systems described above as being at the best a diversion from 'the real world of industry' or at the worst 'dangerous liberal waffle'. But times are changing. The crisis in many of the rigid, hierarchical large organisations is forcing a radical re-examination of much of the given wisdom. At an economic level, the multidisciplinary report on the future of US industry pointed out 'We have tended to treat our workforces as a cost and a liability whereas our major competitors have treated them as an asset whose skills should be ever enhanced'.

Professor Hopwood[37] has highlighted the need for accountants to re-examine their arid thinking. Thus they spend 75% of their time in dealing with direct labour costs but direct labour costs now only constitute about 10% of the total costs if one considers industry as a whole. On the other hand material, which accounts for 50% of the cost is only subject to 10% of accounting effort.

At the technical level the case is even more compelling. In machine-based, hierarchical, Tayloristic systems which relentlessly drive towards wall-to-wall automation, there is now a growing recognition that such systems are extremely vulnerable to disturbance. They are typically good at coping with high frequency, low impact events though bad in dealing with low frequency, high impact ones – e.g. the uncertainty of the real world. Otherwise stated, machine-dependent systems are highly synchronised and coordinated. When one part of the system goes down, the high level synchronisation is turned into its dialectical opposite and one gets massive de-synchronisation rather in the form of catastrophe theory.

In addition to economic benefits, there may be added the long-term advantages in the form of flexibility and strategic capability for innovation. Of equal importance but less easy to quantify are social benefits such as the quality of working life, dramatically improved motivation and the liberating of one of society's greatest assets, the skill, ingenuity and creativity of its people. The resultant flexibility will become paramount in coming years as there are more custom bound, short batch production runs and as an economy of networking becomes widespread. Concurrent or simultaneous engineering will further reinforce the need for systems of this kind.

Overstructuring

> Management is just a bad habit inherited from the army and the church.
>
> *Danny Conroy* – Craftsman

A feature of modern industrial society is its over structuring. This arises within production from a mechanistic, Tayloristic view of optimum organisation. Taylor said on one occasion:

> In my system the workman is told precisely what he is to do and how he is to do it and any improvement he makes upon the instructions given to him is fatal to success.[38]

The United States led in this overstructuring of industry and as a report in the late 1980s from the MIT pointed out, it is seriously debilitating because it treats human beings as liabilities rather than as assets.

Although these lessons are beginning to be understood in industry, the educational system in general and universities in particular still seem determined to pursue teaching forms which are based on factory models. When Henry Ford donated 100 million dollars to an institution which he called The School of the Future, he said:

> I have manufactured cars long enough to the point where I have got the desire to manufacture people. The catchword of the day is standardisation.[39]

The overstructuring reached such a level that even universities are being organised as factories within which the students are referred to as commodities, the examinations as quality control procedures, graduation as delivery and the professors as operators.

The factory model is now all-pervasive. It conditions and distorts every aspect of life in the technologically advanced nations. I am not sure if it was ever true in the Shakespearian sense, that all the world's a stage. It is however certainly true that at the close of the 20th century all the world's a factory and all of nature that surrounds us is seen as inert material for its remorseless production line.

Paradigmatic changes are already at hand and within these we will require people with the competence to cope with ill-defined, loosely structured situations which cannot be defined in a unidimensional way and which embody high levels of uncertainty and unpredictability. At a design level we will have to consider a scientific methodology based on purpose and not only one based on causal explanations.[40]

Educate Not Train

> Any teacher who can be replaced by a computer deserves to be!
>
> *David Smith*

Many of the factory-like universities have ceased any pretence at education and are instead concerned with instruction. In many cases they are so highly structured that even the instruction becomes mere arid training. Training usually provides a narrow explicit machine or systems specific competence which is quickly obsolete with technological change. Education is of a much more durable quality and as one of my German colleagues put it is 'a state of mind'. My hierarchy of verbs in these matters is that you programme a robot, you train an animal but educate human beings. Education in this sense is not just that which occurs in schools or universities where so often there are those who are – as Illich points out:

> Schooled to confuse teaching with learning, grade advancement with education, a diploma with competence and fluency with the ability to say something new.[41]

Education is a subset of the cultural milieu in which it occurs. Thus apprenticeships in the classical sense were the transmission of a culture. They produced the giants who define our European civilisation: Leonardo da Vinci, Filippo Brunelleschi, Giotto. A great skill, as we approach the 21st century, will be to use the new technologies in such a fashion as to build upon the best traditions of those rich learning processes.

The human-centred community questions the given wisdom that daily advises us that we live in an information society. We may perhaps live in a data society whereas what is required is the competence to operate at the knowledge/wisdom/action end of the cybernetic loop (Figure 7). I freely admit that it took us many years and much technocratic rambling to arrive at this rather obvious conclusion. If we had been more fully exposed to the ideas, however ill-structured, of artists or poets, this might have been more obvious. Thus in the case of Figure 7, the basic 'theory' was anticipated by T. S. Eliott in one of his poems:

What wisdom have we lost in knowledge
What knowledge have we lost in information.

Frequently, the big issues in society are prefigured by our poets and our artists and we diminish ourselves as engineers and scientists if we do not interact with them in a multidisciplinary way. The ability to do so may be an important requirement as we approach the 21st century.

Imagination

> It stands almost complete and finished in my mind so that I can survey it like a fine picture or a beautiful statue.
>
> *Mozart*

Industrialisation has in many ways reduced and over concentrated our competencies as human beings. We confer life on machines and diminish ourselves. We are gradually becoming observers of life rather than its active participants. Education in the sense in which I use it above should imbue a sense of excitement, discovery and imagination. We are far too obsessed with narrow facts, details and exams. Exams essentially find out what people do not know rather than what they do know.

There used to be a tradition in some of the older universities that if you didn't like the examination question you were set you simply ignored it and wrote your own question. Life in the widest philosophical sense should be about writing one's own questions and education should facilitate that. It should stimulate and excite our imagination and sense of discovery. The great Einstein on one occasion observed:

> Imagination is far more important than knowledge.

and he went on to say:

> The formulation of a problem is far more important than its solution which may be merely a matter of mathematical or experimental skill. To raise new questions, to look at old problems from a new angle marks the real advances in science.

This, I suggest, is the broader context in which we need to contemplate the skills and competencies for the 21st century.

References

EEC/FAST report, *European Competitiveness in the 21st Century: The Integration of Work, Culture and Technology*. Free from FAST 200 Rue de Loi 200 B-1049 Brussels Belgium.

FAST: Report on each member state available from FAST as in (1).

Cooley, M. *Architect or Bee?: The Human Price of Technology*. Also in German, Japanese and Swedish. Forthcoming in Irish (1993).

Proceedings of *New Manufacturing Imperatives*, London, 1990.

Taylor cited in Cooley, *op. cit.*

Report (1990); *Made in America*, MIT.

Ford, cited in Cooley: *The New Shape of Industrial Culture and Technological Development*. Tokyo Keizai University, 1990.
Cooley, *op. cit.* (3 above).
Rosenbrock, H.H. (1990). *Machines with a Purpose*. Oxford University Press.
Gill, K. S. (editor), *A.I. & Society*, Department of Information Studies, Brighton University, Sussex.
The Technology Exchange. Wrest Park, Silsoe, Bedford, MK45 4HS.

Notes

1 Cooley, M. J. E., *The Knowledge Worker in the 1980s*, Doc. EC35, Diebold Research Programme, Amsterdam, 1975.
2 Braverman, H. (1974). *Labor and Monopoly Capital*. The Degradation of Work in the 20th Century, Monthly Review Press, New York.
3 M. Polanyi, *The Tacit Dimension*.
4 Dreyfus and Dreyfus, *Mind over Machine*, Glasgow, 1986.
5 *The Economist*, 22 January 1972.
6 F. W. Taylor, *On the Art of Cutting Metals*, 3rd edition revised. ASME, New York, 1906.
7 F. Margulies in conversation with the author.
8 W. Fairbairn, quoted by J. B. Jefferys, *The Story of the Engineers*, Lawrence and Wishart for the AEU, 1945, p.9.
9 Eugene Wigner cited page 45 in Cooley, M (1987), *Architect or Bee?*, Hogarth Press, London.
10 N. Wiener cited p. 50 in Cooley, M (1987), *Architect or Bee?*, Hogarth Press, London.
11 G. Nadler, 'An Investigation of Design Methodology Management', *Science*, vol.3, June 1967, pp.642–655.
12 J. Lobell, *Design and the Powerful Logics of the Mind's Deep Structures*, DMG/DRSJ, vol. 9, no. 2, pp.122–129.
13 K. Mark, *Capital*, vol.1, p.174, Lawrence & Wishart, London, 1974.
14 R. S. Silver, 'The Misuse of Science', *New Scientist*, vol. 166, p.956, 1975.
15 S. Rose, 'Can Science be Neutral?', in H. Rose and S. Rose (eds), *The Political Economy of Science*, Macmillan, London, 1976.
16 H. Braverman, *Labor and Monopoly Capital. The Degradation of Work in the 20th Century*, Monthly Review Press, New York, 1974.
17 Dreyfus HL & Dreyfus SE (1986), *Mind over Machine*, Free Press, New York
18 Dreyfus and Dreyfus, *op. cit.*
19 Kantor, Vorlesunfen Uber Geschichte der Mathematik, Vol. 2, Leipzig, 1880.
20 Olschki, Geschichte der neusprachlichen Wissenschaftlichen Litteratur, Leipzig, 1919.
21 A. Sohn Rethel, *Intellectual and manual Labour: A Critique of Epistemology*, Macmillan, London, 1978.
22 T. Bowie, *The Sketchbook of Villard de Honnecourt*, Indiana University Press, 1959.
23 See note. 22.
24 Fairbairn W. quoted by B. Jefferys in *The Story of the Engineers*, Lawrence & Wishart for the AUE 1945, p.9
25 M. Kemp, *Leonardo da Vinci - The Mavellous Works of Nature and Man*, J.M. Dent & Sons Ltd, London, 1981, p.26.
26 Kemp, *op. cit.*
27 M. Polanyi, 'Tacit Knowing: its bearing on some problems of philosophy', *Review of Modern Physics*, vol. 34, October 1962, pp.601–605.

28 Kemp, *op. cit.*
29 M.J.E. Cooley, 'Some Social Implications of CAD' in Mermet (ed.), *CAD in Medium-Sized and Small Industries*, Proceedings of MICAD 1980, Paris, 1980. M. J. E. Cooley, 'Computerisation - Taylor's Latest Disguise' in *Economic and Industrial Democracy*, vol. 1. Sage, London and Beverly Hills, 1981.
30 J. Weizenbaum, *Computer Power and Human Reason*, W.H. Freeman & Co., San Francisco, 1976.
31 Aspinal, Cooley *et al. New Technology, Employment and Skill*, Council for Science and Society, London, 1981.
32 *Shooting Life*, Spring 1987.
33 *Ibid.*
34 *Ibid.* p.11.
35 Taylor, *op. cit.*
36 EEC/FAST report, *European Competitiveness in the 21st Century: The Integration of Work, Culture and Technology.*
37 Prof. Hopgood in Proc. NMI Conference, London 1988.
38 Taylor, *op. cit.*
39 Ford, cited in Cooley, *The New Shape of Industrial Culture and Technological Development.* Tokyo Keizai University, 1990.
40 H.H. Rosenbrock, *Machines with a Purpose*, Oxford University Press, 1990.
41 Ivan D. Illich (1971), *De Schooling Society*, Penguin Books.

Chapter 3

Rosenbrock's Account of Causality and Purpose

A compilation of Howard Rosenbrock's works selected and annotated by Satinder P. Gill*

Introduction

The first extract is taken from Purpose and automatic control, Computing and Control Engineering Journal, March 1992, pp. 88–89, which was addressed particularly to control engineers.

This brief paper is devoted to three propositions:

1. For 400 years, science has explained the world by causal relations, by which the past entails the future, and purpose is excluded. In its own terms, this explanation has been incomparably successful, but it has implications for our relationship with nature, and to one another, which are now forcing themselves insistently upon our attention.
2. The causal description of the world is not something which follows inevitably from observation and experiment. It is something which we impose upon the world before we begin to study and explain it.
3. Practitioners of automatic control are uniquely equipped to understand and appreciate these facts.

These propositions look forward, towards the kind of world which we shall create in the future with the resources of our science and technology. It may therefore seem odd to begin with a reference to the Middle Ages; but ideas have deep roots. It is in medieval thought, and in the scientific reaction against it that we have to look for the origin of our current scientific outlook.

---

* Introductory and explanatory remarks added by Satinder Gill are in italics. This chapter has been compiled, with Howard Rosenbrock's permission, from a number of his published works, as follows: H. H. Rosenbrock, 'Purpose and automatic control', *Computing and Control Engineering Journal*. March 1992, pp. 88–89, IEE (reproduced by permission of The Institution of Electrical Engineers); H. H. Rosenbrock, *Machines with a purpose*, Oxford University Press, 1990 (reproduced by permission of Oxford University Press); Howard Rosenbrock, 'Technology and its environment', *AI & Society Journal*, vol. 7, number 2, 1993, pp. 117–126, Springer-Verlag London Limited.

Some history

Our view of the Middle Ages is obscured by the veil of two centuries, from about 1300 to 1500, when it was in gradual decay. This period of falling harvests, of overpopulation and pestilence, of warfare and intellectual decline, separates us from the two preceding centuries of High Scholasticism. In those two centuries, a renaissance of thought and culture took place. Greek and Roman thinking, with its Arab extensions, was rediscovered in a society where prosperity was increasing and men of the highest intellectual ability strove to reconcile the new learning with religious tradition. It is in these centuries that there were conceived and built the great cathedrals which we, with our immeasurably greater resources, find it difficult even to maintain.

Medieval thought rested on a basis which now seems quite foreign to us. There was a trust in human reason, unaided by observation and experiment, which we no longer share. Authority was vested, not in a direct appeal to nature, but in what had been written in the past, which was endlessly dissected and glossed. Natural history, in the popular 'bestiaries', became a mixture of tenth-hand observation and mythology: 'Bartholomeus reminds us that the weasel alone can attack the basilisk unharmed, provided that it eats rue beforehand'.[1] Underlying these tendencies was the belief in a purpose, manifested in a world which had been created for the benefit and use of mankind.

Between 1500 and 1700, science as we know it now was struggling to break free from this all-embracing outlook. Textual authority was rejected, and reliance was placed on personal observation, carefully recorded, as in William Gilbert's studies of magnetism[2] and Galileo's dynamical experiments.[3] Theoretical explanations to account for the results were not sought in purpose, which tended to answer the question 'why?' rather than 'how?', and to become entangled in theology. Instead, the operation of nature was compared with machinery as it existed at the time, in particular with clockwork, and with the clockwork automata which became fashionable. Kepler wished to regard nature not 'as a divinely animated being' but 'as a clockwork'.[4] Descartes remarked that if we had automata similar to animals and constructed with the same perfection, we should not be able to perceive any difference. La Mettrie in the 18th century extended the same thought to mankind.

The scientific view

In terms of systems theory, the essence of the new approach was to explain the world causally, so that the future behaviour of any system could be predicted from its past history, and from any future inputs from outside. This applies directly to deterministic systems, but it can also apply to certain average properties in stochastic systems. The effect of the past on future behaviour is usually summarised by the 'state', and system behaviour is reduced to a set of dynamical equations governing the evolution of the state. This is the view with which all scientists and engineers are familiar.

We live now with the consequences of this new way of regarding the world. It has brought an immense increase in knowledge, embodied in a science which represents the greatest intellectual achievement of mankind in any age. It has also given us an incomparable mastery of nature through our science-based technology. At the same time it has irrevocably changed our relationship with the world we live in. As Burtt remarks: 'The scholastic scientist looked out upon the world of nature and it appeared to him a quite sociable and human world. It was finite in extent. It was made to serve his needs. It was clearly and fully intelligible, being immediately present to the rational powers of his mind' whereas 'now the world is an infinite and monotonous mathematical machine'.[5] With a deeper and later acquaintance, Monod puts it more strongly: that man 'like a gypsy ... lives on the boundary of an alien world; a world that is ... as indifferent to his hopes as it is to his suffering or his crime'.[6]

Human consequences

Regret for a vanished delight in nature might be dismissed as the price of intellectual maturity, but the implications are more sinister. Together with the changed outlook, there is a change also in the way in which we relate to the natural world and to one another. This can be described most concisely by means of extreme, and therefore unrepresentative, examples; but the application to more ordinary situations is easily made.

The 1840s, the 'hungry forties', were a period of great distress in Britain, and particularly in Ireland, which was met after some delay by charitable endeavours.[7] Lord Brougham responded with a comment in the spirit of the scientific outlook: 'charity is an interference with a healing process of nature, which acts by increasing the rate of mortality, and thereby raising wages.'[8] More recently, the development of the atomic bomb has cast a stain upon science, and a shadow over the future. It was done with a mixture of motives, but would hardly have been possible without a belief that the pursuit of scientific knowledge is exempt from all moral considerations: a view expressed in one physicist's comment: 'Don't bother me with your conscientious scruples! After all, the thing's superb physics!'[9]

A little thought will show the influence of the same outlook in a great many areas of our day-to-day activities. If all plants and animals are machines, we owe them no responsibility. So we exploit nature, to the extent of damaging the environment on which they, and we, depend. Scientific research in medicine has a constant tendency to trespass beyond the bounds of medical ethics: 'in science, man is a machine',[10] and towards a machine we can feel free from any moral obligation. Production engineering regards the human contribution as no more valuable than that of a machine, and demands from men and women a more and more machine-like role.*

An alternative view

If these were the unavoidable costs of our knowledge and competence, we should face daunting questions: whether the price is one we can afford to pay, and whether we can face the penalties of failing to pay it. But the stark choice is only apparent. So far I have used the word 'machine' in its usual sense, as something we conceive to have no purpose, which 'mechanically' follows the laws of nature by which it is governed. In automatic control, however, we have available a different view.

The task of the control engineer is to incorporate human purpose in machines. The purpose is given as an objective to be attained in the future: say, for example, that we wish to take a satellite from one orbit to another at a given time, and to do so with the least expenditure of fuel. We solve this problem by finding the causal relations which will accomplish the objective, and then building a system that satisfies these relations. The system, with its fuel and rocket motors and computers, is certainly a machine. It is governed by causal laws which ensure that the state evolves in the required manner. Otherwise expressed, these laws cause its future behaviour to be determined (absolutely or statistically as the case may be) by its past history and by any inputs which are injected in the relevant period.

Yet at the same time our machine fulfils a purpose. Indeed, in some circumstances the purpose is equivalent to the causal relations, in the sense that these can be deduced mathematically from the purpose by dynamic programming. We have to use the causal description when we build the system, because our science and technology are expressed in causal terms. Yet at a conceptual level, the purposive description is interchangeable with the causal description. We can describe the situation in either way: the device remains the same whichever description we use, and it is we who impose causality or purpose upon it by our choice of description.

A section on 'Variational principles', which is of interest to specialists, is here omitted. It concludes that the usual scientific description of nature in causal terms can be derived from a purpose by methods similar to those of the control engineer, and that this purpose has no theological implications. From the purpose, a 'policy' is obtained, which is a causal description of the actions needed to achieve the purpose. An account of the

* It would be wrong to draw from these facts an accusation against individuals, who invariably temper their views on science and technology with humanity. But anyone who has set out to oppose the anti-human aspects of technology will know that what has to be contended against is what has been described, and that the humanity of those involved in the technological process cannot be engaged effectively against attitudes which are so strongly entrenched in our technological culture.

present state of this theory as applied to special relativity and quantum mechanics is given in H. H. Rosenbrock (1995), A stochastic variational treatment of quantum mechanics, Proc. R. Soc. Lond. A, vol. 450, pp. 417–437.

A world with purpose

Though there are no theological implications to threaten the independence of science, there are strong implications for the way in which we regard the world, and for our relation to it. Let us tentatively adopt the purposive description, not as being the 'true description of the world', but as being scientifically equivalent to the orthodox causal view, though with different human implications. In this way we can obtain an insight into the extent that the orthodox view controls our attitudes and behaviour.

In the first place, human purpose, which in the causal view becomes a phantasm, a mere epiphenomenon, is reinstated. It takes its place among a multitude of purposes coexisting in the world. We can say, for example, in a production system, that the purpose of production is expressed predominantly in the people, and that the purpose of machines should be subordinate to the purpose of those who work with them: they should be tools to be used.

Relations between people will be expressed in terms of their purposes, as has always been the case outside science. This is not in itself a solution to inevitable conflicts, but it does protect us from such aberrations as Taylorism[11] with its treatment of people as (causal) machines. We shall see nature as filled with competing and collaborating purposes, and shall be more inclined to work with and through them than to force them into a different mould.

These and other consequences need to be worked out in detail, for which there is no room here. But by way of illustration a brief view of evolution may be given.[12] In the causal view, organisms replicate themselves, using information passed on in DNA from generation to generation. Occasional errors occur in this process, and most of these mutations reduce the chance of producing offspring. From time to time, a mutation is beneficial, and the mutants compete successfully and become dominant in their population. In this way, species change with time, and occasionally branch to form new species.

Viewed as a causal process, it seems that at each moment evolution offers an unlimited range of possible mutations. Out of this unlimited range, a certain number will occur by chance, and of these usually one at most will be selected by competition. The process appears entirely undirected, dependent only on the chance of a favourable mutation, and the subsequent ability of the mutant to replicate more successfully than its competitors.

Viewed from a purposive standpoint, the process looks different. The organism and its environment form a quantum-mechanical system. This (assuming that the development can be completed) obeys a stochastic variational principle.[13] There is a randomness in its behaviour, but in an average sense the future is determined by an objective which is to be attained: evolution implements a purpose that is inherent in matter itself.

The apparent contradiction between these two views is easily removed. Not every mutation is possible, but only those which obey the prescriptions of orthodox quantum mechanics. These prescriptions are the policy deduced from a stochastic variational principle, so that by following them the system accomplishes a purpose.

The 'Conclusions' of the preceding paper are omitted, and the next extract is taken from Technology and its environment, AI & Society, 1993, vol.7, pp. 117–126, the first section of which is also omitted.

One who has worked both in industry and in academia soon appreciates that there are different kinds of knowledge. There is one kind of knowledge, for example, which is needed by the originator of programs for the computer-aided design of control systems: knowledge of computers, control theory, man–machine interaction, and the kind of problem which the user will wish to solve. The engineering designer who uses the program needs also to understand control theory, but in a way which is more immediately relevant to the systems which he has to control and which impose detailed constraints that he must understand.

Those who will work with the ultimate product, say a computer-controlled lathe, need a still different knowledge: a broad understanding of the machine and its capabilities, in which control will appear mainly through the opportunities it gives and the constraints that it imposes. All of these kinds of knowledge need to be valued equally, because all are essential to success.

A further aspect of this division of tasks and of knowledge is that the designer of the CAD system strongly influences the working life of the engineering designer who uses it, and the engineer in turn influences the working situation of the lathe operator.

Now research into automatic control in Britain in the decade from 1950 to 1960 was strongly influenced by an existing tradition of engineering. This was firmly rooted in experience, and saw theory as a basis from which this experience could be organised and understood. Theory was not seen as a direct source of innovation, but as a springboard from which engineering innovation could take place. Consequently the aim of theory in control was to provide an intuitive understanding of the apparently counter-intuitive behaviour of control systems. In this it was outstandingly successful, and what had for example been a kind of 'black art' in the process industries for three decades, learned only by years of practical experience, became very quickly transparent, and easy to explain and understand. A wide range of control problems, which in the past had been regarded as difficult and doubtful of solution, was shown to be readily solvable in the light of the new understanding.

After 1960, this outlook began to change. In the USA and Russia, research into the problems of rocket guidance was heavily funded. Deep mathematical investigations were undertaken by research workers such as Bellman, Kalman, and Pontryagin, and by a host of later theoreticians. The subject was immensely enriched and deepened, and powerful computer techniques were initiated. The special problems of aerospace yielded readily to these techniques, though applications to industry were less successful.

Yet the invigorating excitement and turmoil of this decade had a darker side, which was more and more evident after 1970. The earlier engineering approach became submerged in a new outlook, which saw theory, not as a way of assisting the engineering designer, but as a way of replacing him. The ideal was no longer that of a designer supported by theory and by computers in making use of his practical experience. It was rather that of a theoretician who took an engineering problem and formulated it in a way that could be submitted to a computer. Then it was the computer which would solve the design problem, by means of an immense algorithm incorporating all that theory and experience could offer.

The tendency in this direction was by no means restricted to control engineering. Similar developments took place in many areas, connected with a debate about 'engineering science' and 'the gap between theory and practice'. Research into 'artificial intelligence' showed similar tendencies in an extreme form.

Now coming from one no longer young, all of this may seem like nostalgia for a simpler past, if not a rejection and undervaluation of the intellect in favour of practical cut-and-try empiricism; but it is far from that. It is based rather on a concern for the relative value which we place, on the one hand on machines and mechanised calculation, and on the other on human beings: human beings with shortcomings that machines and organised knowledge can help to overcome, but human beings also with a tacit knowledge and with unique abilities of observation and ingenuity and experienced response to problems.

Taylorism

The previous section is, in its essence, a protest against Taylorism in its most recent manifestation. F. W.Taylor gave his name in the early twentieth century to a management practice which aimed to remove all initiative from workers: 'Under our system the workman is told minutely just what he is to do and how he is to do it; and any improvement which he makes upon the orders given to him is fatal to success. While, with the old style, the workman is expected to constantly improve upon his orders and former methods'.[14]

This approach to industrial organisation arose from the coalescence of an earlier division of labour with self-acting power-driven machinery. It was primarily a British development, practised by Arkwright in an early form at the end of the eighteenth century, described by Babbage[15] and Ure[16] and later enthusiastically adopted and developed in the USA. It spread to Europe with the dissemination of large-scale industry, but after 1945 was largely rejected there in favour of a more socially consensual view; most notably in the Scandinavian countries and in W. Germany. Japan also passed through a period of Taylorism (the 'second stage') but fairly early in the twentieth century had returned (the 'third stage') to a traditional view emphasising 'harmony unity and solidarity based on the family concept'.[17]

Many reasons can be suggested for the development of Taylorism: the class structure in Britain, the fact that industry in its early stage recruited agricultural workers having no familiarity with machines, and the nineteenth-century faith in scientific knowledge as the key to progress. One may also add, a narrow economic view which failed to see the skill and knowledge of people as a national resource, and regarded their fostering only as a cost. My own view, to which I shall return later, places emphasis on certain beliefs which have underlain our science and science-based technology from about 1600.

In its country of origin, Britain, Taylorism has been responsible for a divisiveness in industrial relations which even now has not been fully overcome. The major criticism of it has arisen, in the past four decades, from social scientists, who have emphasised the damaging effects of fragmented and trivialised work on those engaged in it, and who have taken part in many experimental projects to alleviate these consequences. Techniques of job rotation, job enlargement, job enrichment, and the formation of autonomous groups have been developed and applied, with mixed success.[18] Their major weakness is that the ameliorative effort is usually applied only after the Tayloristic situation has been created and incorporated in the design of the physical plant, so that change is difficult and limited in scope. In Britain, they also have to contend in some cases with adverse managerial views; in Scandinavia the outcome has been notably more encouraging. Much more recently in Britain, Japanese production facilities have come into operation and have been outstandingly successful, and this has prompted the beginning of a process of change in domestic attitudes and practices.

A section on 'The UMIST Project', which was an attempt to give a demonstration on a small scale of an alternative 'human-centred' technology, is here omitted.

Technological convergence

As Cooley has stressed,[19] a uniform technology world-wide, and a uniform management style, are not necessarily desirable. Different countries, with their different histories and cultures, may be better satisfied and more productive with different ways of organising and developing their industrial and commercial activities. One might expect therefore that Japan, Britain and the USA could well follow different paths in the future.

But if one looks at the history of industrial development throughout the world, one is struck chiefly by a strong tendency to convergence. This is commonly explained by economic pressures; by the fact that international trade forces competing nations all to adopt those technical means which give them the greatest advantage. Yet economic calculations are notoriously arbitrary. The neglect to place a value on 'human capital' has already been mentioned, and the neglect to impose a cost on pollution is a second example. Then again, different countries do not all have the same economic objectives.

An extreme example can be found in the socialist countries in their heyday. Their expressed social aims should have made a Tayloristic organisation of work impossible. At the same time their economic calculations were quite different from those of a capitalist economy, and industry was protected from the pressures of competition. A factory making spares for motor cars, for example, would be given production targets for many different items over a future period. The most economic way for the factory to meet its targets was to make all of one item, say windscreen-wiper blades, then all of a second item, and so on. Hence the periodic gluts and famines of goods in the market, the cost of which to the user was excluded from the economic calculation.

Now if the argument from economic necessity has force, one would expect managers, and engineers and other technologists, operating in such different economic conditions to have a different outlook on the way in which technology and industrial organisation should develop. My own experience indicated that this was not so: factories in the socialist countries operated under a Tayloristic regime identical in its intention to capitalist exemplars, even if less well organised. Discussions with engineers developing robots and numerically-controlled machine tools disclosed aims and attitudes no different from those of their contemporaries in the USA, to which they looked as a model. Thus a more backward technology, and a very different social and economic system, did nothing to change the basic outlook on technology and industrial organisation. In his professional life, a competent Russian engineer of 1980 would have had little problem in adjusting to work in the USA or Britain.

A second argument, from technological determinism, can be dismissed more briefly. It is true that, at any time, the feasible range of technology is restricted, and in the early nineteenth century the restrictions were severe. Spinning and weaving machines had to be driven from a central point by line shafting and belts, and working conditions were to an important extent constrained by these and other similar requirements.

Our current technology, however, with electric drives, computer control, and communication systems, is immensely flexible. It will support an organisation based on extreme centralisation, with all initiative removed from the periphery. But with equal ease it will support an organisation in which there is a very high degree of delegation in decision-making, with central co-ordination rather than command.

The scientific belief system

My own view has come to be the following. There is certainly a degree of strength in the economic argument: to the extent that economic costs reflect the effort we employ on a task they will also reflect our tendency to employ as little effort as we may. There is some strength in the technological argument: only a small subset of the total possibilities will fulfil the demands of safety, reliability, economy, low pollution, etc. which we wish to meet, though this subset may be large enough to offer important choices. There is also force in the Marxist argument that one section of society attempts to exploit another, and uses technology as one means to this end. In a democratic society it will always be a problem to contain such attempts by legal and financial and political means.

None of these, however, seems adequate to explain the uniformity of technological practice throughout industrialised nations, and the consistency of thought and experiment directed towards its development. The only activity which generates an equal consistency in all its practitioners seems to be science, and it is here that I should look for the ultimate source of consistency in technology.

Modern science is a European development, which is conventionally dated from about 1600, with Stevin and Gilbert and Galileo, though its roots go far back into the crafts and the Scholastic debates of the Middle Ages.[20] In characterising science, and in distinguishing it from other ways of acquiring knowledge, emphasis is always placed upon the testing of theory against carefully designed experiment. There is another characteristic, however, which is highly significant, yet has become so deeply embedded in our way of thinking that it becomes invisible. This is the insistence that all explanations shall be causal.

Causality is a slippery concept: a cause must never be later than its effect, and the effect must be a necessary consequence of the cause, but beyond this lie endless complexities and difficulties.[21] What is excluded, however, is clear: it is purpose. In science, we are not permitted to give an explanation in terms of purpose, and this embargo is enforced just as rigorously in technology.

For the control engineer, the veto upon purpose imposes a language which is at variance with the way he must think about his task. The natural way to describe what he does is to say that he starts from a human purpose: say that we wish to transfer a satellite from one orbit to another at a given time. He then 'incorporates this purpose in a control system': that is to say, he develops a set of causal relations which, if they are implemented, will fulfil the purpose, and he ensures that the control system obeys these causal relations. Explicitly, certain equations may be solved in a computer, which emits signals to a transmitter, these signals being received on the satellite and causing its rocket motors to fire or to cease firing. Every step in this process can be described causally, and has to be described in this way before the system can be implemented.

The engineer, however, is not allowed to say that his control system implements the purpose which he was given, because purpose is not an acceptable category in a science-based technology. The only explanation of the control system which is allowed is the set of causal relations (computer programs, differential equations, etc.) which it obeys. Yet this causal description will make no sense to anyone inspecting it unless he knows the purpose which it is intended to fulfil.

Beyond this, the engineer in strict terms is not allowed to talk of an initial human purpose, from which he started his design. Human beings, in science, do not have a purpose: 'For the scientist there is only "being", but no wishing, no valuing, no good, no evil; no goal'.[22] At this point, of course, the embargo breaks down. The engineer is in no position to give a set of causes, which determined the utterance of a particular speech, which by a causal mechanism within himself made him design a control system. Even if he could provide such a description in causal terms, it would omit everything which gives human significance to the situation.

Given the impossibility of maintaining a strict exclusion of purpose, the technologist behaves as a human being in his everyday life. But I have contended elsewhere at length[23] that in his professional activity, he is powerfully constrained to observe the scientific rejection of purpose, and to do so not only in relation to machines but also to people. He does this, not from any overt desire to diminish the humanity of those who will be affected by his activity, but because the technological means available to him presuppose that the systems he designs are in their entirety - machines, computers and people - governed by causal laws. And as our commercial and industrial systems become more and more enmeshed with technology - with computers and computer-controlled machines - so this presupposition extends itself to those who manage these systems or bring them into being.

Taylorism is the natural outcome of this underlying presupposition of universal causality, as was recognised by Taylor himself: 'The contrast between Brinley and Davenport's rational, scientific approach to technical problems, and the foreman's capricious, often destructive approach to production problems, including the worker's behaviour, made a strong impression on him. There must be a better method, he reasoned, a way to apply the engineer's scientific perspective to management, as well as machinery. Taylor therefore arrived at the juncture many other engineers would reach in the latter years of the nineteenth century.'[24] The tendency to give a machine-like character to human work follows naturally, and it is not only manual work that can be treated in this way.

The remaining part of the preceding paper is omitted. It is followed by brief extracts from Machines with a purpose, 1990, Chapter 1, pp. 2–3 (Oxford University Press).

'In science, man is a machine; or if he is not, then he is nothing at all.'[25] This was written by a distinguished biochemist and sinologist with wide human sympathies. It defines his view of an imperative imposed by science, rather than uniquely characterising his own beliefs. It was also written in opposition to views which can easily be seen as obscurantist. Yet it contains within it the seed of a highly damaging intellectual development which will be our main concern. It is therefore a convenient starting point for our discussion.

What, first of all, is meant here by the statement that 'man is a machine'? It asserts that all human behaviour, conscious and unconscious, can be explained in causal terms.

A digression on the historical development of the idea of causality is here omitted.

What Needham is asserting is that all biological systems, and all human behaviour, can be explained in terms of cause and effect. A given cause produces a certain effect, and this again acts as a cause to other effects, and so on in an infinite causal chain. The claim is not that we can actually describe every link in such a chain, nor even that we shall some day be able to do so. It is only that in principle such an explanation can be given, and no other is needed: that in fact causal relations express accurately and completely the way the world works, and not only for machines but for people.

The vitalists, against whom Needham wrote, maintained on the contrary that causal explanations were not enough in biology. Machines could be explained entirely by cause and effect. Biological systems required in addition, for their explanation, a reference to purpose. Their behaviour could not, even in principle, be fully explained in causal terms: 'Even in the most elementary and general manifestations of life we have to do with purposive phenomena, quite distinct from all the phenomena of the inorganic world.'[26]

This is a debate which goes back to classical antiquity, and Needham was expressing the view that has predominated during the whole period of modern science, which we can date somewhat arbitrarily from 1600. In this view, purpose does not exist except as an appearance, an epiphenomenon. We seem to perceive purpose in ourselves, in other creatures, or in the progress of evolution. But when we study these scientifically, we find that the idea of purpose is redundant. The phenomena which seem to indicate purpose can all be explained causally.

Our obstinate human belief that purpose is not an illusion arises in this view from our failure to be true scientists:

'Its teleology appears to them [sc. working biologists] as a faint thread ... and only present in them so far as they are imperfect scientists, because they cannot altogether divest themselves of the common and vulgar ways of thought.'[27]

'The former [sc. teleology] is only present in the mind of the scientific worker because he is still to some extent an ordinary man.'[28]

The rejection of purpose in these quotations goes beyond the statement that it is not needed, and becomes a recommendation which we are urged to follow. Teleological explanations are the mark of the common man. Though we all have something of the common man in us, as scientists we must strive to rise above his way of thinking. But if we take Needham at his word, he himself is a machine, explainable in causal terms like ourselves. Between two such machines, what is the force of a recommendation to follow one course rather than another? To do this is to have a purpose, and purpose has already been excluded.

From a later section of Machines with a purpose, Chapter 6, pp. 109–117, is the following extract. In the intervening chapters it has been suggested that a causal description of nature is only one possibility. It is also possible to give a description in terms of purpose, as was done for classical physics in the nineteenth century by Hamilton. This needs to be extended to quantum mechanics, but a start has been made on this (Machines with a purpose, pp. 203–219, and Proc. R. Soc. Lond. A, 1995, vol. 450, pp. 417–437). Then we have two descriptions of nature, which are equivalent in the sense that every observation which agrees with one agrees with the other, and every observation which contradicts one contradicts the other, and we cannot assert that either of them is the 'true and only description of the world'. It is suggested that these two descriptions should be called 'myths', and that only a combination of them, their 'equivalence class', can serve as a scientific theory. To accept either of the myths as the 'true description of nature' is to perform an act of faith.

Technological systems

At the opposite extreme from biological systems are those man-made systems which incorporate a human purpose. The origin of this human purpose must be found in the account given in the previous section *(omitted)*. The way in which the purpose is incorporated in a machine has been described in Chapter 2 *(which is omitted, but in which it is explained how a 'policy' is obtained, a causal description of the way the machine must behave in order to fulfil the purpose: the machine is then built to incorporate these causal laws).* What is the status of a purpose which has been incorporated in this way?

In the causal myth, the purpose effectively disappears. The machine is described causally by the policy derived from the purpose, and the purpose is either denied or ignored. This is the substance of the statement that 'man is a machine', which carries over to man the same denial or ignoring of purpose. In the purposive myth, on the other hand, a purpose which has been incorporated by human action in a machine retains its validity as a purpose. It can be translated into a policy to give a causal description, but equally it can be derived as a subordinate purpose from more fundamental purposes.

The machine will have a purposive description in terms of Hamilton's principle, for the mechanical components, which is subordinate to a description in terms of a quantum-mechanical principle. Electronic devices can equally be described by the quantum-mechanical principle. Subordinate to these descriptions will be the description of a machine in terms of the human purpose which has been incorporated in it. That is to say, this human purpose now exists as a simplified description of behaviour, which behaviour can be deduced from the more fundamental, quantum-mechanical, purpose of the machine.

If, therefore, we describe the machine at a fundamental, quantum-mechanical, level by means of its purpose, this fundamental purpose now has, as a subordinate purpose resulting from it, the human purpose we incorporated. We have, by incorporating the human purpose, modified the fundamental purpose of the machine. This has been achieved by the indirect process described in Section 10 of Chapter 5 *(which is omitted, but where the process of deriving and incorporating a policy is explained)*, because we have no present technique for directly modifying a fundamental purpose in this way.

A first consequence of this account is that moral judgements of human purposes do not cease to be valid when those purposes are incorporated in machines. We may, for example, condemn a human purpose. When this purpose is incorporated in a machine, according to the causal myth, the purpose disappears. We are left with a device which follows certain causal laws, but has no purpose, and so is not subject to moral judgement. Within the purposive myth, the human purpose persists as a subordinate purpose of the machine, as it was earlier a subordinate purpose of a human being, and the same moral judgement is appropriate. Our instinctive moral repulsion from a thermonuclear missile is justified in the purposive myth, whereas it has no force or validity in the causal myth.

Man and computer

A second consequence of our account bears on a conclusion which is often drawn from the alternative, causal, myth. Briefly stated, this is that 'man is a machine'. Therefore anything which a man can do can be done by a machine. In particular, it can be done by the universal machine, the computer, if this is provided with appropriate sensors and effectors. This argument has been put forward particularly by exponents of artificial intelligence, and is very difficult to counter within the framework of the causal myth.[29] The implicit conclusion which is suggested is that a computer can be programmed so that (with appropriate sensors and effectors) it is fully equivalent to a man.

If we believe the purposive myth, human behaviour takes on a different aspect. In principle, science can give a description of the human body based on quantum mechanics. An immensely numerous collection of fundamental particles forms a system which we can identify, with more or less certainty, as constituting the body at a given time. These particles move and interact in such a way as to fulfil a fundamental purpose, described by a variational principle. The description, because of its complexity, will always be inaccessible to us in detail, just as a causal description in terms of the policy derived from the purpose is inaccessible. Nevertheless, the fundamental purpose exists according to the act of faith by which we accept the purposive myth.

Subordinate to the fundamental purpose are many purposes which describe particular aspects of our activity. Some of these subordinate purposes express bodily functions: the repair of damage, the maintenance of structure, the circulation of the blood. Others resemble closely the kind of control system which we can build into machines: for example the mechanism by which body temperature is maintained close to fixed value.

Still other subordinate purposes describe common actions such as walking or running, or other activities which we have learned: driving a car, playing squash, or using a typewriter. All of these we do, once they are familiar, without conscious thought. The purpose is translated directly into action, without intermediate analysis, just as the earth fulfils Hamilton's principle in its passage around the sun without ever formulating or solving the differential equations which express the policy.

Some parts of the purpose are present in consciousness. In driving a car, for example, we shall concentrate our attention when overtaking. We shall have a purpose, which we implement, but if we are experienced drivers we shall do so without any prior translation of the purpose into a policy. We shall change gear, and accelerate, and steer, without conscious thought and often without direct awareness.

The simple and traditional kinds of human work have exactly this character of directly implementing a purpose. When we use a tool, it participates in our purpose. Following Polanyi,[30] for example we may say that when we use a hammer, we 'feel' directly the impact of the head upon the nail. We feel whether the blow is true, whether the nail is driven, whether it is solid, or whether it bends. The purpose of an independently existing hammer, expressed by Hamilton's principle, has been subsumed into our own purpose. The hammer becomes an extension of ourselves.

By the agency of the hammer, our purpose is converted directly into action, without first being translated into the policy. We could, with sufficient labour, and as an intellectual explanation of our actions, deduce the policy which we are implementing with the aid of the hammer. The policy would give a causal picture, showing how the impact of the hammer head results in a reaction between the handle and the palm of our hand, and how by a process we cannot follow in detail, this reaction is converted by the brain into the perception of a blow referred to the point of impact. A similar analysis would show how we manipulate the hammer for the next blow.

This causal picture, representing the policy, is highly dependent upon external circumstances. The size of the nail, the weight and balance of the hammer, whether the wood is hard or soft: all of these details require a different policy to accomplish the purpose. The purpose is constant; the analysis by which it is translated into the policy is highly variable.

But this analysis plays no part in what we actually do. It appeals only because we are so deeply immersed in the causal myth that we ascribe to it a reality going beyond its function as our explanatory description. If we are asked to explain in causal terms how we hammer a nail, we are quite unable to do so, because we do not translate our purpose into the policy in order to fulfil it. The same is true for other kinds of skill, including those which have a strong intellectual content such as engineering design or medical diagnosis.

For the programmable digital computer the case is different. Its actions do not arise directly from its fundamental purpose – a purpose of transistors and electrical circuits. We intervened to incorporate a human purpose, subordinate to this fundamental purpose. In an 'expert system', for example, this has to be done by a process similar to that described in Section 10 of Chapter 5 *(omitted – see previous comment)*. The 'expert' is questioned by a 'knowledge engineer', who asks for a causal description of his behaviour; that is, for the policy by which his purpose is translated into action. The need for this step is indeed presented as an advantage, as a step towards the refinement of our knowledge conceived as expressible only in causal terms:

> 'Yet if artificial intelligence research had done nothing else, it had shown how empty most theories of intelligent behaviour were (likewise theories of creativity, originality, autonomy, and consciousness). When you wanted to make a computer behave intelligently, you had to have a very precise idea of intelligent behaviour in order to specify it to the computer in detail. In neither psychology nor philosophy did such precise models of intelligence exist.'[31]

In terms of the causal myth we do not understand intelligence or creativity or skill until they have been expressed in causal terms. That expert systems force us to do this therefore appears as a contribution to understanding:

'The heuristic [sc. problem solving] knowledge is hardest to get at because experts – or anyone else – rarely have the self awareness to recognise what it is. So it must be mined out, one jewel at a time. The miners are called knowledge engineers ... the expert himself doesn't always know exactly what it is he knows about his domain ...'[32]

'What the masters [of their craft] really know is not written in the text-books of the masters... But we have learned also that this private knowledge can be uncovered by the careful, painstaking analysis of a second party, or sometimes by the expert himself ...'[33]

Yet when we accept the purposive myth, as we do in this chapter, the work of the knowledge engineer appears as a post hoc rationalisation of something that was, in most cases, done spontaneously without conscious analysis. Whereas the causal myth would insist that an analysis was nevertheless carried out, below the conscious level, the purposive myth sees the action following directly from the purpose, without prior translation into the policy. Both views are explanations constructed for human purposes of understanding and control. They are equivalent for scientific and technical purposes, but nevertheless they lead to quite different explanations and different courses of research and development and action.

Accepting the purposive myth we shall see, for example, great difficulties in what a knowledge engineer sets out to do. An expert has a purpose, and without analysis fulfils the purpose by his actions. These actions constitute a policy, but they will be different in different circumstances. And if circumstances arise which were not foreseen when the 'expert program' was developed, the policy generated by the computer may be quite different from the actions which an expert would think to be appropriate.

Because we are so deeply imbued with the spirit of the causal myth, it may be difficult to grasp the distinction which is being drawn between the behaviour of computers and of people. It can perhaps be made sharper by considering the behaviour of faulty systems. When a computer fails, either through a hardware fault or a software error, its behaviour no longer expresses the human purpose which was induced in it, but a different purpose arising from its faulty condition. Failures in the human organism can equally result in aberrant behaviour expressing the purpose of a faulty organism.

The two kinds of behaviour, resulting from human or computer failure, will be quite different. We could perhaps cause the healthy, working computer to simulate human failures, but we cannot make the failed computer behave in the same way as the failed human organism. The computer is like an actor who has learned a part. His 'knowledge base' is the script, together with his experience of acting. On this foundation he creates a simulacrum (but only a simulacrum) of a character whose words and actions are not scripted, but arise spontaneously. A sane actor can represent a madman: a mad actor represents only himself, and is spontaneously mad in his own way.

Expert systems are one special kind of computing system, but what has been said of them can be transferred without much change to systems designed on alternative lines. Algorithmic programs following a set course of calculation, clearly require a causal description of what is to be done. We could program into a computer some variational principle such as Hamilton's, and cause the computer to base its operation on this: but to do so it would first have to generate the policy. Computers can be made to 'learn from experience', but the same remark applies. 'Experience' would have to be incorporated in causal relations before it could be used.

In all cases, it is only necessary to remark that a knowledge engineer would never have the same problem in 'mining' the knowledge of a computer as was described above in the capture of knowledge from the expert. Somewhere within the software of the computer would be found an explicit description of any task which the computer executed.

These comments apply to programmable computers. There are some computer-like digital systems which do not use a program: for example certain devices employed for pattern recognition. The way they behave has been incorporated in them by a process similar to the one described in Section 10 of Chapter 5 *(which has been omitted – see earlier comment)*, but their behaviour is more obviously machine-like (in the ordinary sense) than the computer's. No strong claims to human abilities are usually made for these devices, which set out to imitate, more or less closely, a causal description of the functioning of the brain, rather than the mind.

To summarise what has been said, when we accept the causal myth, 'man is a machine' in the sense that every feature of his behaviour can be described in causal terms. Since anything which can be described in this way can in principle be done by a computer, it becomes hard, if not impossible, to say that any behaviour shown by man cannot also be shown by a computer.

When we adopt the purposive myth, 'man is a machine' in the sense that every aspect of his behaviour arises from an ultimate purpose. Every aspect of the behaviour of a computer equally arises from an ultimate purpose, but this differs from a human purpose. One arises from the physical properties of a vast collection of elementary particles composing a body of flesh and blood. The other arises from a vast collection of elementary particles composing transistors, electric circuits and the like. Human ingenuity can incorporate some parts of a human purpose in a computer. But it exists there as subordinate to an ultimate purpose very different from that of the human being. At some point, therefore, human and computer behaviour will always diverge. The difference of view between the causal and purposive myths stems ultimately from the fact that the former builds up the world from simple, self-sufficient entities, while the second begins from the opposite extreme, from the universal purpose.

The tendency of this discussion is not to suggest that computers should appear less practically useful when we accept the purposive myth. They give us the opportunity to free ourselves from much routine intellectual work, as an earlier stage of mechanisation freed us from much physical labour, and their usefulness can be greatly increased by the new techniques which have arisen from AI. When they do not interact with people, the limited human purpose which can be incorporated in them can be adequate, and need offer no problem.

But where computers and human beings must interact, the purposive myth suggests that computers must be subordinate to men and women. Human purposes can never be incorporated in computers in the way that they exist in people. As the surrounding environment changes, the policy derived from the purpose will also change, and at some point the policy generated by the computer will differ from that which would be generated by the human organism. Unless the computer is subordinate to human judgement, those who interact with it will find themselves helplessly watching as it pursues the consequences of its error.

The role of the computer must be to assist and support the human skill with which we achieve our purposes, for example by suggesting courses of action for human assessment, and by warning of the danger of proposed actions. It must do this in a way that allows human skill to evolve and develop, rather than seeking to replace it. More will be said on this subject in Chapter 9 *(see below)*.

Formulating human policies

The discussion which has been given is somewhat complicated by a fact that we have ignored, namely that we do ourselves, on occasion, translate a purpose into its policy in order to implement it or describe it. Indeed, we regard our ability to do this as one of the most important attributes raising us above the other animals. The chief circumstances in which we make this translation seem to be the following:

i. When the fulfilment of a purpose requires the co-operation of a number of people. If, for example, we are searching for something which we have dropped in the street we might say to a companion, 'I'll take the left-hand side, you take the right.' The resulting action subdivides the purpose of searching the street into two sub-purposes: of searching the right-hand side and searching the left-hand side. Taylorism (Chapter 8, *omitted here*) attempts to take this kind of subdivision to such a point that individual tasks are reduced to a small sequence of simple operations, repeated continuously and exactly.

ii. After the event, we may need to describe and justify actions, which we took in order to fulfil a purpose. After a motoring accident, for example, we may reconstruct our actions for insurance purposes, saying how we saw a car emerge from a side road, and braked hard in a straight line, but then had to relax braking in order to eliminate an incipient skid as we steered to the right. When they occurred, our actions were not analysed in this way - they took place before we were consciously aware of the need for them - and it may demand some effort after the event to reconstruct them. The process by which an 'expert' explains his actions to a computer scientist constructing an 'expert program' is similar. So also is the process by which a skilled driver, unused to teaching a beginner, reconstructs a sequence of actions for the beginner to follow.

iii. When we are learning an unfamiliar task, we may begin by breaking it down into successive actions. When we are learning to drive a car and wish to turn a corner, we may follow a list of actions which we have been given: release throttle, depress brake, depress clutch, move gear lever, release clutch, turn wheel, etc., etc., in a conscious sequence. But after a time we no longer perform the actions in this way, but as a combined motion which fulfils our purpose.
If we believed the causal myth, we should say that the sequence of individual actions is stored somewhere in the brain, but is no longer conscious. As we are adopting the purposive myth, we say that the purpose is now fulfilled without being first translated into the policy. There will always be a description of our actions in terms of the policy, and we can with some effort generate this, as in (ii). The policy will vary with such circumstances as whether the road is dry or wet or icy. But our reaction to these circumstances will not occur by a logical analysis. We shall be aware, for example, of the grip of the tyres by the force needed to turn the steering wheel, and the incipient sliding at front or back. But these statements represent an explicit analysis which we do not make at the time: we 'feel' what is happening between tyres and road, in the same way as we 'felt' what happened at the head of the hammer, and its purpose has been, to some degree, incorporated in our own.

The purposive description of human behaviour illuminates some aspects which cannot be easily understood in a causal description. One of these aspects is the common experience of problem-solving, in such activities as design or the generation of new mathematical theorems. It is generally recognised that these activities, though purposive and highly structured, cannot be reduced to a formal procedure.

A common description of the way that problems come to be solved is that one must worry over them for a long period, usually without much success, but then it is best to put them aside. Some time later, without any further thought, a solution will often present itself. It needs to be checked, because sometimes it is wrong, but often it is correct.

The decisive step is to internalise the problem to the point where one is committed, at a deep level, to solving it. This purpose of obtaining a solution is then fulfilled by the human organism, though we have no conscious knowledge of the way this is done. Nor can we propose any logical procedure by which we can be certain to generate the result; though it is easy to formalise the process of verification and proof once a tentative result is available.

If we accepted the causal myth, we should say that at a subconscious level, some logical procedure was carried out within the brain to produce the required answer. Accepting the purposive myth, we say that our organism fulfils the purpose which we have formed, and which is a subordinate purpose of the whole human organism. Some causal description of what went on can be obtained, in principle, by evaluating the policy obtainable from our purpose, but this description is a description and no more. It is not the mechanism by which the answer was obtained.

Some intervening material is omitted, and the following extract from Machines with a purpose, Chapter 7, pp. 129–134 describes some of the adverse effects of the causal myth.

Science and medicine

Before considering the matter more generally, it is instructive to look at a further example, which is provided by the relation between science and medicine. There is in medicine a long-standing and admirable commitment to an ethical standard of behaviour. There is also a long tradition of scientific study from which great benefits have been obtained. Yet the relation between ethics and science, which here means the causal myth in science, is an uneasy one.

To the physician acting within the ethical tradition, the patient is a fellow human being to whom he can give help and comfort. Nothing is to be done to the patient that is not aimed solely at his own benefit, unless it is done with his own free and informed consent. The medical scientist has a different but equally admirable aim: by increasing knowledge, to permit the more accurate diagnosis and more effective treatment of medical problems.

One might expect that these two aims could coexist without conflict, but there is a tension between them. Science, as always, sets up a distinction between the intelligent and purposive investigator, and the inert, causally determined object of study. Thus Pappworth[34] remarks on 'the significant fact that in most medical reports of patients having been submitted to experimentation, the patients themselves are collectively described as "the material".'

With this outlook, the scientist departs from that attitude to the patient which must underlie and support the ethical requirement. The contrast can become explicit:

'The desire to alleviate suffering is of small value in research – such a person should be advised to work for a charity ... Research wants egotists, damned egotists, who seek their own pleasure and satisfaction, but find it in solving the puzzles of nature.'[35]

It is therefore not surprising that conflicts can arise between the scientific and ethical compulsions, a conflict in which the initiative is ordinarily with science. Some examples will illustrate this situation:

a. A research project[36] in the late 1970s proposed to improve the management of childbirth in the following way. A labour-inducing drug would be continuously injected into the mother while the heart-rate of mother and baby were monitored. If either showed signs of distress, the rate of injecting the drug would be reduced. This would be done by an automatic control system, having the aim of keeping the rate of injection at the highest level which just avoided distress. In this way, the delivery would be made as quick as it could possibly be with safety.

   If it were successfully developed, and made available, a woman might well choose this method for its safety and speed, and there would then be no conflict with the ethical requirement. Nevertheless, the equation of the mother to machine is clear. In a real sense, mother and control system become one machine, and it is this machine which 'gives birth'. Without questioning the aims of the research, one can have strong reservations about the means by which it was proposed to achieve them.

b. Experiments in 1915 were reported by a doctor who

   'made an unsuccessful attempt to discover a cure for pellagra. To do this he produced the disease, which is characterised by diarrhoea and dermatitis, in twelve white Mississippi convicts who became seriously ill as a consequence. Before the experiment was made formal agreements were drawn up with the convicts' lawyers agreeing to subsequent parole or release.'[37]

   Here consent was obtained, but it cannot be considered as freely given. It had been purchased by a reward which might have great value for the prisoner.

c. In March 1962 a forty-year-old married man

   'presented himself ... for operation on a hernia. The previous year, at the same hospital, he had been found to be diabetic, and was put on insulin. The account of this patient contains the following statements:

   At the time of the investigation into the diabetes he was recognised as having a genital deformity about which he was so sensitive that he failed to keep an appointment at which it was hoped to investigate this aspect of his case.

   For this reason he was not questioned about his marital relationship at the time of admission for the hernia operation. Neither he nor his wife volunteered any further information about the matter, and it was not thought to be in his interest to worry him further. It was decided to examine his genitalia very closely under anaesthesia at the time of the hernia operation.'[38]

Some female characteristics were found, the abdomen was opened, and a cystoscope was passed into the bladder to obtain further information. The anaesthetised patient was here treated as an inert object of study. No consent was obtained, and from the account it would clearly not have been given if asked for.

d. The experiments by doctors on inmates of Dachau and other concentration camps are notorious and repulsive. One experiment was intended to discover the best way to resuscitate pilots who had been chilled in the cold waters of the North Sea.

'The subject [an inmate of Dachau] was placed in ice-cold water and kept there until he became unconscious. Blood was taken from his neck and tested each time his body temperature dropped one degree ... The lowest temperature reached was 19° Centigrade, but most men died at 25° or 26°. When the men were removed from the icy water, attempts were made to revive them ... [by various methods].'[39]

Medical ethics here have been forgotten. Men are treated as objects, and objects of no concern. It will be thought objectionable and unfair to attribute the doctors' behaviour to a scientific attitude, since Nazi ideology had already robbed the prisoners of all human consideration. But one can at least say that science, in the mould of the causal myth, does not provide any intrinsic defence against being led in the direction which the doctors took.

The proceedings of the investigating tribunal make it clear that many scientists, not themselves involved in the experiments, listened without protest to accounts of what was done. The two German observers of the proceedings commented as follows:

'Only the secret kinship between the practices of science and politics [sc. Nazi ideology] can explain why throughout this trial the names of high-ranking men of science were mentioned - men who perhaps themselves committed no culpable act, but who nevertheless took an objective interest in all the things that were to become the cruel destiny of defenceless men ... This is the alchemy of the present age, the transmogrification of subject into object, of man into a thing ... '[40]

Lack of a moral argument

The reader will appreciate that the case which is being made against the causal myth in science is only in part the logical one which was made in earlier chapters. It also has a moral and ethical aspect which we have stressed in the present chapter. This being so, it would be helpful if assistance could be obtained from the prevailing world views of the present time. Unfortunately little help is available from this quarter.

In the Western tradition there are three main views to which we might appeal: Christianity, humanism and Marxism. The first and last of these, though fragmented, are explicit and codified, while the second is more diffuse. Christianity in Europe has been in slow decline for two centuries, while Marxism has suffered a severe setback in the last forty years from its economic failures, and its continuing repressive character in those countries which have espoused it. Nevertheless, support from either, or from humanism, would be welcome.

An initial difficulty is that those who accept any one of these views, also in general accept the proposition against which we have argued in Section 3 (*see below, 'The Lushai Hills'*). They accept, that is, that the causal myth is the only basis on which it is possible to generate an understanding and control of nature, having the scope and power of our present science. In consequence, they believe that rejection of the causal myth involves the renunciation of those advantages which arise from scientific knowledge.

Few are prepared to make this renunciation, and the refusal to do so acts as a strong constraint on the conclusions which most will draw from their fundamental belief in Christianity or humanism or Marxism. Without this constraint, quite different conclusions might be drawn, but we cannot easily say what they would be. We therefore have to accept the conclusions as they are at present drawn, pointing out, where we can, the influence which has been exerted by the constraint.

A discussion of the three belief systems is omitted. It is concluded that no effective moral support can be obtained from any of these systems, considered as intellectual structures.

These comments are all directed to Christianity and humanism and Marxism as systems of thought. They are not intended to apply to individual believers in these systems, among whom will be found many of the strongest opponents of the things which have been criticised.

Conclusions

If the analysis in the preceding section is correct, no great assistance can be obtained, from any world view current in Western countries, in making a moral or ethical case against the causal myth and its consequences. The conclusion is not surprising: it is not to be expected that the predominant view, in any industrialised country, will be one which provides a damaging criticism of its scientific and technical foundations. There then remain three arguments which can be used:

i. That the empirical evidence which supports the causal myth also supports, and to the same extent, the purposive myth. This was the aim of earlier chapters. The effect of the argument is to show that the causal myth is a human construct. When we find that it is confirmed by every observation that we make of the world, we are rediscovering what we put there to be found.

ii. That a technology based on the causal myth and incorporating its values is unsound, and less effective than alternatives. We can, for example, argue that the agricultural practice described in Section 2 (*which has been omitted, and in which an extreme form of monoculture is described*) provides ideal conditions for the proliferation of pests and disease. We may control these in the short term by chemical means, but mutant strains will continually arise to penetrate our defences. We are, in effect, pitting our own sole purpose against the purpose of all those organisms which can exploit the conditions we have created; and in any contest of this kind we are likely, in the long term, to be defeated.

   Such an argument has to be made in detail, and case by case. It can be made[41] in relation to the Tayloristic development of production systems using computers. However, the argument cannot be made within the causal framework without implicitly accepting the values which underlie it, and these by a logical development lead back to Taylorism.

iii. That the naive and untutored reaction to many consequences of the causal myth is one of unease, disquiet, and disgust, or desolation. When it is believed that there is no effective alternative, these feelings are held in check. If it can be shown that there is an alternative, then the feelings become significant.

   Examples of such reactions have been quoted, varying from the despair of Monod, to the jaunty dismissal described by Tom Bell, and the magisterial condemnation of Adam Smith (*all three omitted here*). Many others can be found, and together with the argument given in (ii) they form the basis of a growing literature which questions our present technological practice. The argument in (i) adds strong logical support.

The extracts from Machines with a purpose now continue from the early part of Chapter 9, 'An Alternative Technology', pp. 156–176. After stating that the aim of what has been said earlier has not been to replace one myth by another, the extract continues:

It has been to show that there can indeed be equivalent myths, and that these carry values which affect our behaviour. As a case in illustration, Taylorism is a direct response to the causal myth. What we shall attempt in the present chapter is to show how the purposive myth allows us to regard technology in a different light. This will be done in the hope that the insight obtained from this alternative view will release us from the tyranny of the causal view, allowing us to suggest a better kind of technology, better matched to human needs and aspirations.

The scope of technology

The word 'technology' is sometimes restricted to mean the knowledge and skill which underlie our processes of production and distribution and their control. The physical embodiment of technology in systems containing people and machines is then distinguished as 'technics', though this term is not common in English writing. 'Technology' will be used here in a wide sense to embrace the conceptual and physical aspects. Where necessary, distinctions will be made by using such terms as 'technological systems', 'plant', 'machine', 'skill', or 'scientific basis of technology'.

Technological systems may be of the most diverse kinds. A railway is a technological system, comprising lines, rolling stock, signalling equipment, maintenance facilities, and the people who operate, and organise the operation, of this physical equipment. A road haulage company or an airline are similar examples, but ones which make extensive use of independent systems: roads and airports.

Production systems are a major subdivision of technological systems. They may be based upon continuous flow with nearly complete automation, as in an oil refinery. Steel production is only a little less continuous and less automatic. Mass production of consumer goods follows the same pattern as far as possible: closely in the case of food, less closely in such goods as motor cars where final assembly employs much human labour. In all of these systems, computers will usually be found, sharing with people the tasks of control.

Other systems are concerned with providing services of various kinds: shops, banks, insurance offices, hospitals and the like. The military services, army, navy and air force, are technological systems comprising people, machines, communications systems, etc. Government offices are another type of technological system.

A common feature is that specialised organisation, or specialised equipment, is always involved, usually with human assistance but sometimes without. So we should not call an individual lecturer and his audience a technological system; but a university is a technological system through its specialised organisation and equipment. Traffic lights and a completely automated telephone system are both technological systems, which operate without human intervention.

What is, or is not, included in the term 'technological system' clearly cannot be made entirely unambiguous, but one distinguishing feature serves as a useful guide. This is, that underlying the activity which is carried out, there is some feature which is amenable to study in science-based terms: that is, in terms of technology in its abstract meaning.

A telephone system, for example, is designed using mathematical techniques such as queuing theory, and is supported by highly developed theories of electronics, computing, and optical transmission of signals. There are specialised techniques for installation and maintenance, and a specialised organisation for billing. All of these are based upon the application of science in technology; and the scientific basis, as in other technological areas, becomes continually more refined and more powerful.

The purpose of technology

Every technological system starts from a human purpose, from the intention to satisfy some human need or desire. The intention may arise from the aims and actions of government, as nearly always in police or military systems, and sometimes also in many other areas. Alternatively, the intention may arise from the wish to meet a market need, in which case it is subordinate to other aims such as immediate profit for a company, or the fulfilment of a longer-term aim of 'profitable survival'.

However it is formed, the purpose is implemented by a process similar to the one illustrated in Figure 5.9 *(which is omitted here)*. That is to say, a policy which will implement the purpose is generated, and the system is constructed so that it follows the causal laws implied by the policy. This procedure is much less formal in most cases than in the control engineering example of Chapter 5 *(omitted)*. To begin with, the purpose is seldom defined so precisely that a unique policy can be deduced from it. Then the procedure by which a policy is obtained must also be less rigorous, and it usually involves intuitive elements of design and problem-solving.

Nevertheless, a little thought will show that the way in which the technological system is defined follows closely the outline given before. We have certain ends in view, and we generate the means to obtain them. These means will consist of organised groups of people, of machines, and of interactions between people and machines. All of these, if they are to be brought within the scope of scientific study, conceived in the traditional way, must be described causally, because the only science we have is cast in the mould of the causal myth.

If we replace the causal myth by the purposive myth, we at once have to change the way in which we regard the technological system. The people involved in the operation of the system will be seen as sharing in the purpose of the system, and assisting in its fulfilment. Individually, they will have purposes which are subordinate to the purpose of the system; that is, their independent purposes which they strive to fulfil will be consistent with the purpose of the system, but not coextensive with it. The human contribution to the operation of the system will consist of an interlocking network of these subordinate purposes.

The machines will also embody subordinate purposes, but in a more restricted way. Human beings can adopt a purpose at a deep level: we say that they care whether the purpose is fulfilled or not. Because of the limitations of our technology, we cannot implement purposes in machines at this level. Machines do not care whether they fulfil their purpose. We can make a machine behave 'as if it cared', within a restricted repertoire, by making it exhibit the causal behaviour which would go with caring. But this behaviour is superimposed upon an alien purpose (expressed by Hamilton's principle and the like) out of which it does not grow in a natural way. Human purposes can equally be imposed by coercion in an alien way, but they can also be adopted freely.

Within such a purposive view of technology, it would be necessary to consider the interactions of various kinds: of people with people, of machines with machines, and of people with machines. Each of these three kinds of interaction will be considered briefly.

Interactions between people

To regard the interactions of people in terms of their individual purposes is not something new. This is the natural view, expressed in most of our literature and our conversation. The alternative, causal, view is less natural and of later growth: an early and crude form was briskly dismissed by Socrates (Chapter 3, Section 6 - *omitted here*). A current view is that human behaviour can be explained partly by heredity and partly by the influence of our environment (*a brief passage is here omitted*).

It is this outlook which leads to Taylorism, and a manipulative view of human relations in technological systems. A purposive description avoids this danger, and situates the problems of personal relations in a more traditional framework. This does not mean that it solves the problems. It simply restores them to a moral dimension from which the causal account would remove them. Rather than observer and observed, manipulated and manipulator, it shows us moral beings of equal status co-operating or conflicting through their purposes.

This change is particularly relevant in two areas, of ownership and of work. In the causal view there is a justification for ownership in an absolute sense. What is inert and without purpose can be owned by being made absolutely subject to our own will, which we exempt from the otherwise universal rejection of purpose. This conception of ownership applies to our relation with nature, to our ownership of land and plants and animals. It applies also to the ownership by one person of another person's time, in a contract of employment, to the extent of specifying what shall be done to the exclusion of all initiative and independent will.

The relevance of ownership in this sense, which leads in a direct way to Taylorism, is further explored in a passage which is omitted.

The purposive view leads in a different direction, towards a system in which workers share in the purpose which is to be attained. Their own part in this purpose, the fulfilling of a subordinate purpose, must be one which is acceptable and natural to them. It must give scope for the exercise of their own abilities and skills. They in return will have to accept responsibility for their own part in fulfilling the aims of the system.*

Nothing in this is very new. What is different is the justification which could be given, and the likelihood of its acceptance. To argue for such a system in the face of the causal myth constrains us strongly to accept the values which that myth implies. It has to be argued, for example, that a system with these features is more effective than a Tayloristic system in eliciting the desired behaviour from a worker. That is, the justification must be in causal terms; and once the causal view of human behaviour is admitted it brings us round again full circle to Taylorism.

The nature of work

The relevance of the purposive myth to the nature of work is even more direct than its relevance to ownership. Work in modern societies has three functions. It contributes to the production of goods and services. It gives a claim to some part of this production. It is also, for most, a major source of self-esteem through the contribution which it allows them to make to society: it is by breaking this link with society that unemployment can be so damaging.

As the direct contribution of human effort to production continually decreases, these three aspects of work become problematic. Our view of the future then depends very much on whether we adopt the causal or the purposive myth. The first of these suggests that since 'man is a machine', every human contribution to production can ultimately be eliminated by machines. Goods will be produced in the workerless factory. They will be distributed by ever more automated distribution systems, needing progressively less human input. Communication by electronic systems is already highly automated, and will become more so. Other services equally can be provided with less and less human labour.

However enthralling this development might be to the technologist, it is a bleak prospect in human terms. It seems that we shall become less and less necessary to the society in which we live, existing on the margins of a mechanised and automatically functioning world. Our contribution to this activity will hardly be needed, and when it is, the contribution will be as mechanical as the system it serves. Work as an entitlement to share in what is produced will become a source of contention through its scarcity.

---

* It may be that some workers, at some stage of their lives, might not wish to accept this kind of responsibility, or to accept it only to a small degree. This might be true particularly for those who are inured and accustomed to present types of work, or who have strong alternative interests. There would be no difficulty then in providing them with appropriate tasks, particularly if work was distributed among autonomous groups. But with an outlook formed by the purposive myth this would appear as anomalous, as a failure to accept the normal obligations of society: a failure which is to be tolerated if it is strongly desired, but not to be encouraged and never to be imposed.

Work as a source of satisfaction and self-esteem will cease to exist.

Repelled by this prospect, some have suggested a reversion to earlier technology, or a breaking of the machines. But the vision of the future springs directly from the causal myth, and a different future can be imagined if we adopt the purposive myth. This accepts that 'man is a machine', but a machine with a purpose. Those things which we usually call machines, including computers, also have a purpose, but it is very different from the purpose of human beings.

One part of our human purpose is to produce the goods and services we need. To some extent, and increasingly, we can incorporate this purpose in machines. But we can do so only by first translating it into a policy as explained in Section 10 of Chapter 5 *(omitted – see an earlier comment)*. The machine does not become more like the human being from whom the purpose originated; it merely reproduces a desired behaviour. It does so within a limited repertoire of actions, and in a limited range of environments. If pushed sufficiently far outside its repertoire or its range of environments, it will ultimately exhibit a purpose which diverges from the one which a human being would adopt.

By incorporating the human purpose in a machine, we have made this purpose subordinate to the ultimate purpose of the machine – that purpose expressed by Hamilton's principle and its extensions from which, as a policy, all its actions can be deduced. In the human being, the purpose was subordinate to a different ultimate purpose, one of flesh and blood, not one of rods and levers and silicon chips. So although we can expect to incorporate in a machine more and more human purposes, we can do so only in a restricted way.

The human actor 'fulfils a purpose' in the sense that his actions can be derived from this purpose, which is subordinate to an ultimate human purpose. But the actions occur without necessarily being derived explicitly and expressed by a policy; just as the Earth goes around the Sun without any regard to Hamilton's principle or to that policy derived from it which is expressed by Newton's equations. The purpose and the policy are our description of an event which simply occurs. In the same way, the machine fulfils its ultimate purpose naturally, but we have so constructed the machine that it has the human purpose as one of its subordinate purposes: but only in those circumstances where the purpose gives rise to one specific policy or range of policies. Because it is the policy which we impose on the machine, not the purpose.

Despite this restriction, machines in which we have incorporated a human purpose (in the sense defined) will be able to carry out a wide range of simple human tasks. Traffic signals can replace the police who at one time controlled traffic. Robots can put in spot welds on car bodies. Repetitive work of this kind makes no real use of human abilities, and if the time released can be used in more humanly rewarding ways, we can accept it as a benefit.

Those machines which are based upon computers can carry out more complex tasks. If we can, by questioning or observing a competent worker (the 'expert'), define exactly what is needed to accomplish a task, we can build this information into the computer as an 'expert system'. Then the computer can carry out the task, however complicated it may be, if it is provided with the necessary devices for sensing and for manipulating its environment.

The tasks which can be most easily defined in this way are often ones which we regard as prime evidence of human intelligence: calculating numbers, manipulating mathematical formulae, playing chess at a moderately advanced level, some parts of medical diagnosis, and so on. When it is shown, as it has been, that all of these can be done with more or less competence by a computer, we are apt to conclude that 'everything that man can do, a computer can do'. Computers have already shown the ability to do things we regard as difficult, and as indicative of high human abilities: why should they not also be able to perform every other task which is within human capability?

If we accept the causal myth, the conclusion is indeed inescapable, for every human activity is then defined by the actions it entails. But if we adopt the purposive myth, we shall see much human activity as fulfilling a purpose directly, without first translating the purpose into the policy which will fulfil it.

We should then be led to see ourselves as distinguished from the machines by what unites us to the other animals: by the fact that our purposes, with the actions which result from them, are subordinate to the purpose of flesh and blood. We should see ourselves as distinguished from the rest of the animal world by what unites us to machines: the fact that we can implement complicated logical sequences of cause and effect which represent the policy needed to fulfil a purpose. We should see ourselves as distinguished from both, and unique, by our capacity to implement purposes which span these two kinds of activity.

On this view, mathematical calculation is not specifically human, because it can be done by machines. Bodily activities, such as jumping and running, which implement an untranslated purpose, are not specifically human, because they are shared by other animals. The generation of a new mathematical theorem, as the fulfilment of a purpose which is not first translated into causal terms, is a specifically human activity. So is the act, equally fulfilling an untranslated purpose, by which a reverberation is set up between words and their meaning:

> Shall I compare thee to a summers day?
> Thou art more lovely and more temperate

or in a darker mood,

> And this is the manner of the Daughters of Albion in their beauty.

So also is the activity of the painter or sculptor, and at a more accessible level the exercise of craft skills such as those of the carpenter or blacksmith, and the practice of engineering design.

All of these are activities which implement a human purpose, and do so without first translating it completely into the causal actions which will be needed. They are all activities, in consequence, which resist formalisation and the reduction to method, but must rest on a basis of ability developed through experience. In the tradition of the causal myth, they are therefore ignored, or mentioned only briefly in passing. In Popper's account of the scientific method, for example, the process by which a new theory is generated is outside the logical framework, and is left undefined, though it is the most specifically human part of the whole activity.

What is most distinctive in human work, on this view, is that activity which fulfils a human purpose in its untranslated form. The purpose must involve more than the simple activities which we share with other animals, such as the co-ordination of hand and eye in muscular activity. It must be more than the logical, causal sequence of actions which we share with machines, and which Taylorism would see as the only content of work. Both of these may be components of a truly human kind of work, but there must also be that element of directedness, of aiming at a goal, and achieving it by judgement and skill based upon knowledge and experience.

In these terms we can imagine a different future for human work. Instead of the workerless factory and the marginalising of all human contribution, we can imagine a technology in which the human contribution is central: where this contribution is assisted by machines, but where it retains the quality of directly fulfilling a purpose. A technology conceived in this spirit could provide a truly human kind of work, and could produce goods and services which met more amply the needs of those for whom they were provided.

The knowledge and the skill required by such a technology will be at different levels in different kinds of work, and will be appropriate to different kinds of human ability. But no human work should have the machine-like triviality and aimlessness of the lamp plant (Section 5 of Chapter 7 – *an example of extreme Tayloristic organisation which has been omitted in these extracts*). This offends by equating people to automata, fulfilling a purpose appropriate only to machines. It also offends in another way, by subordinating people to machines.

Machines and people

The distinctions just made can be illustrated by Henry Ford's introduction of the assembly line. This was accompanied by an extreme subdivision of the work, expressed in his dictum, 'The man who puts in a bolt does not put on the nut; the man who puts on the nut does not tighten it.'[42] In such a working situation, all human content has been eliminated by an ultimate trivialization of the task. The work is machine-like, and men are treated as though they were machines. This would be true regardless of the other aspects of their work: for example if they were assembling components which they picked from a bin. In these circumstances the workers might be subject to pressure to increase the rate of production, and the situation could be unsatisfactory to varying degrees.

The production line, however, goes beyond this. When he introduced it, Henry Ford regarded it as a way of bringing the work to the worker, and eliminating the time otherwise lost in moving from one car to the next. Its effect is more profound. It links all the separate parts of the assembly process into one whole. The assembly plant becomes one vast machine, in which certain actions are indeed carried out by machines, while others, equally machine-like, are carried out by workers.

Here the whole system has a purpose, which was incorporated into it by its designers. They did this by incorporating a purpose into the machine components, and requiring workers to conform to the purpose of the machine. When the production line speeds up, workers must also speed up. When it slows down, they must also slow down. A car is presented to them, and they have so many seconds to carry out their task. The machine, the production line, defines what they are to do and how fast they are to do it. All control is taken from them, and they are required to subordinate themselves as servants to the purpose of the machine.

This inverts the traditional relationship, in which a craftsman may use machines, but they are subservient, as tools to assist him in accomplishing his purpose. The purposive myth suggests that this traditional relationship is the correct and humanly satisfactory one; but it is extraordinarily difficult to defend it or promote it within the framework of the causal myth.

A first difficulty is that the causal myth rejects purpose. It does not allow us to say that machines have a purpose, or that the purpose of a worker is subordinated to the purpose incorporated in a machine. Definition of the problem becomes impossible, because the terms needed to discuss it are disallowed.

Secondly, human purpose itself tends to disappear. An unsatisfactory working situation may lead to problems – to poor quality of work or to a high labour turnover. The difficulty may then be studied, and perhaps in a sympathetic spirit, in order to provide a more satisfactory working situation. But if this is done in a 'scientific' way, which means in the light of the causal myth, the problem has to be stated in the form, 'what changes in the working situation will cause it to appear better in the eyes of workers?' Even if the tendency to manipulation is resisted, the question will not lead to the reply, 'the machines should be made subordinate to the purpose expressed by workers in their tasks'. The attempt to design a more humanly satisfactory technology therefore meets a basic difficulty, because the terms in which it would have to be done contradict the orthodox understanding of science, upon which our technology is based.

The relation of machine to machine

Of the three types of relation which were described earlier, we have said that the relation between people is situated by the purposive myth in a traditional framework, pre-dating the causal myth and still persisting. Different people, and different groups of people, have purposes which sometimes coincide and sometimes conflict. Little can be said about this that is not already incorporated in the traditional account, which has always been expressed in terms of purpose.

The relation between people and machines, by contrast, cannot be treated in a satisfactory way without a change in the usual terms of discussion, which are based on the causal myth. It is this change which has been our main subject. The third item, the relation between machine and machine, offers no fundamentally new difficulty.

Machines can relate to one another in two ways, either by the passage from machine to machine of material for processing, or by the interchange of information. When passage of material is accomplished manually, the important interaction is between machines and people, and this has already been discussed. Where it takes place automatically, the linked machines become one larger machine, as in the car assembly line. The important question is again the relation between this larger machine and the people whose work it either controls or assists.

This leaves the flow of information as the new factor to be considered. There are two different extreme forms which this flow can take, with an unlimited number of intermediate forms. At one extreme, all information can be sent to a single, central point. Decisions can be generated on the basis of this information, and transmitted back to initiate action. At the other extreme, decisions can be taken at the point where the information arises, in accordance with guidelines laid down by a central co-ordinator.

The first model is one which fits most naturally into the centralising, Tayloristic, causal mode of thinking. Some human interaction or supervision of the decision-making activity will usually be required: by centralising it, the interaction can be made at some high level in a hierarchical management structure, where knowledge and authority are assumed to be concentrated, and where the denial of purpose is relaxed. Decisions, also, can be taken on a basis of total information, rather than of partial, localised information.

There are practical difficulties in such a system, which have been at least partly recognised. Gathering and transmitting and processing information centrally can introduce delays which degrade the operation. Human interaction in the centralised decision-making can be hampered by an overload of data, and by the absence of contextual information which is needed to assess the relevance of isolated facts. The chief objection, however, is that if there are other human actors in the system, they become subordinate to the centralised decision-maker interacting with them through the machine.

As seen by the worker, the system gives no information, and allows no control. It issues instructions, the reasons for which are unknown. It embodies Taylor's system under which 'the workman is told minutely just what he is to do and how he is to do it', and it provides this control, moment by moment. This may not have been the intended aim when the system was designed, but the structure of the information system, based upon the causal, Tayloristic view, will ensure that it occurs.

To the worker, the system of interacting machines will constitute one larger machine to which he is subordinate. This is not admissible on the view which has been put forward. We are brought back again to the relation between machines and people: what seems at first sight to be a relation only between machines, implies also a relation between machines and people.

'Assisting' and 'replacing'

The preceding discussion can be summarised in the following way. On the causal view, machines are designed to replace the human contribution to production. They seldom do so completely, and those workers who are left are required to behave like machines: to stand in place of machines which it has so far proved impossible or uneconomic to produce.

On the purposive view, machines can embody a part of the human purpose of production, but in an imperfect way. Machines are not to be designed to replace the skills and abilities of people, but rather to assist these skills and abilities and make them more productive. The machines should allow existing skills to be relevant, but should not attempt to preserve them in an unchanged form. Scope should be given for skills to change and develop as technology itself develops. Above all, people should never be subservient to machines, but machines should be subservient to people.

These aims, which have been associated with the purposive myth, will no doubt be accepted as desirable by many. Indeed, they are often claimed, at least in part, to be the aims of those who develop new technical systems. For example, 'The expert [computer] system is a high-level intellectual support for the human expert, which explains its other name, intelligent assistant.'[43] As one comes closer to application, however, the emphasis changes:

> 'This is a revolution, and all revolutions have their casualties. One expert who gladly gave himself and his specialized knowledge over to a knowledge engineer suffered a severe blow to his ego on discovering that the expertise he'd gleaned over the years, and was very well paid and honored for, could be expressed in a few hundred heuristics [computerised 'rules of thumb']. At first he was

disbelieving; then he was depressed. Eventually he departed his field, a chastened and moving figure in his bereavement.'[44]

This episode reveals no attempt to produce an 'intelligent assistant', working under the direction of the expert. Still less is there an attempt to provide scope for the expert's skill to develop, in such a way that he can perform better with the assistance of the computer. 'Assisting the expert' has become 'replacing the expert', and the process by which this change takes place passes unnoticed.

Upon analysis it is easy to see that 'assistance' will always become 'replacement' if we accept the causal myth. The expert's skill is defined to be the application of a set of rules, which express the causal relations determining the expert's behaviour. Assistance then can only mean the application of the same rules by a computer, in order to save the time and effort of the expert. When the rule set is made complete, the expert is no longer needed, because his skill contains nothing more than is embodied in the rules.

It is this destructive metamorphosis of good intentions into ill effects which makes the causal myth so damaging in the context of work. It will be illustrated below in more detail by means of a demonstration project, supported by the ESPRIT programme of the EEC, which had the aim of developing and demonstrating an alternative 'human-centred' technology. The demonstration was successful, but it raises the question whether such a development can resist subversion by the causal myth. (*The description of this demonstration is omitted*).

Recapitulation

This concludes the project upon which we embarked. To avoid misunderstandings, it will be helpful to say what the book has not been intended to do.

First, it is not an attack upon science. Rather, it is an attempt to show how the benefits of science can be obtained without those highly undesirable consequences which have so often arisen from it. We need the honesty and the patience in the face of evidence which are characteristic of the best science. We need also the coherent intellectual framework, which it provides more strongly than anything previously achieved. We do not, on that account, have to accept with it the preconceptions of the causal myth with their damaging human consequences.

Secondly, it is not a plea for the rejection or restriction of technology. We need the insight which it gives us into man-made systems, not least to undo the damage which past technology has inflicted. It is rather a plea for a different kind of technology which is not antagonistic to people or to the environment. Because technology is so firmly based upon science, the change in technology will have to begin with a changed outlook in science.

Thirdly, it is not intended to propagate a belief that the purposive myth incorporates truths about reality which the causal myth does not. Both myths are descriptions which we impose upon nature, and both carry with them presuppositions for which there is no warrant in the physical evidence. The evidence supports the equivalence class, the theory, to which both myths belong. It does not support any feature in which one myth differs from the other.

From these negatives follow the more positive aspects of our analysis. To be truly consistent, we ought to use the theory and never any of its myths. But the theory is an abstract and tenuous entity, and it is beyond our intellectual capacity to use it in a practically constructive way. For scientific purposes, we therefore select one representative myth, and use this as though it were the theory. The choice can be made on grounds of convenience and simplicity and fruitfulness, and we shall never be led into scientific error in this way.

The danger is that if we choose the causal myth for its scientific convenience, and forget that it is a myth and not a theory, we are likely to end by acting as though the myth expressed fundamental truths about the world. We shall then see it as natural and right to exploit nature, and exploit the future, and exploit society, to the extent of our ability. Much of what has been written in previous chapters has been intended to establish this connection.

Now it is true that if we are free to choose a representative myth for its scientific convenience, we are equally free to choose a different myth when we consider how we ought to behave, and the ethical problems that this question must raise. The purposive myth has been used in this way in what was written earlier, and it acts as a corrective to the causal myth.

We regard the world as fulfilling a purpose, which explains every observation we can make as completely as the causal myth, being indeed equivalent to it. But a world impregnated with purpose (even if a purpose which we have ourselves constructed to explain our experience) is different in its effect upon us from a causally determined world. Our own purpose finds a natural place among many. We are not given the world to coerce it to our will, but to live in as an organic part of it.

This has been the conclusion to which we have tended, and if the foundations of this different view are not complete, they are well begun. One doubt must remain. Adam Smith remarked that 'the understandings of the greater part of men are necessarily formed by their ordinary employment'.[45] If we continue to use the causal myth for the purposes of science, and therefore also technology, will it not continue to influence the development of our technology-based society in the same way as before?

This is what Monod foresaw in a future dominated by the causal myth: 'the choice of scientific practice ... has launched the evolution of culture on a one-way path ... what we see before us is an abyss of darkness.'[46] But if we cannot be sure of the final outcome we have at least a more promising situation than Monod imagined. He saw one unchallenged view of the world, leading inevitably to his abyss. We have suggested a struggle for influence over our behaviour of two views, one leading to the abyss, but the other to a different and more welcoming future.

This concluding section of the discussion is followed here by two extracts from Chapter 7 of Machines with a purpose, pp. 122–124 and 127–128.

The Lushai Hills

Four hundred years of belief in the causal myth have given us a technology, including our agricultural technology, which embodies the values incorporated in the myth. We have great difficulty in imagining a different technology which could give equivalent benefits, or in believing that such a technology is possible. What is, seems to be all that could ever be. Only by showing an example of a better technology, or at least the basis on which it could be developed, can we hope to overcome the disbelief in its existence.

This will be attempted, for one particular technology (*the ESPRIT project referred to above, which has been omitted*). Meanwhile, we can at least try to show how what has been done in the past four hundred years has come to seem the only possible course which could have been followed. To this end, consider a metaphor which has been used before.[47]

The Lushai Hill Tracts, in the early 1940s, were almost untouched by any human activity. Long ranges of hills ran north to south, rising gently at their western boundary to a few hundred feet, then falling again almost to sea level in a river valley. Range upon range succeeded to the eastward, each ridge climbing higher, but falling again into deep valley, until at Blue Mountain the highest peak reached 7000 feet.

Scattered settlements were confined to the crests, partly to escape the malaria of the lowlands, and partly from a tradition of defence; the Lushais had only recently given up their immemorial practice of head-hunting. Their villages contained perhaps thirty or forty inhabitants, living in bamboo huts, and growing hill rice upon land fertilised by burning off the brushwood. After a few years, the poor soil would be exhausted, and the villagers would move on, taking with them their chickens, their few pigs and goats, and their dogs.

For the most part, the hills were covered in jungle. Almost everywhere were tall trees providing a canopy, with dense undergrowth, with here and there occasional patches of close-growing, unshaded bamboo. The war mostly passed by the area, with only forays by small parties, or feuding between tribes attached to either side.

It was a region of great natural beauty. Following one of the major rivers by canoe, in the brilliant sunshine of noonday under a cloudless sky, there would be a complete silence, broken only by the distant barking call of monkeys. On either side the dense foliage would rise up steeply, covering the hills to their summit.

Travelling north to south along the valleys was easy, but not so the passage west to east. There were occasional tracks or game trails, but often the only practicable routes were along the streams which cascaded down the hillsides, forming low tunnels under the overhanging foliage.

Climbing upwards in this way, one would reach a fork where two streams joined, and a choice had to be made. No reliable information could be obtained from the map, and no general overview was possible to guide the choice, which must be based only on what could be seen within a few yards, or on any general predisposition to go towards the right or the left.

For a while, it was possible to reverse a decision if it proved unfortunate; either by going back to the junction or perhaps by cutting across from one stream to the other. But very soon this became impracticable, because of the great effort needed. One was committed to the chosen stream, for better or worse.

Having climbed high up the side of the valley, one would pause and camp for the night. Looking back to the west, one would see range after range of hills, falling away to the plain. The red of the setting sun would cast a glamorous light over the country from which one had come, covering the hills with a purple haze and disguising the heat and the malaria and the leeches from which one had escaped.

Then it was possible to feel a sense of achievement: to have climbed so high and to be able look back over the lower country out of which one had come. And it was easy to believe that all the choices which had been made along the way were justified by the outcome, and were the only right choices to be made.

This self-congratulation might of course have been quite unwarranted. Some other route might have led to still higher ground, and done so more easily. But if so, the knowledge was hidden, and the complacency uncontradicted.

This is an image of the way our technology has developed. We have climbed by our efforts out of a past in which the relative ineffectiveness of their labour condemned the great majority to unremitting toil and poverty; and looking back there is even a glamour upon this past which conceals some part of its harshness. In our progress, we have continually had choices before us. We have been ignorant of their ultimate consequences, and have taken our decisions by the light of the causal myth.

The advantages we have gained cannot be given up lightly, nor can we, if we wish, go back again and start from where we began. We arrived where we are by the route that we chose and, knowing no other, are persuaded to believe that it is the only one we could have followed with success. Yet Monod's 'abyss of darkness' which opens before us must cause us to ask whether this appearance of inevitability arises only because we chose one particular guide in making our decisions; whether we might not have climbed as high by some other route, but arrived in a more welcoming countryside.

Recapitulation of the argument

Before proceeding it will be useful to recapitulate the argument which is being developed in this chapter. This is, that our long tradition in science accepts no explanations except those in terms of cause and effect. The result is that we regard everything outside ourselves as a machine, and a machine without purpose.

In principle, we also regard ourselves as machines, but the strain of doing so proves in practice to be too great. So we accept our own purpose without reservation; we have also a social life within which we set aside the need to reject purpose. Yet in all serious matters of science and technology and commerce we act as though our own purpose were unique.

The examples used above in illustration (*some of which have been omitted*) are taken from personal observation. In agriculture we treat the natural world more and more as though it were an inert, purposeless machine, governed by causal laws which we can elucidate, and which we use to attain our purposes. We take the same view of animals, treating them also as machines. We design our production systems so that human activity is made to resemble as closely as possible the activity of a machine. If these statements seem exaggerated and extreme, a consideration of tendencies in the last few decades will confirm their accuracy. Some fringe reactions apart, all movement has been in the direction suggested. Where the statements do not apply, we usually find a survival of previous conditions, not yet amenable to the mechanising tendency.

In the naive (but not thereby less perceptive) observer, the more extreme examples of this tendency will raise misgivings, some of the bases for which have been indicated above. In agriculture it produces a landscape which repels all human feelings of closeness, of belonging, of delight: a dreamlike landscape as of some other planet. It sets a gulf between ourselves and the animals, severing a bond which we feel though we cannot easily articulate. Between those who design production systems and those who work in them, it sets up a destructive antagonism, in which a human purpose of production is achieved by denying purpose in humanity.

These feelings are probably becoming more widespread and stronger, but to be effective, rather than merely destructive, they need an intellectual basis from which something more satisfactory can be evolved. The difficulty of providing this basis is considered in Section 8 (*see 'Lack of a moral argument', above).*

It is also necessary to reply to the objection that all the progress we have made in alleviating poverty and hunger and disease has been achieved by exactly that scientific outlook which is in question. If we reject it, we reject at the same time the advantages which it brings. A partial reply has been given in Section 3 (*see 'The Lushai Hills', above*): that the success of one path to the acquisition of knowledge does not prove there cannot be other, and perhaps better, paths.

Notes

1 Anderson, M. D. (1963), *Drama and imagery in English medieval churches*, p.55 (Cambridge University Press).
2 Dijksterhuis, E. J. (1986). *The mechanisation of the world picture*, pp. 391–396. Princeton University Press.
3 Reference 2, pp. 333–359.
4 Reference 2, p. 310.
5 Burtt, E. A. (1932). *The metaphysical foundations of modern physical science*, p. 116. Kegan Paul.
6 Monod, J. (1971). *Chance and necessity*, p. 160. Collins.
7 Trollope, A. *Castle Richmond*.
8 Shaw, C. (1977). *When I was a child*, p.113. Caliban Books.
9 Attributed to E. Fermi in Jungk, R. (1960). *Brighter than a thousand suns*, p. 184. Penguin.
10 Needham, J. (1927). *Man a machine*, p. 93. Kegan Paul.
11 Rosenbrock, H. H. (1990). *Machines with a purpose*, pp. 136–154. Oxford University Press.
12 Rosenbrock, H. H. (1992). 'Science, technology and purpose', *AI & Society*, vol. 6, pp. 3–17.
13 Rosenbrock, H. H. (1990). *Machines with a purpose*, pp. 203–219. Oxford University Press.
14 Taylor, F. W. (1906). *On the art of cutting metals*, p. 55. American Society of Mechanical Engineers.
15 Babbage, C. (1835, reprinted 1963). *The economy of machinery and manufactures*. Kelley, New York.
16 Ure, A. (1836, reprinted 1970). *The cotton manufacture of Great Britain*. Johnson Reprint Corp.
17 Satofuka, F. (1992). 'Forum: Some aspects of the debate on scientific tradition in Japan (III)', *Historia Scientiarum*, vol. 2, pp. 66–67.
18 See, for example, Herzberg, F. (1966), *Work and the nature of man* (World Publishing Co.); Drake, R. I. and Smith, P. J. (1973), *Behavioural science in industry* (McGraw-Hill); Klein, L. (1976), *New forms of work organisation* (Cambridge University Press); Legge, K. and Mumford, E. (eds.), (1978), *Designing organisation for satisfaction and efficiency* (Gower Press); Kelley, J. E. (1978), 'A reappraisal of sociotechnical systems theory', *Human Relations*, vol. 31, pp. 1069–1099.
19 Cooley, M. (1987). *Architect or bee?*, p. 147. Chatto and Windus.
20 See, for example, Burtt, E. A. (1932). *The metaphysical foundations of modern physical science* (Kegan Paul); Clegg, A. (1979), 'Craftsmen and the origin of science', *Science and Society*, vol. 43, pp. 186–201; Dijkterhuis, E. J. (1986), *The mechanisation of the world picture* (Princeton University Press).
21 See, for example, Bunge, M. (1955), *Causality* (Harvard University Press); Lerner, D. (ed.), (1965), *Cause and effect* (Free Press, New York).
22 A. Einstein (1950), *Out of my later years*, p.114. Thames and Hudson.
23 Rosenbrock, H. H. (1990). *Machines with a purpose* . Oxford University Press.

24 Nelson, D. (1980). *Frederick W. Taylor and the rise of Scientific Management*, p. 34. University of Wisconsin Press.
25 Needham, J. (1927). *Man a machine*, p. 93. Kegan Paul.
26 Rignano, E. (1926). *Man not a machine*, p. 10. Kegan Paul.
27 Reference 25, p. 43.
28 Reference 25, p. 45.
29 Dreyfus, H. L. (1972). *What computers can't do*. Harper and Row.
30 Polanyi, M. (1958). *Personal knowledge*, p. 55. Routledge.
31 Feigenbaum, E. A. and McCorduck, P. (1984). *The fifth generation*, p. 52. Pan Books.
32 Reference 31, pp. 104, 114.
33 Feigenbaum, E. A. 'Themes and case studies of knowledge engineering', in Michie, D. (ed.), (1979), *Expert systems in the micro-electronic age*, p. 8. Edinburgh University Press.
34 Pappworth, M. H. (1967). *Human guinea pigs*, p. xi. Routledge.
35 Reference 34, p. 11. Pappworth is quoting Dr. Szent-Gyorgi.
36 The account is based on conversations with a member of the research team.
37 Reference 34, p. 61.
38 Reference 34, pp. 92–93.
39 Lord Russell of Liverpool (1979), *The scourge of the swastika*, p. 164. Corgi Books.
40 Mitscherlich, A. and Mielke, F. (1949). *Doctors of infamy*, p. 152. Henry Schuman, New York.
41 Brödner, P. (1989). 'In search of the computer-aided craftsman', *AI & Society*, vol. 3, pp. 33–46.
42 Ford, H. in collaboration with S. Crowther (1923). *My life and work*, p. 83. Heinemann.
43 Feigenbaum, E. A. and McCorduck, P. (1984). *The fifth generation*, p. 86. Pan Books.
44 Reference 43, p. 115.
45 Adam Smith, (1977, reprinted 1976). *The wealth of nations*, V.i.f. 50, pp. 781–2. Glasgow Edition.
46 Monod, J. (1970, translation 1971). *Chance and necessity*, pp. 158–159. Collins.
47 Rosenbrock, H. H. (1979). 'The redirection of technology', IFAC Symposium *on Criteria for selecting appropriate technologies under different cultural, technical and social conditions*. Bari, Italy, 21–23 May.

Chapter 4

Culture, Mind and Technology: Making a Difference

Eunice McCarthy

Introduction: Culture and Technology

> Difference which occurs across time is what we call change.
>
> *Bateson (1972: 452)*

While culture is a concept which has a long history in anthropology, it has nevertheless been described in many diverse ways. Margaret Mead in her comprehensive book *Cultural Patterns and Technical Change* (1954: 12) observes that:

> A culture is a systematic and integrated whole ... culture is an abstraction from the body of learned behaviour which a group of people who share the same traditions transmit entire to their children, and, in part to adult immigrants who become members of the society. It covers not only the arts and sciences, religions and philosophies, to which the word 'culture' has historically applied, but also to the system of technology, the political practices, the small intimate habits of daily life.

This approach to culture is systemic, in that it is based on the assumption (itself drawn from field-work among many kinds of societies), that a change in any one part of the culture will be accompanied by changes in other parts and that only by relating any planned detail of change to the central values of the culture is it possible to provide for the repercussions which will occur in other aspects of life. Mead referred to this as 'cultural relativity', that practices and beliefs can, and must, be evaluated in context in relation to the cultural whole.

The wholeness of a culture is not a statement that all cultures are integrated in the same way, or that all are equally integrated. Some cultures are so tightly integrated that any change threatens the whole. Others are flexibly or loosely integrated which makes it easier to introduce particular changes without disturbing the whole way of life. Mead (1954), argues that whether a society is tightly or loosely integrated, the different parts will still be interrelated in systematic ways. This systemic or patterned quality of culture is a function of the integrated character of human beings, who as they incorporate culture traits and behaviour, organise them into a viable way of life.

The core concepts and issues highlighted by Mead (1954), in her cross-cultural studies of aspects of technical change, and the mental health implications of technical change, (sponsored by the World Federation of Mental Health), are equally of significance for us to keep in mind when exploring the meaning of 'culture and technology' in our modern-day cultures, societies and organisations. These are as follows:

- the interdependence of culture and technology;
- the need to consider the introduction of new technology in relation to the cultural whole - cultural relativity. The evolution of new technology in context;
- the systemic nature of change (Bertalanffy, 1950);
- the degree of integration in a culture, society, organisation. For example, tightly-coupled systems or loosely-coupled systems;
- the use of values - central and peripheral values, that are impacted on by technological change;
- mind and technology - the relationship between technology and our way of seeing the world, reality, and beliefs, thoughts, our construction of reality.

The Interdependence of Culture and Technology

The interdependence of culture and technology is as old as human kind and civilisation. Both culture and technology are embedded in an environment, which is transformed by the human person and also transforms the person. From the earliest of times, we can see that technology has been evolved to extend the human person, much as a spider's web expands the spider. Further, it becomes evident on closer examination that the earliest tools and artefacts created by the human person are magnified copies of something biologically natural. Recent technological developments, however, bring with them a shift in perspective from the physical to the mental, from a consideration of physical processes to mental processes. For example, human machine/computer compatibility is connected with the limits and tolerances of fatigue, memory, attention, vigilance, mental workload; in other words, with identifying the conditions that allow some kind of synergy to be established between human capacities and work tasks and demands.

It has been argued by Cooley (1989, 1995), that designers should set out to design systems to celebrate and enhance human skill and ingenuity rather than to modularise it. There is also an emerging trend towards a growing humanisation of technology. An impetus for change has been the vulnerability of complex technological systems, made evident by recent technological catastrophes. For example, the Three Mile Island nuclear accident (TMI) in 1979, as well as the Chernobyl disaster in 1986, were linked with deficiencies in the human-machine interfaces in the control room, together with organisational dysfunction. An analysis of these nuclear plant disasters highlighted the importance of organisational variables and training. One aspect of the humanisation process stresses user-oriented design, user-friendliness of hardware and software. A machine is described as 'user-friendly' if it is compatible with some physical or mental characteristics of the user, and also if it can interact or communicate with the person using it. It is now becoming evident that humanisation of the workplace is not a soft option but rather that it is intrinsic to effective deployment of human resources and the realisation of healthy and safe work settings. Further, the need to integrate understandings of human and technological systems so that they achieve optimal performance is recognised as necessitating more careful thought and analysis across a wide range of high technology work and societal systems, (see McCarthy and Tiernan, 1987). Recent research on nuclear power plant safety, has identified three domains that seem central to developing new directions in effective safety resources management. These include organisational analysis and self-analysis, balancing procedurisation and professionalism, and communications and learning. These domains suggest that there is a need to understand the nature of relationships involved in operational activities, to find ways to produce a better balance between the value placed on the expertise of risk handlers and that of risk estimators. These frameworks suggest that communication and learning activities need to be regarded as operating issues in that they affect actors' capacities to reduce risk; (Carroll and Perin, 1995).

Perin (1992), further highlights dimensions to risk handling in high technology workplaces, that are often ignored by risk-estimation techniques. Briefly, these embrace the following:

- the *communication and cognitive* dimensions which consist of extraordinary demands made on workers to be aware of hundreds of technical details;
- the *cultural dimension* refers to schedule pressures in what is often a physically hostile work environment;
- the *social and organisational* dimension of risk arises from a doubling and tripling of the on-site work-force during critical phases of maintenance development;
- the *knowledge* dimension of risk refers to the disturbed relationships of operating systems.

These four dimensions, which are critical to understanding problems relating to outages and outage risks in nuclear power plants, provide a useful framework for examining expectations relating to technological demands and the activities and practices that ensue in other high technology work settings. It was established by Carroll and Perin (1995), that while work schedules set up expectations of sequential activities, in practice, activities are non-linear and recursive. Thus, a scheduling of linear exchange between persons and technology may be inappropriate to the demands of the situation and exclude and inhibit flexible and appropriate recursive responses, embodying complex feedback learning loops. The framework outlined by Carroll and Perin (1995), draws our attention to deficiencies in systemic understanding and reflection, as well as in communicative practices and organisational learning. It was further demonstrated that although the literature on cognitive process and training effectiveness is large, there appears to be little research into experiential learning and technology transfer specific to high-hazard production systems. Organisational learning is new to those organisations that resist self-study and self-reflection, and to those that are dominated by cognitive engineering concerns, e.g. control operations and maintenance task training.

Communication and Culture

> The medium, a process of our time - electric technology - is reshaping and restructuring patterns of social interdependence and every aspect of our personal life.
>
> *Marshal McLuhan*
> The Medium is the Message, (1967)

Communication is the process by which culture is developed, integrated and maintained. Information, the content of communication, is the basic source of all human intercourse. Throughout history, information has been embodied and communicated in an ever-expanding variety of media, including spoken words, graphs, art, music, dance, written text, film, recordings, computer hardware and software. The communication of information permeates the cultural environment and is essential to all aspects of social life.

The new information and communication technologies provide many opportunities to impact on our culture(s) by expanding the infrastructure for information sharing and exchange. Communication can be used to generate greater amounts of information and new cultural forms to make this knowledge more accessible, and to provide it in more convenient and suitable ways. If, into the future, people are to become active in creating their own cultural environments, efforts will need to be made not only to assume that people can assess a broad variety of information and cultural content, but also that they have the skills and resources necessary to create, package and distribute information.

Communication and the Individual

Emerging technologies promise on the one hand to provide individuals with opportunities to increase their personal autonomy, enhance their sense of connection to others, and, in general, enable greater accomplishment and self-fulfilment. On the other hand, the same technologies could produce the opposite outcomes, contributing to personal isolation, increased dependency and loss of privacy.

How new technologies will affect individuals will depend in part on the rules that governments adopt to govern access to information and the new communication technologies. In order to take the greatest advantage of new technologies, people will need to be more technically skilled and have access to better 'navigational tools' (means to help people to access systems – i.e. analogous to today's TV guides etc.).

At the organisational level, how new technologies impact on people and on the organisation is tied more to the management ideology and ways of organising work, than on the technology. New technologies will not only affect how people access information, but also how information impinges on people's lives. The pace of technological change has created confusion about the appropriate standards of information use. While embracing new ways to access information for their own use, many individuals may find it difficult to cope with the fact that others, in turn, now have much greater access to them.

The opportunities for people to participate in cultural, political and economic life depend on their ability to access and use communication sources. Individuals need skills and tools to create the communication pathways and information in an appropriate form. Unequal access to communication resources leads to unequal advantages, and ultimately to inequalities in social and economic opportunities. Moreover, the people most likely to be adversely affected are those whom the new communication technologies could help most – the poor, the educationally disadvantaged, the geographically and technologically isolated, those in need of primary health care and the struggling small and medium social businesses.

Social Vision

Communication as societal infrastructure emphasises the linkages between communication, human activity and social structure. It focuses on the relationship between access to communication and services and access to power, wealth and position in society. Hence, in weighing communication policy choices, it places great weight on equity. From this perspective, the role for government is to ensure not only that needed technologies and communication services exist, but also that they are available to everyone and will serve all social purposes on an equitable basis.

New technologies create new potential and new opportunities that change our notions and expectations about what is possible and what is not. Historically, the creators and developers of new technologies did not foresee the massive impact these technologies would make – e.g. the potential of the telephone was not widely appreciated, the computers role in society has far exceeded the expectations of its early creators.

The gap between expectations and actual experience with new technologies can be explained in part, by our limited understanding of the relationship between technology and society, technology and the person. Attempts to depict this relationship have typically been uni-dimensional, focusing either on:

- technology as the driving force (technological determinism); or,
- on a particular set of social forces that has determined the evolution of technology.

Lacking an adequate understanding of technological development, we have been unaware of the realm of choices available. Thus, 'decision-makers', have been unable to channel technological development in the most positive directions - to do this there is a need to improve our analytical basis for assessing their development.

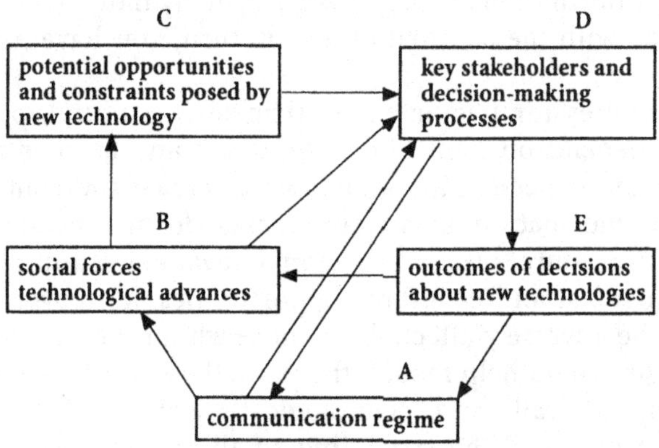

Figure 10: Interactive model of communication and society

A conceptual framework for analysing communication issues proposed by U. S. Congress Office of Technology Assessment (1990) identified the critical points at which choices about technology might be made, as well as suggesting key questions that need to be raised about new communication technologies (Garcia, 1990). This model of the relationship between technology and society takes into account technological developments, social forces and the values and roles of individuals and groups who have authority to make decisions about technology (see Figure 10).

People choose their definition of technology to suit the questions they are asking and the problems they must solve. Scientists and engineers, for example, may have less need to consider human factors, thus their definitions concentrate on machines and on physical structures, such as airports and nuclear reactors. A purely mechanical definition of technology would be inadequate for analysing, for example, how technology might affect communication and communication systems.

Understanding how technologies might affect communication systems or the processes in which individuals and groups come together to formulate, exchange, and interpret information, requires a definition of technology that embodies the interface/intersection of physical objects and people - e.g. human-computer interaction. This in turn involves a focus on apparatus, technique and social arrangements. The interaction of technological advances and social forces also creates new communication needs and desires and changes actors/stakeholders and perceptions of their interest.

A. *Communication Regime:*

1. norms, values, goals and roles that sustain and maintain communication within a given realm;
2. communication infrastructure that supports and facilitates communication processes;
3. decision-making processes and the rules and regulations that govern how the communication regime is managed and regulated.

B. *Interaction of social forces and technological advances:*

1. to create new ways of carrying out economic political, cultural and social activities as well as new opportunities and constraints.

C. *Potential opportunities and constraints engendered by new technologies:*

1. while technological advances might give rise to new economic opportunities for some people - these same advances might establish constraints for others.

D. *Decisions about technology will be made in a variety of arenas:*

1. the scientific/technical community;
2. the marketplace, the social/organisational, political and cultural arenas.

These decisions will be determined by, and reflect, the preferences of those, who within the relevant context, have the authority and/or the resources to structure the choices of others.

E. *Outcomes of decisions about new technological opportunities, will affect all elements of the model, selling the entire complex of interrelated changes into motion once again.*

Source: adapted from US Congress and Office of Technology Assessment Report - *Critical Connections* (1990).

A Post-Positivist Paradigm and Organisational Change

Recent discussion on the limitations of traditional human-computer /machine interaction frameworks e.g. ergonomics, cognitive ergonomics, cognitive engineering (Gorayska and May, 1995; Gill, 1995), as well as new research on high technology risk factors (Carroll and Perin, 1995), illustrates the pervasiveness of the mechanistic paradigm of the world of work. It further highlights the problems that can arise when new technology is enmeshed in a mechanistic production culture. In other words we can ask where is the *human* in human-centred systems, and how to move further on the idea of designing the cultural interface rooted in human diversity (Gill, 1995).

The need to overcome the limitations of the machine model of organisational systems which diminish the human side of enterprise (at the level of the individual, group, organisation and environment), is still a major challenge facing organisational theorists in the social sciences. There is now a need for greater cross-fertilisation of ideas across disciplines to facilitate the generation and evolution of enriched frameworks for understanding the meaning of human technology/computer systems. In what follows, the post-positivist paradigm of organisation developed to increase sensitivity to the complexity of the human person in context and to the human side of enterprise and technology will be delineated. This paradigm draws on different theoretical stands from the natural and the social services. It promotes a framework for the realisation of a new view of the person in the human-computer technology system and content.

Complexity and Change

One variant of the post-positivistic approach called the naturalistic paradigm, proposed by Lincoln and Guba (1985: 37), draws together core issues that impact on theory and research not only in psychology but also in the other social sciences and the natural sciences. These can be summarised as follows:

- the nature of reality – viewed as multiple, constructed and intrinsic;
- the relationship between the knower and the known is interactive and inseparable;
- with reference to the possibility of generalisation – only time and context bound working hypotheses (idiographic statements) are possible;
- the possibility of causal linkages – sees all entities in a state of multiple, simultaneous shaping, so that it is impossible to distinguish causes from effects;
- inquiry is viewed as value-bound rather than value-free.

These characteristics isolated as intrinsic to the naturalistic paradigm can further be built on to display coherence for the task of naturalistic inquiry. Naturalistic inquiry, grounded on the naturalistic paradigm, offers a post-positivist approach for working with system and technological/organisational change.

Systems concepts introduce a focus on relationships between organisations and their environments (the environment was neglected by the closed mechanical model), on interdependencies between parts of the organisation (sub-systems), on optimal relationships between the parts of the total system, on feedback mechanisms and on complexity (see Katz and Kahn, 1978). Systems theory, as applied to organisations, introduced biological models to organisation theory. The metaphor of organisation as organism began to take root. Exploring the parallels between organisms and organisations produced different theories and explanations that have had very practical implications for organisations and for members of organisations. An important strength of the 'organismic' metaphor rests in its contributions to the theory and practice of organisational development (or organisational change, with the help of human resource professionals) - see Morgan, 1986.

Complexity, Chaos and Change

Scientific investigations in the past sought and found many predictive elements in nature: they found order and regularity. Typically, when they accordingly came upon disorder or chaos (complexity) they spontaneously rejected it as too bizarre (Breuer, 1985: 471). There were, however, critical exceptions. Separate streams of inquiry by leading edge scientists in the natural sciences went beyond the positivist paradigm and proposed new questions, new methodologies and new interpretations. Their explorations tapped into core issues such as:

- Is science objective or subjective or both?
- Is the observer the observed? (Bohm, 1980)?
- Can all properties of a system be known exactly? (Heisenberg, 1958).
- Do micro-structures and macro-structures evolve together? (Jantsch, 1980).
- Are we part of a self-organising universe? (Jantsch, 1980).

Paralleling these developments in the natural sciences, which posed fundamental challenges to the classical paradigm, researchers in the human and social sciences were also faced with many paradoxes arising from the range of phenomena they chose to consider and the methodologies they chose to adopt under the umbrellas of their respective disciplines. As we have seen, it is now becoming clear that those questions which grew out of dilemmas and paradoxes faced primarily by physicists, chemists and biologists, are also of paramount significance for probing the human and social dilemmas and paradoxes faced by people as they live out their lives in family, work, organisational and societal domains. One theme central to all disciplines is the phenomenon of change. Bohm

(1980), has developed a theory that seeks to understand the universe as a flowing and unbroken wholeness. He views flux and change as fundamental, arguing that the state of the universe at any point in time reflects a more basic reality. He calls this the implicate (or enfolded) order manifested in the world around us. The implicate order is viewed as a creative process, which like a hologram has everything enfolded in everything else (see Morgan, 1986). For Bohm, the world unfolds and enfolds from moment to moment. This process creates the appearance of continuity in the midst of change.

Bohm's theory encourages us to see the world itself but as a moment in the fundamental process of change, rather than seeing change as an attribute of reality. The implicate order proposed by Bohm, as the dynamic underlying the flux of change, is analogous to the unconscious in the theories of Freud and Jung. At the level of the organisation it resonates with the concept of culture which can be understood as the carrier of shared meanings, shared understandings and shared sense making in the organisational setting (Schein, 1990). Weick (1979), speaks of the proactive role that we unconsciously play in creating our organisational models as a process of enactment. In modern organisations, which are ever-frequently described as complex systems in turbulent environments, the search for new frameworks, with which to encounter the hidden, the uncertain, the unexpected, the disordered is ever-pressing. It is proposed here that newly-emerging concepts, metaphors and insights derived from the newly-evolving field of chaos and complexity research can open up a creative debate for those who are seeking to understand, inquire into and become part of the process and dynamics of technological and organisational change.

Chaos research has also created special techniques of using computers and special kinds of graphic images, pictures that capture a fantastic and delicate structure underlying complexity. For some physicists chaos is a science of process rather than of state, of becoming rather than of being. Further, it is becoming evident now that the scientists who set the description of chaos in motion, showed certain sensibility in the following ways:

- They had an eye for patterns, especially patterns that appeared on different scales, at the same time.
- They had a taste for randomness and complexity, for jagged edges and sudden leaps.
- They speculated about determinism and free will, about evolution, about the nature of conscious intelligence.
- They spoke about the universal behaviour of complexity.
- They feel they are turning back the reductionist trend in the natural sciences – the analysis of systems in terms of their constituent parts.
- They believe they are working for the whole (see Gleick, 1988).

In all, they believe that chaos theory cuts away the tenets of Newton's physics and that it comprises the third greatest revolution in the physical sciences in the twentieth century (besides, relativity theory and quantum mechanics). Of the three, chaos research applies to the universe we see and touch (clouds, ecology, snow flakes): to objects at a human scale.

Chaos and Organisational Change

The lead that chaos theorists within the domain of natural science are giving in integrating chaos as a fact of life, provides a strong impetus for researchers in the human and social sciences to see chaos as belonging to our world picture. Natural science has come to see the world as being in a state of constant movement, change and growth. Nothing is static and stable. Within such a universe chaos plays a major role. Chaos and order are no longer to be viewed as mutually exclusive. They can and do exist in the life processes of natural organisms. Of fundamental interest is their applicability to humanity. In what follows here, the focus will primarily be on the applicability of chaos concepts to the world of work, or organisations and of organisational and technological change and dynamics. I will first present some core scientific concepts and principles relating to chaos and order before going on to suggest what they might look like in the realm of human experience in organisational contexts.

Chaos Concepts

The tenet that there is order in chaos pushes us towards a multi-perspective in looking at human phenomena in organisational systems. The problem is to find the proper perspective to be able to perceive the order in chaos. We could generate new modes of inquiry that are sensitive to multi-perspectives and meta-analysis and experienced paths from chaos to order could be charted. 'Deterministic chaos' is a term which has been coined to describe seemingly disordered states which, however, reveal a certain degree of order. Such chaos can be defined by a limited number of core variables as outlined below.

Order Out of Chaos

This principle derives from the work of Ilya Prigogine who won the 1977 Nobel prize in chemistry for his theory of 'Dissipative Structures' which describes a new perspective on change. Prigogine and Stengers (1984), in their demanding book *Order out of Chaos*, point out that our vision of nature is undergoing radical change: towards the multiple, the temporal, the complex. 'Dissipative Structure' is a concept which refers to new structure which emerges out of a simple structure. They argue that order and organisation can arise spontaneously out of disorder and chaos, through a process of self-organisation.

Synergetics

The phenomenon of 'self-organisation', a synergetics identified by

Prigogine has been found to apply in many disciplines, including ecology, sociology and Jungian psychotherapy. The phenomenon whereby systems pass from one stage of order through a chaotic or disordered phase to arrive at a new and more complex stage of order, is central to 'self-organisation'. Self-organisation is a phenomenon which also emerges in group and organisational life. Symbiotic relationships of human–technology interfaces, could also be explored from a self-organisation perspective.

Non-Linearity
The thermodynamics system to which Prigogine and Stengers (1984), refer to are all non-linear systems. Change takes place within them in a non-linear fashion. Their sum is more than the total of their parts. In such systems, complicated feedback mechanisms are at play: simple laws of cause and effect do not suffice to explain the changes one observes.

The Butterfly Effect
According to this principle, tiny differences in input could quickly become overwhelming differences in output – called sensitive dependence on initial conditions. The Butterfly Effect, captures this phenomenon, with the notion that a butterfly stirring the air today in Beijing can transform storm systems next month in Dublin. A tiny input difference in a computer program can yield widely different outcomes.

Discontinuity
Discontinuous phenomenon has had no place in the geometry of the past 2000 years. The shapes of classical geometry are lines and planes, circles and spheres, triangles and cones. They represent a powerful abstraction of reality and they underpin a powerful philosophy of platonic harmony. For understanding complexity, these turn out to be the wrong kind of abstraction.

The new geometry mirrors a universe that is rough, twisted tangled, intertwined. Mandlebrot's (1977) work made a claim that odd shapes carry meaning.

Sensitivity and Flexibility
Biologists such as Maturana and Varela (1980), have pointed out that natural organisms are uniquely sensitive to their milieu. In other words, they are in a continual process of adapting and re-adapting to their outer and inner worlds. Flexibility is viewed as the key to survival in an ever changing environment.

Applications to Organisational Change
Gleick (1988), in his analysis of chaos research concludes that scientists wanted to know where there are forms in nature – not visible forms, but shapes embedded in the fabric of motions – waiting to be revealed. The Jungian analyst Wieland-Burston (1992), contends that chaos is the starting point for most psychoanalysts. When clients are in mental and

emotional turmoil they turn for professional help to psychotherapists. At a human level, Freud, since the turn of this century, had already identified the turbulence of defence mechanisms and saw his major mission as the unravelling of the secrets of the unconscious - the region of psychic chaos. Contemporary Western society is marked by a fear of chaos, a denying of the validity of chaos as an aspect of experience and is characterised by a relentless search for order (Wieland-Burston, 1992). Chaos research, suggests that both order and chaos are two aspects of living and experience that have to be accommodated.

Order and Chaos

One cannot be considered without the other. There is order in chaos and chaos in order. The classical machine model of work organisations, as explored by social scientists has clearly unearthed and delineated unanticipated consequences of the machine model, i.e. the 'chaos' of the bureaucratic model. Typically, these consequences have been negative given that many human cognitive and social emotional elements are suppressed and not integrated into the dominant organisational model. The creative aspect of the organisational underworld is typically not recognised and frequently rejected by organisational management. Lessons from mythology and ancient societies have shown that once the element of fear retreated into the background, chaos became the domain of creative experimentation (Weiland-Burston, 1992). In the history of the Greek usage of the word 'Khaos' - it was used to describe the world before it was ordered with a universe. The process of creation followed the 'formless void'. New possibilities of vision, of action and interaction can arise from chaos as a field of experimentation.

The humanist approach in psychology, in contrast with the scientific approach (see Kimble, 1984), has typically argued for space, for human values, for creativity and unpredictability in human experience rather than for a fixation on control and predictability. Organisational psychologists, who have driven the post-positivistic paradigm in organisational change, have strongly identified with the humanist culture of psychology and social science in general. The challenge is to channel the energy released by this approach creatively.

Attention to chaos also leads us away from the repetitive and the universal to the specific and the unique. In psycho-dynamic terms, one could argue that the collective unconscious is stirring and rising to compensate for the controlling dominant structure and values in our workplaces and society. Feminist researchers (see Wilkinson, 1986), have, for some time now, championed a reframing, a refocussing of the way we view reality. In particular, feminist researchers seek to evolve new forms of self-organisation and group-organisation that will lead to a new coherence: new forms of communication in the major domains of living, home, work, community and across traditional domains, within different societal contexts. Chaos research also places emphasis on 'becoming', on the quality of experience. This transition to a new description can find resonances in organisational change around principles of quality, equality and health (see McCarthy, 1988, 1995). Thus, the principle, order out of chaos, provides a welcome impetus to organisational-change agents to seek to understand the processes of self-organisation or *autopoiesis* (Maturana and Varela, 1980). Fluctuations within a system interact with one another in qualitative ways causing wholly new structures to arise. The implication of this process is important to sensing new possibilities in work design and new forms of work and technology, and generating a culture of creativity and entrepreneurship.

Building on the insights derived from chaos research it is important to consider the modes of inquiry and the modes of action and learning that can create conditions to allow the new chaos paradigm to flourish. It is not sufficient to understand phenomena, but it is also essential to evolve modes of inquiry and action that are sensitive and flexible to the phenomena and experiences laid open by the new paradigm. Within the social sciences four approaches can be isolated as offering fruitful pathways in the transition to the new approach. These four include:

- the naturalistic paradigm proposed by Lincoln and Guba (1985), for post positivist research in the social sciences (see discussion above);
- the action learning model of Morgan and Ramirez (1984).

This approach is concerned with creating the conditions through which self-organising systems can develop. The objective is to create systems that can learn from their own experience, to modify their structure and design to reflect what they have learned. This work draws on the ideas of single-loop and double-loop learning as proposed by Argyris and Schön (1978), and Bateson (1972, 1979). In his recent book *The Fifth Discipline: the Art and Practice of the Learning Organisation*, Senge (1992), argues that organisations must be able to learn in order to survive. In other words, learning disabilities threaten many organisations and prevent them from recognising threats and opportunities; to succeed, they must relearn how to learn. Cooperative learning is a cornerstone of the learning organisation. A challenge for the future is to evolve ways of integrating cooperative learning with competitive and individualistic learning activities. It is proposed here that one of the most important elements of cooperative learning is 'positive interdependence' (see Deutsch, 1993). Managers need to understand that it is to their organisation's advantage if their work-force learns well, and that it is to their disadvantage if workers learn poorly - a recognition of the power of mutual interactive learning is required. Learning and working are interdependent. Learning facilitated by a mutually respected interdependent culture can yield effective and satisfactory outcomes both in performance and in attitudes. This can be achieved in a number of different ways:

- through the generation of goal interdependence and mutual goals;
- through a creative division of labour (task, job, skill interdependence) thus defusing traditional segregation – e.g. by class, gender, race etc.;
- through the sharing of resources and information among group members (resource interdependence);
- by giving joint rewards (reward interdependence).

It has been demonstrated (see Deutsch, 1993) that co-operation learning can have very favourable effects on learners. For example, learners developed considerably greater commitment, helpfulness and caring for the other – regardless of differences in ability, ethnic backgrounds, gender, social class or physical disability. They developed skill in taking the perspective of others emotionally as well as cognitively. The organisations evolving today, and into the future, using integrated computer and communications technologies will increasingly be defined by collaborative networks linking small, medium and large groups of people together. There is a need in the work context for new conversations between human resource managers/consultants and technologists who are engaged in mapping out salient underlying conceptual understandings of self-learning, group learning and organisational learning.

Feminist critiques of research on gender and the social sciences
Research on differences between the sexes, gender roles, gender role socialisation, gender and technology and on modes of inquiry which has been conceptualised within a strong positivist framework, has been challenged by feminist researchers (see Hare-Mustin & Marecek, 1990; Unger, 1979, 1989, 1990). Unger (1990), in her discussion on 'imperfect reflections of reality' argues for alternative modes of inquiry into feminist research, which include the individual's story, an emphasis on 'conscious subjectivity' (Wilkinson, 1986), the need to embrace diversity and uncertainty and for multiple perspectives.

Networking and Network Cultures

Since the early 1960s there has been an intensification of effort by the powerless in nations around the world to organise themselves to effect social structural change. No matter what the 'cause', the goals, or the beliefs, and no matter what type of movement it is – political, social ecology, women's, religious – there is the same basic structural form and mode of functioning. Wherever people organise themselves to change some aspect of society, a non-bureaucratic but very effective form of organisational structure seems to emerge. This structure does not involve boxes arranged in a hierarchical order with the controlling box at the top as in the typical bureaucratic system. It would look like a knotted fishnet, with a multitude of nodes or cells of various sizes, each linked to all the others either directly or indirectly. Some of those cells within the network would, in themselves, be hierarchically organised bureaucracies recognised by the public as regional, national or even international organisations; but the majority of cells are local groups, some organised according to the conventional mode, others *ad hoc* groups that are here today and gone unrecognised tomorrow.

The dynamics of networking is concerned with new balances between structures and processes. The web-like structure of networks permits new forms of communication, participation, leadership and decision making to evolve. In all, these processes are both person- and group-oriented, designed to tap potential and build a culture of shared learning, decentralised, boundaries are fluid rather than fixed, members relate to one another as equals rather than as subordinates responding to superiors. Procedures are people-oriented rather than institution-oriented. The structures are polycentric rather than monocentric. In conventional organisations people are typically slotted into hierarchical boxes, while the organisational picture (chart) of a networking system is more like a 'fishing net' or a spider's web, with a multitude of nodes and cells of various sizes each linked to all the others either directly or indirectly (see McInnis, 1984).

Network Groups

Networks are our oldest social invention. They can be viewed as the webs of life. Features of networks in an organisational context, which were poorly understood and which can have ambivalent effects (particularly for those outside the dominant networks) are becoming more transparent in recent times – e.g. a recognition of the subtle power of the 'old boys' network in workplaces. The spider's web is distinguished by its interconnectivity, which maximises flexibility and minimises vulnerability. What is new about today's networks is that groups such as ecology groups, women's groups, whether based in the home (full-time or part-time), in formal workplaces or in voluntary groups, recognise that they can invigorate the home, the community/neighbours, workplace and society by a sharing of their resources. For example, women's network groups have been evolving on two fronts – informal and formal. Women's networks, sometimes called support groups, have grown out of women's needs and anxieties in coping with the demands and stresses of the modern world.

Women who have been marginalised from the dominant power structures of society, such as work, education, politics etc., have experienced a strong urge to come together, to learn together, to pool their resources and to shape their own destinies and roles. These networks range from formal professional/technical groups (e.g. women in management networks etc.) to less-formalised community groups who are beginning to identify the dynamics of their present roles and future possibilities. Authority in a network structure tends to be decentralised, resulting in individuals and groups who have appropriate knowledge and information rather than in those who occupy formal assigned positions. Grassroots women's groups are beginning to discover the force of shared power. The sharing of power, enhancement of empowerment and the facilitation of constructive interdependence characterises networking. The shift is away from dependence to interdependence. Policies and boundaries tend to be fluid rather than fixed (see McCarthy, 1994).

Table 1: Values of network and conventional work systems. Core characteristics and some guiding values of network systems compared with conventional organisations

Network Systems	*Conventional Systems*
Complex - high interconnectivity power-sharing, empowerment	Simple - low interconnectivity power at top
Decentralised	Centralised
Policies - boundaries fluid	Policies - boundaries fixed
Equal power relations	Power distance, high-authority laden
People-oriented	Institution-oriented
Polycentric structure	Monocentric structure
Flexible	Rigid
Trust	Mistrust
Context for learning	Inhibits learning

Table 1 delineates core characteristics and some guiding values of network systems compared with conventional organisations.

Networks that are built on values of cooperation, mutual trust, sharing, learning together, teaching one another with dignity, listening to other voices, incorporating diversity have the potential to consolidate an inner strength, elasticity and robustness. Such networks are inoculated to cope with the traditional blasts of stereotyping and discrimination that exude from patriarchal systems. Networks which carry ECO-feminist values build strength with others at the individual, group, organisational and community-societal levels. This leads to a re-humanisation of working and living for those who participate and share in this evolutionary process. This process involves commitment to basic and shared assumptions. Thus, basic assumptions grounded on caring, valuing the other, respect for the self, for society, for nature and the ecosystem can catalyse the unifying of disparate groups and individuals. Networking is a qualitatively different way of organising to that so typical of our everyday workplaces and institutions. It brings a shift from 'position power' to 'knowledge power'. It involves 'having a say' and having an impact on outcomes.

It is now recognised that technology - telecommunications, computers etc., - has got the potential to create network cultures as delineated above, which in turn have the potential to generate a new symbiosis between the person and technology. In this way, new forms of power-sharing which enhance empowerment through the facilitation of constructive interdependence can be created. Thus the shift is away from dependence to interdependence.

The modes of analysis and inquiry emerging within these four perspectives outlined above are demonstrating features that resonate with the characteristics of chaos and research. These emergent modes of action and social images stress a shift from the simple to the complex, from the mechanical to the holographic, from linear to multiple causality, from the objective to the subjective and perspectival. They also include a concern with meaning, with the small, and with self-organisation. An understanding of organisational change from the perspective of chaos and complexity research draws one into the hidden recesses of systems and allows one to embrace uncertainty, recursiveness and diversity. While Prigogine and Stengers (1984) and Gleick (1988), carefully chart ways of seeking order in chaos, it also behoves researchers and practitioners who are working with systems that appear ordered, to seek 'chaos in order' e.g. nuclear power plants and to work with chaos towards revealing creative organisational change.

Virtual corporations and network systems are integrally dependent on mind and the evolution of structures and processes which are sensitive to the interdependence of mind and context. The evolution of the 'virtual organisation', which Bleecker (1994), describes as a new forum of cooperation that uses information technologies to collapse time and space will be paralleled by collaborative and consultative networks. Collaborative networks will input the drive for higher quality products and the ability to draw on internal resources as needed regardless of where they are and regardless of who owns them - supplier or customer. They will focus on the interoperability of processes and supporting systems which, in turn, will be driven by mobile knowledge workers. The 'virtual office' feeds on spontaneity and flexibility. Knowledge makers process information and assignments in parallel, interweaving bits of one problem with pieces of solution to another. The sharing of information does not diminish it: as people develop skills in sharing ideas, feelings, and information, the more skilled they become and the more empowered. Conventional managers typically lack awareness that the most valuable and transforming resource of any human system is internal and non-depletable, that is, the human mind which thrives on being renewed and is self-renewable.

This new work paradigm demands a set of organisation behaviours: trust, individual empowerment, personal accountability, cooperation and teamwork. Organisation systems which adopt these behaviours will also nurture a fundamental improvement in the individual's role in the work context. Thus, information intensive, fluid environments will be anchored on the potentials and abilities of both individuals and groups (teams).

Mind, Culture and Technology

Jung points out that there are two voices of explanation or voices of understanding, the pleroma and the creatura. In the pleroma, the world of things, there are only forces and impacts. That is the voice of the hard sciences. The voice of the creatura is the voice of communication and organisation. In the creatura there is difference. It is argued by Bateson in his book *Steps to an Ecology of Mind* (1972: 481) that the difference is an idea. Drawing on the insight of Kant (1781), who, in the *Critique of Judgement*, states that the primary act of aesthetic judgement is selection of a fact. There are, in a sense, no facts in nature or, rather, there are an infinite number of potential facts in nature, out of which the judgement selects a few which become truly facts by that act of selection.

Bringing the insights of Kant and Jung together, Bateson (1972), argues that we create a philosophy which assesses that there is an infinite number of differences in any object (e.g. piece of chalk), but that only a few of these differences makes a difference. Bateson (1979), asserts that is the base for information theory. The unit of information is the difference. In fact, the unit of psychological input is difference. Thus, the energy structure of the pleroma, the forces and impacts of the hard sciences have evaporated so far as explanation within creatura is concerned. The letter you did not write can precipitate an angry reply. Zero can be a cause. Sameness can be a cause, because sameness differs from difference. The creative effects which do not occur in the pleroma, have yielded notions which come together to give us a broad conception of the voices in which we live – a new way of thinking about what a mind is. Characterisation of a system called mind, according to Bateson, includes the following:

- the system shall operate with and upon differences;
- the system should consist of closed loops or networks of pathways along which differences and transforms of differences shall be transmitted. (What is transmitted on a neuron is not an impulse, it is news of a difference);
- many events within the system should be analysed by the respondent part rather than by impact from the triggering part;
- the system shall show self-correctiveness in the direction of homeostasis and/or in the direction of runaway. Self-correctiveness implies trial and error.

For Bateson, mind is a necessary and inevitable function of the appropriate complexity, whenever that complexity occurs. Further, as determined by Bateson (1979), in his seminal work *Mind and Nature*, that complexity occurs in a great many processes besides the inside of the human head. In answering the question whether a computer thinks, Bateson observes that what 'thinks' and engages in 'trial and error' is the human, plus the computer plus the environment. The lines between the person, the computer and the environment are fictitious lines. What thinks is the total system which engages in trial and error, which is the person plus the computer and the environment. The lines that matter are pathways along which information or difference is transmitted. Bateson's holistic approach challenges the Darwinian theory of natural selection, in which the unit of survival was the family line of the species or sub-species. For Bateson, the unit of survival is organisation plus environment. Within this framework, history in the widest sense turns out to be the study of the interaction and survival of ideas and programmes (i.e. differences, complexes of differences etc.) in circuits. For Bateson, the separation of mind from the structure in which it is imminent, such as human relationships, human society or the ecosystem, leads to a fundamental error, which in the end will result in hurt. Imbalances occur, when certain basic errors of our thought become reinforced by thousands of cultural details.

Recent discussion has been channelled into a research programme called 'Cognitive Technology', (CT), which is spurred by a need to distinguish the classical HCI approach which focuses either on the mainly engineering aspects of information technology, or draws attention to the strictly human problems involved in that technology. CT purports to integrate these two trends. It is proposed here that the distinction between creatura phenomena and pleroma phenomena as developed by Bateson (1972, 1979), would provide a useful framework for integrating core assumptions and concerns in HCI – including cognitive ergonomics, cognitive engineering and cognitive technology. Metaphors of mind which extend the computational – information processing model (see Sternberg and Wagner, 1994), to 'Mind in Nature' as determined by Bateson (1979), offers a new perspective. The creatura elements of both the person and the computer in context can be distinguished from the pleroma elements which are a primary focus of engineers. While the distinction between software and hardware resonates with creatura and pleroma issues, the latter concepts, however, embrace a rich and deeper mosaic of phenomena, including not only the 'cognitive' domain of the human person, but also the social and emotional domains. Not only the human part of the process but the computer as well are embedded in a cultural context, which affords meaning to the environment and is in turn challenged by the questions emanating from such an environment.

Mind and Technology: The Dynamics of Intensive Care

The relationship between mind and technology can be mapped into by a detailed consideration of work settings that embody sensitive and complex relationships between humans and technology. The impact of new medical technology (NMT) on nurses and patients in hospital intensive-care settings provides a rich domain within which to explore, issues of stress and work, the role of NMT on task performance and on both the 'caring' and 'curing' dimensions of intensive-care nursing work.

The interplay between the demand that the job makes on an employee and the demands that an employee requires of a job underpins numerous studies on the well-being of both organisational members and of organisations (French *et al.*, 1984). Arising from the thrust of recent studies, it is now recognised that the attainment of healthy work environments requires an understanding of the effects of both physical (including chemical forces) as well as psycho-social stressors on employees (Cooper, 1983).

The dominant ideology that prevails in a hospital is the medical model. This interpretation is the major influence on the way we view the treatment of an individual in hospital. The model is a traditional model handed down through the evolution of hospital care. It is an ideology of medicine subscribed to by the majority of physicians. Since physicians have largely been the policy makers in hospital life, the way a hospital is run (i.e. the organisation) is influenced by the medical model.

The medical model as a prototype views the doctor as the central authoritarian figure with a nurse as dependent on the instruction of the doctor and the patient as a passive object to be 'cured'. Since the doctor specifies how treatment should be given, this would be in line with the medical model. This mechanistic interpretation minimises human contact in treatment. Since the patient is given a passive role, active participation in his/her own treatment is not encouraged. This is the environment which prepares the way for the introduction of new technology. The basic triadic model of the interaction between doctor, nurse and patient has gradually been augmented in the history of hospital care to include a technical relationship where the communication between these three groups can become less personal and more technical (see Figure 11).

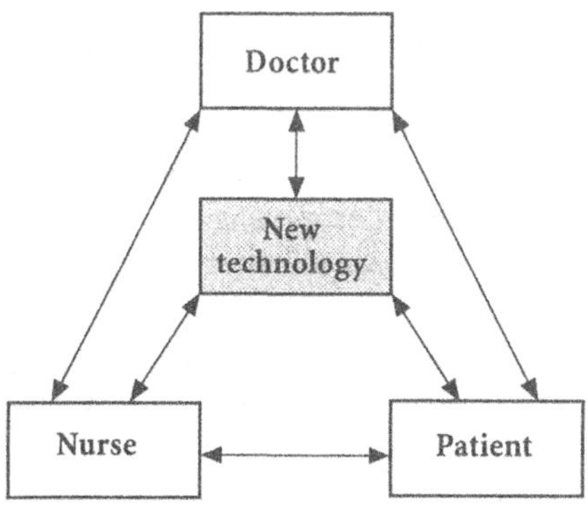

Figure 11: Model of interaction between doctor, nurse, patient and new technology

Recent research on the nursing profession is critical of the medical model, which has dominated the training and evaluation of nursing, and stresses the importance of a holistic or systems approach in nursing (Jacobson and McGrath, 1983; Sobell, 1979). This approach translated into nurses' work emphasises the significance of the 'caring' aspect and the danger of it being eroded by undue concentration on the 'curing' dimension. The impact of NMT on nurses and patients requires the context of the work setting to be attended to. Cognitive ergonomic aspects of NMT, as well as psycho-social aspects of organisational situations, need to be included in an analysis of the transforming role of NMT on task performances (Welford, 1978), and on nurse and patient satisfaction.

In recent years, it is becoming more and more understood that 'efficiency' is not the sole criterion of performance at work, but that human needs as related to performance also require attention. These needs are associated with health and well-being, with safety, comfort, as well as with the quality of working life (QWL). It has been observed, particularly with the transition into high-technology workplaces, that some types of organisations and work are vulnerable. For example, Caplan *et al.* (1980), following their comprehensive review, suggested that workers in health care organisations had a greater probability of experiencing excessive stress than others.

It has further been noted that occupational stress is not uniformly distributed across departments within organisations and that research needs to be occupation-, department-, or unit-specific. The evolution of medical technology in specialist hospital units brings with it not only a need for increased skills and competence on the part of the medical team (nurses, doctors), but equally a need to isolate sources of stress for those employed in such units, and to develop organisational mechanisms which would prevent and ameliorate those 'stressors' identified.

Intensive Care Units (ICUs)

Intensive care units have evolved in response to clinical needs, which reflect rapid advances in medicine and medical technology since the 1950s, 1960s, 1970s, and 1980s (Taylor, 1975; Stehle, 1981). In particular, key concerns for nurses arising from the development of specialised intensive care units in hospitals are tied to (1) the training of staff, and (2) the use of medical technology with an emphasis increasingly on new technology (NMT). Moriarty (1978: 160), stated that intensive care units (ICUs) are designed:

> To diagnose, treat and maintain patients with immediate acute but potentially reversible life-threatening impairment of single or multiple vital-organ systems.

The advantages of ICUs isolated by him include the following:

- highly trained medical and nursing personnel;
- able to concentrate, expensive monitoring and intervention equipment and clinical investigations can progress efficiently because of the high concentration of equipment and trained personnel.

Disadvantages were viewed primarily as:

- the danger of infection;
- high cost;
- legal factors; and,
- psychology issues.

The advantages of ICU work as outlined by Moriarty (1978), reinforces the centrality of highly skilled and well trained personnel (nursing and medical), the significance of expensive monitoring and intervention equipment and the cognitive-psychological demands on ICU staff, i.e. ability to concentrate, ability to respond effectively to machine cues and initiate appropriate action.

Psychological Disadvantages

The psychological disadvantages noted by Moriarty (1978) focus on the ward atmosphere as a 'tension charged stratified war bunker with sounds of hissing machinery, flashing monitors,' etc. Thus it unfolds that both the advantages and disadvantages of ICUs can be a source of stress and strain for ICU nurses. This can arise in particular where:

- training does not match job demands;
- knowledge of the NMT and equipment is inadequate; and,
- where the cognitive psychological requirements of concentration, alertness, judgement, accuracy and sense of responsibility may not always be appreciated or understood by those working outside the ICU settings.

By 1983, Bishop in her research on staff stress in ICUs in the USA observed that 'the emotional consequences experienced by nurses working in these units are barely mentioned,' (p. 181).

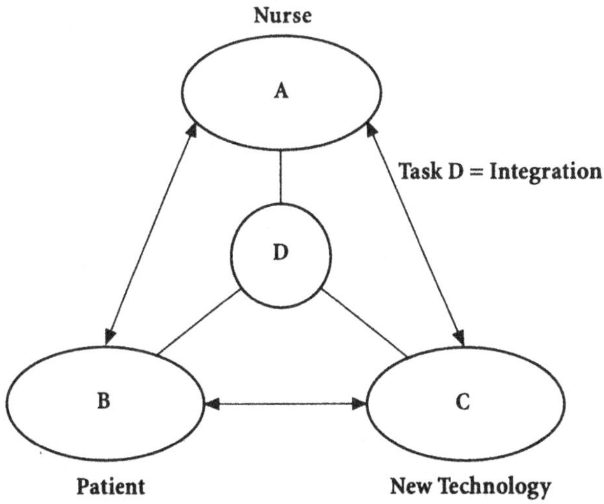

Figure 12: Relationships between ICU nurse, patient and new technology

Figure 12 delineates the systematic interactive nature of the relationship between the ICU nurse, the patient and new technology. The task of ICU nursing involves task integration (D) of salient interrelationships between A (nurse), B (patient) and C (Medical Technology), i.e. the arrowed, solid lines, AB, AC and BC.

Research conducted by McCarthy and Tiernan (1987; see also McCarthy 1989a, 1989b), was designed to tap the sensitive and intricate relations between new medical technology and nurses' experience of NMT in their work in hospital intensive care units. A sample of 52 nurses (all female) in intensive care units (ICUs and coronary care units (CCUs) in five Dublin hospitals participated in a study designed to tap the relationship between NMT and stress and coping at work. Both qualitative and quantitative methods were adopted to explore the salient elements and issues central to the nurse–patient technology interactions (see Figure 11 and Figure 12). Qualitative analysis included observations of ICUs/CCUs, in-depth interviews with nurses and case study profiles, while the quantitative approach included the development of questionnaires tailored to the central research questions and essentially tapped the perceptions of nurses towards NMT, the stress experienced and its relevance for the 'caring and curing' components of their work.

An underlying concern was the relationship between NMT, which detailed patient psychological data and nurses perceptions, judgement of working both as 'carer' and 'curer' focusing on the 'hard' information transmitted by means of the NMT. The first phase of the study identified nurses' perceptions of:

- the most important aspect of their work; and,
- the most demanding aspect of their work.

Both of these questions draw on the significance of the ICU work for nurses and on the demands both mental and physical that emanate from that work. The most important aspect of ICU work highlighted a holistic orientation as follows:

> The total patient care and satisfaction when they are well enough to leave the unit.

Combined with this was an emphasis on nurse–patient–technology-related responses. A third emphasis was on maintaining a supportive intensive care unit culture, while promoting collaborative teamwork between nursing and medical staff and a caring culture for patients. The most demanding aspects of intensive care work for nurses highlighted the following:

> Coping, with a variety of different responsibilities emerged as the most demanding aspect of ICU work. Perhaps it is the extensive variety of responsibilities which actually makes ICU work difficult, coupled with the ultimate responsibility – that of the patient's life. One of the most demanding aspects of ICU work for nurses is being short-staffed or being relieved when

short-staffed by student nurses or nurses not experienced in ICU work. Dealing with medical staff, distraught relatives, the various mood of different staff members while trying to ensure continuity of patient care can make nursing work stressful. Equipment can be an added demand if it has not been properly sterilised or if it breaks down or is faulty.

Nurses are expected to carry out instructions exactly and often have to cope with being given many different instructions at the same time. Nurses are aware that patients are ultimately dependent on their observations for their lives. Coping with life and death situations and emergencies often means that nurses take on more responsibilities than is advisable in an attempt to maintain high quality care.

Problems With New Medical Technology

Problems with new medical technology were identified through open-ended questions in a specially developed questionnaire administered to the nurses. Responses were content analysed to obtain trends in nurses' thinking and experience. A variety of problem areas were identified, including:

- patient-related problems;
- technical problems;
- staff-related problems;
- organisational problems.

Patient-Related Problems

As would be expected, the problems of new medical technology in relation to the patient varied with the unit specialisation. The problems that units seemed to have most in common included the observation that patients tend to become dependent on equipment they are linked to, thus requiring considerable psychological support and reassurance from nurses. Patient anxiety was perceived as adversely affecting the rate of recovery of the patient. A further perceived anxiety for patients relates to the intrusive nature of some technology, such as the cardiac output machine and the Swan Ganz machine, which are not vital life-sustaining equipment and which seem to cause considerable distress to patients.

Ethical problems concerning patients also came to light because of the new medical technology. Basically, these are twofold. Firstly, patients' lives can be prolonged almost indefinitely with ventilating equipment and so forth, and doctors and nurses may not be able to decide how best to treat the patient who has little hope of recovery. As one nurse put it: 'It seems that people are now more interested in the new technology than in the patient.' Secondly, problems arise, particularly for intensive care and coronary care nurses, when one patient must be given priority treatment over another patient or when the decision about who is to be admitted to a ward is determined by the availability of life-sustaining equipment.

Physical problems have also been created for patients because of new technology. Some patients develop allergies to electrodes and the risk of infection is high because most equipment requires a trans-venous link between the patient and the machine – as one nurse stated: 'The more lines you have going into a patient the less likely they are to recover.'

Some problems emerged which were peculiar to the neonatal units studied. One of the principal problems is that new technology tends to leave little time for caring for the neonates. Other problems identified suggest that babies who previously would have died are now being kept alive and could have a low quality of life. New complications are arising as more and more babies are surviving. One of the main ethical problems for nurses is the knowledge that early delivery of some neonates means that they run a higher risk of brain damage.

Technical Problems

Breakdown of equipment was the most common technical problem encountered by nurses in all specialised units. Each of the hospitals studied had a back-up set of equipment available in the event of an emergency, a technician on call to repair equipment when it was needed and also a specialised 24-hour servicing facility outside the hospital to cover periods when technicians were off duty. However, breakdowns still occur and nurses find they must be able to make small repairs to faulty equipment themselves – as one nurse observed: 'I've become more of a technician than a carer.' Other technical problems identified by nurses include the fact that not everyone has the same understanding of the functions and uses of the equipment in a ward and this, in turn, can lead to mistakes and breakages. Machines sometimes break down because of the large number of people handling them. A final technical problem identified was that some monitors are not suitable for particular types of patients, especially in neonatal units.

Staff-Related Problems

In the view of nurses, technology tends also to create a variety of interpersonal problems between staff. The problem most consistently identified here was inexperienced or relief staff not being able to handle equipment because of lack of training. Senior nurses found themselves having to 'train in' new staff and this added to an already heavy workload.

The level of cognitive knowledge about machinery varied between members of staff and this meant that equipment could be misread or misinterpreted. New staff may not be able to handle the equipment because of inexperience or may be reluctant to learn about machines. After all, not all staff perceived themselves to be mechanically minded. Further, it was observed that medical staff tend to rely more on the new technology than on the judgement of nurses and they also tend to expect nurses to be able to fix equipment themselves if things go wrong.

Organisational Problems

Machines are often ordered without consulting the nursing staff who are the principal users. This tends to vary from hospital to hospital. In hospital A, nurses were not consulted about the technology nor was their advice sought when the equipment was installed. Nurses have found this frustrating since many simple problems with technology could be overcome or avoided if use was made of their knowledge and expertise in this area.

Hospitals C and B had a policy of involving senior nursing staff in all decisions about technology and the level of satisfaction with it was more positive in these hospitals.

Further difficulties identified by nurses focused on training. Where the training in new medical technology is inadequate, nurses tend to be blamed for misreading or misinterpreting the machine readings. Other problems include missing breaks because of emergencies and such like - breaks which are necessary unwinding periods for a nurse. Monitoring equipment is perceived as more demanding than caring for the patient and nurses sometimes are fearful because of their lack of knowledge about the intricacies of machines.

Perceptions of New Technology - SD Scale

The data presented in Table 2, below, clearly show that positive perceptions of NMT in intensive-care nursing is closely tied to its new role in facilitating effective task performance. It emerges that a majority of nurses agree that NMT has special advantages - e.g. 'helpful' (86%); 'enables me to do a good job' (73%); 'comfortable' (67%); 'adequate' (64%); 'very good' (61%); and 'not difficult to handle' (58%).

On the other hand, in the view of these nurses, NMT also present difficulties. These cluster around the special demands NMT exerts on them. For example, it demands being 'alert all the time' (69%); it is 'not challenging at all' (73%); 'routine' (48%); 'not relaxing at all' (42%); 'takes my attention away from the patient' 92%); 'stressful' (19%). It is worth observing that nurses also expressed a certain degree of ambivalence regarding the impact of NMT on their role in intensive care nursing - for example, with regard to the degree of 'stress' incurred (50%), and the extent to which NMT 'makes me feel competent' (40%). Throughout the interviews it became evident that the pull between technology and patient is one that the nurse has to resolve consistently throughout her working day.

In all, the overall mean score of 2.64 for the total S-D scale suggests that while NMT makes a positive contribution to ICU/CCU work, it nevertheless generates many complex work activities that are demanding and that require technical skill, alertness, and a constant balance between the cure and the care dimensions of the work.

Table 2: A semantic-differential analysis of nurses' perceptions of new technology – percentages, mean scores and standard deviations (N = 52).

	S-D Item	*Scale Range*			*Mean Score**	*Standard Deviation*
		1/2 (+ve)	3	4/5 (-ve)		
1.	Helpful – unhelpful	86	10	2	1.71	0.72
2.	Enables me to do a good job – prevents me from doing a good job	73	23	10	1.86	0.85
3.	Very good – very bad	61	36	10	2.02	0.86
4.	Comfortable – uncomfortable	67	25	6	2.08	1.01
5.	Adequate – inadequate	64	29	4	2.10	0.98
6.	Efficient – not efficient	63	25	10	2.20	1.60
7.	Not difficult to handle – difficult to handle	58	21	19	2.37	1.11
8.	Reliable – not reliable at all	52	34	10	2.42	1.01
9.	Not rushed – rushed	46	31	19	2.57	1.15
10.	Very important for curing the patient – unimportant for curing the patient	54	15	30	2.59	1.33
11.	Gives me confidence in my job – does not give me confidence	46	33	17	2.60	1.43
12.	Allows time for attention to patient – takes my attention away from patient as a person	44	29	25	2.69	1.26
13.	Not stressful at all – stressful	29	50	19	2.82	0.91
14.	Makes me feel competent – does not contribute to my competence	31	40	27	2.98	1.27
15.	Relaxing – not relaxing at all	31	25	42	3.20	1.24
16.	No routine at all – routine	19	29	48	3.46	1.64
17.	Challenging – not challenging at all	10	15	73	3.90	1.08
18.	Does not require me to be alert all the time – requires me to be alert al the time	12	17	69	3.92	1.21
	Overall mean score index				*2.64*	*0.48*
					Alpha score = 0.74	

*Mean score ranges = 1 (positive) to 5 (negative)

A symbiotic physical relationship between medical technology and patients in intensive care is paramount. Added to this is a psychological and emotional relationship, reflected in beliefs and feelings expressed by patients regarding the significance of the technology for patient well-being and recovery. Technology generates precise quantitative information graphically displayed regarding the status of vital bodily signs including temperature, blood pressure, heart rate etc., which can be viewed as an extension of the physical self. The responses of patients selected from the Mended Hearts Association (N=18) capture this form of sensitive symbiosis between patients in intensive care and medical technology:

- 'My life depended on the machinery to which I was linked' (65% agree);
- 'Staff were reassuring about the equipment', (83% agree);
- 'At no time did I feel that a change in my condition would go unnoticed' (69% agree);
- 'I feel that the machinery was my life support' (47% agree);
- 'I am grateful that this high-technology machinery is available for health care' (93% agree).

The relationship between the patient and technology in the case of intensive care can be conceptualised at two levels. First and most significant is the function of the technology as an extension of the physical vital signs of the patient, while the cognitive and emotional domain of the patient is reflected in his or her attitudes, beliefs and feeling relating to the centrality of the technology in maintaining the patient in a healthy state. The nurse plays a significant role in accurately recording the medical information provided by the technology, in observing the clinical symptoms of the patient and in making valid judgements regarding the patients progress and the nursing care required. The nurse–technology interface reflects a state of mind. Further, the nurse plays a salient role in creating a culture of nurturance and care for the patient which in turn enables the patient to relate to the technology and the treatment received with a positive attitude and expectation. In an attempt to further explore the extent to which medical technology 'attracted' the nurse towards a 'curing' model for nursing with an emphasis on technical efficiency or a 'caring' model which is patient-centred, a set of 'care–cure' questions were included in the research. The overall patterns which emerged when the 'care–cure' scale was correlated with stress and perception of medical technology was of interest. A thirty-item care–cure scale with a care–cure-technology focus was developed to highlight the extent to which nurses exhibited a 'caring' orientation towards their work in intensive care or a more 'curing' orientation. The following care–cure scale items highlight the interface dilemmas between medical technology and the nurses role (care–cure approach):

- I have become a technician because of the new technology rather than a carer.
- The new technology frees me to have more time with patients or for other work.
- I have more time for patients' other needs because of new technology.
- New technology has made my relationships with patients less personal.

The overall results demonstrated a greater orientation towards the more holistic 'caring' approach than on the 'curing' approach. In other words, these nurses did perceive the new medical technology as clearly contributing to the 'caring' dimension of the ICU–CCU work. Further clarification of the meaning of the care–cure scale demonstrated a positive significant correlation with a measure of overall job content ($r = .81$; $P < .001$) (modified Karasek (1984) Scale).

The meaning of this highly significant correlation suggests that where the nurse views her job as high on cognitive skills such as discretion skill, and overall psychological skills such as meaningfulness and value of work in relation to life goals, psychological workload, positive on recognition, she scored high on the care–cure scale, or strong on the 'care' dimension.

The analysis further set out to explore links between the care–cure scale and a stress index. The emergent pattern demonstrated a strong positive correlation ($r = .62$; $P < .001$) between the care–cure scale and the stress index. This suggests that where stress is low, caring can occur in intensive care work, while on the other hand where stress is high, the emphasis appears to be on the curing dimension.

This association presents a very interesting configuration of the impact of stressors on the care–cure dimension. Further, it emerged that where the requirements and demands of technology are experienced as low-stressors then the nurse-caring approach reigns, while on the other hand where technology demands generate high stress then a nurse-curing approach is adopted. In addition, workload stress as well as unit/staff stress correlated highly with the care–cure dimensions.

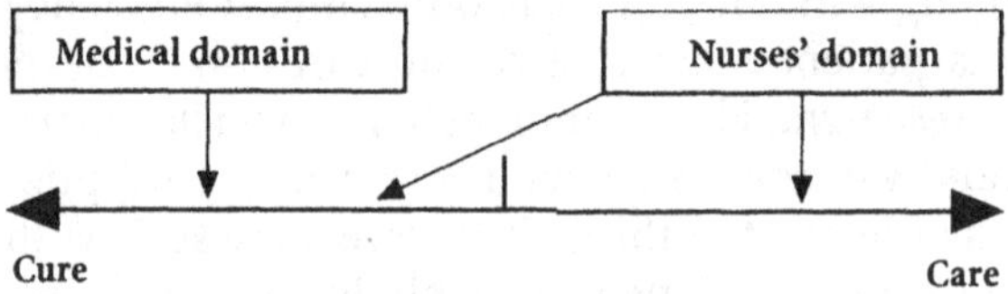

Figure 13: Traditional and holistic care–cure model of nursing in intensive care

Taken together this interesting profile emerging from this correlational matrix clearly indicates that the 'caring' aspect of the nurse's role and identity is sensitive to the stressors that can emerge from workload, and machine/technology requirements. Further, it can be inferred from this pattern that the 'caring' dimension of nursing or a holistic approach, may require a more organic, supportive and caring organisational workplace for a nurse in intensive-care work-settings to effectively realise the 'caring' aspect of her work. Additional explication of these data suggests that rather than viewing care–cure as a bipolar scale (see Figure 13), a two dimensional model (see Figure 14) might have more heuristic value.

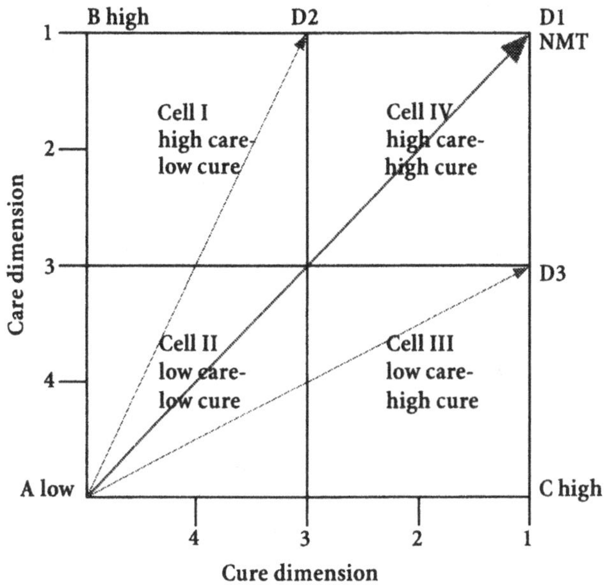

Figure 14: A two-dimensional care–cure model of intensive-care nursing: 1 = high, 4 = low

Thus, rather than viewing 'care' as the opposite of 'cure' (traditional model), I am proposing that both 'care' and 'cure' are two independent dimensions. As shown in Figure 14, nurses' beliefs may be categorised into 4 key beliefs:

- high-care/low-cure (Cell I)
- low-care/low-cure (Cell II)
- low-care/high-cure (Cell III)
- high-care/high-cure (Cell IV)

New medical technology can further impact in an independent and qualitatively different manner on each of the two independent dimensions of the nurses' role – i.e. the 'caring' dimension and the 'curing' dimension.

Cells I and III represent the classical medical model with nurses concerned with 'care' (Cell I) and doctors with 'cure' (Cell III). However given that the nurse is embedded in a hospital setting permeated with a 'cure culture', (Figure 12) the caring dimension was frequently subsumed in the cure ideology. Thus, the traditional nurse drifted to the cure model of nursing (see Figure 14). The two-dimensional model as shown in Figure 14 graphically highlights other options Cell III - could be described as a 'negative' approach with a low concern for either 'caring' or 'curing'. The holistic approach is typified by Cell IV, which reflects a nursing process style which embodied concern for both the caring and curing dimensions of the nurses role. New medical technology (represented by the arrows AD1, AD2 and AD3) introduces a third critical variable. The ideal situation is one where in new medical technology both the caring and curing dimensions of intensive work can be sensitively and effectively integrated.

The care-cure grid provides a conceptual framework for exploring the multi-faceted relationships between advanced medical technology and the cognitive-emotional concerns of competent and successful intensive care work. The research outlined here draws our attention to the mosaic of interactions between technology, patient and nurse in high-technology hospital settings which necessitates a comprehensive conceptualisation of the mind-technology issues that impact on the 'delivery' of effective and satisfactory life nurturing outcomes for both patients and nurses (as well as other medical professionals). The relationships between culture, mind, technology and work in intensive care, can be understood as a symbioses between human (patient)-new technology-human (nurses)-human patient. Figure 12 highlights this relationship which goes beyond a merely human-machine symbiosis to a more complex cycle of human/machine-human/human interactions and attachments as illustrated by C-A; A-B; B-C interfaces, leading to overall system task integration (D).

Concluding Remarks

To conclude, the complex interdependence between the human-person (mind and body) and technology (as illustrated in the relationships represented in Figure 12), yields a truly symbiotic exchange.

It is important to remember however, that the results intrinsic to such interdependence are high (e.g. death of a patient in the case of ICU work), and requires sensitive adaptation to 'the edge of chaos', (see Lewin 1993). That these relationships are fully enmeshed in an overall culture - societal value-laden milieu also underpins the complexity of the symbiotic relationships between culture, mind and technology.

> we draw the boundaries, we shuffle the cards, we make the distinctions
>
> *James Keys* (1972)

References

Argyris, C., Schön, D. A. (1978). *Organisational Learning: A Theory of Action Perspective.* Addison-Wesley, Reading, MA.

Bateson, G. (1972). *Steps to an Ecology of Mind.* Ballintine Books, New York.

Bateson, G. (1979). *Mind and Nature: A Necessary Unity.* Fontana Paperbacks.

Bertalanffy, L., Von (1950). 'The Theory of Open Systems in Physics and Biology', *Science*, 3: 23–29.

Bleecker, S. E. (1994). 'The Virtual Organization', *The Futurist*, March–April, pp. 9–12

Bohm, D. (1980). *Wholeness and the Implicate Order*. Routledge and Kegan Paul, London.

Breuer, R. (1985). 'Das Shaos', *Geo*, 7 (July).

Caplan, R. D., Cobb, S., French, S. R., Harrison, R.V. and Pinneau, S. R. (1980). *Job Demands and Worker Health: Main Effects and Occupational Differences*. Institute for Social Research, Ann Abbor, Michigan.

Carroll, J. S. and Perin, C. (1995). *Organising and Managing for Safe Production: New Frameworks, New Questions, New Actions*. The MIT International Programmes for Enhanced Nuclear Power Plant Safety: Final Report June 1995. Report no. NSP 95-005. MIT Centre for Energy and Environmental Policy Research.

Cooley, M. (1989). *European Competitiveness in the 21st Century*. FAST, EC, June 1989.

Cooley, M. (1995). 'Visions and Problems of the Post-Industrial Society'. Invited lectures in Gill, K. S. (ed.). *New Visions of the Post Industrial Society*. SEAKE Centre, University of Brighton.

Cooper, C. L. (ed.), (1983). *Stress Research: Issues in the eighties*. Wiley, London.

Deutsch, M. (1993). 'Educating for a Peaceful World'. *American Psychologist*, May, pp. 510–517.

French, J. R. P., Caplan, R. D. and Van Harrison, R. (1984). *The mechanisms of job stress and strain*. Wiley, London.

Gorayska, B. and Mey, J.L. (1995). 'Cognitive Technology' in Gill, K.S. (ed.) *New Visions of the post-industrial society, the paradox of technological and human paradigms*. University of Brighton, SEAKE Centre.

Garcia, L. (1990). Personal Communication. US Congress Office of Technology Assessment.

Gill, K.S. (1995). 'Co-Development: Towards a new vision of the Post-Industrial Society' in Gill, K. S. (ed.) *New Visions of the post-industrial society, the paradox of technological and human paradigms*. University of Brighton, SEAKE Centre.

Gleick, J. (1988). *Chaos: Making a New Science*. Sphere Books Ltd, London.

Hare-Mustin, R.T., Marecek, J. (1988). 'The Meaning of difference, Gender Theory, Postmodernism and Psychology', *American Psychologist*, vol. 43, no. 6: 455–464.

Heisenberg, W. (1958). *A Physicist's Conception of Nature*: Hutchinson, London.

Jacobson, S. F. and McGrath, M. (eds), (1983). *Nurses under stress*. Wiley, London.

Jantsch, E. (1980). *The Self-Organising Universe: Scientific and Human implications of the emerging paradigm of evolution*. Pergamon Press, Oxford.

Kant, Immanual, (1781). *Critique of Judgement*. (Trans, 1987 L. La Fleur). Hacket Publishing Co, Cambridge.

Karasek, R. A. (1984). *The Job Questionnaire*. Columbian Heart Project.

Katz, D. and Kahn, R.L. (1978). *The Social Psychology of Organisations*. John Wiley, New York.

Keys, James, (1972). *Only two can play the Game*. Julian Press, New York.

Kimble, J. A. (1984). 'Psychology's Two Cultures', *American Psychologist*, 39: 833–39.

Lincoln, Y.S. and Guba, E.G. (1985). *Naturalistic Inquiry*. Sage Publications, California.

Lewin, R. (1993). 'Order for Free', *New Scientist Supplement*, Feb., pp. 10–11.

Mandlebrot, B. (1977). *The Fractal Geometry of Nature*. Freeman, New York.

Maturana, H. and Varela, F. (1980). *Autopoiesis and Cognition: The realisation of the Living*. Reide, London.

Mead, M. (1955). *Cultural Patterns and Technical Change*.

McCarthy, E. (1988). *Transitions to Equal Opportunities at Work in Ireland: Problems and Possibilities*. Employment Equality Agency, Dublin.

McCarthy, E. (1989a). 'Stress: New Technology in Intensive Care Units – Implications for Training'. *Nursing Review*, vol. 7, nos. 3 & 4: 1–7.

McCarthy, E. (1989b). 'The impact of new technology on health workers in intensive care – a psychosocial analysis', *Journal of Health and Safety*: 3: 21–32.

McCarthy, E. (1993). 'Chaos and organisational Change', *Thornfield Journal*. vol. 16. 25–32. (Special issue 25th Anniversary Edition, The Psychological Journal of University College Dublin, Ireland).

McCarthy, E. (1994). 'Work and Mind: Searching for our Celtic Legacy', *The Irish Journal of Psychology*. 15, 263, 372–389. (Special Issue: A. Halliday & K. Coyle (ed) *The Irish Psyche*. 15.263).

McCarthy, E. (1995). *Changing Contexts: Women and Work in Ireland*. Social and Organisational Psychology Research Centre, University College Dublin.

McCarthy, E. and Tiernan, J. (1987). *The Impact of New Technology on Workers and Patients in the Health Services*: Ireland European Foundation for the Improvement of Living and Working Conditions.

McInnis, N. (1984). 'The Tao of Networking', *The Futurist*, June, pp. 9–10.

McLuhan, M. (1967). *The Medium is the Message*.

Mead, M. (1954). *Cultural Patterns and Technical Change*. The New American Library.

Moriarty, D. (1978), 'Intensive Care Units', *Irish Medical Journal*, 31.

Morgan, G. and Ramirez, R. (1984). 'Action Learning: A Holographic Metaphor for Guiding Social Change', *Human Relations*, 37: 1-28.

Morgan, G. (1986). *Images of organisation*. Sage Publications, London.

Perin, C. (1992). 'British Rail: The Case of the Unrocked Boat', Network Workshops, *Managing Technological Risk in Industrial Society*. Bad Hamburg, Germany, May. Unpub ms.

Pribram, K.H. (1986). 'The Cognitive Revolution and Mind/Brain Issues', *American Psychologist*, 41(5): 507–520.

Prigogine, I. and Stenger, I. (1984). *Order Out of Chaos*. Heinemann, London.

Schein, E.H. (1990). 'Organisational Culture', *American Psychologist*, 45(2): 109–119.

Senge, P. (1992). *The Fifth Discipline: The Art and Practice of the Learning Organisation. Century Press*, London.

Sobell, D. S. (1979). *Ways of Health*. Harcourt Brace, Ivanovich; New York.

Stehle, J. L. (1981). 'Critical Care Nursing Stress: the findings revisited', *Nursing Research* August, vol. 15, no. 4.

Sternberg, R. and Wagner, R.K. (1994). *Mind in Context*. Cambridge University Press, Cambridge.

Taylor, R. (1975). *The anatomy of a malignant technology: medicine out of control*. Sun Books, Melbourne.

Unger, R.K. (1979). 'Towards a Redefinition of Sex and Gender', *American Psychologist*, 34: 1085–94.

Unger, R.K. (1989). 'Sex and Gender and Epistemology' in Crawford, M. and Gentry, M. (eds.) *Gender and Thought*, pp. 17–35. Springer-Verlag, New York.

Unger, R.K. (1990). 'Imperfect Reflections of Reality', in Hare-Mustin R. T. and Marecek J. (eds.) *Making a Difference: Psychology and the Construction of Gender*, Yale University Press, New Haven.

U.S Congress, Office of Technology Assessment (1990). *Critical Connections: Communication for the Future* Washington, DC: US Government Printing Office.

Weick, K.E. (1979). *The Social Psychology of Organising*. Addison-Wesley. Reading, MA.

Welford, A. T. (1976). Skilled Performance in Perceptual and Motor Skills, Glenview III: Scott, Foresman.

Wieland-Burston, B. (1992). *Chaos and order in The World of the Psyche*. Routledge, London.

Wilkinson, S. (ed.), (1986). *Feminist Social Psychology Developing Theory and Practice*. Open University Press, Milton Keynes.

Chapter 5

The Social Construction of Human-centredness[1]

Lars Qvortrup

Introduction

Social Systems as Self-Observing Systems

In his book from 1947, Herbert A. Simon writes that '... administration is not unlike play-acting', i.e. that the effectiveness of the administrative process will vary with the effectiveness with which an organisation's members play their parts (Simon, 1947, p. 252). Twenty years later, Simon raises a significant question: How is it possible to construct a theory of administration – and, more generally, of social systems – if such systems are systems of play-acting? The problem is that if social systems are based on play-acting they are also based on self-observation, i.e., on humans observing themselves playing as themselves. In a theoretical context the problem is that if such systems work through self-observation, this includes self-observation through the theories constructed by, and of, themselves. For Simon, the problem is, insofar as the theory becomes part of its object, the basic precondition for constructing an empirical theory will be obstructed. Any such theory will contain nothing but 'normative rules of good acting, (Simon, 1969, p. x). Simon's observation and conclusion can be confirmed – insofar as many administrative and organisational theories are unreflected catalogues of good practice, i.e. circular arguments of normativism. However, I cannot support Simon's implicit conclusion that the phenomenon of self-observation should be avoided in administrative and organisational theories.

Quite the opposite, self-observation (including self-observation through social and organisational theories) is an indispensable fact of social systems. Consequently, we should not try to *avoid* self-observation and self-reference, but should make self-reference a significant object of and a reflected condition for our theory. One dominating set of normative rules in current social theories is human-centredness. It is difficult to think of modern analyses of social phenomena which do not base themselves on the idea that the human being is, or at least ought to be, the centre of the political system, the organisational system, the computer system, etc.

However, all too often what is actually *meant* by 'human-centredness' is not reflected. It is rather considered an indisputable precondition for, the *ex ante* of social theories. In this way, social theories forget about – and sometimes even deny – their own self-referential structure with the implication that unavoidable self-reference transforms into unreflected dogmatism. Consequently, the aim of this paper is to analyse a particular normative concept: human-centredness or antropocentrism, in order to reinforce its status as a self-observational – and not a dogmatic – concept. In other words I try to support self-reflection as an important part of human-centred systems theory. I do so by reconstructing the social construction of human-centredness in 15th century's Europe.

The Current Status of Human-centred Systems Theory

The roots of the current theory of human-centred systems and technology can be found in 1979, when the first copies of Mike Cooley's *Architect or Bee?* were disseminated through informal networks (cf. Cooley 1987 p. 1). In direct inspiration, in 1984 Karamjit Gill opened his SEAKE Centre in Brighton. Actually, the *concept* was first used in 1987, when the revised edition of *Architect or Bee?* was published, coining and defining the concept of human-centredness (cf. Gill, 1990, p. 6), with practical reference to the English LUCAS plan and to the European ESPRIT project *Human Centred CIM System*, and the same year the first volume of the journal *AI & Society* with the significant subtitle *The Journal of Human-Centred Systems and Machine Intelligence* was available. In 1988 *The Human Centred Systems Book Series* was launched in Japan. And in 1992 the *International Research Institute in Human Centred Systems* was inaugurated. Evidently, the efforts toward developing the concept of human-centredness had both practical and theoretical aims. Human-centredness provides a practical alternative to the Tayloristic approach to production, as well as a theoretical alternative to the Newtonian–Cartesian mechanistic paradigm (cf. Gill, 1990, p. 4 and p. 11). It is evident that the pioneering document concerning human-centred technology is Mike Cooley's book. However, when reading the book it is interesting to note that much is said about the concrete social conditions of human-centredness, i.e. the socio-political process leading to the articulation of a human-centred perspective in production and use of technology, while relatively little is said about the semantic contents of human-centredness. This constitutes the first issue of this paper: what does 'human-centredness' actually mean?

My second issue concerns the social and cultural context of human-centredness. Obviously, it is not a universal concept, and its values do not have universal or global significance. On the other hand, the idea of installing the human[2] as the centre of the world is not a 20th century phenomenon. For me, it is important for current human-centred programmes and 'movements' to know that they are part of a long and influential tradition. Thus, I will trace the historical roots of human-centredness.

My third issue concerns the ethical characteristics of the concept; or, as one may say, the 'ethical range' of the concept. Does 'human-centredness' primarily constitute a perspective, i.e. an attitude, or is it a well-defined moral concept from which world-views and directives for good practice can be derived?

On the one hand, Mike Cooley seems to support the view that human-centredness is a perspective. He abstains from giving a detailed definition, emphasising that human-centredness should develop out of concrete experience and practice, rather than from academic definitions. On the other hand, quite far-reaching human-centred principles and characteristics can be found in his book. Here, there are at least four dimensions of human-centredness: a technological, an economic, an organisational, and an ecological dimension. According to Cooley, human-centredness is opposed to Taylorism (cf. Cooley, 1987, p. 147) and de-skilling. It reflects such values as 'motivation, self-activism, dignity of the individual, concern for quality etc.' (*ibid.*). It also opposes an economy based on money and profits: 'The product and processes should be regarded as important more in respect of their use value than their exchange value.' (Cooley 1987 p. 155). This is the economic dimension of human-centredness. It has an organisational dimension, representing participation as an alternative to delegation. And, finally, it has an ecological dimension, supporting sustainability.

I will not discuss the specific relevance of these attitudes. In the present context I am only concerned with the question, whether they can be derived from human-centredness as their 'mother' concept. Let me briefly refer to a couple of examples. Literally speaking, 'human-centredness' means that the world is centred around the human. How does this imply an ecological dimension of sustainability? Has not nature been destroyed as a result of our efforts to satisfy human needs? Don't we have a latent conflict between 'human-centrism' and 'eco-centrism'? Furthermore, why are certain principles of social interaction better – in the sense of 'more human-centred' – than others? Of course, one can discuss the pros and cons of different ways to coordinate complex economic systems, but I am not convinced that human-centredness in itself is a valid argument. Thus, in addition to the semantics and the history of human-centredness I am concerned with the normative potentiality of the concept. Is it possible – and appropriate – to use human-centredness as the keyword of a moral code, from which rules and principles can be derived? Here, for me it is important to specify a difference between human-centredness as a *normative* code, i.e. as a perspective, vs. human-centredness as a *dogmatic* code, i.e. as a moral code with a specific and indisputable set of behavioural implications.

I will discuss these three issues in an, apparently different, historical context, i.e. the 15th-century European renaissance. I do so, firstly because it is a well-known experience that we often see our own problems more clearly if we use a picture of somebody or something else as our mirror. Secondly, I do not feel that the gap between our century and the 15th century is so wide that it cannot be bridged. On the contrary, the relationship is so close that although it *seems* that during the latest 15 years we have experienced the birth of a 'new technological tradition' (cf. Gill, 1987, p. 4), in my opinion we have rather experienced a *re*-birth, a *renaissance* of human-centredness, consciously or unconsciously building on very important traditions in European cultural history.

Consequently, in this paper I will look at the historical and cultural roots of human-centredness, going back to the very beginning of this kind of thinking in the 15th-century's renaissance. Particularly, it is interesting to analyse the social and cultural context in which the human was moved into the centre of the world. It is my hypothesis that human-centredness – the idea that the world is centred around the human – is a social construction, and that the values attached to human-centredness are the result of social and cultural processes, i.e. that they are social values, rather than natural (i.e. universal) values. This means that the current human-centred 'movement' is oriented towards culturally constructed values; this is important to keep in mind in order to avoid that kind of fundamentalist intolerance which is a result of the belief that one represents natural values. Human-centredness is – and could not avoid to be – normative, but it should avoid to be dogmatic.[3] My approach also emphasises that the current human-centred 'movement' is concerned with issues and problems rooted in centuries of cultural discussions and elaborations, and that the contemporary human-centred movement can, and should, learn from tradition. Thus, in order to identify our historical roots, rather than focusing on the renaissance of human-centredness, my paper examines the human-centredness of the renaissance.

The History of Human-centredness

As mentioned above, in our century the concept of human-centredness was coined by Mike Cooley. However, in order to understand what we mean by 'human-centredness' we have to be familiar with its history. It is my hypothesis that human-centredness is not a concept invented in this century, but is based on the tradition of modern European philosophy rooted in the culture of 15th century Europe.

The Construction of Human-centredness in Renaissance Painting

To illustrate the historical construction of human-centredness I will refer to a number of paintings from the 14th and 15th century.[4] To eliminate other sources of influence I have chosen paintings of the same motive, i.e. the last supper, most of them using the same technique, the fresco painting. The order of presentation is not chronological, but represents the order of ideological development.

1. From Denmark, on Zealand, in the church of Mørkøv, painted around 1450.
2. From Denmark, on Funen, close to Odense, in the church of Bellinge, painted in 1496 by Ebbe Olsen and Simon Petersen.[5]
3. From Crete, Church of Our Lady in Kritsa, from the 14th Century, painted by an anonymous artist.
4. From Italy, the Arena Chapel, Padua, year 1305, painted by Giotto del Bondone.
5. From Italy, in the Cenacolo di Sant'Apollonia in Florence, 1445–50, painted by Andrea del Castagno.
6. The famous fresco painted by Leonardo da Vinci in the refectory of Santa Maria delle Grazie in Milan during the years 1495–98.
7. And finally, a painting from 1592–94 by Jacopo Tintoretto in San Giorgio Maggiore in Venice.

Although depicting the same motive, these seven pictures are so stylistically different that it is not exaggerated to say that they represent totally different cultural eras. While the first three pictures are clearly pre-modern, the last three pictures are from 'our age', number 4 being 'in-between'. The significant difference is the difference of perspective. While the pictures by Andrea, Leonardo and Tintoretto are painted according to the rules of the linear perspective, the frescos from Denmark and Crete are, as we would say, 'out of perspective'.

Even because it is out of question that there should have been any contacts between the Danish and Cretan artists, the striking similarities between these three frescos are all the more interesting. In all three pictures the disciples are seen *en face* and with no spatial distribution, the table is depicted both from the side and from the top, and the objects on the table are seen from their most characteristic angle: plates and knives from the top, bottles and glasses in profile. There is no specific point of observation, the scenery is seen from everywhere, i.e. by an observer who can see everything (regarding the analysis of the Danish fresco cf. Johansen, 1979, p. 52).

In comparison, the painting of Giotto represents an interesting intermediary phase between pre-modern and modern painting (cf. John White pp. 57–71). The human figures are clearly distributed in space, and there are signs of linear perspective in the seat and the building. However, it is quite evident that Giotto does not fully master the rules of perspective: the flatness of the painted architecture is quite eye-catching, and he has difficulties in managing the nimbuses which look like solid plates sometimes placed in the faces of the disciples. It is obvious, that at the time of Giotto there was still '... no fixed theory of perspective construction. There were no set rules to be obeyed or broken' (*ibid.* p. 60).

Turning to Andrea del Castagno's picture, 'at first sight', as says John White, it seems to be a perfect example of linear perspective. However, a closer inspection unveils some significant problems (cf. White, p. 198): there are six square marble panels on the back-wall, and they seem to be repeated (i.e. as squares) on either side-wall. Consequently, the whole room should be based on a perfect, foreshortened square. However, in the decorative frieze under the ceiling there are thirty-four loops on the back-wall, while there are only seventeen loops in depth. Also, the roof does not seem to fit with the rest of the building.

Finally, the table is painted in a very simple way, in silhouette, avoiding the complications of the plates and glasses, and the disciples appear as rough and motionless figures (cf. Leinz, 1979, p. 76). In particular, instead of separating Judas from the other disciples by his psychological characteristics, Andrea has chosen the simpler solution to place him at the opposite side of the table. Of course, these violations of the rules of the perspective *might* be made on purpose, but it is difficult to find a convincing artistic reason. Therefore, this fresco might also be taken as an example of a renaissance artist being on his way towards the full mastery of linear perspective.

Leonardo's fresco of the Last Supper[6] in the refectory of Santa Maria delle Grazie in Milan is one of the most famous renaissance paintings, and here the illusion of perspective seems to be perfect. The linear perspective of the room is supported by the atmospheric perspective of the landscape, and the painting is convincingly integrated into the refectory, almost as a *trompe l'œil.* Moreover, the perspective is used to support the 'message' of the fresco: the vanishing-point being exactly behind the face of Christ. Finally, the figures are psychologically characterised as individuals, and in order to achieve the highest level of realism, the nimbuses have been avoided – or have been replaced by the light and the vault over the head of Jesus. Still, not everything is perfect. Most importantly, there is a contradiction between on the one hand the continuity between the real architecture (the refectory) and the virtual architecture of the fresco, and on the other hand the 'correct' point of observation being 4.6m above today's floor (cf. Kubovy, p. 141; Kemp, p. 49).

Some arts historians have said that this – and other ambiguities – is a result of the technically insoluble conflict between the existence of one and only one correct viewpoint and the fact that real viewers move around in the room. As a result, Leonardo:

> ... has done his ambiguous best to preserve the illusion, but even he must admit ultimate defeat.
>
> *Kemp, p. 49*

However, I would rather support Kubovy's analysis that the inconsistency between the *trompe l'œil* effect and the correct viewpoint creates:

> ... a 'difficult' work of art, one that forces you to engage in mental work to overcome the obstacles ...The result is a vibrantly tense work full of foreboding. Leonardo used perspective to elevate the viewer to an extraordinarily high center of projection, thus achieving a feeling of spiritual elevation.
>
> *Kubovy, pp. 145 and 148*

My final example is painted almost 100 years later (1592–94) by Jacopo Tintoretto, perhaps the greatest of the so-called Mannerist painters. His 'Last Supper' is one of his last pictures. While going from Castagno to Leonardo the observer was, so to speak, 'elevated' in order to face Jesus and the disciples – with the effect of feeling subordinate, considering the actual point of observation. In Tintoretto's painting the implicit position of the observer has totally changed; now he is clearly superior. Moreover, the observing 'camera' has turned to the side of the motive, and an oil lamp throws a dramatic light from the back of the scenery, shrouding the scenery in smoke from which angels seem to materialise. Evidently, there is no 'background' in the picture, but only a:

> ... tunnel of darkness without end ... Space here is not a measurement, not an artefact of perspective, but a kind of infinite music ... With Leonardo, Christ was still center; here He is caught up in the force which sets the world spinning ...
>
> *Wolf and Millen, p. 114f*

Stylistically speaking, this painting reminds one of the contemporary painting of the Last Supper by Federico Barocci (1535–1612) in the cathedral in Urbino with its complex scenery, materialising angels and infinite numbers of persons (a copy of the painting can be found in Hauser, illustration no 188). However, while the latter painting has evident signs pointing toward a baroque idyll with Christ in the centre, looking with sentimental confidence toward the heaven, the painting of Tintoretto is chaotic, Christ being neither the sacred figure looking 'everywhere' as in Leonardo's fresco, nor the sentimental person looking toward the heaven, but a very concrete human, concentrating on serving his bread and wine. Thus, I can fully support Hauser in his conclusion:

> Tintoretto 'fühlte sich aber von der Welt, mit der sich Tizian so leicht versöhnte, noch immer entfremdet, und fand auch im Glauben noch nicht die heitere Ruhe, die der Barock vom Geiste der sieghaften Gegenreformation auf die Kunst übertrug'.
>
> *Hauser, p. 213*

In Tintoretto's painting the artist is so fascinated by his own skills of observation that he almost forgets about the religious scene that he is painting, or he is so influenced by the development initiated by the early renaissance that he cannot return to a harmonious religious faith in God. The latent message in the fresco of Leonardo, that anthropocentrism and deocentrism cannot be reconciled, has been fully realised by Tintoretto.

The Ethical Dilemma of Human-centredness

As a consequence of human-centredness the human being is put into the centre of the world. This is reflected by renaissance painting in the sense that the world is seen from the single-human's perspective. However, in entering the centre of the world, the human being replaces the former central figure, i.e. God. Thus, human-centredness is closely related to secularisation, and thus to making the human being divine. Thus, an unavoidable ethical dilemma of human-centredness emerges which is of significance also to contemporary human-centred systems theory. Again, the problem is best illustrated within the field of renaissance painting which I have chosen as my primary example.[7] For many years, the construction of the linear perspective has been interpreted as the discovery of an already existing reality. The linear perspective is a *fact of realities* which has just waited for being discovered by humans.

Particularly, the linear perspective was then seen as an expression of progress due to the accomplishments of science and technology. As an example, Piaget saw the linear perspective as an historical achievement which again is repeated in each individual's personal development. The linear perspective is an expression of maturity, and consequently one can say that the world matured during the 15th century.

However, recently an alternative approach has been proposed. In a paper called 'Visual perception in people and machines' in a book about artificial intelligence and the eye, Ramachandran sees the linear perspective as one out of many potential ways to organise or construct reality, that is as one *trick* among a number of visual tricks (Ramachandran, 1990). Here, the linear perspective does not represent abstract progress, but is a deliberate *choice*, one way to see the world which has been chosen for particular reasons.

If the latter approach is true, the question is of course, *why* this structuring principle was chosen. What is the *meaning* of the linear perspective? Or, to put it differently, what is the significant difference between the first two and the last three paintings presented above?

The basic difference concerns the point of view. Every picture is communication and communication about communication (cf. Watzlawick, Beavin and Jackson, 1967). It says *what* it is, and it says *how* it is what it is. Every utterance is also the utterance of somebody, every sign is the sign of somebody. In the case of the paintings, its aesthetic organisation implicitly says whose painting it is.

Whose signs are the early paintings of the last supper? These paintings are the signs of God – or, more specifically, they are the signs of a *non-localisable* observer. As it has once been said, a straight line is part of an infinitely great circle. In such a circle the centre is non-localisable. Similarly, in the pre-modern paintings, all figures have the same size, irrespectively of their position, and the objects are not structured according to a specific organising principle. These paintings have no localisable centre of observation. The non-localisable observer is: God. He or she for whom everything is observable, and for whom everything has the same size and the same importance.

The alternative to the non-localisable observer is of course the individual observer, the individual painter. And, as an observer of the picture, one should place oneself in the position of the painter – in what the arts theory of linear perspective calls the 'correct point of observation', which can be reconstructed by analysing the painting in question. This then represents the ideological revolution of the 15th century: *the construction of human-centredness.*

This is, of course, a revolution not only in the art of painting. On the contrary, human-centredness is a basic ideological issue of our modern society. Niccolo Machiavelli (1469–1527) placed the human in the centre of *politics.* Politics was made by and for the person in power, not by or on behalf of an external God. *Science* developed into the construction of truth according to humans' observation and reason, not according to God's creation of the world. In *the judicial system* the construction of justice according to human-centred legal rules, not according to God's will, became the fundamental concern.[8]

There are innumerable implications of this ideological revolution. One particularly important, unintended implication is that the artist puts himself in the place of God. He or she becomes divine.[9]

For me, this is particularly evident in the above-mentioned Last Supper of Leonardo da Vinci. As already demonstrated, in this fresco one has a perfect example of linear perspective, and of people characterised as psychologically-differentiated individuals. All the technical skills which were achieved during this period are represented by this painting. But there is an even deeper dimension in Leonardo's Last Supper. His painting depicts the very moment in which Jesus stands up and says: 'One of you shall betray me.' We all know that the betrayer is Judas. The tradition has it that he must therefore be given a special mark or indication, and one way of doing this is to place him at the opposite side of the table, as did not only the pre-modern frescos, but also still the picture painted by Andreas del Castagno.

In the first drafts of his Last Supper, Leonardo chose the same solution but obviously this solution did not satisfy him – cf. two studies for the Last Supper to be found in Venice, Academy, in Santi p. 46)[10] and at Windsor, Royal Library (in Santi p. 48). Consequently, he 'integrated' Judas into the group of disciples, only indicating who is Judas by the shadow in his face, caused by his bending on his elbow over the table. Thus, he is both *in* the group and *outside* the group. This is important, because Judas is not a *particular* person, but could be *everybody.*

This is the real drama in Leonardo's painting, that potentially everybody is Judas, the betrayer of Jesus. This even includes Leonardo himself as the painter and the audience as observers, i.e. as modern humans. The implicit indication of the fresco is that having replaced God in his position as the universal observer, and having changed the way the world appears accordingly, *the modern human is the betrayer.* Therefore, Jesus does not address himself to any specific person, but to everybody, including the artist and the observer. To me, this is what differentiates this particular painting from all the others depicting the Last Supper: it does not just show an important moment in the life of Jesus, benefiting from all the technical achievements of renaissance arts, but in doing so it reflects a basic dilemma of human-centredness, that the cost of making the human being the centre of the world is that he or she replaces God.

Actually, this radical change of the position of the artist which can be identified as the latent contents of Leonardo's fresco, and which later on developed into a clear change in the artist's societal position, was indirectly articulated in Leone Battista Alberti's classical book on painting, *De pictura,* from 1435 by two revealing metaphors: 'Painting possesses a truly divine power', he writes in §25, followed by §26: 'The virtues of painting ... are that its masters see their works admired and feel themselves to be almost like the Creator' (Alberti, 1972, p. 61). Here we see the first signs of the artist as the celebrated centre of the modern world, the institutionalised divinity.

As a consequence of this change in the arts of painting, we cannot anymore talk about an 'objective' world, or a 'neutral' point of view; on the contrary, we always have to choose among different perspectives. If this is forgotten, a new dogmatism is constituted, but this time not in the name of God, but in the name of one or the other human-centredness. Again, this is a general problem in the modern era: the politician becomes divine; the artist becomes a God; and with the best of intentions, the architect, the engineer or the programmer, creating human-centred systems, becomes he or she who knows best. Thus, there is a fundamental inherent ethical dilemma in the principle of human-centredness.

From now on the piece of art is not given by God, but is a question of choice – a choice which must be articulated by the piece of art itself, such as Leonardo does so radically. Also, as an aesthetic choice art must reflect itself. When it doesn't represent God, it must speak on behalf of itself, and in speaking in this way it must speak within a specific code. Consequently, codes such as 'beauty', 'harmony', 'sublimity' – i.e. codes which only refer to and legitimise themselves – enter the scene of arts.[11] In the publications of the most influential contemporary arts theorist, Leone Battista Alberti, this is explicitly discussed. He replaces the former principle of 'verisimilitude' – the painting must resemble its external object as much as possible – by the principle of 'convenienza' or 'concinnitas', which is an internal principle, i.e. a principle of internal correspondences within the single piece of art. The painter:

> ... must take pains, above all, that all the parts agree with each other; and they will do so if in quantity, in function, in kind, in color, and in all other respects they harmonize (corresponderanno) into one beauty.
>
> *Alberti, op. cit.*[12]

The Role of Science and Technology in Human-centredness

Developments of Contemporary Science and Technology

We have already seen that the modern artist ousts God as the undisputed observer, and that the construction of the world changes accordingly. However, this trend is not only found in the artistic world, but is equally evident in science and technology.

In my opinion three parallel tendencies can be observed. First, there is an increasing focus on *how* humans observe. The study of the way in which we see is called *optics*. In the long period which Ulrich Beck calls 'simple modernity' (cf. Ulrich Beck, 1992 [1986]), science seems to have forgotten that this is what optics is about. Instead, scientists have thought that optics alone deals with improving the way – in self-confidential singular – in which we see. Second, there is a growing interest in the way in which the world is organised according to human-centred observations. One important example of such studies – the study of the organisation of the universe such as we see it – is called *astronomy*. Again, it may be noticed that modern science has since been hit by collective amnesia. Scientists have thought that astronomy is about the way the universe *is* – and not that it is about the way the universe appears within a specific perspective, i.e. the perspective of modern, anthropocentric human beings. Finally, our ability to observe is being improved by developments of observational techniques. This again reinforces the interest in optics and astronomy.

In order to summarise this context we have to go back in history to the Greek astronomer, Ptolemy. According to the English science historian James Gingerich, Ptolemy's the *Algamest* (the Arabic name of 'The Greatest') which was written in Alexandria around 150 AD *is* the greatest surviving astronomical work from antiquity. What Ptolemy has done is to show for the first time in history (as far as we know) how to convert specific observational data into numerical parameters for planetary models. (cf. Gingerich, p. 4) This was revolutionary because:

> ... more than any other book, it demonstrated that natural phenomena, complex in their appearance, could be described by relatively simple underlying regularities in a mathematical fashion that allowed for specific quantitative predictions.
>
> *Gingerich, p. 5, cf. also p. 74*

As a matter of fact, Ptolemy almost accomplished the scientific revolution of replacing the traditional geocentric cosmology by the modern heliocentric cosmology which was not achieved until 1400 years later, when Copernicus in 1543 published his *De revolutionibus*. Ptolemy realised that according to his observations it would be simpler to have the Earth spinning daily on its axis than the entire heavens rotating about the Earth, but he gave up this idea as it would contradict the Aristotelian terrestrial physics: If the Earth moved:

> ... animals and other weights would be left hanging in the air, and the Earth would very quickly fall out of the heavens. Merely to conceive such things makes them appear ridiculous.
>
> *Gingerich, p. 5*

However, Ptolemy's most important contribution is not to be found in his results, but in his methodological innovation, i.e. in the belief that the human through his/her observations and reasoning can reduce the complexity of the external world and that the proper name of the resulting construction is 'scientific truth'. The latter is evidently a basic source of the development of human knowledge ever since.

This also meant that observational theories (theories based on systematic observations) and theories of observation (theories about the way humans observe the world) were two equally important and mutually interrelated scientific activities. Not only was is evident that astronomy was based on observations, but it was equally clear that the scientific results were derived not only from an 'objective external world', but also from the way in which humans observe. Astronomy always reflects both the movements of the observed objects and the conditions and movements of the observer: the position and movements of the earth, the atmospheric reasons for the apparent size, form and colour of the observed objects, etc. Consequently, theories on human vision – i.e. internal and external conditions for the way in which human beings observe (optics) – were complementary to theories on the external world (astronomy), and thus, quite logically, the second known major work of Ptolemy was his *Optics* which is a treatise (from approximately 175 AD) about human vision. Although the original Greek version was lost, the book survived in an Arabic translation, and in 1158–1160 this Arabic version was translated into Latin by admiral Eugenius of Sicily (cf. Albert Lejeune's 'Introduction historique' to Ptolemaeus p. 132).

According to Albert Lejeune, Ptolemy's *Algamest* was translated from Greek into Latin at the same time, i.e. around 1160 (cf. his 'Introduction historique' to Ptolemaeus p. 9), but the full text was not available in print until 1515. Consequently, there is no direct connection between Ptolemy and, e.g., Copernicus in the sense that one book causes the next one. Rather, Copernicus was influenced by his stay in Northern Italy in the beginning of the 16th century during which period the world view was undergoing major changes, although from this period it took approximately 30 years before Copernicus, in 1543, published his *De revolutionibus*, presenting for the first time a heliocentric cosmology,[13] thus drawing the conclusions which already were latent in the work of Ptolemy (Gingerich, 1993 p. 145). In chapter 10 of his *De revolutionibus* Copernicus proclaims: 'In the center of all rests the Sun... as if on a kingly throne, governing the family of stars that wheel around.' (Quoted from Gingerich p. 34; see a similar quotation, Gingerich p. 180).

Thus, according to recent science history the connection between Ptolemy and 15th-century scientific revolution is only indirect, one important link being Arab astronomy. The Ptolemaic–Aristotelian cosmology clearly fascinated Islamic astronomers around 1000 AD, and they adopted it as their world view. However, criticism of Ptolemy emerged in the 11th century in the work of the most influential Arab astronomer, Ibn al-Haytham (*c.* 965–1040). In his detailed commentary on Ptolemy's *Algamest* (cf. A. I. Sabra's introduction to *Ibn al-Haytham*, p. xxxiv), among other things he discussed the optical problems of astronomy. A more extreme criticism was launched in the following century by the Cordoban astronomer and philosopher, Averros (1126–1198), who wrote:

> The astronomical science of our days surely offers nothing from which one can derive an existing reality. The model that has been developed in the times in which we live accords with the computations, not with existence.[14]
>
> *Gingerich, 1993, p. 140*

The Scientific and Technological Influence on Culture

Following upon his critical commentary on Ptolemy's *Algamest* within the period 1028–38[15] Ibn al-Haytham finished his major work, the five volumes *Optics* or *Kitab al-Manazir*, which was rendered from the Arabic into Latin in the late 12th or in the early 13th century (*ibid.* p. lxxiv) under the (in our present context) significant title *Perspectiva* or *De aspectibus*. Still, at that time, 'perspective' was a phenomenon closely related to optics, i.e. to the way we observe, it was not, as is commonly believed today, a phenomenon in the external, 'objective' world. Perspective was a matter of observation, rather than a matter of reality.[16]

Ibn al-Haytham opens his book by presenting two apparently contradictory approaches. Looking for a theory on optics which can make his empirical astronomic observations calculable, he is handicapped by the fact that the very nature of vision is disputed: does vision come from the outside, or does it come from the eye? When looking at an apple, does the image of – or, in order to better understand the dilemma, the concept of – the apple come from the external object into the eye, or does it move from the eye out onto the external fruit on the tree? Ibn al-Haytham's contemporary natural scientists claim that '... vision is effected by a form which comes from the visible object to the eye ...' (Ibn al-Haytham chapter 1, section 3, p. 4), a theory which is in harmony with Ibn's own approach, but which does not provide him with tools for calculation. Mathematicians, for their part:

> ... agree that vision is effected by a ray which issues from the eye to the visible object and by means of which sight perceives the object; (and) that this ray extends in straight lines whose extremities meet at the centre of the eye ...
>
> *Ibid.*

From this approach Ibn can make calculations of straightlined pyramids etc., but as a theory of vision it contradicts his own ideas. Thus, the problem is that on the one hand an optical theory can only be based on the idea of rays extending in straight lines, but on the other hand Ibn al-Haytham cannot accept the idea of rays of vision going out of the eye. After a long discussion and an empirical inspection of the eye, he concludes that the two approaches - the centripetal theory of the natural scientists and the straight lines - but centrifugal - theory of the mathematicians - *can* be combined:

> ... the eye does not perceive any of the forms reaching it expect through the straight lines which are imagined to extend between the visible object and the centre of the eye ...
>
> *Ibid., chapter 6, section 60, p. 82*[17]

Although his book was not printed until 1572, quite a number of copies circulated in Europe, and according to Sabra, Ibn al-Haytham - and not Ptolemy - became the principal source of information for the optical writings produced in the 1260s and 1270s by Roger Bacon, John Pecham, and Witelo (cf. *ibid.* p. lxxiii). It is thus evident that the 'revolution' of the renaissance was not a sudden incident, as is often believed, but that there was a long line of development during the so-called 'dark' middle ages. Also, the sources of influence were not only Greek, but the Arab world played an important role.

As I have argued, the theory of the way in which we observe - the theory of optics - became a significant scientific and philosophical subject, but so did the development of tools for increasing the potentiality of observation and of reasoning. In order to develop a scientifically-based astronomy, the questions on how to increase one's ability to see, and how to improve one's ability to reduce the complexity of what one has observed, became major concerns.

A famous example of how to operationalise the theories of optics and of light into practical tools is provided by the achievements of the Italian renaissance architect Brunelleschi. Not only did he use the optical theories in order to improve his architectural design instructions, but according to his contemporary biographer, Antonio Manetti, between 1401 and 1409 (others have set the date to 1413, or to 1425, cf. Kubovy, 1986, p. 32) he painted two panels of buildings in Florence in accordance with the principles of vision as the product of straight lines of light, forming visual pyramids with the top in the middle of the eye, and furthermore he arranged the panels to be viewed through a small hole, i.e. in a peep-show, creating a perfect illusion of a three-dimensional construction.

In 1435, according to most art historians, the painter, architect and scholar Leone Battista Alberti transformed Brunelleschi's experiments[18] into a practical guide for painters.[19] In Alberti's instruction *On Painting* the reader is provided with 'a full-grown, mathematically based theory of vision applied to artistic representation', a theory: '... which is founded on a simplified theory of vision, in which rays of light travelling in straight lines convey information from surfaces of objects to the eye.' (Cecil Grayson's introduction to Alberti , 1972, p. 11).

It has been disputed whether the idea of the linear perspective developed directly from optical science through applied architecture to painting,[20] and for instance Masaccio's famous *La Trinita* was created independently of Alberti's book as it was painted 10 years before the publication of the book. Consequently, the use of linear perspective in painting was rather a question of broad scientific and cultural trends combining the principles of optics with the belief that the human is a rational observer, who is able to report what he sees, and who is confident that what he sees is 'the truth'.[21] All these trends are combined in the famous words of Albrecht Dürer, saying that:

> 'Perspective' is a Latin word, meaning 'looking through'.
>
> *Elkins, 1992 p. 209*

They were operationalised in constructions of countless numbers of projection methods, stereometric diagrams, measurement techniques, perspective machines, observation nets, etc. (cf. Kemp, 1990).

The Cultural Influence on Science

As we have seen, science and technology influenced the development of culture and thus the cultural articulation of the principle of human-centredness. Consequently, it is misleading to think that human-centredness exists *ex ante*, before technology, as an indisputable universal fact which just has to be realised by scientists and technological designers. Antropocentrism is a social construction, and science and technology have contributed to this construction. However, the relationship between science and culture goes both ways. As we have seen, there is a long and windy, but still a clear line from Ptolemy to renaissance painting, and a similar connection between Ptolemy and renaissance science – cf. the astronomical achievements of Copernicus. However, this line of development cannot explain why Copernicus chose a heliocentric model and there is apparently no reason to believe that the scientific world, through increasingly accurate observations, came closer an closer to an 'objective truth' hidden in 'the external world'. In this respect Thomas B. Kuhn is correct in saying:

> ... Copernicus's radical cosmology came forth not from new observations but from insight. It was, like Einstein's revolution four centuries later, motivated by the passionate search for symmetries and an aesthetic structure of the universe. Only afterward the fact, and even the crisis, are marshalled in support of the new world view.
>
> *Thomas B. Kuhn, 1962, pp. 67–68*

Although this explanation is correct, it is however not sufficient. Although we can only guess, according to Gingerich, the reason for Copernicus proposing a heliocentric model was what we today would call irrational. The reason for choosing the heliocentric model seems to be that Copernicus's observations of the movements of the planets contradicted his belief in the existence of solid-crystal planet spheres or spherical shells. If Copernicus did not make his revolutionary rearrangement, he would have had to change the theory of crystal planet-spheres, simply because according to his observations one sphere would smash another. As we know, the Danish astronomer Tycho Brahe chose a geocentric (or rather a geo-heliocentric) model. Today, it seems obvious that Brahe was the conservative astronomer, while Copernicus was the revolutionary scientist, but actually there was no clear difference in radicalism. Copernicus maintained the idea of solid spheres, while he radically accepted a heliocentric model. Brahe maintained the geocentric model, radically giving up the idea of crystal spheres. The problem was only solved with Isaac Newton's gravitational theory.[22]

Why then did Copernicus choose his heliocentric model? I think that Gingerich is correct in saying that the introduction of a revolutionary new astronomical paradigm in the 16th century was heavily influenced by Copernicus's socio-cultural context, rather than being an answer to a scientific crisis within its own narrow scientific field, as claimed by Thomas B. Kuhn (Gingerich, p. 200).[23] Although being born and later on returning to Poland, Copernicus achieved his scientific education in the intellectual world of North Italy, dominated by high-renaissance culture, where there was a burning intention of seeing things from a human-centred perspective. Actually, to simplify the situation, the observer was more important than the observed, the 'cogito' more important than the 'sum', the organisation of the world in accordance with a single-human's rationally-based sense of the world more important than the world 'an sich', innovation more attractive than tradition.

As mentioned above, tools for observation (peep-holes, windows for observation, mathematical theories of linear perspective) played an increasing role in painting, being tools that strengthened the position of the observing artist at the expense of the status of the observed object. As an example – as we saw in the first section – the main figure in Leonardo's

Last Supper is the observer, and the total aesthetic arrangement of the fresco is made not only from the viewpoint of the observer, but also so as to challenge the observer, i.e. to see things differently. Actually, the inherent intention of this fresco (and of many contemporary pictures) is for the observer to observe himself *through* the painting. This unlimited belief in the ability of the single-human observer to construct the world in accordance with his senses and his intellect, even at the cost of a traditional world view, was the cultural environment in which Copernicus was educated, and, I would assume, a basic reason for his scientific achievements.[24]

The Normative Status of Human- centredness

Design as Codification of Human-centredness

During the 15th century, the human moves into the centre of the world. This is the result, not of a single person or a single field of study or practice, but of a broad cultural change including arts, science, philosophy, politics, etc. Even the field most closely related to what we today define as systems design, i.e. architecture, plays a role in the development. Actually, the most ambitious publication by Alberti, best known for his books on painting and sculpture, is his ten books on architecture, *De re aedificatoria,* which was presented - probably in an early and rudimentary version - to Pope Nicholas V in 1452, but which was not published until 1485, 13 years after his death. It was translated into many languages, including an English version which was published in 1726, and reprinted in 1739 and 1755.[25] In the ten books[26] a great number of aspects concerning architecture are analysed, both regarding design, which, according to Alberti, is 'produced by the thought' and thus can be directly formed by the architect's plans, and regarding matter, which, although being produced 'by nature' has to be carefully prepared and chosen by the architect (cf. Alberti, 1965, p. xi). Consequently, the first book is concerned with design, while the second is devoted to material. However, everything in the ten books is based on the fundamental principle of human-centredness, i.e. that design must be derived from human needs:

> Him I call an architect, who, by sure and wonderful art and method, is able, both with thought and invention, to devise, and, with execution, to compleat all those works, which, by means of the movement of great weights, and the conjunction and amassment of bodies, can, with the greatest beauty, be adapted to the uses of mankind.
>
> *Alberti, 1965, p. ix*

Here, it is clearly illustrated how human-centredness is being developed from a philosophical idea, a scientific and technological principle, and an aesthetic ideal, into a concrete and practical principle of design. Indeed, one might add that with his book about architecture, Alberti is the first person to present the basic working principles of human-centred design. But as soon as this has been realised, the problem occurs: it seems as if Alberti establishes a normative school of unambiguous principles of human-centred design, based on a positive knowledge concerning the nature and needs of human beings; a codification of human-centredness with a direct connection to our century's functionalism.

Human-centredness as Self-Observation

Personally, I would say that although the above-mentioned dogmatism has been a strong tendency of human-centredness, not only would I not support the idea, I also do not find it in agreement with the basic ideas of human-centred philosophy.

The point is that exactly at the moment in which the human being is put into the centre of the world – feeling almost like the Creator, as Alberti said with a slip of his tongue – the indisputable status of the nature of humans disappears. This was indirectly realised by the contemporary Italian renaissance philosopher Giovanni Pico della Mirandola (1463–1494). In his book from 1486, *Speech on Human Dignity*, God speaks to Adam in his Paradise, and among other things God says to Adam:

> You are not restricted by any insurmountable barrier, but shall decide upon your nature in accordance with your own free will to whose care I have confided you.

Thus, in our modern era, human is not formed by God or by nature, but it is formed by him- or herself, in and by the modernity of modern humans. As a result, human-centredness represents an inherent contradiction. On the one hand, the human-being moves into the centre of the world. On the other hand the definition of 'human' evaporates. The human is, and ought to be, the centre of the world; this is one dimension of the new, modern concept of human.

However, the other dimension is that the positive answer to the question, what *is* human – that this positive or dogmatic answer disappears. The 'nature of human' is a *construction* of human, and in this respect human nature becomes an artefact. Here, we have the important difference between normativism – human-centredness as the basic ideal of the modern era – and dogmatism, being the positive knowledge concerning the nature of human. As I have demonstrated, the latter is – or ought to have been – made impossible by human-centredness.

This, of course, has wide implications, and the discussion can been found in many contexts, as for instance in the political context where many ideologies in our century have been based on the self-contradictory combination of human-centredness and dogmatism. If we turn to the field of human-centred technology, this means that such technology is not created according to unambiguous, 'scientifically' identified human qualities. No, this approach exactly represents the latent danger of making the engineer a new God: he who knows what is good for human, i.e. he who knows the nature of human needs.

In order to illustrate how this dilemma of human-centredness is articulated in the art of painting, I will refer to two additional renaissance paintings: The first one is Luciano Laurana's *La città ideale*, in Palazzo Ducale, Urbino.[27] Laurana (*c.* 1420–1479) was primarily an architect, and among other things he designed the beautiful main court of the Palazzo Ducale in Urbino. The second is Masaccio's *The Trinity*, in the church of Santa Maria Novella in Florence. This fresco was painted in 1425–26.

The painting of the ideal city is an example of the construction of the world according to an unambiguous representation of the qualities of the human. Actually, this can be seen both in the contents of the painting and in its aesthetic form. In its contents, the painting indirectly represents the demands to architecture made by Alberti in his above-mentioned books on architecture, that architecture must reflect the needs of humans. Actually, when looking at this painting, it is hard not to make associations to modern functionalist suburbs in Northern Europe: straight lines, everything planned in advance, no ambiguity, every detail of course formed in accordance with asserted human needs. In short, the perfect design.

Also regarding its aesthetic form, the painting follows the rules of the linear perspective, i.e. the principle of human-centred observation, in every detail. So, in both respects, as a semantic representation of reality, and as a pragmatic representation of the relationship between observer and object, this is – as the title says – the ideal city.

However, it is also awfully boring and sterile, and – more importantly – it does not understand the fact that when humans install themselves as the creator of the world, they also necessarily become self-constructive, and self-observing. When we look at a painting, we are not interested in seeing a perfect and unambiguous world, but we are interested in seeing ourselves 'becoming involved', as one says, by the painting, such as it was exemplified above by the fresco by Leonardo da Vinci.

This is clearly demonstrated by the other example, the fresco of The Trinity, painted by Tommaso di Ser Giovanni Masaccio in 1426. The fresco shows us the crucified Christ, behind him God the Father, and just in front of the cross to the left Mary, and to the right St. John. Kneeling in front of the chapel are at both sides the two donors, a couple of man and wife, and under them and the cross there is a grave with a skeleton. The fresco is commonly admired to be the oldest surviving painting exhibiting unambiguously an almost perfect adherence to the rules of linear perspective (Mey, 1993, p. 1). Particularly, the representation of the columns and the vault has been celebrated because it is actually very complicated to transform such constructions from the three-dimensional space to the two-dimensional fresco.

However, most analyses soon realise that something is wrong. First, any reconstruction of the linear perspective must make certain assumptions, for instance that the vaulted area is a perfect square (this argument can be found already in J. G. Kern's classical article from 1913: 'Das Dreifaltigkeitsfresko von S. Maria Novella, eine perspektivisch-architekturgeschichtliche Studie'; cf. Mey, 1993; cf. also Kemp, 1990). Thus, if we admire that the painting is a perfect example of the linear perspective, we do so by implicitly basing ourselves on specific cultural values (e.g. that a square is to be preferred for a rectangle).

Second, however, still something is wrong. Let me mention three obvious examples:

1. The people in the painting are not foreshortened (Kern 1913, cf. Mey 1993). Of course, the people in the foreground should be larger than the people in the background, but here it is just opposite: God the father is the biggest person, then comes Jesus, then St. John and Mary, while the sponsoring couple in the front are clearly the smallest people in the painting.
2. Some of the most important elements in the painting are not placed in accordance with the three-dimensional order of space, but in accordance with the two-dimensional order of the surface of the picture. One such example is Christ who is placed exactly in the middle of the two-dimensional trapezium defined by the four visible Ionic capitals (cf. Mey,1993,p. 6). Here, the observer is definitely forced to change perspective.
3. At least according to Kemp (1990), the two major parts of the picture are organised in accordance with two different points of observation. In the bottom we see the grave and the sponsoring couple: they represent the mortal world. In the top we have The Trinity, i.e. the spiritual realm inhabited by God, Christ, Mary and St. John.

How should these observations of the aesthetic composition of the painting be explained? A very prosaic explanation is offered by Martin Kemp. Presenting similar results of a close analysis of the fresco, he (1990 pp. 17–21) concludes that:

> ... the sheer sophistication of the construction, unmatched in Masaccio's other works, raises the possibility of Brunelleschi's active collaboration.
>
> *Ibid, p. 21*

The hypothesis seems to be that while the building elements have been constructed in accordance with the mathematically-based design principles of Brunelleschi, Masaccio has added the figures in accordance with aesthetically-based artistic principles. Personally, I do not support this idea. On the opposite, as actually pointed out by Martin Kemp himself, it is the use of *different* principles of representation which makes this fresco the work of a genius. Consequently, I would rather support the conclusion of Marc de Mey, that Masaccio is playing tricks with the observer, partly in order to challenge the observer, partly in order to avoid that the rules of linear perspective and their inherent principle of human-centredness are made a new, indisputable and holy principle. Masaccio is using the principle of linear perspective in order to praise God, not to challenge or replace him. Thus, ambiguity becomes a main point. Masaccio does not aim at combining mathematics and aesthetics into a unified principle, he rather displays a *conflict* between the mathematics of three-dimensional representation and the rhetoric of aesthetic representation. He does not aim at subordinating the celebration of the religious motive under the new, rational principles of 'scientific truth'. The painting represents a *differentiation* of organising principles, and in this respect it reminds us of contemporary pictures of an artist like Escher (cf. Hofstadter, 1980). Actually, one finds a clear parallel to Masaccio avoiding to become the tool of scientific progress in post-modern architecture and design, reacting against the scientific uniformity of functionalism by playing games and by representing an ironic attitude. By looking at the painting, and by being challenged by the painting, one becomes aware of oneself seeing in just one of a number of potential perspectives. Thus, the painting is a painting of self-observation.

Design as Support of Self-Observation

In one of the sections of his book *The Conscience of the Eye* Richard Sennett[28] argues that in its best works, renaissance town-planning and architecture represented exactly the idea of self-observation and that it was realised by 'playing tricks', i.e. by demonstrating that the modern human represented one out of a large number of possible perspectives.

In the renaissance towns the:

> ... centers created through perspective were places in which, it was thought, people would keep moving and look searchingly around them.

This effect was achieved because the architecture supported men and women to:

> ... design ways to 'see outside themselves'.
>
> *Sennett, 1990, p. 153 and p. 152*

One example is the church of St. Lorenzo in Florence designed by Brunelleschi. In the interior, the vanishing point is not the altar, and none of the ritual or prayer places. On the contrary, '... the vanishing-point seems indeed not within the building at all; it is suggested ... to lie somewhere invisible beyond St. Lorenzo's farthest wall. That invisible vanishing point reinforces the sense, within the church, of where one is standing. *One becomes aware of oneself seeing in perspective.*' (Sennett 1990 p. 156, my italics).

Conclusions

In this concluding section I will summarise the most important points of this paper:

1. In the paper I have tried to show that human-centredness is not a construction of our century aimed particularly at the design of technology and artificial systems. Human-centredness is a basic construction of modern, European civilisation with its roots in the 15th century's renaissance. This of course does not reduce the significance of human-centred technology and design, but rather emphasises the fact that it is part of a cultural construction. Human-centredness in technology and computer systems should not be taken less, but more seriously.

2. However, being part of a historical tradition we should of course learn from history. I have tried to demonstrate that in its analysis of human-centredness and its implications, renaissance philosophers and artists achieved results which certainly have not been rendered superfluous.

3. Furthermore, we can learn from history that the achievements in one field of work (e.g. design) is influenced by other fields. The achievements of renaissance painting was a result of scientific and technological developments and of internal aesthetic discussions. On the other hand, the scientific achievements were heavily influenced by current developments in arts and literature, i.e. by new ways of seeing the world.

4. However, history provides us with many examples of what could be called the dialectics of human-centredness, i.e. that human-centredness transforms into its own negation. So-called human-centred design may well develop into a dogmatic design methodology in which deo-centrism (or in our century: techno-centrism) is replaced by a similarly dogmatic and unambiguous human-centrism. Human-centredness represents a certain point of observation, in which the observer is always also observing him- or herself as an observer. This should not be confused with 'human-centredness' as a guiding principles for global world-views and behavioural commands.

5. Therefore, it is important to emphasise that human-centredness is not a set of principles achieved by the identification of a universal order of the world and of a set of given human qualities existing independently of the observer, just waiting to be discovered by the continuous progress of science and technology. Quite the opposite, 'human nature' is itself one of the constructions of modern humans, and the order of the world is as much a construction of the observer.

6. Thus, the ideal of human-centredness is not to make unambiguous designs (of buildings, information systems, interfaces, etc.). The inherent ideal of human-centred products is to challenge the user, to represent different perspectives, to play tricks, hereby supporting the observer's self-observation.

References

Alberti, Leone Battista (1972 [1435 and approx. 1466]) *On Painting and On Sculpture*, ed. Cecil Grayson, Phaidon, London

Alberti, Leone Battista (1965 [1485]) *Ten Books on Architecture*, facsimile reprint of the 1755 English translation, Alec Tiranti, London

Beck, Ulrich (1986): *Risikogesellschaft: Auf dem Weg in eine andere Moderne*, Suhrkamp Verlag, Frankfurt a. M. English translation (1992): *Risk Society. Towards a New Modernity*, Sage Publications, London *et al.*

Benesch, Otto (1965) *The Art of the Renaissance in Northern Europe*, Phaidon, London

Bo, Carlo (1991) *Der Herzogspalast in Urbino*, Manfred Pawlak Verlagsgesellschaft, Herrsching

Cessi, Francesco (1978): *Giotto. La capella degli Scrovegni*, Editoriale Arsuna, Firenze

Cooley, Mike (1987) *Architect or Bee? The Human Price of Technology*, Hogarth Press, London

Elkins, James (1992) 'Renaissance Perspectives', in *Journal of the History of Ideas*, 53, pp. 209-230

Foerster, Heinz von (1981) *Observing Systems*, Seaside, California

Gill, Karamjit S (1990) *Summary of Human-Centred Systems Research in Europe*, Brighton Polytechnic, SEAKE Centre

Gingerich, Owen (1993) *The Eye of Heaven. Ptolemy, Copernicus, Kepler*, The American Institute of Physics New York

Hauser, Arnold (1964) *Der Manierismus*, Verlag C. H. Beck, München

Hofstadter, Douglas R. (1980) *Gödel, Escher, Bach: An Eternal Golden Braid*, Penguin Books, London

Ibn al-Haytham (1989 [approx. 1030]) *The Optics of Ibn al-Haytham*, translated with Introduction and Commentary by A. I. Sabra, The Warburg Institute, London

Johansen, Broby (1979) *Med Broby på Nord- og Midtfyn*, Hamlet, Copenhagen

Kemp, Martin (1990) *The Science of Art*, Yale University Press, New Haven and London

Kubovy, Michael (1986) *The Psychology of Perspective and Renaissance Art*, Cambridge University Press, Cambridge *et al.*

Kuhn, Thomas S. (1962) The Structure of Scientific Revolutions, Chicago

Leinz, Gottlieb (1979) *Die grosse Zeit der italienischen Malerei*, Verlag Herder, Freiburg *et al.*

Luhmann, Niklas (1988) *Erkenntnis als Konstruktion*, Benteli Verlag, Bern

Luhmann, Niklas (1990) *Soziologische Aufklärung 5. Konstruktivistische Perspektiven*, Westdeutscher Verlag, Opladen

Niklas Luhmann *et al.* (1990): Unbeobachtbare Welt. Über Kunst und Architektur, Bielefeld

Marani, Pietro C. (1986) 'Leonardo's Last Supper', in Marani, Pietro C., Roberto Cecchi and Germando Mulazzani: *The Last Supper and Santa Maria delle Grazie*, Electa, Milan 1986

Mey, Marc De (1993) 'Masaccio's Bag of Tricks', paper presented at the XIXth International Congress of History of Science, Zaragoza, Spain

Mirandola, Giovanni Pico della (1990 [1486]) *Tale om menneskets værdighed (Speech on Human Dignity)*, Museum Tusculanums Forlag, Copenhagen

Panofsky, Erwin: (1969 [1960]) *Renaissance and Renascences in Western Art*, Harper & Row, New York *et al.*

Pirenne, M. H. (1970) *Optics, Painting & Photography*, the Syndics of the Cambridge University Press, London

Ptolemaeus, Claudius (1989 [approx. 175 AD]) *L'optique de Claude Ptolémée*, dans la version latine d'apres l'arabe de l'émir Eugène de Sicile, édition critique et éxégétique augmentée d'une traduction française et de compléments par Albert Lejeune, E.J. Brill, Leiden *et al.*

Qvortrup, Lars (1991) *Kedsomhedens tidsalder (The Age of Boredom)*, Odense Universitetsforlag, Odense

Ramachandran, V. S. (1990) 'Visual perception in people and machines', in Blake, A. and Troscianko, T. (eds.) *AI and the Eye*, Wiley, Chichester, pp. 21-77

Santi, Bruno (1990) *Leonardo da Vinci*, Scala, Florence

Sennett, Richard (1990) *The Conscience of the Eye*, Faber and Faber, London & Boston

Simon, Herbert A. (1947) *Administrative Behavior*. New York

Simon, Herbert A. (1988 [1969]) *The Sciences of the Artificial*. The MIT Press, Cambridge, Massachusetts

Steinberg, L. (1973) 'Leonardo's *Last Supper*'. in *Art Quarterly*, pp. 297-410

White, John (1967 [1957]) *The Birth and Rebirth of Pictorial Space*, Faber and Faber, London (first edition 1957)

Watzlawick, Paul, J. H. Beavin, and D. D. Jackson (1967) *Pragmatics of Human Communication*, New York

Wolf, Robert Erich and Ronald Millen (1968) *Renaissance and Mannerist Art*, Harry N. Abrams, Inc., New York & London

Notes

1 This paper is based on a lecture I held at the European Workshop in Human Centred Systems, University of Brighton 10–13 July, 1994.

2 In order to avoid the sexist term 'man', in the following I use 'human' as the concept for the modern citizen, bourgeois and homme. Thus, 'human' is a sociological concept and should not be mistaken for the biological concept 'human being'.

3 This important difference was made clear to me by Jürgen Friedrich from University of Bremen.

4 The illustrations can be found in the annex.

5 Normally, these pre-modern frescos were anonymous, but here the artists have written: 'haec pictura completa fuit per manu ebonis olai et simonis petri', 'ebonis olai' being the Latin version of the Danish Ebbe Olsen, while 'simonis petri' stands for Simon Petersen.

6 This fresco has been analysed again and again. My brief summary is based on the very detailed analysis of Steinberg (1973), and on Kubovy (1986) and Kemp (1990). I have also benefitted from the very good illustrations in Santi (1990).

7 This may have the broader point that one often can understand one's own problems better by looking outside one self.

8 This particularly means that 'human-centredness' implies conflict. If we act according to the will of God, this is a way to treat everybody equally, i.e. as God's children. But if we act according to human-centredness it means that somebody has to define what a human is, and from whose particular perspective the world is seen.

To treat human-centredness as a matter of consensus is, in my opinion, misleading and may well be the first step towards a new dogmatism, this time not on behalf of God (religious dogmatism), but on behalf of a particular definition of the human being and its 'natural' needs presented as if they were universal.

9 Consequently, today churches have turned into museums, while arts museums have turned into churches where pieces of art - representing the Gods and saints of the great artists - are worshipped.

10 By some, this drawing is suspected to be a counterfeit but by others to have been reworked by another artist, cf. Marani 1986.

11 Niklas Luhmann has discussed this in a number of works, cf. e.g. Niklas Luhmann *et al.* (1990).

12 Here, I have quoted from Panofsky, Erwin: (1969 [1960]) p. 26.

13 Actually, the first indication of a heliocentric arrangement can be found in Copernicus' *Commentariolus* or 'Brief Treatise', cf. Gingerich p. 28 and p. 163.

14 Ibn al-Haytham is also known as Alhazen which is a 'European' version of al-Haytham. Averröes is also known under the name Ibn Rushd which simply means 'grandson'. It was Averröes who stated that a sentence may be scientifically true even though it is theologically incorrect. He thus contributed to the secularisation of science and thus to the foundation of the closed, differentiated system of modern science based on the code 'true/false'.

15 It is difficult to exactly specify when *Kitab al-Manazir* was finished, cf. Sabra *op. cit.* However, the first Latin translations were based on Arab copies from 1083–84. In the following I indicate the date of Ibn al-Haytham's *Optics* to 'approx. 1030'.

16 This type of basic considerations have found a renaissance today with so-called second-order cybernetics, cf. Heinz von Foerster (1981), and Niklas Luhmann (1988 and 1990).

17 From here there is a direct link to theories from our century believing that the human mind is a passive 'reflection' of the 'real world'. Thus, the dilemmas of modern theories of knowledge can be traced back to pre-renaissance discussions, and particularly the alleged innovations of so-called 'virtual reality' theories build on very old traditions.

18 Actually, this idea of a straight line of development from Brunelleschi through Alberti to practical artists has developed into a very popular story - almost a myth - in arts history, cf. for instance John White (1967) pp. 113–121, Pirenne (1970) p. 182, Kubovy (1986) pp. 32–52 and Martin Kemp (1990) pp. 11–14.

19 Cf., for instance, John White (1967) pp. 121–126, Kubovy (1986) pp. 10–20 and Martin Kemp (1990) pp. 21–25.

20 As writes Cecil Grayson, ... there is no real evidence to support this view beyond the experiments described in his biography 'Life' by Manetti, and even if they should be true, they do not offer instructions on how to represent three-dimensional space on a two-dimensional surface. Grayson, in Alberti (1972) p. 11. An interesting critical discussion can be found in James Elkins (1992).

21 At dette igen hænger sammen med den fremvoksende norditalienske købmandskapitalismes understøttelse af en abstrakt rationalitet har bl. a. Peter Larsen, *op. cit.* fremhævet.

22 As a matter of fact gravity as an 'action-at-a-distance' was considered to be occult. In public, Newton desisted from presenting any hypothesis regarding the nature of gravity, but privately he wrote to Richard Bentley: 'That gravity should be innate, inherent and essential to matter so that one body may act upon another at a distance through a

vacuum, without the mediation of anything else, by and through which their action and force may be conveyed from one to another, is to me so great an absurdity, that I believe no man who has in philosophical matters a competent faculty of thinking, can ever fall into it.' (quoted from Gingerich p. 151).

23 Actually, in his epilogue Gingerich confirms that this is probably the most important finding in the book: its papers – articles written by Gingerich in the period between 1972 and 1989: ' … examine in several ways the idea that Copernicus's intuitive leap to a heliocentric cosmology was a decision of mind's eye, against the apparent evidence of the senses and proposed in the absence of any physical proof. In other words, they furnish a sustained argument against an empiricist view of the astronomical revolution of the sixteenth century.' (p. 423). For the same reason, a closer examination of the socio-cultural context of his revolutionary results is important.

24 Although both the artistic and the scientific developments had a major religious significance, I have not discussed the reactions of the church. Often, the inherent implications which I have pointed out were not clear for contemporaries. However, as soon as it was realised that the heliocentric cosmology was a model of reality it was condemned. 'As long as heliocentrism was regarded as a convenient fiction, its study posed no threat to the churchmen. However, Decree XIV of the Holy Congregation of the Index, issued 5 March 1616, placed the *De revolutionibus* on the prohibited list *donec corrigatur*, 'until corrected' (Gingerich p. 273f).

25 Here, I am referring to the facsimile reprint from 1955 of the original English 1755 version.

26 Here, Alberti seems to copy the structure of the classical book by Vitruvius: *Ten books on Architecture*. The book was still of much influence and had been revised and copied in 1415, but it was explicitly disliked by Alberti.

27 According to Bo (1991), it is disputed who is the painter. Some say it is painted by Piero delle Francesca as an illustration to his book *De prospectiva pingendi*, others think it is a draft of a set piece.

28 Thanks to John Götzsche for drawing my attention to this book.

Chapter 6

Human-centred Methods of Social and Technical Design

Lauge Baungaard Rasmussen

Introduction

'Design' is a concept which occurs in many contexts: graphic design, engineering design, product design, corporate design, fashion design, architectural design, software design, design methods. It is not immediately obvious that a common essence underlies all these different usages. The word 'design' causes ambiguities because it has more than one common meaning. It can refer to a *product* (a sketch, a model, a plan, a designed good) or it can refer to a *process* (the act or practice of designing). This chapter is mainly devoted to the process of design and the paradigms underlying that process. As J. C. Jones points out:

> The whole point of transformation, the central part of the design process, is to change what already exists, and this includes both theories and practices. Each should influence the other.
>
> *Jones, 1984*

The main objective of this chapter is to present and reflect a particular paradigm: *human-centred* design, and the particular kinds of processes that such a paradigm may provoke.

In the first section general ideas of human-centredness are presented. Questioning the mechanistic paradigm so powerful in Western societies so far, the human-centred approach is defined as an individual and collective learning process including at least four interrelated perspectives:

- A *dialectical* development orientation focusing on how to experiment with alternatives generated by tensions or contradictions between interpretation and interaction of the actors of a living culture.
- A *shaping* perspective focusing on reuniting the separated 'worlds' of social and technical design.
- A *dialogue* perspective integrating 'producer' and 'user' viewpoints in the design process.
- A social *sustainability* perspective focusing on the interconnectedness between individuals, societies and nature.

The central point is that the design process should reflect all the four perspectives in common, if it is to be termed human-centred.

The second section deals with different kinds of human-centred design processes related mainly to Scandinavian research experiences, though many of the methods discussed are more universally approached. The aim is to show examples of the human-centred design process and thus provide the reader with an insight into how different human-centred design processes may reflect different design traditions and still be part of the same fundamental paradigm. Dialogue is considered as central to the human-centred design. The general condition of the dialogue process is defined as a symmetric process which supports mutual relations and human development. Different methods of creating dialogues, such as prototyping, metaphors, organisational plays and integrated interaction and interpretation models, are presented and discussed in the context of different Scandinavian design projects.

In the third section the relationships between the methods and the paradigm of human-centredness are reflected more explicitly. The process of dialogue is reflected in terms of human development as well as a scientific approach of a particular kind stressing different criteria for guiding the design process. The central point of the section is to stress that human-centred design dialogue is itself a *development process* and not just a method of development. Dialogue is developmental in the sense of a continuous learning process toward a more sustainable dynamic balance between nature, society and individuals. This dialogical perspective influence the design in a particular direction. It also influences the participants during the design process towards an individual and collective process of experimentation, providing alternatives of shaping human development. In the fourth section the concept of *learning*, as a very central part of human-centred design, is discussed more in detail. Different kinds of learning models are presented. The central point of the section is that learning of a human-centred process is not only a means to reach an end, but is also a continuous reflection of rationality towards that end. This is precisely the difference between single- and double-loop learning. The 'double-loop' comprehends reflections of the appropriateness of 'means' as well as of 'ends'.

In the fifth section a model of technical and social design perspectives is presented. Several ways of integrating these perspectives are discussed and further elaboration of the human-centred design approach is suggested. Underlying all the sections in this chapter is the conviction that human-centred design presupposes *cultural diversity* as well as a continuous exchange of viewpoints and practices, reflecting different kinds of patterns of interpretation and interaction related to different cultures.

Thus human-centred design is fundamentally a process of instrumental, reflective and utopian processes confronting and reuniting the viewpoints and practices of different design professions and practitioners towards a more sustainable, dynamic balance. As such it is a never-ending process of continuous learning of how to improve human capabilities of self-reflection and realising utopian visions in concrete practices of shaping and reshaping cities, buildings, machines, furniture, social institutions and life.

Background and Assumptions

What is human-centredness about? How did it develop as a scientific approach? Behind the approach lies a questioning of the paradigms and ways in which Western societies have been influenced by a mechanistic world view: that is the imagination of a universe of solid matter made of fundamental building blocks which, by definition, are indestructible. In this universe time is uni-dimensional, flowing from the past through the present to the future. It resembles a gigantic super-machine governed by linear chains of causes and effects. Similar to the imagination of the universe, the human beings are viewed as biological machines. During the 19th century, attempts were made to codify and promote the ideas which could lead to the efficient organisation and management of workers as if they were machines.[1] It was not until the 20th century, however, that the mechanical approaches were synthesised in a more systematic way.

Thus the German sociologist, Max Weber, observed the parallels between the mechanisation of industry and the emerging bureaucratisation of organisations. Bureaucracy was defined as a form of organisation that emphasises precision, regularity, speed and efficiency achieved through the creation of fixed-division tasks, hierarchical structure and detailed rules for activities. The parallels to machines were clear and deliberate. The mechanistic model proved to be extremely powerful and emerged even stronger during the last four decades, leading to the integration of information technology in everyday life.

This development is like the double-faced Roman God, Janus: one of the 'faces' shows higher productivity, an enormous increase in total volume of production and consumption and new possibilities of autonomy from traditional ways of thinking and behaving. The other 'face' shows several kinds of interlocking crises. Nature has been depleted and polluted. Groups of people have been alienated from culture-based experiences. Societies have been disintegrated. It is as if we have been influenced by the '*Lushai Hill Effect*' as Howard Rosenbrock has pointed out:

> Our biggest danger lies in what I may perhaps call the Lushai Hill effect. Resting in the evening and looking back over the lower hills, it is easy to say, at every fork we took the right-hand branch, and see how high we have climbed. Taking the right-hand must be the only way. Though, if we had sometimes taken the left-hand branch, we should in all probability be just as high and perhaps in a region which was in other ways richer, more friendly and more fertile.
>
> *Rosenbrock, 1980*

The human-centred approach questions the thinking of 'the only way' or 'the one best way' (F. W. Taylor). The point is not always to take the 'right-hand branch' (or the 'left-hand branch'). The point is never to go back and start again from scratch. The point is if and how we can avoid the 'Lushai Hill effect' in the future. Are we able to learn how to reflect not only on our single steps one-by one, but on the rules underlying these - steps? The human-centred approach is foremost an individual and collective learning process of how to change steps as well as rules underlying the steps. This learning process is not restricted to work life. It includes a much wider range of interrelated perspectives.

A. Development Orientation

Our habit of 'thinking along straight lines' tends to hamper our ability to think dialectically. Consequently, we fail to appreciate how the seeds of future development are enfolded in the contradictions within a given culture. A dialectical viewpoint invites us to recognise that every action has a tendency to produce a movement in the opposite direction. It has major implications for the interpretation of social and technical change. The dialectical perspective helps one to focus on how experimentation with alternatives is influenced by tensions or contradictions between paradigms of different cultures as well as different patterns of social life inside a given culture.

Experimentation presupposes that actors are part of certain institutions and are able to transcend the existing state of these institutions in some sense. Essentially, the dialectical interplay of 'being part of' and 'transcends' is one of the most essential aspects of understanding a culture bounded by differences of learning processes. The focus on how this interplay is carried out in different cultures is central for the human-centred approach.

B. The Shaping Perspective

The dialogue perspective is fundamentally rooted in the human-centred approach. The accentuation of the shaping aspect, rather than effects or consequences, demands a broader concept of innovative processes than the mechanical world view offers. Shaping does not adapt very well to linear chains of causes and effects. It has much more to do with mutual exchanges of perspectives, imaginations, theoretical knowledge and practical experiences. The perspective is to confront and re-unite separated worlds of social and technical design orientations.

C. The Dialogue Perspective

The shaping perspective differs sharply from the 'objectivistic' approach dominating mechanistically-oriented approaches in which individuals are perceived as objects of modelling like anything else. From a human-centred viewpoint individuals are not perceived to be passive objects who are acted upon. They are perceived as subjects which are also able to act and participate in the process of dialogue between different kinds of designers as well as between producers and users.[2] The possible users of a prospective technology or social organisation are to be involved in the shaping process, not only at the stage of adaptation, but also in the initial development-process.

D. Social Sustainability

Environmentalists warning us of the irreversible consequences of continuing environmental exploitation are developing a paradigm emphasising the interconnectedness between individuals, society and nature. The question is how different concepts of sustainability can be reflected as part of an integrated frame of reference of human-centredness? Both individuals and social organisations contain the potential for self-renewal and self-transcendence. *Self-renewal* refers to the ability of organisations and organisms to continuously renew and recycle their components, while maintaining the integrity of their overall structure. *Self-transcendence* refers to the ability to reach out creativity beyond physical and mental boundaries in the processes of learning and development (Corbett *et al.*, 1991: 44).

If the potential for self-renewal and self-transcendence are prioritised in the shaping of the prospective technologies and societies, the objectivistic paradigm which considers design as a mapping of reality in a context-free manner has to be abandoned. A new paradigm emphasising human beings as active participants in the design of socially sustainable technologies and societies includes new ways of shaping processes too.

In the following sections I intend to introduce and discuss some Scandinavian examples of human-centred processes related to different shaping perspectives and different contexts of practice. In doing so, I intend to demonstrate the *strengths* as well as the *weaknesses* of different kinds of methods used within the frame of reference of human-centredness. It is by no means a complete overview of possible methods. It nevertheless contains central aspects of understanding how the human-centred paradigm may influence different participants in designing new technologies or new social patterns of interaction.

Methods of Social Shaping

Cultural Diversity: Scandinavian Background

The human-centred approach has rediscovered the importance of subjectively bounded knowledge and experience developed through face-to-face communication and practical learning-by-doing. Methods of social shaping meet the challenge of transcending the barriers which exist between theoretically and practically experienced persons regarding doing collectively innovative processes together. New methods are needed. Though the last decades of research and development have contributed fruitful insights, we are still far from establishing universal principles for human-centred methods. This may be fortunate, because the idea of universality in this respect is perhaps non human-centred in itself? Methods of social shaping may be more culture bound than we would like to admit? Given the cultural diversity it may still be relevant to exchange methodological experiences, if we do not claim superiority of our methods independent of the cultural context in which they are supposed to be used. The following examples refer to a Scandinavian context which is far from homogeneous itself. Nevertheless, some common culture-bound characteristics of the Scandinavian societies may help to explain why the conditions of social shaping from a human-centred perspective have been particular favourable in this cultural context.

By virtue of their social position, the Scandinavian labour movements and social democratic parties have achieved a relatively large influence on the industrial and technological development in comparison with most of the other Western industrial nations. They have thus tried to counter the potential disadvantages of the technological development within the working life by an active labour market policy. They have introduced laws which formally ensure the basic rights to establish something like workplace democracy, although the actual effect of these laws is debatable.[3] In our context, however, they illustrate the extension of and relatively strong interest for workers' participation in technological development in the Scandinavian countries. Since the beginning of the 1970s, the universities have developed an openness, allowing students and researchers to be involved in participatory learning and research projects. Finally, Sweden, Norway and Denmark have rich historical traditions for grassroot-movements and local cultural activities centred around folk high-schools and local centres of culture.

In general, the Scandinavian context for social shaping and dialogues between researchers, practitioners and ordinary people has been rather favourable compared to most other industrial societies. This does not mean that is has been ideal. In the Scandinavian context there exist a number of serious barriers too. But the cultural context makes it possible to deal with and overcome the barriers.

If we examine more specific sources for the emergence of the Scandinavian tradition for user-oriented working life research, we have to go as far back as to the 1960s. At that time, the socio-technique gained ground in Scandinavia resulting in the first working life research involving the users. The socio-technical approach was developed in the 1950s by members of the Tavistock Institute of Human Relations in England. The basic idea can be characterised as (Bjørn-Andersen, 1980):

- job design includes other individual needs of the employees besides pay (hereby clearly distinct from Taylorism);
- social and technical needs and goals are considered equally important;
- instead of specialised individual tasks, the focus is on autonomous, self-steering groups.

In the early 1960s the socio-technical approach was taken up and further developed in Scandinavia. The Norwegian Industrial Democracy Project was initiated around 1960, approximately ten years after the initial Tavistock experiments. The ambitions were high:

> Changes at shop-floor level were to create new production relations and release forces which in the end would force the traditional organisational design to be replaced by a new design. The emergence of a new type of organisational design were to guarantee new forms of participation - even within national politics.
>
> *Gustavsen, 1987*

The hope was that Norway would serve as a model for this revolution in the organisation of work. And the changes were to spread to the central industrial nations. However, the socio-technical experiments initiated by the Norwegian employers and the Norwegian Federation of Trade Unions suffered a sad fate.[4] The ideas were opposed and died when they started to involve the lower levels of management. Consequently, the workers soon lost interest in the projects and the labour market parties withdrew their support because of the lack of results.

In Sweden, however, the socio-technical inspiration from Norway found a ready market, although both the objective and the method were reformulated. The keywords in Norway had been 'industrial democracy and participation' but in Sweden they became 'job satisfaction and productivity' (Qvale in Ehn, 1988). Like Norway, the Swedish Federation of Trade Unions received the ideas of socio-technique very positively. They found a possible solution to a critical period of illegal strikes and unrest among their members as a consequence of changes in the production process. But in spite of the positive interest among trade unions, the employers were the prime initiators of the new ideas. They organised and implemented a number of development projects. For them, the ideas about semi-autonomous groups and improvements of the conditions at shop-floor level were the answer to the problems of recruiting and maintaining labour during the boom. Better working conditions were to ensure stable labour which would make it easier to plan and improve efficiency and product quality. However, the employers' concept of the socio-technique meant:

- that it was not researchers but the managers of the employers' association who were in charge of the innovative experiments;

- that these managers adapted ideas for humanisation of the socio-technique while the user-involving method was reduced to asking the user questions and keeping him informed. The changes were implemented top to bottom fashion. A genuine form of participation was not applied in which users developed proposals for humanisation of the work and exerted influence on the changes;
- the changes were confined to shop-floor level while the vertical division of labour was left unchanged (Ehn, 1988).

The negative experiences from Norway and the employers obvious control of humanisation attempts in Sweden became an important motive for establishing an alternative school within the science of work. It was decisive to this development that it took place in the very same year as the students' protests inciting both their fellow students and the researchers to rebel against traditional science and pursue new ideas. The objectivity of research was strongly contested through criticism of ideology that castigated the social doings and actions of science. In Scandinavia, the criticism of the experts penetrated especially in Denmark and Norway. Neither did the new socio-technical science of work in Sweden escaped criticism: 'science of work' and 'humanisation of work' were stamped as ideological veils of mist behind which people were being manipulated and tamed (Christianson *et al.*, 1971). In Denmark, slogans such as 'research for the people' became the headline of a comprehensive cooperation between students and workers about work environmental issues (Christianson *et al.*, 1971). In Norway, however, *Action Research* became the keyword for methodological discussions and a large number of user-oriented projects within different areas of society (Axelsen and Finset, 1973).

On the basis of socio-technical experiments during the 1960s and the student movements and revolts by the turn of the decade, a new tradition was established within the science of work in opposition to the socio-technique. In reality, the new *collective resource tradition* and socio-technique had much in common methodologically as well as with regard to content. But the Scandinavian experiments had, as mentioned above, made the socio-technique appear as a tool by which the employers could legitimise ideological manipulation with the workers. Moreover, the result of ideological critical studies of the father of the socio-technique, Enid Mumford, was that it was stamped as an anti-democratic and anti-union strategy (Ehn/Kyng, 1987).

Instead of the focus of the socio-technique on, 'how do we design systems to fit people', the collective resource tradition asked 'how do we make it possible for people to design their own systems themselves'? (Ehn, 1988: 270). The collective resource tradition continued to use and develop the socio-technical methods which have been important for the human-centred approaches too. In the following sections some of the methods are described in more detail.

Prototyping in General

Prototyping in the present context means a learning process which involves several early practical demonstrations of relevant parts of the desired software or tool.[5] Prototyping is a constructive method to involve people with different experiences, providing more precise ideas about what the target system should be like.

- *Exploratory prototyping* aims to support communication between researchers, designers and users. Alternative solutions are sketched in preliminary shape, thus enabling potential users to assess more adequately the changes in their work consequent upon the introduction of a new tool or system. Moreover, the problem solutions may help users as well as designers to develop tangible ideas. Very early on, the immediate experience of operating the prototype version enables customers and users to render their objectives more precisely and, if necessary, to revise them in a way that is close to practical experiences. Exploratory prototyping is of strategic importance at the transition from requirements analysis to design.
- *Experimental prototyping* presupposes a comparatively stable objective or requirement description. It may adequately help to realise existing specifications. Here, too, prototyping encourages participatory development: the experiment considerably facilitates designers and users grappling with, and assessing of, potential solutions.
- *Evolutionary prototyping* is the far-reaching kind of prototyping. It assumes that technology development can only adequately be a part of constantly-changing organisations. The boundary between technology development, service and maintenance has been blurred.[6]

An Example of Using Prototyping

In the Danish part of the ESPRIT Project 1217, human-centred CIM systems, the three kinds of prototyping mentioned above were used in different stages of the project. The objectives of ESPRIT project 1217 were fivefold (Corbett *et al.*, 1991):

1. Establish criteria for the design of human-centred CIM systems.
2. Establish their economic and commercial competitiveness.
3. Achieve a high level of flexibility and robustness in CIM systems.
4. Define the training for a new type of multi-skilled worker.
5. Demonstrate at a number of production sites that there are better means of organising manufacturing, especially suited to Europe.

The project was carried out by three separate but interlinking national groups between 1986 and 1989: a computer-aided manufacture (CAM) group in the UK, a computer-aided design (CAD) group in Denmark, and a computer-aided planning (CAP) group in Germany. The following discussion focuses on the prototyping experiments carried out by the Danish CAD group of the ESPRIT 1217 project.[7]

The remit of the Danish group was to optimise the alternative design methods to bring the best of the designer, accommodating each designer's preferred method of working. During the first six months of the project, the social science and engineering groups with the overall CAD group worked separately within very few established lines of communication and discussion. The engineers began work by evaluating state-of-the-art flat-panel displays, graphic drivers and digitisers. The Hannover Fair was visited to closely examine products similar to the electronic drawing board concept which lay at the heart of the CAD work. Quite independently, the social scientists established three usergroups comprising technical school teachers, industrial designers and industrial draughtsmen.

At the first meeting between the social science and usergroups, the project objectives were introduced and mutual expectations and benefits were discussed. At subsequent meetings, the positive and negative attributes of conventional CAD systems with regard to the design process were debated, in addition to possible principles for developing prospective job structures and work organisation. During the first six months of the project, 12 all-day meetings were conducted with these three usergroups. It is worth noting that the project engineers did not participate in these meetings. They expected that the social science group would transform the results of the usergroup discussions into an operational form for subsequent software development.

The meetings produced a detailed number of criticisms of existing CAD systems together with a list of suggestions for their improvement. This list was presented to the technical groups at an early stage of the project, although little notice was taken of these suggestions as the engineers perceived their primary task at this stage to be the development of a drawing board prototype and a CAD system interface.

The first meeting between the social science and engineering groups was held in an atmosphere of distrust, scepticism and misunderstanding. Hence, the social science group feared that the engineers would develop their technical specifications and models too quickly, thereby making important technical decisions before the concept of human-centredness and its related criteria were developed.

On the other hand, the engineers feared that the social scientists and the user groups would only generate unrealistic demands concerning technical specifications and software development. One engineer viewed human-centred CAD merely as something different to systems currently on the market. Another engineer defined the concept as a question of being more or less user-friendly. The general feeling appeared to be that once human-centredness was confronted by the criteria of technical feasibility it would not differ much from the conventional automation approach.

Moreover, the engineers feared that the social scientists would play the role of controlling their work without clearly specifying the criteria of human-centredness. Therefore they continuously asked the social scientists to precisely define these criteria. During the early stages of the project the social scientists were not able to do this, thereby increasing the prejudice of the engineers towards a belief that human-centredness is scarcely more than a smart way of getting money.

After six months of independent endeavour the engineers presented their ideas at a meeting to which all usergroup participants were invited. The technical groups had considered six different possible designs and presented the 'best solution' to the interdisciplinary group meeting. This consisted of a more advanced drawing board design (see Figure 15).

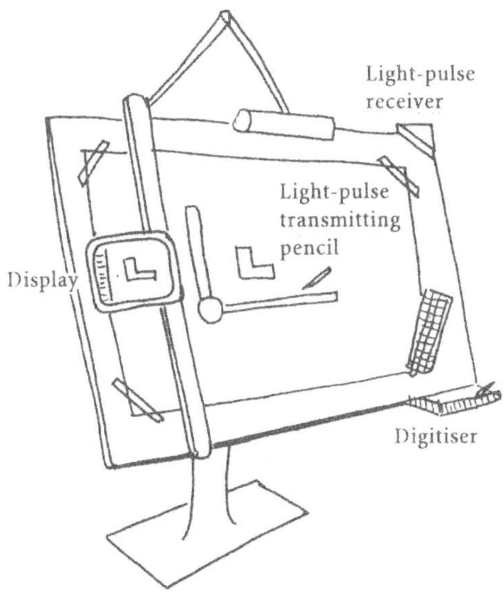

Figure 15: An advanced drawing board design

This drawing board design includes a drawing machine, a display and a digitiser. The display is of the flat-panel type showing the menu as an item and a text field, together with the geometrical contour of the drawing. The pencil is wireless so as not to disturb the user's freehand movements. A digitising eraser makes it possible to erase any part of the drawing. The drawing board is intended for use in three modes: freehand drawing, digitising and keying input from which the designer is free to choose his or her preferred method of working. Whilst drawing, the display shows the window of the drawing at the tip of the stylus (on a 1:1 scale if desired). The idea is to reduce the 'keyhole' effect associated with some conventional CAD systems.

The users' response to the design was not very positive, partly owing to misunderstandings of the engineers' intentions, and partly because some of the users did not approve of combining CAD and drawing boards in the same system. It was therefore decided to arrange a usergroup meeting with only one representative from each of the three usergroups present in an effort to go more deeply into the problem. At this meeting the representative from the designers' usergroup put forward an idea that was to have a significant impact on the CAD group. This designer suggested that the CAD group should develop an Electronic Sketch Pad. This would be portable to allow the designer to discuss aspects of a product's design with people elsewhere in the organisation. Such a sketchpad would greatly facilitate this process of interpersonal communication and discussion. Furthermore, the designers' usergroup representative seemed interested in testing out the idea in a practical organisational setting.

Figure 16: The Electronic Sketch Pad

This suggestion was an important catalyst to the work of the CAD group for two reasons. First, it offered a specific user-oriented tool as a clear goal for the engineers to work towards, thus motivating them to develop prototype software and to find appropriate hardware to achieve this goal. Second, it improved the relationship and collaboration between the social scientists and engineers. The presentation of a clear product triggered off an intensive discussion between the different disciplines concerning the technical and economic feasibility of the sketchpad idea, its role as a medium of face-to-face communication and how the research and development work should progress and be organised.

The design 'shift' occurred in September 1987 during an all-day meeting between the leaders of the three CAD groups, where it was decided that the Electronic Sketch Pad suggestion put forward by a representative from the industrial designer usergroup represented the ideal design solution for the CAD project groups. After this meeting, the collaboration between the Danish groups began to develop in a positive and more constructive manner. The CAD group decided, in September 1987, to focus on the possibilities of keeping informal communication between the designer and the pattern maker intact, whilst at the same time simplifying the means of exchanging ideas and the duplication of effort regarding the transfer of the chosen solution to the CAD system. Furthermore, the sketchpad would simplify learning-by-doing.

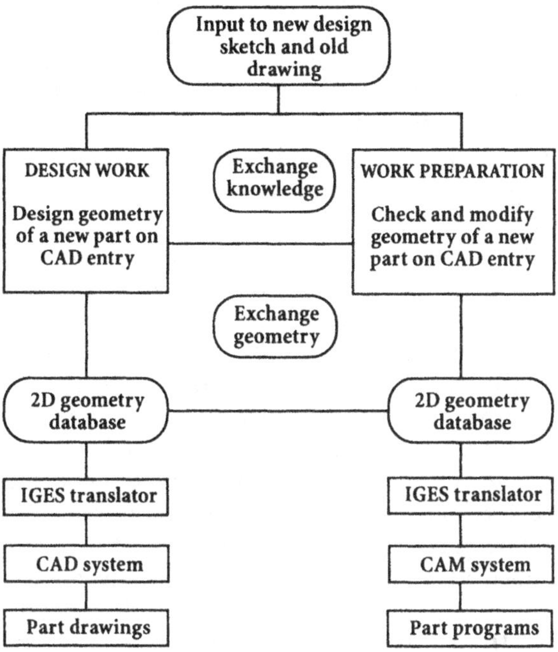

Figure 17: CAD system data-flow and structure

As illustrated in Figure 17, the suggested CAD entry medium makes it possible to use sketches, old drawings and IGES files as inputs from which the design of a new component can be made. The geometry would be calculated and edited either by the designer or the pattern maker and exchanged between them. Because the CAD entry system is the same for both, they can easily communicate and change the design in a common language. The CAD entry is portable, thus facilitating informal interpersonal communication.

The CAD system development process made provision for a series of design iterations and prototype experimentation. This process has a number of objectives, namely:

1. To evaluate how the prototype supports personal communication between users and planning/manufacturing personnel.
2. To refine the CAD entry equipment.
3. To observe the possible changes in the work process and the reactions of users.

4. To evaluate the work process from a human-centred perspective regarding productivity, learning and work quality.

The experimental prototyping took place in the Development Department in a Danish industrial company between May and December 1988. During this period five company employees (designers and pattern makers) participated in the experiment, including the designer who had originally suggested the sketchpad concept. Before the experiment began, the usergroups and the CAD research group discussed how to structure the process. It was agreed that the designers should try to use the sketchpad in carrying out their normal daily work tasks, and report back their feedback to the group after a couple of weeks. Where necessary, the group would initiate any changes to the system based on the feedback.

This cyclical try–evaluate–change process worked well on a number of levels. A number of suggested improvements arose from the process. Although resource limitations made it impossible to realise all these suggestions, a number of important modifications were made. The 'spin-off' effects to the process were equally important. During the evaluation of the sketchpad, a creative and very stimulating atmosphere developed in the group (two designers, one engineer, and two social scientists). A number of ideas on how to improve work organisation and educational opportunities were discussed during the six-month prototyping period.

So far the tool perspective dominated the project. The Electronic Sketch Pad was developed very closely to meet the 'needs' for handy tools in a *given* social context of the users in question. During the prototyping phase of the Electronic Sketch Pad a number of issues about social worklife in the drawing office came to our attention. At first we considered dealing with them as an appendix to the discussion of the Electronic Sketch Pad, but this proved to be unsatisfactory. Many of the issues were significant in themselves, with more far-reaching implications than their technical contribution to facilities of communication and creative work activities. We considered continuing the prototyping in a wider frame of reference which not only includes efforts to improve the functions of the Electronic Sketch Pad, but also to highlight certain aspects of social worklife in the actual and prospective drawing department.

Would it be possible to shape fruitful practice-oriented ideas by means of an interchange of the practical experiences of industrial designers, the research-based knowledge of the social scientists and the technical knowledge of the systems developer or engineer? We decided to try.

Organisational Development by Using Metaphors

A *Metaphor* is an attempt to describe an object or an event as if it resembled another object or event. If we, for instance, say 'he is brave like a lion' then 'lion' is used metaphorically to recall certain images about the man's character. By means of metaphors it may be possible to learn to view the concrete task from different perspectives and then evaluate each perspective from a kind of synthesis or juxtaposition resulting from different insights into the metaphorical analyses.

By using a metaphor we are not taking our point of departure in abstract theories and concepts. Instead, we can start with immediate images and gradually involve the relevant concepts and theories as if they become relevant to the understanding and interpretation of the given problem. The consideration of this point is important in cases in which people work together based on rather different theoretical and practical knowledge and experience. For example, practical experienced people may be more attracted to behave in a creative manner if they can reflect their practical experience from non-conventional perspectives instead of being 'taught' how they should understand and behave by 'experts'. This was the assumption we made in the case of looking at how to change the industrial design office from a human-centred perspective.

Before I describe how the metaphor was used in the case of the ESPRIT 1217 project, it may be informative just to make explicit some impressions of social life patterns in a typical Danish industrial Design Office. These are based on the experiences of two Danish industrial designers (A and B). If allowed to do so, industrial designers tend to develop their own individual style of work. Some designers like to sketch and draw at the outset of a project, using drawings as creative tools. Others prefer to withdraw and think about the design before making any drawings at all. Still others find inspiration in creative talks with colleagues, which designer A describes as follows:

> Sometimes, we use the brainstorming principle, if we have some problem or other which we can't solve. Then we get hold of two or three men, who we know have had something to do with that area, and then I raise the problem and we sit and talk about it and then I go back to work on it.

He goes on to describe the intense and often enthusiastic atmosphere of such a situation:

> You're sitting there with two or three others and drawing for everything you're worth and then somebody says, 'We'll do it this way'. Then I draw it like that ... that's the way we operate.

The point here is, that the enthusiasm experienced in such situations is *situational and not planned*. It grows out of informal, spontaneous communication between colleagues engaged in a common problem. When such enthusiasm arises, it can have enormous creative potential. Such crucial phases in the design and development of products may depend heavily on this kind of direct personal engagement and informal face-to-face communication. The existence of the above mentioned situations depends upon the broader context of the social life in the drawing department. As designer B describes it:

> An important part of working life in the drawing department is the atmosphere. You can describe it on the one hand by the product development,

where we know that what we do is very important, that we initiate a lot of other processes. But on the other hand we don't take ourselves so seriously that we don't have time to joke around. That's important and it also inspires you to do things. You feel comfortable and you have time for a good laugh once in a while. That's reflected in the products in one way or another.

This statement illustrates why it can be difficult to separate creative and non-creative aspects of design work. Parts of the social life in a drawing department consists of the opportunities it offers for designers to regulate their activities in accordance with their needs. As designer B expresses his variable situation:

On the one hand you need peace and quiet. During certain periods I find myself thinking about a deserted farm somewhere in Sweden. But in the midst of the peace and quiet you also need a sparring partner ... I also need to talk about other things, too. Just a few minutes where you don't think about the product you're working on at all. So you talk to someone and then you go back to the work-at-hand.

Sometimes they need to work alone in silence and at other times they need professional conversations with colleagues. Often, an informal chat about matters unrelated to the work-at-hand may serve to stimulate creative activities immediately afterwards. Even when engaged in totally non-creative activities, the mind can be simultaneously engaged in creative thinking. Then, too, a break in intense, creative activity can motivate and enhance further creative work. As part of the social and physical work environment, tools and technical systems play an important part in social work life. As designer B explains:

It's important that we develop tools with man in the centre. It has to do with our job satisfaction and happiness, generally, as well as our relationships, our way of communicating with one another. I think that our surroundings at our place of work have a profound influence on our lives. The kind of potential our surroundings offer us.

During our discussion with designers A and B, they described how the atmosphere of the drawing department might be improved, so increasing individual involvement, creativity and quality of work. They named the following aspects as important in this context:

... more playful, experimental working conditions.

... greater frankness and tolerance about taking risks, being in doubt, making something which is different.

... better chances for 'using' and 'pushing' one another in relation to assignments and on the personal level as well.

... greater energy reserves.

At this stage of the dialogue process a number of ideas come up. They were, however, still rather disintegrated. In order to integrate different ideas of developed so far, we decided to look metaphorically at the design office as a 'human brain'.

The Design Office as a 'Brain'

The following question was posed at the workshop: 'Is it possible to organise a design department with characteristics which are just as flexible and self-regenerating as the human brain? While this is unlikely, we can imagine that the design department could 'borrow' some of the characteristics of the human brain which might conceivably provide us with ideas for organisation, cooperative activity, tools etc.[8]

Many images of the brain can be envisioned. For example, the brain can be conceived of as a control system similar to a complex computer, or as a kind of television system, as a complex system of chemical reactions, as a linguistic system, or as a mysterious 'black box'. Most recently, the brain has been compared to a holographic system, in which memory is distributed throughout the brain and can thus be reconstituted from any one of its parts (Morgan, 1986: 79). The patterns of connectivity by which each neuron communicates with multitudes of other neurons, create a far higher degree of interconnection and exchange than is actually needed at any one time (*ibid.*: 95). This connectivity allows the brain to:

- develop both generalised *and* specialised functions;
- create redundancy and excess capacity which permits the development of new functions;
- develop flexibility of operation and the capacity for self-organisation.

Using this metaphorically, participants in the workshop attempted to envision whether the worklife of future drawing departments could be structured to implement such principles.

The principle of parallel development of both generalised and specialised functions was interpreted as:

> A principle for creating a positive interchange between the social life of a department and the individual's work. On the one hand, there must be 'room' for individuals. As far as possible, the individual should be able to choose his own method of work and technology. Particular abilities should be respected and specialisation encouraged. On the other hand, the department should encourage cooperative endeavour and a mutual utilisation of particular abilities in the common endeavour.[9]

The *principle of flexibility* in operation and *capacity of self-organisation* means that:

> ... there is a continuous process of learning and renewal in the department. By organising leeway for experience, problem-solving and imagination, the working methods of the department can be continuously revised and developed.[10]

The principle of redundancy and *excess capacity* was discussed in greater detail. Firstly, it was related to the dependence of the excess psychological capacity of the individual designers upon such factors as personality, family relations and other circumstances outside the workplace.

Secondly, an atmosphere of 'openness', 'tolerance' and 'resources of self-organisation' in the drawing department itself were soon to be important ingredients or prerequisites for creating excess capacity. Excess capacity does not mean:

> ... that there are always a couple of people sitting around drinking beer, waiting for a peak period. We should like to see it realised as a practical consideration for what is energising and exceptional in the work process.[11]

Ideas of how to promote and facilitate excess capacity in the latter sense were subsequently elaborated during the following workshops. An important characteristic of the management/leader role of a project is *attentiveness*, that is, management's ability to:

- be able to initiate communication;
- be able to initiate creative teamwork;
- be able to orchestrate project teamwork;
- have good insight into individuals in the 'orchestra';
- be personally engaged in the individuals of the department ... for example, by providing positive and negative feed back (Morgan, *op. cit.*).

Product development should be organised in autonomous project groups consisting of representatives from all relevant departments right from the outset of the development process. An example of the make-up of such a multi-functional team consists of representatives from the development department, plastic department, metal department, purchasing, dispositions, planning, quality control department, sales department, production planning and production. Instead of a fixed division of tasks, a multi-functional team collectively possesses skills and abilities needed in the development of a product in a pattern of *overlapping* skills and knowledge which is illustrated in the following diagram (Figure 18).

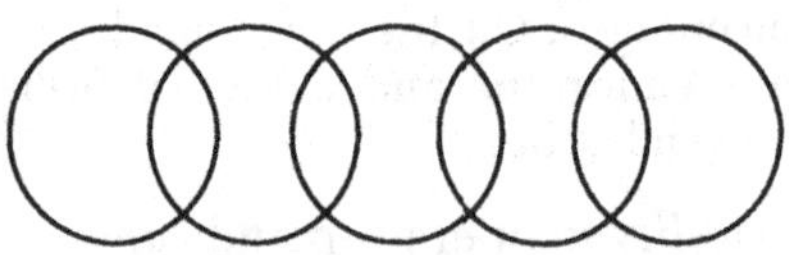

Figure 18: Overlapping skills

Moreover, this kind of organisation may *transcend* traditional, hierarchical divisions between development, work preparation and production, and thus create a more flexible structure for a direct communication network. The project group should assume responsibility for all decisions regarding detailed time-scheduling, expenditure, and the purchase of relevant tools, materials and technical equipment. As regards *technical equipment* the following pragmatic concept was put forward by the industrial designers:

> Each individual could have their own screen at their desk (mounted on a swivel-arm so that it could be pushed aside). It could be used on-line like a telephone screen, as a work bench, as a word processor and so on. It would be connected to a little printer for making hard copies. It should not have many complicated systems, but rather resemble a personal multi-tool. In addition, there should be a drawing board or a table where drawings can be placed. It should also have a vice and a few other implements.[12]

Moreover, as mentioned earlier, a portable Electronic Sketch Pad was envisioned as part of the equipment. Though the video-telephone and electronic-mail system may increase opportunities for quick communication, the importance of *direct face-to-face communication* between members of the project group was heavily emphasised:[13]

> ... so you wouldn't have to write everything to each other.
>
> ... it would be a much more direct and immediate form of communication.
>
> ... it also has to do with connecting with each other mentally, you know?, like sending and receiving on the same wavelength.
>
> ... then we could get rid of the negative impression we have of the other departments.

Communication mediated by computer applications may constitute a supplement to face-to-face communication but not its substitute. Rather than reducing face-to-face communication by means of technical equipment, they prefer an organisation which allows ample time and opportunity for this kind of activity. At the same time, they expressed the need for improved methods and means of communication in order to employ more time-saving procedures; but where such methods also safeguard the freedom to select the most feasible method for exchanging ideas, feasibility is measured in terms of the concrete social content of the work.

In the present case only one metaphor was used. It proved to be rather fruitful. The industrial designers were certainly able to associate the 'brain metaphor' to their own visions and the other way around too.

Though this example of a metaphorical process is not complete (we only decided to use the brain metaphor due to our limited time schedule), it may illustrate tentatively how the shaping process may take form and develop from rather scattered suggestions to a gradually more and more firm vision.

Still the vision is one thing, how to transform it to practice may be quite another matter. In this case another method may be fruitful to implement as a learning process.

The Organisational Play

The idea of the 'organisational play' is to imitate a *theatre* as a means towards new organisational design. Users are involved not only by the means of ordinary discussions using words. Their senses and abilities to participate in different kinds of play have a high priority in this case. Just like a play in a theatre, the organisational play contains a prologue and several acts and sometimes an epilogue. An example of this is described by the Swedish FoU project (1989).[14]

Prologue
The roles are defined, sometimes by the researchers and sometimes by researchers and users in common. To each role there is a description of the skills and anticipations attached to it. The participants of the play discuss the roles and make their choices according to what they feel suits them and expresses their personal ambitions. Then the play board is elaborated, containing a schematic description of the work situations of the organisation. The group accepts or rejects this frame of reference or clarifies it until they reach a workable degree of consensus (which is not necessarily identical with total agreement).

Act 1
The play starts. Researchers and participants have constructed a number of action cards (e.g. 100) describing small and large problems occurring during the work. When a card is put on the table, the problem must be solved. Everybody participates and a consensus on how to solve the problem must be reached. One of the players must assume responsibility for the decision by undertaking the task. But he or she can make it as a condition that further training is offered. The researchers summarise the process by presenting revised and elaborated role descriptions. Furthermore, they construct a new version of the play board. The participants select work tasks for the next phase of the play. Each of them formulate, for example, 5 action cards related to these work tasks.

Act 2
The play continues as in Act 1 by using action cards. In this act, however, the problems have to be solved in accordance with the role descriptions, the agreements and organisational initiatives formulated in the act. On the basis of the problem arising, the researchers formulate a number of developmental questions. The emergence of the problem gradually outlines the patterns for a new way of working.

Act 3
The point of departure is now the image of a new way of working. The question is how to carry it into effect. The group makes it tangible by working out an action programme containing demands for product planning, development of knowledge, technical facilities and organisational support.

Epilogue
Why establish such a play? The actors are asked to reflect on the roles and the processes which have taken place. The rules as well as the actual

happenings are evaluated in common and improvements are suggested.

From a *human-centred perspective* the play may offer a favourable tension between the roles of the play and the factual jobs in reality. Ideally, the users become more aware of the work roles in the company and of the alternatives to the existing stage. Moreover, they are invited to be active in a creative manner and still be able to associate to their personal practical experience. The researchers do not have *a priori* introduce fixed concepts. They try rather to catalyse the learning process than control it. The learning process is reciprocal instead of uniform.

The methods of organisational play may help to transcend traditional mental borders. It is, however, far from being the only ingredient of the social shaping process. A more elaborated example integrating several kinds of prototypes, metaphors, organisational play, methods and future creating workshops are described in the following sections.

Integrated Interaction Model of Social Shaping: Multidisciplinary Approach

Premises of the Model

The integrated interaction model of social shaping is based on rules of dialogue and cooperation between those involved in the shaping process. Instead of just using a particular kind of dialogue it describes how different kinds of dialogue can be combined in an integrated shaping process. The model consists of two levels:

1. *Metarules* – that is a set of rules which the participants are obliged to follow during the shaping process.
2. A *scenario* of a possible social-shaping process describing how these rules may be translated into actual processes of integrated social shaping.

It should be stressed that the following example is *not* the model for social shaping from a human-centred perspective. Rather, it is a prospective case for illustrating how the principles of social shaping may be carried out in an interdisciplinary context.[15]

Metarules of the Model

- **Metarule A:** before the process is initiated an assignment is formulated in a manner comprehensible to all participants, and on which agreement has been reached.

According to the human-centred learning perspective the assignment is *not rigid* in the sense of a specification defining all the sub-objectives and sub-procedures of the process. As illustrated in the process description later on, these are supposed to be open for discussion during the social shaping process. The assignment must allow a gradual change of objectives and means as part of the innovative aspect of the social shaping process. The general purpose of the assignment is to secure that some procedures are agreed upon regarding how to handle conflict situations or actual/potential crises during the shaping process.

- **Metarule B:** at regular intervals, particularly at the end of defined work stages, the project group reflects upon the method employed and the quality of the intermediate result achieved, deciding on the form and content of the subsequent project stage.

The alternation between action and reflection-in-action is an important process in any kind of shaping. Stages of reflection and interpretation have to be integrated into the overall process. Otherwise, shaping projects stand in danger of evaporating into meta-communication and of eventually failing. Practising and applying Metarule B should be subject to time limits agreed upon in advance by the project group. Though these time limits may be overrun during the actual shaping process, it is important for the project as such to have a common understanding of the time perspective before the shaping process is initiated. Metarule B is a *collective participative* process of reflection and decision rather than a rigid hierarchical decision structure. As the project group is supposed to be interdisciplinary, a single person appointed as the leader would not be in accordance with the principle of equality of the different groups participating. If the group consists, however, of more than six persons a coordination committee may be a practical solution (see Metarule D later on).

- **Metarule C:** the participants involved are encouraged to contribute to the shaping process those forms of expression that correspond to their experience, abilities, interests and wishes.

It is important to secure openness regarding work and communication methods in an interdisciplinary design group. Some may prefer to write systematically, others to draw or present their ideas orally or by the means of physical models or techno-scientific instruments. Because everybody is supposed to be an expert in his/her field, each of the expert methods of communicating or work are the best possible conditions for communication and participatory/interdisciplinary shaping. This pluralism of methods is simultaneously a potential barrier for understanding each other. How to cope with barriers of communication will be discussed more in detail later on.

- **Metarule D:** if the group exceeds more than six participants a coordination group has to be chosen among the participants.

The coordination group has two main functions:

1. Integrating and systematising the contributions to the project made by the groups involved.
2. Organising the collective reflection workshop in order to enliven up the process of self-evaluation.

The establishment of a Coordination Group (CG) is not intended to re-establish a hierarchical structure. It is rather intended to follow the Metarule B: both moderation of the dialogue and cooperation have to occur collectively. The Coordination Group may play an important role in initiating the formulation and progression of the assignment. Moreover, it is important to explicate problem solving horizons of the groups involved (it is described in the following process example).

A Scenario of a Possible Social Shaping Process

The following scenario is a prospective case illustrating how the principles of social shaping may be carried out in an interdisciplinary context. It should be stressed that the following example is not a model for social shaping processes. It is rather one of more possible models. Nevertheless, as an example in this context it may illustrate how it is possible to implement the Metarules and the general principles of human-centredness in a systematically considered process of social shaping.

The following example presupposes a rather complex task, i.e. it is more than just a prototype of a tool or software system for a stand-alone human–machine system. It is a combined change of social and technical art. The example, however, is not restricted to the industrial companies. Service centres or local communities may just as well be in focus. Nevertheless it is assumed that at least three aspects are to be combined:

- Some technical or physical relationships are supposed to be changed.
- The existing pattern of communication is supposed to be changed.
- A learning process is necessary in order to secure the changes in the long run.

Preparation of the Shaping Process

At least three kinds of experts are supposed to join the process: a group of end-users (U), a group of social scientists (S) and a group of technical specialists (T).

The U-group may consist of employees from certain factories or service centres or selected people from a local community. It is important to avoid a hierarchical structure in the group, i.e. a chief and a subordinate in the same group, because it may disturb the creative process in the group.

The social science group may include people from different social science disciplines (e.g. psychology, sociology and ergonomics) but members should be encouraged to dispel any disciplinary parochialism or narrow-mindedness. The technical group may include people from different technical disciplines in order to cover the different problem areas within the project.

Finally, there should be sufficient time to discuss the meaning and possible contradictions surrounding the concept of human-centredness both within and between the groups in an effort to encourage the formation of a mutually agreed definition of, and philosophical basis for, human-centredness. Each group should choose one of its members as their representative on a Coordination Group (CG). This group of three has a number of important functions to perform during the design process (see Metarule D). Before the groups begin working, the CG defines the tasks and the problem-solving horizon of the project in order to secure comparative standards across the groups. These standards must be agreed upon by the three groups.

Future-Creating Workshops

During the initial stages of the project, it may be fruitful to carry out a number of rather open and loosely structured discussions. 'Future-creating' workshops incorporate both a critique of existing conditions of work and technology and creative ideas concerning new conditions and suggestions for the realisation of human-centred work and technology.

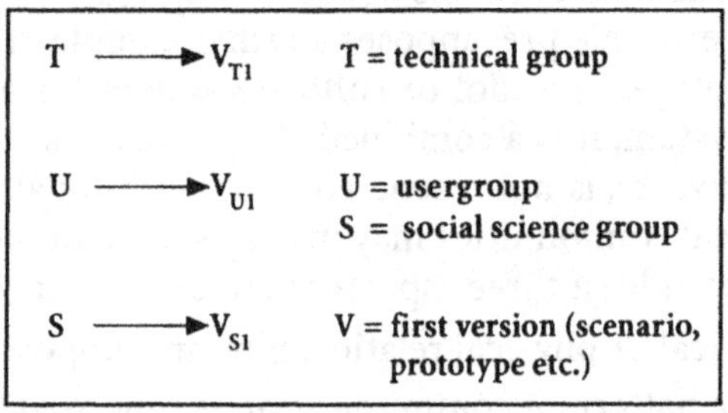

Figure 19: A future-creating workshop

Each of the three groups (U, S and T) separately describe their initial versions of an organisation scenario or equipment prototype in their preferred language (e.g. written word, physical models, drawings, role playing). The process is illustrated in Figure 19.

The reason for separating the groups according to language and experience at this stage of the project is to create opportunities for each group to develop their own vision *before* they are influenced by the ideas of the other two groups. Thus a much wider set of design options and ideas may be shaped if the three groups work in parallel at the same task using their own preferred language and method instead of a mixed group process.

Shaping Workshop

Depending on decisions made after the future-creating workshops (see Metarule B), the process may continue in different directions. In this case we suppose the next step to be a 'shaping workshop' of mixed groups; that is, each group consisting of at least one social scientist, one engineer and one end-user. Compared with the future-creating workshop, the shaping workshop is firmly structured from the beginning as the following procedural example indicates.

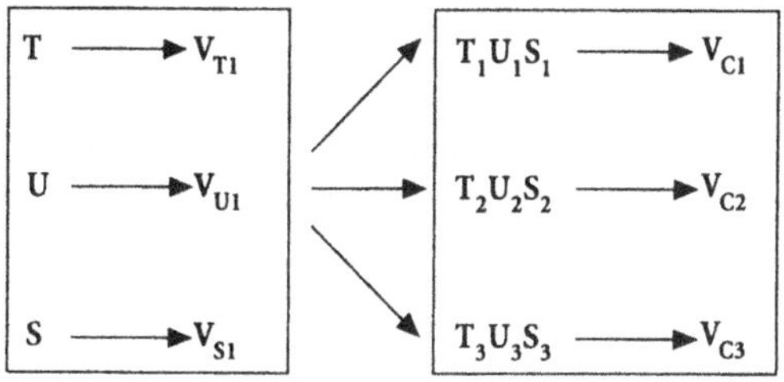

Figure 20: A shaping workshop

- *Step 1.* The aims and rules of the workshop are presented by a person outside the group who has been chosen to guide the workshop.
- *Step 2.* A user representative presents his or her results from the previous future-creating workshop (V_{U1} in Figure 19) to the shaping workshop participants.
- *Step 3.* Discussion of the technical and social implications of V_{U1}.
- *Step 4.* The social science representative presents the results from the social science future-creating workshop (V_{S1} in Figure 19) to the group.
- *Step 5.* Discussion of the technical and practical implications of V_{S1}.
- *Step 6.* The engineering representative presents the results of the engineers' future-creating workshop (V_{T1} in Figure 20).
- *Step 7.* Discussions of the social and practical implications of V_{T1}.
- *Step 8.* Comparison and discussion of the similarities and potential contradictions, or differences, between V_{U1}, V_{S1} and V_{T1}.
- *Step 9.* Integration (as far as possible) into a common version (V_{C1}).

The process thus far is illustrated diagrammatically in Figure 20. Hence we see that the mixed design groups (T_1 U_1 and S_1, T_2, U_2 and S_2, T_3, U_3 and S_3 in Figure 20) are multidisciplinary in order to enable the translation and transformation of the design options derived from the separate future-creating workshops into a common design option or specification which includes technical, social and practical considerations. In this case, the products are more or less three different versions (V_{C1}, V_{C2} and V_{C3}) produced by three parallel multidisciplinary working groups.

Coordination of the Results of the Shaping Workshops

The coordination group (CG) should receive the three 'products' of the shaping workshops (i.e. V_{C1}, V_{C2} and V_{C3}). This is followed by discussions and decisions are then taken regarding how to continue the process. In this case, I suppose that the CG produces a design option based on the juxtaposition and integration of these three products.

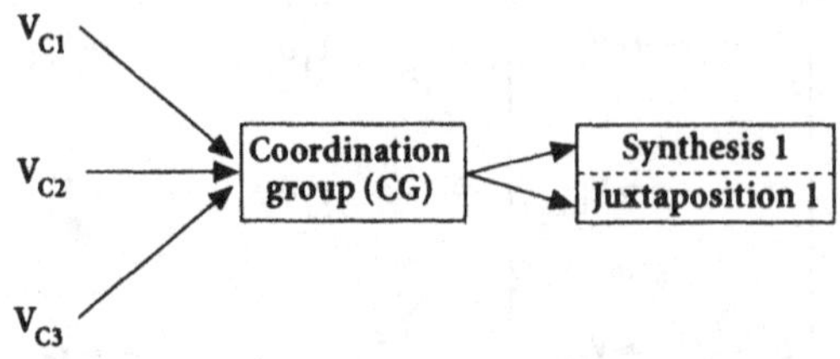

Figure 21: Coordinating the results of the shaping workshops

Moreover, the CG may organise an evaluation procedure in an effort to improve methodology and overcome conflicts and misunderstandings.

The Second Shaping Workshop
In this case, I suggest a rearrangement of the groups back to their original format (i.e. and engineering group, social science group and end-user group). The reason for this change relates to the specific task of the second shaping workshop: namely, the comparisons of the coordination groups' deliberations and common version with the original versions produced by the groups in the future-creating workshops (V_{T1}, V_{U1} and V_{S1}) in order to criticise and produce new ideas. These are delivered back to the coordination group who develop a new synthesis or juxtaposition (S_2 and J_2). This process is illustrated in Figure 22.

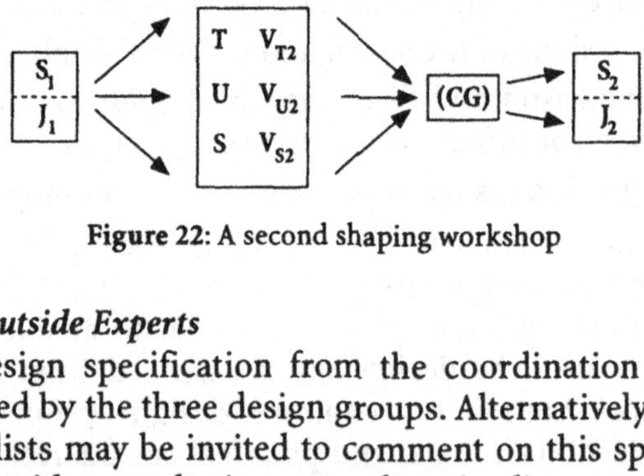

Figure 22: A second shaping workshop

Seminar With Outside Experts
The refined design specification from the coordination group may be further discussed by the three design groups. Alternatively, a few carefully selected specialists may be invited to comment on this specification. The advantage of outside consultation stems from its distance and detachment from the design process. In this case, I propose that outside experts are invited to a short series of seminars of this kind to enable the integration of a new specification by the coordination group.

Subsequently, the procedure is revised and reappraised. In this case I propose that a third shaping workshop is held with mixed groups of engineers, social scientists and end-users. The objective of this workshop is to prepare an exhibition or workshop for the public. This public airing of the design specification (presented in the form of models, film, prototypes, etc.) is suggested for a number of reasons. First, it might stimulate the project participants. Ideas communicated thus far in the design process are at a relatively abstract level. Through preparation for an exhibition, these ideas may become more concrete.

In the present case, it is intended that the exhibition is viewed as a part of the continued development process rather than as an end product. A second reason for incorporating a public exhibition into the development process stems from the desire to encourage a more open and democratic procedure of technology development. People from outside get a chance to 'get inside' the project and even to contribute to it. Third, the project group may receive stimulating feedback from such an exhibition if it is shaped in such a way as to facilitate meaningful participation.

Following this exhibition, the coordination group may take the initiative for another internal, self-reflective seminar. Depending on these discussions, the development may again take different routes. In this case, we suppose that some of the ideas are implemented within a company as experimental prototypes.

Simultaneously, it may be decided to carry out educational experiments in order to achieve a balance between prospective techniques, organisation and educational initiatives.

Summary

Our initial questions concerning the mediation between the abilities moulded by the industrial culture and attitudes of staff on the one hand and the development of sustainable procedures on the other is not merely, a matter of reflecting and respecting, but also that of transcending. The existing forms of work and industrial culture are largely characterised by Tayloristic structures. Set roles and an internalised working practice counteract innovations in all departments and at all levels of the company, even though such innovation may be in the staff's existing converging interests. The transformation of both management and existing converging interests of staff in reducing Tayloristic structures of work and technology into their personal interests and corresponding willingness to innovate is a key problem faced by those R&D projects that are concerned, directly or indirectly, with organisation development. R&D projects with a focus on technology run the risk of intensifying old work and factory structures by means of new production technology. Here, collision may occur between conservative management's interest in maximum supervision and deterministic production control and corresponding staff interest in preserving the status quo. In the search for new forms of humane, yet efficient work and technology for the 'new' factory, there is a great need for intelligent basic ideas and for committed experimenting behaviour. The prototyping, the metaphorical process, the organisational play and the integrated interaction model of social shaping may all serve to strengthen the opportunities of human-centred development but only in close relationship to the paradigm of human-centredness, and not as methods in themselves.

Relationship Between Methods and Paradigms of Human-centredness

Dialogues as Human Development

What is the common paradigm underlying the methods described so far? It is the aspect of human development assigned to them. They are in this context neither just methods of getting data nor exchanging viewpoints. They are not just methods of development. They are developmental in the sense of learning how to realise the individual self as well as realising collective objectives of groups. Differing from economic growth, human development comprehends material as well as mental and spiritual development.

By focusing on human development as the 'centrepiece' rather than economic growth, several new problems of definitions and priorities become visible. How do we evaluate whether a given development contributes to human development or not? Of course, one could claim it depends totally on the culture in question. Part of the human-centred paradigm is to support cultural diversity. It may be tempting to describe development in this way. It is, however, not very satisfactory from another perspective of human centredness: the assumption that every human being is a subject with basic rights and abilities to act and participate actively in the processes of self-realisation and collective relationships. Cultures oppressing individuals or groups preventing them from realising themselves in a sustainable manner may just as well be a barrier to human development. As pointed out by Iris Young (1993: 251):

> Agents, whether individual or collective, have the right to sole authority over their actions only if the actions and their consequences: (a) do not harm others, (b) do not inhibit the ability of individuals to develop and exercise their capabilities within the limits of mutual respect and cooperation, and (c) do not determine conditions under which other agents are compelled to act.

Such limitations of activities may narrow the freedom of agents too much or make any action nearly impossible, one could claim. However, this is not true. The principle is simple (though often difficult to practice): wherever actions affect a plurality of agents in the ways I have specified, all these agents should participate in deciding the actions and their conditions (*Ibid.*: 251). Inspired by Iris Young's distinction between ***autonomy*** and ***empowerment*** the participation aspect of human-centredness can be developed more precisely. According to Young, '***autonomy***' is:

> ... a closed concept, which emphasises primarily exclusion, the right to keep others out and to prevent them from interfering in decisions and actions. Autonomy refers to *privacy*, in just the sense that corporations are private in our current legal system.
>
> *Ibid.: 251*

Empowerment is:

> ... an open concept, a concept of publicity rather than privacy. Agents who are empowered with a voice to discuss ends and means of collective life, and who have institutionalised means of participation in those discussions, whether directly or through representatives, open together onto a set of publics where none has autonomy.
>
> *Ibid.: 251*

By pointing at *empowerment, tolerance of group differences* and *cultural diversity* in a broader frame of reference, Young advocates a 'third way' between the liberal ideal of 'individualism' and the socialist ideal of 'community'. The problem of liberal political theory is that it conceives the person as an atomistic, separated individual determined only by their own private desires. Such a consumer-oriented conception of human nature reduces social relations as instruments to achieve individual desires, and not as having intrinsic value in itself. This atomistic conception presumes competition as *the* characteristic mode of social interaction. Other persons are reduced to 'instruments' or 'barriers' in such a competitive conception of social life. Furthermore, liberal individualism denies cultural diversity by bringing all separated individuals under a common measure of rights.

These criticisms of liberal individualism do not necessarily imply that the socialist ideal of the community is the only alternative, as Iris Young points out:

> Too often contemporary discussion of these issues set up an exhaustive dichotomy between individualism/community. Community appears in the oppositions of individualism/community, separated-self/shared-self, private/public. But like most such terms, individualism and community have a common logic underlying their polarity, which makes it possible for them to define each other negatively. Each entails a denial of difference and a desire to bring multiplicity and heterogeneity into unity, though in opposing ways.
>
> *Ibid.: 228–229*

The ideal of community expresses a longing for harmony, consensus and mutual understanding among persons. This is what Foucault calls the Rousseauist dream of

> ... a transparent society, visible and legible in each of its parts, the dream of there is no longer existing any zones of darkness, zones established by the privileges of royal power or the prerogative of some corporation, zones of disorder. It was the dream that each individual, whatever position he occupied, might be able to see the whole of society, that men's hearts should communicate, their vision be unobstructed by obstacles, and that the opinion of all weigh over each.
>
> *Foucault, 1980*

Such an ideal of the transparency of subjects to one another denies the difference within and between subjects. Each understands the others in the same way as they understand themselves. People become identical and mutually sympathetic, understanding one another as they understand themselves. This commitment tends to value and enforce homogeneity among subjects and groups. It implies a model of a good society as consisting of decentralised small units. It privileges face-to-face relations as the only alternative to alienation and bureaucratisation of governance in existing mass societies.

Opposed to both the ideal of community and the ideal of liberal individualism, Iris Young suggests a third way of what she pronounce *'city life'* as:

> ... a form of social relations which I define as being together of strangers. In the city persons and groups interact within spaces and institutions they all experience themselves as belonging to, but without those interactions, dissolving into unity or commonness.
>
> *Young, op. cit.: 237*

She extrapolates four virtues of a normative ideal of city life which have much in common with the paradigm of human-centredness as I conceptualise it. The four virtues are:

1. *Social differentiation without exclusion* - that is, different groups do stand in relations of inclusion and exclusion, but overlap and intermingle without becoming homogeneous.
2. *Variety* - that means the urban spaces should include a diversity of activities (e.g. parks, restaurants, clubs, offices, residences) instead of strictly separated areas for each activity.
3. *Eroticism* - by this Young points out the pleasure in being open to and interested in people and aesthetics of the material shapes of the city we experience as different from our self:

> The erotic meaning of the city arises from its social and spatial inexhaustibility. A place of many places, the city folds over on itself in some many layers and relationships that it is incomprehensible. One cannot 'take it in' one never feels as though there is nothing new and interesting to explore, no new and interesting people to meet.
>
> *Ibid.: 240*

4. *Publicity of group diversity* - instead of trying to standardise or homogenise every aspect of group differences the public should be:

> ... heterogeneous, plural, and playful, a place where people witness and appreciate diverse cultural expressions that they do not share and do not fully understand.
>
> *Ibid.: 241*

Young's four virtues of a normative 'city life' are not necessarily be restricted to 'cities'. The underlying value of tolerance and mutual interests of cultural differences or diversity is a more general character of human life. Furthermore, it is at the *threshold* between different cultures that most innovations or social learning take place.

When two or more cultures collide or interfere, new ideas or habits may be shaped. Change itself often arises from the meeting of different cultures along their edges. Moreover, the cultural plurality of peoples inside a given space and time may create one of the most important conditions of social shaping of mutually inspiring design processes of possible futures. This is not to claim that cultural diversity is without problems such as conflicts and violence.

The causes of violence, however, should not be derived from cultural diversity itself, but from social and political dreams of identification of some and exclusion of others.

Dialogues as a Scientific Approach

The overall purpose of dialogue is human development as defined in the last section. It makes dialogue different from conventional positivistic thinking of 'methods'. One could ask if dialogues are scientific methods at all or just a hidden way of political work? The answer depends of course on the scientific paradigm of the evaluator. The dialogue is obviously unscientific from a positivistic criterion of 'reliability', if the dialogue follows the human growth paradigm. As the Norwegian Sociologist Johan Galtung points out:

> ... if the dialogue is really good the participants are no longer the same persons after they have been through the process, so how could they replicate the process – particularly if both the social scientists and others have undergone changes.
>
> *Galtung, 1988*

Nor does the dialogue method fulfil another positivistic criterion: 'objectivity'. As mentioned the researcher is forced to involve his personal judgement and abilities in the dialogue process, so how could he be 'objective' in the positivistic sense?

Instead of trying to fulfil the classical ideals of natural sciences, it may be more fruitful to define other criteria in order to evaluate if a given dialogue process is 'good' or 'bad' research. In general I will suggest the following four criteria:[16]

1. *Explicitness*: the researcher must be able to explain what he/she is doing/has been doing and why.
2. *Coherence*: it is possible to see a logical relationship between problem stating and the empirical method used.

3. *Transparency*: it should be possible for scientists not participating in the process to follow the dialogue process step-by-step just by reading the analysis and evaluation.
4. *Validity*: criteria of validity are explicitly formulated and reflected in the evaluation of the dialogue process.

The first three criteria mentioned above will be rather agreeable to most scientific paradigms. It is the validity which may be the point of disagreement. Let us then go a little more in detail with this aspect. The question is what kinds of validity criteria can be stated in general and what kinds may be more context dependent? Johan Galtung suggests the following criteria of validity:

> ... First, there is a dynamic, a dialectic to the discussion which serves to unearth what normally is dominant, deeper, even hidden. Second, dialogues serve to create attitudes and insights in the way such things are created in social reality; they are less artificial. Third, it is a group process, not a lonely reflection. And fourth, it can be tied to a project so that the approach would be a part of an action research program, insights immediately or relatively quickly being translated into practice and vice versa.
>
> *Galtung, op.cit.: 83*

Galtung's definition of 'validity' however does not answer the question regarding the relationship between policy and science. It is obvious that his concept of validity is political, but is it scientific too? His own answer '... it is a different kind of research (than the survey method – LBR) and hence a different kind of politics' (*ibid.*) is not sufficient. It does not answer the question, of whether the political result is the criterion for evaluating the success of the process as science or not? If the growing knowledge fostered during the dialogue process does not result in practical change, does this influence the evaluation of the scientific part of the process? Or the other way around: if the growing insight or consciousness is transformed into practical actions, is the success of the transformation sufficient to evaluate the process as 'good' research? I am sceptical about Galtung's interpretation in this case. Just as the positivists are trying to make a total separation between science and politics, Galtung tries to make a total integration. A more promising, but also more difficult road may be the following: the process of dialogue is political as far as it tries to change social relationships. It remains, however, non-scientific until new experiences or knowledge is created in a way that 'external' persons (i.e. persons not participating in the given dialogue process) can accept the experiences as new. This experience needs not to be expressed in a systematic theory or hypothesis, which can be tested in a broader frame of reference. The experiences can be deeply rooted in a particular situation of the group or persons involved and therefore of limited generalisation value. It may have something to do with the external persons success in moving from the surface level towards deeper-level patterns in the particular context. It may also happen that researchers have learnt to link the dialogue process in a more fruitful way and thus have improved the dialogical method. Even if the new experience cannot always be formalised or generalised, other researchers must be able to acknowledge it as new experience. Therefore the criteria of explicitness, coherence and transparency is important. Moreover a further criterion of validity may be claimed: the researchers' interpretation of the actions of participants or meanings should be comparable to the participants' own interpretation. This is not the same as saying that they necessarily need to agree upon the action or meaning in question. As pointed out by the Norwegian sociologist Harald Enderud (1979):

> ... a certain disagreement may result in a dialectical tension between the participants and thus create a further dialogue and development of new viewpoints.

The interpretation of dialogue as a particular scientific approach following particular scientific criteria makes it necessary to reflect on the role of the researcher in more detail.

The Roles of Researchers in the Dialogue Process

In the study of innovative processes of social networks or movement in which the researcher participates as part of the dialogue process, this participation must be part of the reflection itself. Any research practice which requires an intervention in the dialogue process creates an artificial situation which must be explicitly acknowledged. As Alberto Melucci (1992: 50) points out:

> The researcher cannot pretend that phenomena impacted by his or her practices are natural, but must be capable of moving the level of observation and communication. It is necessary to introduce into the field of research a capacity for meta-communication, regarding the relationship between the observer and the observed.

Acknowledging the limited rationality which characterises patterns of interpretation and interaction, the researchers and other participants of the dialogue process can no longer just apply a criteria of truth and ethics defined *a priori* outside the participatory relationship. As part of the development process, researchers must also reflect the limits of their own methods and paradigms. Therefore neither the role of total 'identification' or total 'distance' between the researcher and actor are acceptable as a scientific approach of social science from my viewpoint.

The role of *identification* is often preferred by the 'empathetic' oriented researcher. The process of identifying him/herself with the actor either by imagination or action may give the researcher a kind of insight not accessible using conventional methods of surveys or interviews. There is a risk, however, that the researcher believes that he is able to overcome the 'gap' between interaction itself and the reflection of the interaction. Even a well-trained empathetic researcher would never be able to overcome this 'gap'. One reason is that identification between the researcher and actor would never be totally even if it seems to be rather close. Therefore, reflection is always necessary in order to understand the limits of the identification process. Another reason is that the 'gap' between the interaction and the reflection of the interaction is a very fundamental human condition of being able to 'be within' and 'transcend' a situation mentally and/or physically. This is not a particular attribute of the researcher role. It is rather a condition of human growth as such. The researcher, however, may be particularly aware of this dialectical process, because of his explicitly double role of being 'actor' as well as 'interpreter' of the effects of his own and other's actions on the dialogue process and the results of the dialogue. The role of *distance* is often preferred by the 'objectivistic' oriented researcher. I have already argued why 'objectivity' or 'neutrality' of the researcher is an illusion.

The dialogue researcher inevitably enters into the 'game' of questions, expectations and interests. Thus he cannot meet these challenges without a strong sense of self-reflection which becomes gradually more explicit as part of the research process itself, a sort of situational epistemology of being 'inside' as well as 'outside' at the same time. In other words, the researcher must be able to acknowledge the limitations of the points of view of the participants and of 'meta-communicating' it.

Following the human-centred paradigm, the purpose of the research is *not* to give the researcher a monopoly of knowledge and decision-making abilities. It is rather to offer the actors including the researcher himself an opportunity to develop their capacity of human growth and at the same time to reflect on his/her own way of learning as part of the dialogue process in a consistent and transparent manner. Thus the relationship between the researcher and the other actors becomes a 'terrain of responsibilities' of 'the space for a cognitive and ethical contract between the researcher and actor' (Melucci, 1992). By making more explicit or visible the relationships linking the elements of a conflicting or harmonious field, the researcher can contribute to a better understanding and handling of opportunities and constraints of human growth. She/he may provide a wider perspective of the possibilities to act, because the dialogue in our sense favours interaction as well as meta-communication (Bateson, 1972, 1979; Melucci, 1992), about the dynamic or static relations between the actors in question. The researcher interested in innovative processes and human growth should never forget that the core of the analysis is not the actors themselves, but the interactions and interpretations they are making before and during the dialogue process. In order to get at this meaning the researcher has to try to enter the actors' inner world and to reconstruct from this process the frame as the actors have experienced and lived it themselves (Melucci, 1992). Then the researcher must be able to 'step' outside this world and analyse the subjectively bounded experiences as part of a given context of the field. During this process then the researcher alternates between at least two positions: the 'empathetic' understanding of the actors' interior world and the critical reflection of an external interpreter.

Relationships Between Power and Communication Patterns of the Dialogue

The relationships between power and communication are very important to reflect on all human activities. It is, however, particularly essential in situations in which the participants are supposed to be 'equals' such as in a symmetric process of a dialogue. What does it mean to be 'equals' in a dialogue? The participants are supposed to be 'equals' in the sense that each is a source of knowledge and imaginations for the other participants. It does *not* mean that each participant exert the same influence on the direction or the content of the dialogue process. On the contrary, even if democratic metarules and intentions are guiding the dialogue process some of the participants may possess much stronger mental models or resources than others. What do I mean by 'mental model' in this case?

> A mental model is a pattern of interpretation which is used to produce preliminary answers to issues which the model is assumed to represent.
>
> *Bråten, 1983*

The prototyping process is based on a series of such models. More generally, a 'model-building process' or 'model-destroying process' is carried out as soon as mental perspectives of the participants are 'crossed' during the dialogue. In a more or less conflicting way an actor's inner mental auxiliaries for filtering or initiating innovations are related to other actors' auxiliaries during the dialogue. If the actors differ regarding the strength of inner conviction or the abilities to use symbolic or material resources a power relation may be implemented in the dialogue even if the participants want to be as 'equal' as possible. The actor applying the 'weak' model or resources is more or less implicitly forced to adapt his/her contribution to the premises of the actors applying the strong model. Well known examples of this is the authoritarian relationship between the teacher and the pupil or between the doctor and the patient. The strong-model actor is able to integrate and utilise the information from the weak-model actor. The weak-model actor will try to get access to parts of the strong-model actor's knowledge and thus be more and more adapted to his universe of understanding. Formally, Bråten describes the *relationship* of *model power* as:

> 1) In order for actor A to control 'X' the precondition is that 'X' operates on a model developed on the premises of A (i.e. on the basis of A's concept).
>
> 2) It is a precondition for starting a dialogue between the two actors A and B that they have access to the subject of the model referred to.

The consequences of this, in the situation of interaction between the strong model A and the weak model B are that:

3) The weak model B tries to gain control over the models of the strong model A.

The consequences of 1, 2 and 3 are:

4) The more successful B is in gaining control over the models of A, which are developed on the premises of A, the more B will be controlled by A.

Bråten, 1983

The question is whether inequality of the power model destroys the conditions for practising research based on dialogues? In other words: is the power model contradictory to the fundamental principles of dialogue? Bråten's answer is 'yes'. At the same time he perceives the power model as a relationship which can be discouraged or diminished. He suggests several ways of doing that. *Firstly*, he suggests to *move the limits of the issue* in such a way that it contains elements outside the competencies of the strong-model actor and within the universe of the weak-model actor. I agree this is necessary, if each participant of the dialogue is to fulfil the criterion of being a particular resource of knowledge. The question is, however, *how* the issue is moved in order to be in accordance with this principle? The strong-model actor may urge the weak-model actor to present or elaborate his/her imaginations *before* his/her own ideas are presented. The question, however, is if this is sufficient to prevent a power relation to be implemented? I do not believe so. Of course, it may be a help for the weak-model actor to be aware of his/her own position. Nevertheless the strong-model actor does possess a number of other possibilities to exercise his/her model power, even if he/she does not want to.

Secondly, Bråten claims that *being aware* of the model may in itself be a way to reduce it. At least it seems to be a minimum condition to begin to work with the problem. As such a growing awareness of concrete forms of the model in the dialogue process may be considered as part of the learning process. The micro-cosmos is a mirror of how power relations should or could be dealt with in a wider context. As such, the particular power relations in the dialogue group may be handled as part of the learning process towards a growing autonomy of the model actors. Awareness itself is of little help, if no actions are initiated to change the situation.

Though I agree with Stein Bråten's diagnosis of power model, I disagree regarding how to cope with the phenomenon. Instead of trying to avoid it, I suggest the relationship between power and communication patterns during the dialogue process should be in focus as the most important part of the collective learning process.

Learning as a Human-centred Process

Learning as Human Development

The fundamental issue of learning from a human-centred perspective is the development of human abilities to take care of social and ecological sustainability and thus contribute to human development in a qualitative meaning. An essential part of the learning process from this perspective is a focus on how to learn not only to follow the existing rules but also to reflect on them critically and change them, if necessary. Reflection and action are part of the learning process rather than just knowledge. Of course knowledge is necessary. One has to understand in order to reflect and change. The most important aspect, however, is not knowledge' as such, but how to make a dynamic relationship between knowledge, reflection and 'action'.

Learning Models of Reflection-in-Action

'Knowing-in-Action' and 'Reflection-in-Action'

A part of the Western design tradition has coined the concept 'reflection-in-action' which is opposed to technical rationality as the dominant epistemology of practice. Instead of focusing only on the instrumental adjustment of 'means' to 'ends', the 'ends' are continuously reflected and perhaps changed too.

In our spontaneous, intuitive performance of the actions of everyday life, we show ourselves to be knowledgeable in a special way. If we try to describe it we find ourselves in a situation in which the produced descriptions are obviously inappropriate. Our knowing is ordinarily *tacit* in our patterns of doing or feeling. Michael Polanyi (1967), who invented the phrase *'tacit knowing'*, draws examples from the recognition of faces and the use of tools. If we know a person's face we are able to recognise it among thousands or millions of other faces. We are usually, however, not able to tell exactly how we recognise a face we know. Similarly, when we learn to use a tool our initial awareness of its impact on our hand is transformed to its effect on the things to which we are applying it. This process is essential to the acquisition of a skill. The initial awareness becomes internalised in our tacit knowing. We are often unaware of having learned to do these actions. We simply find ourselves doing them. They become internalised, a routine which may be more and more perfectly performed as we get more and more practical experience. But the perfectly performed routine may have *ambivalent* consequences on our learning abilities.

A very deep internalisation of a given knowledge or actions may obstruct the consciousness and the body to change perspective or practice. This is well-known among psychologists and historians. It is common in the industrial work relationship too. Employees who have been doing the same actions for decades are usually not very motivated to change their routines even if it may give them increasing responsibility and job enrichment. Thus 'knowing-in-action' is different from 'reflection-in-action'. A good example of reflection-in-action is given by Donald A. Schön (1983: 55):

> When good jazz musicians improvise together, they also manifest a 'feel for' their material and they make on-the-spot adjustments to the sounds they hear. Listening to one another and to themselves, they feel where the music is going and adjust their playing accordingly. They can do this, first of all, because their collective effort at musical invention makes use of a schema a metric, melodic, and harmonic schema familiar to all the participants – which gives a predictable order to the piece.
>
> Improvisation consists in varying, combining and recombining a set of figures within the schema which bounds and gives coherence to the performance. As the musicians feel the direction of the music that is developing out of their interwoven contributions, they make new sense of it and adjust their performance to the new sense they have made.

The musicians are reflecting-in-action on the music they are collectively making and on their own personal contribution to it. They do not reflect in the medium of words but through 'a feel for the music'. Similarly a football player may have a 'feeling for the play'. Usually there is no time to discuss when the play is going on. The reflections are transformed to physical action as soon as possible.

Reflection-in-action is often initiated by surprise. Something unanticipated happens and starts a reflection of 'why'? If the unexpected happens several times, the reflections may continue and increase in strength. In these situations changing of actions may not immediately follow the reflections. At the beginning the reflection may just be like a 'silent wonder' encapsulated in the consciousness of the person him/herself. This 'wonder' may gradually be communicated to colleagues or specialists and thus initiating a more or less collective reflection process. At this stage new aspects of interpreting the unexpected events may be added to the initial ones. It might even be connected to other unexpected events. The person who initiated the wonder may, however, also be oppressed by the colleagues if he threatens the dominant paradigm.

This has happened throughout the history of science as well as in religious matters or in the cross between science and religion. Historical examples of major paradigm shifts are the transition from the geocentric astronomy to the heliocentric system of Copernicus and Galileo and, most recently, from the Newtonian mechanics to quantum-relativistic physics. A paradigm may, for a period, be extremely powerful: clearly defining not only what reality is, but also what it is not and cannot possibly be. The extreme example is behaviourism – an attempt to eliminate the element of consciousness as a legitimate object of scientific interest. In his *book The structure of Scientific Revolutions* Thomas Kuhn (1970), proclaims a kind of on-going 'fight' between existing and new scientific paradigms. At first, all research challenging the dominating paradigm tends to be suppressed. The scientist who generates controversial ideas or data is criticised, isolated or even accused of cheating. Usually, many attempts to provoke or transcend the existing paradigms are prevented in developing further at this initial stage. If new data or ideas, however, hold in subsequent studies and are further confirmed by other researchers too, the discipline in question moves into a paradigm crisis. Out of the chaos of different alternatives, more or less fantastic theories, one of these finally emerges victorious as the new paradigm. The sequence of events is repeated again and again, Kuhn proclaims.

On a smaller scale, similar processes may be observed in other fields. Innovative processes in industrial areas, for example, may follow similar patterns of change. Change in pedagogical paradigms of education and training is another example.

The process of reflection-in-action comprehends an extensive span of processes ranging from radical changes in world-views to usual everyday experiences of ordinary people. Not all kinds of reflection result in changes. The important matter, however, is focusing on how the reflection in question may be learned and practised as a possible way of human growth. In order to be able to do this a more detailed interpretation of the reflective processes is necessary.

Single-Loop and Double-Loop Learning

During the last decades cyberneticians have discussed the distinction between the process of learning and the process of learning to learn. For example a house thermostat is able to detect and correct deviations from predetermined norms. But it is unable to question the level of temperature appropriate to meet the preferences of the inhabitants of a room. More advanced systems such as the human brain and body have this capacity. The essential difference between these two types of learning is usually identified in terms of a distinction between 'single loop' and 'double loop' learning. Morgan (*op. cit.*) has illustrated this by distinguishing between 'single loop' learning and 'double-loop' learning (Figure 23).

Single-loop learning rests on an ability to detect and correct error in relation to a given set of operations norms. *Double-loop* learning depends on being able to take a 'double-look' at the situation by questioning the relevance of operating norms:

1. The process of sensing, scanning, and monitoring the environment.
2. The comparison of this information against operating norms.
   - 2a. The process of questioning whether operating norms are appropriate.
3. The process of initiating appropriate action.

Of course, the double-loop learning seems to be more advanced than single-loop learning. Including the extra loop (2–2a) at least makes an opening for reflection and change to a certain level. Innovative processes sometimes start from such an extra loop. Just like a 'spark' can make a 'fire', questioning whether operating norms are appropriate (2a) may initiate further reflections (and thus further loops) resulting in innovative processes of a smaller or bigger scale. Perhaps the 'fire' has not been made and no change will be visible, before a new 'spark' ignites the 'fire'.

If the human learning process was to follow single-loop learning four sets of key qualifications would need to be taught:

- *First*, the persons must learn the capacity to sense, monitor and scan significant aspects of the situation in question, whether it is a CNC (Computer Numerically Controlled) machine or a social relationship in a group.
- *Second*, the persons must learn to relate this information to the operating norms, role expectation or rules that guide the machine system or the patterns of interpretation and interaction.
- *Third*, they must learn how to detect significant deviations from the norms or rules of behaviour.
- *Fourth*, they must learn how to initiate corrective actions when discrepancies between expected and deviating behaviour are detected.

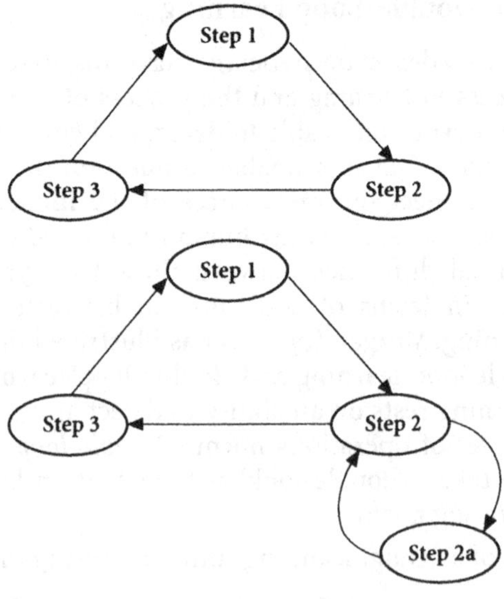

Figure 23: Single-loop and double-loop learning

Thus, single-loop learning covers most of the school curriculum at the primary and secondary level as well as traditional industrial shop-floor training. The single-loop principle can be associated with a certain level of socialisation of children and adults in general in a given frame of reference. But it can also be related to a special job situation, for example operating a CNC machine. The great span of possible perspectives does not make it useless as an analytical model if the concrete circumstances or context in question are specified. The problems occur, if the model itself is taken to be a real learning process. It is no more than a certain way of distinguishing between two ways of reflecting in a learning process. In general, it would not be correct to claim that double-loop learning is always 'better' or more 'advanced' than single-loop learning. Again, it depends on the context and the quality of the 'second loop' and how the 'second loop' is related to the 'first loop'.

From a human-centred viewpoint the cybernetic models of learning are far too simple to comprehend the complexity of human learning processes. Nevertheless the explicit focus on the 'second loop' as a (self-) questioning way of learning point at a central issue of the reflection process. Consciousness at every moment includes the possibility of reflection and opens onto the world of imagination and knowledge. But not every reflection is a learning process in itself. Reflection must be related to a certain context of understanding and action before it can be evaluated as a learning process. On the one hand, learning may take place within an agreed, accepted and unchallenged context. In this view it is clear what should be the result, the end of the learning process and what should not be achieved. The learning process becomes an instrument to reach this end. On the other hand, learning may take place on the threshold between two or more different cultural contexts. In this view there is not a general agreement of the ends. The continuous reflection of the ends themselves are part of the learning process. For example, take the case of human growth as the end of a human-centred development. Even if there exists an agreement of human growth as a likely overall objective, there would certainly be disagreement about what it is and how it can be realised in practical interactions. The existence of multiple criteria of evaluating human growth makes the ends/means distinction ambiguous, if not useless. The dialogue is not a means to realising the end: human growth. It *is* human growth. It is human-centred development itself. The cybernetic 'loop- thinking' may contribute to transcending the mechanistic 'straight-line thinking'. But the 'loop-thinking' is not *the* new paradigm of human-centred development. It is anything but an analytical perspective of distinguishing between different levels of learning which may be human-centred or not.

Liminal Learning and Cultural Diversity

As pointed out by Raymond Williams, it is possible to distinguish between at least three levels of culture:

A: There is the *lived* culture of a particular time and space, only fully accessible to those living in that time and space.

> B: There is the *recorded* culture, of every kind, from art to the most every day facts: the culture of a period.
>
> C: There is also, the factor connecting lived culture and period cultures, the culture of *selective tradition.*
>
> *Williams, 1961*

It is the process of selective tradition that makes it possible to reflect on the dynamic aspects of culture as a continuously selective and interpretative process of learning, communication and production. Culture is not just an absolute body of things and habits of a particular time and space, nor is it only value patterns recorded from a particular period. It is also, and most important for our purpose, a continuous process of accepting and rejecting areas of what was once a living culture as well as a process of shaping new kinds of relationships between learning, communication and production. The concept of selective tradition is linked to William's dialectical perspective of change. While the linear perspective of change reduces our understanding of change as more or less predetermined development of cause and effect, the dialectical perspective invites us to embrace contradictions and oppositions as defining features of culture.

All phenomena generate latent oppositions that tend to change themselves. Oppositions are intertwined in a state of tension that also defines a state of harmony and wholeness. Some cultures have concepts that explain this more explicitly than others. For example, the relationship of oppositions is clearly reflected in the Japanese concept of 'omote' and 'ura' as expressed by Takeo Doi:

> Omote and ura belong together although opposing concepts ... Even when we use them separately, one term implies the other: omote-muki refers to that which is public, open, official, ura-muki suggest something private, closed, personal ... Even when we use them separately, one term implies the other: to speak of ura is to speak of omote.
>
> *Doi, 1986: 23*

According to Doi, institutions, the omote of a society, express the characteristic cultural features of that society. To the extent that the individual as ura is expressed by *omote* institutions, these institutions also belong to the individual. Institutionalisation is like a multidimensional continuum where the scope as well as the content of role expectations and integration of different institutional sectors constitute both conflicting and homogenous situations.

If a given lived culture is relatively homogeneous and isolated from the influence of other cultures the learning process takes place within an agreed, accepted and unchallenged framework of symbols and habits. Learning within such a culture may be characterised as mainly *adaptive* to the given role expectations with few if any processes of selective changes taking place. If two or more lived cultures interfere or collide several things may happen. One of the cultures may completely dominate the others. Or a situation of *cultural diversity* with two or more different frames of reference may survive side by side for a shorter or longer period. The interesting point here is: what happens when learning is taking place on the *threshold* between two or more different frames of reference? I will term this kind of learning as *liminal learning* in contrast to *learning within a frame.*[17]

The thresholds between two or more different cultures may be particularly interesting from a learning perspective. Persons placed on thresholds may be more able and motivated to do reflection-in-action or double-loop learning than persons firmly placed within the frame of a certain culture. Thus cultural diversity may create better conditions for liminal learning processes. An *accepted* cultural diversity may be a promising point of departure towards more reflective and deeper learning processes even from a social and ecological sustainable viewpoint. Instead of perceiving diversity as only a potential or real threat against cultural survival, which it might be too, it may be a better option to valorise and make advantage of cultural diversity in certain kinds of learning and development.

Integration of Social and Technical Design

A Model of Technical and Social Perspectives

The purpose of the shaping perspective is not a romantic binding to the past. Nor is it a blind, unreflected homage to the 'progress of enlightenment'. The real challenge comprehends a combined transformation of separated technical and social perspectives of design into a new kind of collective skill formation and imaginations of possible futures.

The accentuation of the 'shaping' aspect rather than its effects, consequences and so on, demands a double view of the concept of design. Ideally, design is supposed to be the union of the technically possible and the socially desirable as illustrated in Figure 24.

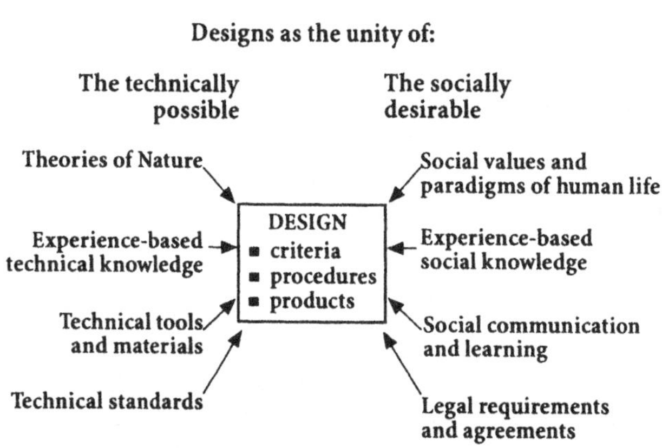

Figure 24: Perspectives of design

The figure illustrates the necessity to reflect upon the following questions. What are the cultural backgrounds behind the separation of the technical and social perspectives? How is it possible to reunite the technically possible and the socially desirable from a human growth perspective?

How is it possible to use the cultural diversity of interests, knowledge, methods, tools and practices in a creative and human-centred way?

Multidisciplinary Approach

The continuous mutual exchanges of perspectives, so essential for the shaping approach, involve a multidisciplinary approach. Thus we are encouraged to learn how to think of situations from different viewpoints and how to confront or combine these viewpoints in a dialectical or loop-oriented process. For example, how to cope with the engineering viewpoint stressing efficiency, precision, functional determinism and use of technical tools in the same design approach in which sociologists are focusing on conflicts of interest and different means of social power exercise? Or another example: how to deal with the cybernetic perspectives of connectivity, redundancy and double-loop learning in the same design approach in which anthropologists stress the symbolic meaning of day-to-day rituals, cultural traditions and family networks?

Taken separately, every viewpoint highlights certain interpretations of design criteria, processes or products. At the same time it tends to force other viewpoints into a background role. The opportunity to achieve a united shaping process depends on the learning situations and the subjectively bounded learning abilities of members of the design team. The different kinds of dialogue processes of social shaping (prototyping, metaphors, organisational theatre, and integrated interaction) are possible ways of initiating the shaping process of a multidisciplinary design team. They are, however, context dependent. A given method may succeed in one situation and fail in another. The context of our actions is made up of a succession of situations that we walk into, and to which we respond.

Insofar as the alternatives of action that the cultural context provides are seen as constraining the social imaginations of the individual a tension or opposition begins to develop between the person(s) in question and the others bounded by the cultural traditions. These kinds of opposition may be handled in different ways. If the fundamental paradigm of the design approach is a shared system of culturally established symbols and meanings which should not be questioned the 'deviators' will soon feel the pressure to (re)adapt their imaginations of transcending values or ideas. In this case cultural diversity is viewed as a threat against established order. In contrast to this view a more open interpretation may be the following: the continuity or stability of culture does not depend on a stable body of shared meanings, but on our ability to be 'within' and 'transcend' traditions in a dialectical process.

Integration of social and technical design is not a set of skills, techniques or procedures but should be thought of as a series of mutual interpretations and interactions occurring simultaneously rather than in a linear sequence. For example, engineers are instrumentally oriented and would often prefer the social scientists or the users to present their wishes or imaginations through means which the technical specialists could use to find possible technical solutions. Another example may be that the engineers having decided upon the main functions of the innovation, wants social scientists to convince users to accept the innovation. A third example may be the following: social scientists create an ideology which they believe is the best solution for the users. The engineers are then asked to shape the technical tools or standards appropriate for the realisation of the sociological imaginations. None of the above mentioned examples are simultaneous as the original meaning of this concept. Nor are they symmetrical in structure, that is characterised by an equal power relation. These examples illustrate different possibilities of domination patterns opposed to a simultaneous process.

A simultaneous process is occurring *when* the technically possible as well as the socially desirable interact as open-ended 'worlds' rather the predetermined, closed systems. The principle of cultural diversity as a criteria of integrated design is essentially stressing the necessity to keep several open-ended 'worlds' interfering or colliding with each other without seeking the dominance position at the cost of the other worlds.

Nevertheless, oppositions between technically- or socially-oriented designers of the same group may take on the character of *contradictions*. How does one handle internal contradictions in a design group between different kinds of paradigms? I have already discussed some possible ways of handling them by using a special coordination group or inviting persons from outside the group to evaluate or judge the internal contradictions of the design group. Nevertheless, it may turn out to be a failure too. The 'gap' between the different positions in question may be too fundamental. These kinds of fundamental gaps may result in dissolution or a bitter 'fight', but if the dialogue principles are taken sufficiently seriously and respected by all the participants of the group, the gaps may be the focus of the common endeavours to deepen the understanding of the culturally- bounded thinking and acting of the design team. In such a situation, double-loop learning may occur in the technical as well as in the socially-oriented design process.

Both parts of the design group may begin to practice reflection-in-action. By means of organisational theatre, prototyping and/or future workshops the synthesis may be elaborated at a higher level, even if the participants do not agree internally. Thus cultural diversity may be viewed as an opportunity to develop new products and processes of design instead of being perceived as a constraining factor to common design actions.

Conclusions and Perspectives

One of the reasons why integrated design approaches may become a focus of attention during the forthcoming decades is the irreversible consequences of continuing environmental and human exploitation. New paradigms emphasising the interconnectedness between individuals, societies and nature are becoming more widespread throughout the Western world, even though the mechanistic world view is still very powerful both as a scientific paradigm and as a basis for decision making at a global level. The designers have been as heavily implicated as anyone in issues concerning the environment and human exploitation.

Pollution problems and ecological threats that were first recognised during the 1960s and 1970s were often solved with quick 'technological fixes'. Many solutions of these kinds only masked or intensified the problems. Acid rain is a case in point. Some companies installed scrubbing equipment and afterburners on their smokestacks, thus limiting pollution at the sources of elimination. In the 1980s several approaches of 'cleaner technology' began to be a promising strategy of 'greening' the industry. Environmental aspects should be part of the industrial designers considerations regarding his/her choice of tools, methods and materials. It seems that integrated design approaches combining ecological, technical as well as economic perspectives are growing rapidly at least in some of the most technologically advanced countries. However, this raises the question of whether such approaches are sufficient to deal with environmental problems or are they just a 'technological fix' on a higher level?

A human-centred design approach demands that we establish at what *level* of complexity the problems belong. Are we, for instance, dealing with a process that must be redesigned, or should we rethink of the product itself or even the needs behind the product in question? In the case of ecological and human exploitation the socially desirable production must be independent of concern for the Gross National Product. The paradigm that we must buy more, consume more, waste more and throw more away might have been consistent with human development for a particular historical period, but may not be valid for the future. Neither purely technical design 'solutions' nor purely social design 'solutions' are sufficient. Fundamental changes of social needs and values must be integrated with the new way of shaping technological tools and products.

In order to place the level of design problems in the appropriate frame of reference the *historical* perspective of the problem must be envisaged. Thus the historical perspective of reuniting different design traditions may be viewed as a shift of paradigm from mechanical to human/ ecological world view. Or the mechanical world view may be extended into a new level as the more sceptical analysts might claim. Whatever is the case, the historical perspective helps the analyst to create a necessary frame of reference regarding culturally bounded patterns of design alternatives.

Different kinds of inhibitors or perceptual, emotional, intellectual blocks that keeps us from solving a given task in a certain way may become more understandable if we are aware of our particular cultural heritage compared with other cultures. Furthermore, a historical perspective of the problem in question may contribute to the understanding of not only the 'surface' or 'symptoms' of the problem, but the deeper levels too. Thus, an early warning can be recognised against too superficial solutions, such as for example the above mentioned 'technological fixes' of coping with acid rain.

Furthermore, integrated design must place the problem in its *social* perspective. For example, factory automation may release workers from boring labour functions, as well as it may release them from industrial labour at all. What is the social perspective of such a development? Taken step-by-step no single design approach can envisage the social consequences. The redundant workers may find more interesting jobs or be more-or-less permanently unemployed. Furthermore, automation above a certain degree has proved to be counter-productive even from a strict operating economic viewpoint. Nevertheless, automation, as such, is not the 'devil'. It is the one-dimensional design approach that focuses on just one particular factor: cost reduction on a plant-level, without reflecting the design solution in a social perspective.

If the social perspective is integrated in a given automation approach the point of departure may be taken in the combination of skills of the worker, the long-sided perspective of competitiveness of the industry as well as the possible new ways of shaping jobs in new areas of regional or urban development. In such an integrated design project the list of design criteria may be more comprehensive and therefore more complex to handle in practice. However, if the outcomes are more comprehensive design solutions, additional costs may be worth considering in particular if the paradigm of economic growth is substituted with the broader reference of human development. Integrated design is comprehensive: it attempts to take into consideration multiple factors and perspectives which influence the conditions for human development.

The most important objective of social shaping is to transcend the deterministic viewpoint that the world-of-tomorrow will be like the world-of-today, except for 10% increase or decrease of everything. The ultimate objective of social shaping of human development is to transcend the 'straitjacket' of the mechanistic world view of only one probable future.

If the deterministic perspective of *the* future is rejected and several futures are perceived as possible, human-centred design approaches have to comprehend *utopian* as well as *instrumental* and *reflective* aspects of social sustainability. The relationship between instrumental and reflective aspects has already been discussed as a matter of double-loop learning. By introducing the *utopian* element, a further aspect of the design process as social learning in a wider sense is becoming visible.

The German philosopher Ernst Bloch (1986), makes an important distinction between abstract and concrete utopias. *Abstract utopias* are transforming and wilful in the sense of driving forward to action and a real, possible, transformed future.

The challenge then is to include the development of utopias as part of the human-centred design approaches in such a way that utopian images are strengthening and not weakening social change toward a more socially sustainable society. How does one cope with such a challenge?

In the case of using *prototyping* to develop new software techniques or new job profiles of an organisation, the utopian element may be included in several ways. During the phase of *explorative prototyping* several images of prospective use-situations (if it is a product) or prospective communication patterns (if it is an organisation) may be developed reflecting not only the existing state or context but also other possible social contexts that may fit even better to the perspective of increasing social sustainability. Development of different kinds of concrete utopias (here in a small scale) may help the design team to widen the scope of possible solutions and ways of reaching the solutions. In practical design approaches such utopian thinking is sometimes an integrated element of the initial stages of development, and sometimes not. Often, the process of explorative prototyping is reduced to new ways of introducing a 'smart' technique or method without asking the question of the possible social context in which such a technique should be used. In this case both the utopian and reflective aspects of the design process have disappeared and only the instrumental aspect is left. The utopian aspects of using 'organisational play' or to imitate a theatre as a means toward new organisational design become more visible, if the instruction of the play allows or even deliberately comprehends different kinds of social shaping during the play. The instructor should not introduce fixed roles *a priori* but let the actors develop them during the play. Part of the play may include shaping different concrete utopias and another part of the play may focus on how to realise or not realise the developed images of prospective organisations.

The *integrated interaction model of social shaping* (*ibid.* page 223) very adaptable regarding mixing utopian, reflective and instrumental aspects of social learning. By means of 'future-creating workshops' end-users as well as social scientists and technical specialists from the beginning may shape their own concrete utopias. By the first 'shaping workshop' of mixed groups the original versions are reflected and maybe further developed or changed completely. During the 'second shaping workshop' the original groups of separated end-users, social scientists and technical specialists have another possibility to alter the versions or to develop them further towards an operational action level. This cyclical process may continue until different versions are made sufficiently concrete in order to initiate actions and change in a given social context, either an organisation or a local community. From a human-centred perspective the question is not *if*, but *how* utopian thinking should be part of integrated design approaches. The examples above are just a few of many other possibilities. The point, however, is not just to call for more and better utopias, images or maps of possible futures, *but* to do this in close interrelationship with possible points of interventions and agents of change of the existing state of circumstances. Utopias without reflected actions are escapism. Utopias related to reflected actions are a promising way of transcending the determinism of the mechanistic world view.

What kind of changes inside and between different institutions of the industrial societies may help developing these kinds of integrated design practices?

From an *educational and research* perspective, multidisciplinary design centres including humanists, social scientists, media professionals, artists and engineers are obviously an opportunity to strengthen the *integrative* design approach, but no guarantee in itself! The crucial point is not the spatial closeness of these different design traditions *but the quality of dialogues* they are able or willing to develop. Instead of just establishing such a multidisciplinary design centre from the beginning and then let it develop by itself, depending on the different practitioners engaged at the centre, it might be better to make a loose connection between already established multidisciplinary groups and then gradually building an *open* centre in a particular regional context with close contacts to the public outside the centre as well as to similar centres in other regions. A centre like this may have a better point of departure with regards to including the empowerment of the local people in the design approaches and thus establish a mutual social learning process inside as well as outside the centre. Regionally-based networks including design specialists as well as actors from industry, social services and local communities may be the most promising way of strengthen human-centred social and technical design approaches in the future.

References

Axelsen, T., Finset, A. (1973) Aktionsforskning i teori og praksis. Cappelens forlag, Trondhjem.

Babbage, D. (1832). *On the economy of machinery and manufactures*. Charles Knight, London.

Balzert, H. (1986): Programmierung. Reihe Informatik 50: 37–51 Bjerknes, G, Ehn, P., Kyng, M.(eds.), (1987): Computers and democracy: a Scandinavian challenge. Avebury Books, Aldershot.

Bateson, G. (1972) *Steps to an Ecology of Mind*. Ballintine Books, New York.

Bateson, G. (1979). *Mind and Nature: A Necessary Unity*. Fontana Paperbacks.

Björn-Andersen, N. (ed.), (1980). Artikelsamling til SociotekniskSystemboustruktion. IFA Publications, Copenhagen.

Bloch, E. (1986) *The principle of hope*. Basil Blackwell, Oxford.

Bråten, S. (1983) Dialogens Vilkår i Datasamfundet, Universitetsferlaget, Oslo.

Brödner, P. (1985). Fabrik 2000: Alternative Entwicklungspfade in die Zukunft der Fabrik. Sigma-Verlag, Berlin.

Brödner, P. (1986) 'Skill based manufacturing versus unmanned factory: which is superior?', *Int. J. Ind. Ergnomics*,1: 145–153.

Christianson, L. *et al.* (1971) Kunsten at dressere mennesker. Hans Reitzels forlag, Copenhagen.

Cooley, M. J. E. (1987). *Architect or bee? The human price of technology.* Hogarth Press, London.

Corbett, J. Martin, Lauge Baungaard Rasmussen and Felix Rauner (1991). *Crossing the Border.* Springer Verlag, London.

Doi, T. (1986). *The Anatomy of Self.* Kodanska, Tokyo.

Dreyfus, H. L., Dreyfus, S. (1986). *Mind over machines.* Basil Blackwell, Oxford.

Ehn, P. (1988). *Work oriented design of computer artifacts.* Arbetslivscentrum, Stockholm.

Ehn, P., Sjögren, D. (1986). 'Typographers and Carpenters as Designers' *in Proceedings of Skill-based Automation.* Karlsruhe.

Ehn, P., Kyng, M. (eds.), (1987). *Computers and Democracy.* Avebury, Aldershot.

Ehn, P. (1988): *Work-oriented design of computer artifacts*, Arbetslivscentrum, Stockholm.

Enderud, Harald (1979). Kvalitetskriterier for teoriudvikling og praksisorienteret forskning i organisationssociologien. Copenhagen, Nyt fra Samfundsvidenskaberne.

Fayol, H. (1949). *General and industrial management.* Pitman, London.

Floyd, D. (1984). 'A systematic look at prototyping' in: Budde, R. (ed.) *Approaches to prototyping.* Springer-Verlag, Berlin Heidelberg New York.

Foucault, M. (1980). *Power/knowledge.* Pantheon, New York.

Galtung, J. (1988). *Methodology and development.* Chr.Ejlers Forlag, Copenhagen.

Gill, K. S. (1989). 'Reflections on participatory design', *AI and Society*, vol. 3, no. 4.

Gill, K. S. (1990). 'AI and Social Citizenship: an anthropocentric approach'. *Tenth European meeting on cybernetics and system research.* EMCSR 90, Vienna.

Gill, K. S.(1990): 'The computer and the Social Citizen: Technology for participation', Proceedings of the international conference on 'Computer, Man and Organisations I. Bruxelles.

Gustavsen, G.(1987). Reformer på arbetsplatsen och demokratisk dialog i Odhnoff,J. & van Otter,C.(red): Arbetets rationaliteter, Arbetslivscentrum, Stockholm.

Habermas, J. (1981). Theorie des kommunikativen Handelns, Suhrkamp, Frankfurt/Main.

Jones, J. C. (1984). *Essays in Design.* John Wiley & Sons, Chichester, p.7.

Koslowski, K. (1988). Unterstützung von partizipativer Systementwicklung durch Methoden des Software Engineering. Westdeutscher Verlag, Opladen.

Kristensn, K.M. and Nielsen, B.S. (1979). Arbejder-studenter-samarbejdet i Kontext nr.38.

Kuhn, T. (1970). *The structure of scientific revolutions* (2nd edn). Chicago University Press, Chicago.

Læssøe, Jeppe, Lauge Baungaard Rasmussen, Poul Tøttrup (1989). *The Electronic Sketch Pad.* DTH, Copenhagen.

Læssøe, Jeppe, Lauge Baungaard Rasmussen (1989). *Human-Centred Methods.* DTH, Copenhagen.

Melucci, A. (1992). 'Liberation or Meaning?' in Jan Nederveen Pieterse: *Emancipations, modern and postmodern.* Sage Publication, London.

Mooney, J. C., Reiley, A.P. (1931). *Onward industry.* Harper and Row, New York.

Morgan, G. (1986): Images of organisation. Sage Publications,London.

Noble, D. F.(1977). *America by design.* Alfred Knopf, New York.

Noble, D. F. (1984). *Forces of production: a social history of industrial automation.* Alfred Knopf, New York.

Numinen, M. I. (1988). *People or computers: three ways of looking at information systems.* Studentlitteratur, Lund.

Oppermann, R. (1983). Forschungsstand und Perspektiven partizipativer Systementwicklung. Oldenbourg, München.

Perrow, C. (1983). 'The organisational context of human factors engineering', *Administrative Sci*, Q 28: 521–541.

Pieterse, Jan Nederveen. (1992). *Emancipations, modern and postmodern.* Sage Publication, London.

Polanyi, M. (1957). *Personal knowledge.* Routledge and Kegan Paul, London.

Polanyi, M. (1967). *The tacit dimension.* Anchor Books, New York.

Rauner, F. (ed.), (1988a). *Gestaltung: Eine neue gesellschaftliche Praxis.* Verlag Neue Gesellschaft, Bonn.

Rauner, F. (1988b). 'Die Befähigung zur (Mit) Gestaltung von Arbeit und Technik als Leitidee beruflicher Bildung' in: Heidegger,G., Gerds, P., Weisenbach,K. (eds) *Gestaltung von Arbeit und Technik: Ein Ziel beruflicher Bildung.*

Rauner, F., Rasmussen, L. B., Corbett, J. M.(1988). 'The social shaping of technology and work: human-centred computer integrated manufacturing systems', *AI Soc.*, 2: 47–61.

Rosenbrock, H. H. (1977). 'The future of control', *Automatica*, 13: 389–392.

Rosenbrock,H. H. (1980). 'Human resources and technology' *in Proceedings of the sixth world congress of the international economic association on human resources, employment and development.* Mexico.

Rosenbrock, H. H. (1983). 'The social and engineering design of a flexible manufacturing system' in Warman, E. A. (ed) *CAPE '83 Part 1.* North-Holland, Amsterdam.

Rosenbrock,H. H. (ed), (1980). *Designing human-centred technology.* Springer-Verlag, London.

Sandberg, T. (1982). *Work organisation and autonomous groups.* Liber Forlag, Uppsala.

Schön, D. A. (1983). *The reflective practitioner.* Basic Books, New York.

Smith, A. (1776). *The wealth of nations.* Stratton and Cadell, London.

Suchman, L. A. (1987). *Plans and situated actions: the problem of human-machine communication.* Cambridge University Press, New York.

Taylor, F. W. (1911). *Principles of scientific management.* Harper and Row, New York.

Thorsrud, E. (1972). *Worker participation in management in Norway.* Institute for Labour Studies Publications, Geneva.

Trist, E. L., Bamforth, K. W. (1948). 'Some social and psychological consequences of the Longwall method of coal getting', *Hum. Relations*, 4: 3–38.

Ulich, E. (1978). 'Über das Prinzip der differentiellen Arbeits-gestaltung', *Industrielle Organ*, 47: 566–568.

UTOPIA (1981): Report Number 1. Swedish Centre for Working Life, Stockholm.

Weizenbaum, J. (1976). *Computer power and human reason.* Freeman, San Fancisco.

Williams, R. (1961). *The long revolution.* Chalto & Windus, London.

Young, Iris Marion. (1993). *Justice and the politics of difference.* Princeton University press, New Jersey.

Notes

1 Thus, Adam Smith's *The Wealth of Nations* (1776), praising division of labour, was followed by Eli Whitney's public demonstration of mass production in 1801. Then, in 1832, Charles Babbage, inventor of one of the earliest forms of mathematical computers, advocated a scientific approach to organisation (Babbage, 1832).

2 This statement is developed more in detail in: J. Martin Corbett, Lauge Baungaard Rasmussen and Felix Rauner (1991). *Crossing the Border*. Springer-Verlag, p. 45 ff.
3 In 1973 all three countries passed laws ensuring the employees' right to be represented on the Board of Directors.
4 Between 1964 and 1967 four experiments on work organisation were carried out in different Norwegian organisations.
5 This definition differs from the traditional engineering one in which a prototype means the first of a kind or type where the aim is to mass-produce goods of the same type.
6 See note 5.
7 A detailed description and evaluation of the whole project is presented in: J. Martin Corbett, Lauge Baungaard Rasmussen and Felix Rauner (1991). *Crossing the Border*. Springer-Verlag, London.
8 Paper presented at the workshop held in ESPRIT 1217 on 23 January 1989.
9 Paper presented at the workshop in ESPRIT 1217 on 23 January 1989.
10 Paper presented at the workshop in ESPRIT 1217 on 23 January 1989.
11 Paper presented at the workshop held in ESPRIT 1217 on 23 January 1989.
12 Summary from a workshop in ESPRIT 1217 on 8 February 1989.
13 Summary from workshop held in ESPRIT 1217 on 23 January 1989.
14 The present presentation is based on the experiences from a Swedish project: 'Forandring och Utvickling' (Change and development) carried out by Ehn, Møllerud og Sjögren. 1987–90.
15 The model is based on positive and negative experiences within ESPRIT project 1217 'Human-centred CIM Systems'. This description is a revised version of the model described in: J. Martin Corbett, Lauge Baungaard Rasmussen and Felix Rauner (1991). *Crossing the Border*. Springer-Verlag, London.
16 Based on an article: 'Research roles and quality of research in collaboration with actors', written by Jeppe Læssøe and Lauge Baungaard Rasmussen in *Deltagelse i teknologisk udvikling*, (Cop. 1992), (Participation in technological development).
17 Turner, Barry, A. (1991) 'Organizational learning and Safety Management', draft for discussion at the Copenhagen Workshop.

Chapter 7

Information Systems Design: a User-Involved Perspective

Siv Friis

Introduction

The user-involved design model, the PROTEVS model, is based on research on active user-involvement in information systems development (ISD) of computer-based information systems (CBIS). It is directed towards development and change where the future users are in control of the design process. User-controlled information systems development (UCISD) is thus defined as:

- when the future users take the responsibility for the design, i.e. when they make binding decisions about the design project;
- when the users are regarded not only as the problem owners, but also as the problem solvers.

The PROTEVS model was developed during the course of a number of case studies on 'design in practice'. These involved participating actors in all stages of the process of information system development. Prototyping is the main technique in this model and is used throughout the experimental design, testing, use and evaluation. The aim of the development process was to enhance active learning, communication and active user participation. Participation is seen as the democratisation of the processes of the development of information systems. This kind of democratisation is described as: the solution-generating ability of people in the area of problem solving and the change process (Emery & Thorsrud, 1976; Emery & Emery, 1989; Gustavsen, 1985, 1990).

The research on 'design in practice' involved four cases studies called the Lund, Malmö, Falkenberg and Factory case studies. These case studies highlighted a number of similarities arising in different case study experiments. For example, the experiments showed a temporary change of user roles arising from the learning progress of the information systems development model. They informed of blindness of user perspectives and taken-for-granted perspectives of different actors involved in the case studies; and they illustrated a discrepancy between the organisational model for information systems development and the model in actual use (cf. Argyris & Schoen, 1978; Salaway, 1987).

The knowledge acquired throughout these case studies offers a basis for new ways of action for future users of computer-based information systems acting as participants in their design, evaluation and change. Apart from the resulting PROTEVS model the experiences gained during the development of the model form the basis of locally developed theories for a suitable environment for user controlled ISD, an environment described as a 'local design shop'. In a local design shop the PROTEVS model may be one possible approach to future 'local theories' about change in work organisation, resulting from computer-based information systems.

Frame of Reference

Workplaces are managed with the aid of organisational information systems (IS). 'Organisational' is here interpreted as having both an organisational level and a functional level. Olerup (1982), defines IS as:

> ... activities for collecting, storing, processing and distributing information about an organisation and its environment.

Today, some of these processes are automated by means of computers, the new tools of information technology, and many workplaces are managed with the aid of computer-based information systems. These new tools, however, are used for processing only the formalised parts of an information system.

For the purpose of further clarifying the concept of computer-based information systems and information systems research I will use the proposed framework of Ives, Hamilton and Davis (1980). They define the concept of management information system (MIS) as:

> ...a computer-based organisational information system which provides information support for management activities and functions.

It is further claimed that:

> ...MIS may also be called an organisational information system, a computer-based information system or information system (IS)

This framework termed 'information systems' includes both transaction systems and management-oriented systems.

A complete information system is thus defined as 'a collection of subsystems (ISS) defined by functional or organisational boundaries' (Ives, Hamilton and Davis, 1980). An ISS is also referred to as an application system. One might add that an information system also comprises activities of communication between actors within and between functional organisations in a workplace (Nissen, 1976). Information systems development for larger organisational systems are traditionally divided into two processes:

1. The design process, controlled by professional systems analysts, aimed at producing a requirements specification and/or a systems specification.
2. The construction (programming) process, controlled by programming experts, aimed at producing both the final system (application) and the security system for that application.

The design and construction process of information systems development for small personal and/or departmental systems, may become an interwoven process controlled by the future users of these systems (Friis, 1984). When desired, these small systems may be integrated into larger organisational systems by the programming experts, with all the necessary compatibility, quality, and security arrangements.

Prototyping

A prototype system has been defined as:

> ... a hard/software artifact developed by programmers with the purpose to evaluate a future computer based information system.
>
> *Rzewski, 1983*

I will go one step further and claim that a prototype system should be designed by the users of a future CBIS because a limited model of a known work-routine is useful for achieving a better communication between data processing (DP) experts and the users. Hence, prototyping is not only about involving users in a first try of a final system, as described in Rzewski (1983). User prototype systems may be thrown away as recommended by Floyd (1984), and Friis (1984). They may also be used as learning vehicles for newcomers, i.e. the next generation of users to evolve continuous experimental design (Friis, 1984).

Today, prototyping is one of the main techniques where user participation is part of the information systems development process. Two main forms of prototyping can be distinguished and applied in information systems development and CBIS research:

1. *Rapid prototyping*, where the DP experts create the technical design of a prototype system. An initial prototype should be developed very quickly, almost overnight (Naumann & Jenkins 1982), hence the notion of rapid. According to Floyd (1984) this prototyping is defined as: '*Processes which involve an early practical demonstration of relevant parts of the desired software on a computer, and which are to be combined with other processes in systems development with a view to improving the quality of the target system.*' The rapid prototyping technique is mainly performed during the requirements specification phase in an information systems development process (Naumann & Jenkins; 1982, Jenkins, 1983). Rapid prototyping may also be used for software evolution (Luqi, 1989).

2. *User prototyping*, where the future systems-users, often in collaboration with DP experts, are the main designers of the prototype systems. User prototyping is defined as a technique that allows even the future users to develop a model of a future CBIS. This model can be designed manually or on a computer (Friis 1984). A model, as discussed in Smith (1985):

> ... deals with its subject matter at some particular level of abstraction, paying attention to certain details, throwing away others.

A user-designed prototype system would probably pay attention to certain requirements of a new or changed IS. A requirements specification for a future information system is defined as either a document, a manual prototype system or a computer-based prototype system, illustrating the future needs and requirements as seen from the future users' point of view, and expressed by them. The requirements specification may not necessarily comprise all the requirements, but it is important that the most essential requirements are illustrated. Thus a requirements specification is not necessarily a finished product.

Theoretical Perspectives

In practice, there is a need for both human knowledge and technical knowledge in a workplace situation which involves both humans (intentional people) and formal logical machines (e.g. computers), and data processing techniques (Nissen *et al.*, 1982). One way to achieve this symbiosis of knowledge is to create an alternative approach to information system development. Such an approach works in dialogue and involves not only the professional DP experts who have the technical knowledge, but also the people who act in the work situation. They are all subject to change, i.e., they are people who have the human and contextual knowledge and the personal knowledge that may be either tangible or 'tacit' (Polanyi, 1958, 1983).

Taking its human aspects seriously, the discipline of systems development belongs to the domain of social science. Hence, when searching for a theoretical perspective in order to understand the nature of knowledge and gain insights into the research process, I have looked to the social sciences. I have looked for guidance to the 'interactionistic field' in order to understand the documented data and actions of my research. The discussion on 'grounded theory' by Glaser and Strauss (1967), belongs, according to Cuff and Payne (1979), within 'interactionism', and borders the interpretative sociology. I am not a sociologist, and even though Glaser and Strauss (1967), clearly say that the development of a grounded theory is strictly for sociologists, I find their theoretical discussion relevant to CBIS research concerning user-controlled information systems development. According to the work of Glaser and Strauss (1967):

> The theory arrived at must be understandable to all participants in the research area, both researchers and laymen ... is derived from data and then illustrated by characteristic examples of data ... Generating theory from data means that most hypotheses and concepts not only come from data, but are systematically worked out in relation to the data during the course of research ... comparative analysis can facilitate the discovery of grounded theory ... This strategy involves the systematic choice and study of several groups...

The grounded theory is induced from contextual data. The theory requires related properties: it should have a close fit to the problem area; it must be understandable to both laymen and scientists; it must have sufficient generality for similar situations within the problem area and it must allow the user control over the daily situations.

The PROTEVS model concerns information systems development that have the characteristics of context sensitive processes. Hence it does not state *a priori* assumptions about the area of change, but it necessitates collaboration with people from the research area, e.g. both management and white-collar workers in an office workplace or blue-collar workers in an industrial workplace. There is thus a fit between grounded theory and research about user controlled information systems development. However, Glaser and Strauss (1967), aim at a grounded theory for sociologists and do not involve the layman in the construction of the explanatory grounded theory. In my view this line of thought should be extended to include the laymen as well, and according to Whyte (1991):

> ... It is a mistake to assume that only scientists have theories. It is impossible for anyone to behave in an orderly fashion without developing some kind of theoretical framework to provide a context for guiding actions in search of particular consequences.

With the grounded theory by Glaser and Strauss as a starting point, Elden (1983), argues that not only the sociologists but also the laymen should be active in the discovery of local theories and actions arising from these theories, because:

> ... not only shall the scientists decide what is the theory arrived at in a participative work research situation, but also the layman. The scientists shall, in dialogue, help synthesize the theories that will result in some kind of action.

The intentions behind the PROTEVS model stand in accordance with such a perspective. From this perspective, the development of a user-controlled working model for information systems development should involve only those people who are concerned with a specific work situation, and they should be the designers (the insiders) in collaboration with DP experts or researchers (the outsiders). The key factor is dialogue. This dialogue may also take place as communication in writing. The insiders may, together with researchers and experts, undertake an inquiry aimed at diagnosing a work situation. The result of the diagnosis may then be used as an explanation of why things are as they are. This explanation was first termed 'local theory' by Phil Herbst in the mid-1970s. The insiders/outsiders theory should then be tested in action (Elden & Levin 1991).

Dependent upon the research perspective, new kinds of knowledge may be more important than the knowledge traditionally acquired in social science research areas. In the area of 'user controlled information systems development' one may claim that research questions about systems quality and efficiency are most relevant. With the aim of supporting future users of CBIS as participants in their design, evaluation, and choice, other questions also become at least equally relevant, e.g., questions about learning and understanding.

The Problem Area

Recently, the practice of user participation in systems development has gained considerable momentum, at least up to the programming (and implementation) part, i.e. with the work of requirements specifications. However, the level of this practice may depend on the fact that user involvement traditionally means that DP experts approach the users and enquire about the requirements for a new information system.

Although the work with systems development and requirements specification is undertaken using well-defined and conventional systems lifecycle models (often adapted by experts to the specific organisation), the new systems have often been unsatisfactory. Consequently, the users call out for help in expressions like: 'This is not what we meant'. In my experience the reason for this is that there is poor communication between the users and the data processing (DP) experts. The same dilemma has been discussed by Naumann & Jenkins (1982), and Alavi (1984). Boland (1978) mentions that even when users and DP experts work very closely together they do not really communicate in the sense of understanding each other.

When working with computer-aided information systems one soon discovers that there is need for a specially formalised computer language, a language with only 'either/or' forms, because the computer does not accept a 'maybe' form. Because computers are machines driven by binary truth-value logic - i.e. true or false - it cannot accept ambiguities like 'maybe' forms. Sandström (1985), has reported a case where this had undesirable consequences in an IS application for the medical profession.

Today's development of experts systems, e.g. those drawing upon Zadeh's (1965, 1973), 'fuzzy set theory', has introduced a virtual 'maybe' form. However, this still cannot handle the 'fuzziness' of common language. The majority of business data systems will probably continue to be based on 'true' or 'false' statements as being the only alternatives for a considerable time.

Information phrased in the everyday language of a workplace thus still has to be narrowed into either/or forms to fit the computer. We formalise it. This means the creation of a different sub-language from the common one being forced upon people in work situations where computers are employed. A spoken or written language represents in itself a first normalisation of human perceptions and ideas. A computer language must be formalised further. The computer can not handle inconsistencies in input – see *Explorative Case Studies of Requirements Specification Work*, (starting on page 267) where I describe the concept of 'estate' used as input in a real-estate declaration system. The social security number furnishes another example of rigour. A person's name is not unambiguous because other people may have the same name, hence every person is given a unique number. Thus computer language is forced upon you.

In order to make users understand a computer language one often has to use metaphors. When I try to explain a computer-based system or even a prototype system, I use metaphors for illustrations.

For example, in my empirical work with blue collar workers (see *The Factory Case – Blue-Collar Workers*, starting on page 282) I found the word 'card-box' (kortlåda in Swedish) useful to illustrate a HyperCard system. After this illustration the users named HyperCard (and the illustrating prototypes made by myself) as 'Siv's-card box', using my name, Siv, as their metaphor. The level of abstraction in a computer programming system like HyperCard needs metaphors for people to understand the computational concepts and their use (cf. Järvinen, 1987, and Kensing & Halskov Madsen, 1991).

A formalised language used in order to enable computer processing of messages employs a much more limited set of concepts and words than professionals use when speaking about their work areas. The context to which these words belong is peeled off when used for data processing. The people who are not familiar with the new computer language will not always understand it or its implications.

We must realise that users on the one hand, and systems analysts including DP experts on the other, belong to two different professional worlds or cultures, and they have their own frames of references. They are both experts in their own fields with their own professional languages (cf. Greenbaum, 1990).

We are not born with a ready-made language, we must learn it from scratch. George Herbert Mead (1930, 1934), teaches us that we are socialised into society by learning the language used in that society. First you learn the language of the family and later you learn the languages used in different kinds of societies like for instance: national languages, scientific and professional languages, jargons and so on. Each time you are confronted with a new language in your daily life, whether it is jargon or a professional language, you must learn its characteristics. You must do this in order to be able to participate in that reality which is affected by this language and in which it exists. Berger & Luckmann (1967), express it in the following way:

> The common objectivations of everyday life are maintained primarily by linguistic significations. Everyday life is, above all, life with and by means of the language I share with my fellow men. An understanding of language is thus essential for any understanding of the reality of everyday life.

An office routine is an everyday reality for many people with its accompanying professional language and knowledge. Like many such realities it is exposed to changes, partly because the surrounding world changes, and partly because new tools for administrative management are developed. A new reality is then created for the people involved, which they must internalise. Otherwise they can not fully participate in it. Those who have the ability to learn the new language first are those who, by externalising themselves in it, will shape the new objective reality.

The computer is a new technical tool and a very effective tool when it comes to reshaping everyday reality since messages to be processed by it have to be written in a special language, a new formalised computer language. The participants of this reality are the users of the new language, the tool, and hence they are the ones who must either accept that reality or change to another.

To my knowledge these users meet the formalising problem for the first time when they are asked to specify their requirements for a new information system. How can these future users know how far to go in formalising their everyday language? They may even go too far. As a researcher of user-controlled information systems development, I ask myself the questions:

> How can the users state their requirements for a future computer-based IS, when they do not have the knowledge (and hence the language) to give good translations of their intentions with the system? How do they know what to ask for?

The users meet the formalising problem again when they use the new computer-aided information system in their daily work. They might fail to recognise the old routines and then also fail to notice the implications of the new technique. The new language contains words and concepts that are fitted for the formalised representation of the computer. The everyday reality is reinterpreted from these new suppositions. This means that professional language and knowledge may have to make way for a new language. The old language may even be experienced as an obstacle in the development of the new technique.

If you do not understand the new language you either need further learning or an *interpreter*. I have studied the working process of some existing requirements specifications (described in the section *Explorative Case Studies of Requirements Specification Work*, starting on page 267) with a view to demonstrating the necessity of an interpreter to help the users when they work with requirements specifications. Some of the questions I was interested in were:

- How far do the users believe they can go in the normalisation of natural language to computer language, and how far do the users want to go?
- What are the implications of a formalised computer-aided information system?
- How did the users participate in the work on requirements specification?
- How would they have liked to participate?

The last question is due to my conjecture that *users are both able to and willing to participate* actively in the design of an IS. The main problem raised in this work is:

> How can the future users' competence, knowledge, obligations and commitment be safeguarded, and how can the competence and knowledge be increased, during design and change of computer-based information systems?

Research Method: Action Research

In a traditional functionalistic perspective the researcher strives for objective results and sees the actors in a research area as objects to be observed and experimentally manipulated. In an action research perspective, the researchers have a more equitable view of the research objects. They do not manipulate objects, but 'collaborate with subjects' (Torbert, 1981). The latter is the perspective of the work presented here. This is close to the 'participative research' described by Jenkins (1984), as a methodology which 'allows researchers to become part of the research – to be affected by and to affect the research'. The objective is not the testing of hypotheses but instead the 'realisation of human creative potential'.

Kalleberg (1990), defines action research as constructive social science where the researchers, in order to learn from the reality and to influence it in a certain direction, participate actively in the projects. The researchers are not only observers, they participate in the work.

In the studies reported here I have worked in close cooperation with people, mostly from office environments, when testing the PROTEVS model in order to develop and refine it. The participants are considered as co-researchers. In the case studies and practical experiments their questions and our mutual experiences have shaped the final model. Hence, they have also contributed as evaluators of the research process, experiences and findings.

This cooperation is not without problems. As argued by Sandberg (1981), it is all very well if all participants are considered as researchers, but it also requires academic researchers willing to share their theoretical and methodological experiences with co-researchers, so that this knowledge may be further increased. There is also a possibility of over-reacting on the side of the co-researchers. It has also been my experience with action research, that disappointment is sensed in the group when the expectations of results have not been fulfilled. Often these high expectations are caused by the high ambitions of researchers.

In order to capture the perspective of the co-researchers and their current evaluations of activities during the practical experiments, I have used 'open diaries'. These have had the form of both ordinary notebooks and/or loose notepaper stacked in a folder. In the first planned case study I was the only person who kept a diary. However, the book was kept open for all participants to read. Mathiassen, Nielsen and Jepsen (1986), define diaries as private notebooks which are not shown to other people. Since my kind of 'notebook' was open to all participants in the research projects, I decided to call it a *logbook*.

In the empirical studies (see the section *Empirical Tests for the Development of the PROTEVS Model*, starting on 272) we all used logbooks. One for each researcher and a shared one for the participating users. The logbooks have been there all the time and in case of evaluation, negative as well as positive, the participants wrote it in the logbooks. The *logbooks have thus functioned as instruments for documentation, communication and evaluation*. The logbooks of the researchers have been the main source of recapitulation and planning in the different practical experiments.

In each project I had initial problems with getting the participating users to write in the logbooks. They claimed they did not know what to write, but as soon as the first politeness between us had worn off, they became bolder and quite 'talkative' in the logbooks – see the example below (author's translation):

> Nu botaniserar vi igen. Det måste hända något som är arbetsrelaterat – *We're leaving the path again. Let's have some work-related action!*

Hence, the documentation during the practical experiments was also done in cooperation with all the actors in the project. Both researchers and co-researchers made notes in each other's logbooks. Most references to the different projects in this work are taken from the logbooks.

Many of the meetings involved group discussions concerning the project process, results, and findings. These meetings were mostly documented by one of the researchers during discussion. Hence, many of the findings were discussed, documented and evaluated during the same meeting. Sometimes, though, the discussions continued over more than the period of one meeting. The *documentation undertaken by the researchers was always open for reading and correction* if needed, and for further discussions.

I argue that the collection of problems should be seen as the collection of data in a participative research situation. In my research, case study participants have also collaborated in evaluating the validity of the data. Thereby they have also contributed to validating the data gathered earlier.

There have been cases in field-work in which the author's interpretation of earlier data has become corrected during the evaluation. The generality of the findings is thus argued for in two ways:

1. In order to contribute to the generality of the findings, different groups of users have been studied, particularly by also including blue collar workers (Factory case), (cf. Glaser and Strauss' (1967) recommendation of theoretical sampling).
2. The validity of the findings has also been corroborated by the users and by other readers of the case reports and other papers based on these cases. Also the generality of the findings have been corroborated by the users and other readers due to open field data, and data understandable to others than the researchers (see Figure 25 below).

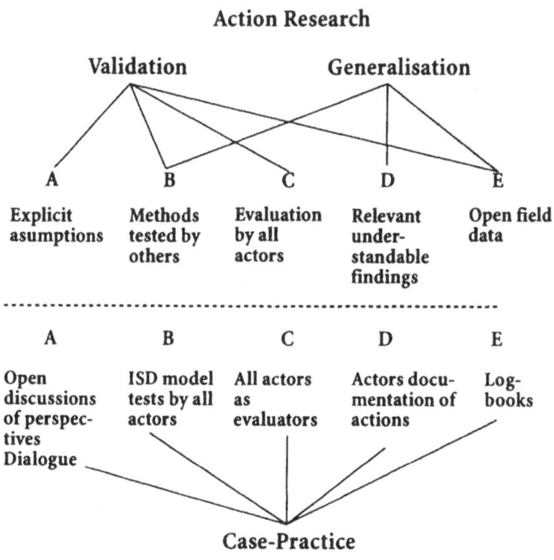

Figure 25: Criteria for validation and generalisation in action research methodologies and practice concerning ISD research

Beard and Easingwood (1989), are critical of the data collection and analysis approach for the development of 'grounded theory' recommended by Glaser and Strauss (1967). They describe it as very time-consuming and laborious, due to large quantities of qualitative data. To some extent I share this view. Glaser and Strauss recommend that only the scientists should do the data analysis.

As I see it, the data analysis should be done together with the actors supplying the qualitative data, as recommended in Elden (1979, 1983), and Elden & Levin (1991). The results of the data analysis in an area of change are documented as 'problem identification tables' and 'organisational maps', showing the relationships between causes and effects of a problematic situation. Based on these maps, 'local theories', i.e. solutions with inherent action plans, are developed.

In the case of information systems development, the insiders (users) and outsiders (researchers/DP experts) collaborate in dialogue in order to develop a local theory of the situation. Problem identification tables are helpful techniques used to gather the situational knowledge (data) from the insiders in accordance with Elden (1983). The problem inquiry is done in collaboration by researchers (outsiders) and people from the work situation (insiders). The problems are grouped in jointly accepted categories for mutual satisfaction.

According to Whyte (1991), local theories are not always written out or explicitly stated by the users (workers):

> A researcher can deduce their general outlines through observing consistent patterns of behaviour and/or through interviewing and examining written communications.

The problem identification tables presented in the section *Empirical Tests for the Development of the PROTEVS Model* (starting on page 272) have been designed in collaboration with the people at the work situation for the Falkenberg and Factory cases. In the Falkenberg case, I have also used data from three questionnaires given to the participants during the project. All the data have been confirmed by the participants in this case.

For the Lund and Malmö cases I designed the tables myself. These tables are based on the notices that were documented in the research collaboration with respective participants.

Hence, the joint documentation with the co-researching users of our findings constituted one way of data analysis. The joint use of 'organisational mapping' and construction of 'local theories', as described by Elden (1979, 1983), constitutes another way, which is the main analytical tool used in this work. The 'local theories', based on the maps, were designed partly together with the co-researchers and partly by referring to the respective logbooks.

These categories are then used to develop preliminary sketches of local theory, showing the cause and effect relationships between the problems listed. Arriving at a local theory is then not only a presentation of the causes and effects of a problematic situation, it is also an action plan towards a solution of the problem. This work is done by both management and workers. It is argued and shown by Elden (1983), that the white- or blue-collar workers' problem identification is both richer and more multifaceted than that of the management.

I have extended the problem identification technique to comprise also the *possibilities* seen by both insiders and outsiders during collaborative analysis work. To see problems as possibilities is inherent in the PROTEVS model.

The Research Process

This section is organised as a chronological narration of both the technological studies of techniques suitable for a user-controlled information systems development, exploratory studies for the development of the conceptual PROTEVS model, and the uses of the model in case studies and practical experiments.

Explorative Case Studies of Requirements Specification Work

Before I started the practical experiments with user prototyping for requirements specifications I undertook a study of requirements specifications processes in three large organisations. I was mainly interested in the questions stated on page 263. Also, I wanted to check my conjecture that the users really wanted to participate actively in ISD.

A Case Study of an Economy System in a Ceramic Industrial Plant

Early in 1980 I had the opportunity to follow the process of user-developed requirements specification in a very large ceramic plant. The intentions of the management of the organisation were to rebuild the current economy database in order to make it more accessible. As in many organisations that have relied on computer-aided information systems for the last decade, this organisation had quite a large amount of data in a mainframe computer environment, about 300 to 400Mb. This data was stored in different kinds of files with varying quantity definitions and with different retrieval facilities. This resulted in a great risk for errors in the statistical reports produced. If these errors were not discovered by chance they were very difficult to correct at a later time.

In this organisation, we were interested in undertaking a survey of the database for the economy system and, if possible, in building a report system together with the EDP staff. Firstly, we wanted to make a survey of the different functions of the economy system, so that we would be able to determine expectations and requirements of the future database. In one case, of the expense account routine, we had difficulties in understanding all the concepts and definitions. We did not understand what data were needed or where they originated from. Also, we could not remember all the different account numbers mentioned in the interviews. The users within the organisation could not understand these difficulties because they were all familiar and transparent to them. Through routinised experience their knowledge had become tacit.

Back at the university we decided to build a small limited model of what we thought was the expense account routine. The more we worked with the model the more it came to resemble a prototype for a new system though this was not really our intention. We had built a small menu-based retrieval screen. Still we decided to invite the users to see the prototype and to evaluate it. The prototype system sparked a very lively discussion among the users. First they corrected us where we were mistaken about the transactions of the data and the data media used, then they started to evaluate the structure of our model. They said it made them understand how a computer worked, so we called it the 'interpreting prototype'.

We told the users how the prototype was built, what programming language we had used, how it worked, and how the data structure was organised. In a very short period of time the users started to ask technical questions that amazed us because we thought they could not have gained all that knowledge in such a short time. The users even volunteered to build a model by themselves with a new structure for a more advanced prototype. We indicated that they should bear the time scale in mind when they started their work with the requirements specification, and have that document replaced with a prototype system.

Here, almost by accident, we learned that prototyping as a technique is quite useful for users' understanding, communication and learning. With reference to the example of the economy system, I discovered that the need for a prototype as an interpreter is important for both DP experts as well as users. Only a superperson can be an expert in all fields. So, why must DP experts be supermen? Why not use the expertise of the users?

I also learnt that prototyping furthers the possibility for a more process-oriented systems design. Once the users get to know the possibilities of the new very high-level programming language it is quite possible for them to participate more actively in the design of new computer-aided information systems, even to build their own systems, at least single-user systems and/or systems for smaller work units (cf. end user computing (EUC) in Davis, 1984; Gunton, 1989). This case also indicated that the users do want to try out prototyping by themselves.

A Case Study in Malmö Town Council – Real-Estate Department

Below I shall present the results of a case study of a real-estate declaration (revenue) system used in the real-estate department of the Malmö town council. The organisation needed this system to collect information about their real-estates, partly for the County Government taxation and partly for their own real-estate accounts. The collection and revision of the information needed for about 15000 estates was done manually. The information had to be revised to suit the declaration form issued by the Government. This was considered to take too much time: the form must be completed in two months from the date the organisation receives the notice from the County Government Board.

All the estates were regularly declared every fifth year, and the changes in the estates were updated every year. There were approximately 4000–5000 changes each year. The collection of information was done in many different offices, and sometimes the clerks had to go there themselves to fetch it. It was a rather difficult job to collect the right information and it took a clerk about three years to learn how to do it.

When the decision was made to replace the manual information system with a computer-aided system, it was also decided that the requirements specification should be done by the users, with the aid of a DP expert. Two users were selected to do the job. In the end it turned out that only one user - the head of the department - and the DP expert were to do the specification. According to the DP expert the reason for letting them participate only in the work on requirements specification was that 'the users do not have sufficient knowledge of computers and similar things'.

What they did was 'discuss' the possible requirements and on the basis of these discussions the DP expert made program fragments that were sent to the EDP department as specifications. When I asked this user if she felt in any way influenced by the expert during the work with the specification, she promptly said no. The DP expert had merely pointed out some of the possibilities and limitations of the computer. In principle she was allowed to work on her own.

There were additional discussions about some of the concepts though. Many of the concepts were hard to define and many had double meanings. For many users in the organisation the concept of 'estate' meant different things. It might be a house, a piece of land, a flat, and many other things. It would depend on what you were dealing with - accounts, sales, rentals and so forth. The output of the system caused no trouble but the input did. The new computer-aided system must have consistency in the input, i.e. an input concept must be unique in use. It meant that some of the estate concepts had to be renamed, hence a specially formalised language for person-to-person communication mediated by a data processing computer system was produced, (cf. Nissen, 1976).

There are cases today when even the declaration form is not sufficient for all input purposes. The clerks must then write a special note which is attached to the form. With the new system one could not write extra notes with the computer. According to the user there was no extra room for it. In the users' own words (*but the author's translation*):

> Because, if you did put it in, you would not be able to find it afterwards because the new system is menu-based and there is no function for that kind of retrieval. If there had been we could have used the system for many more tasks. Still, the new system will make the work with the information collection much easier. No more running about, you will get the information on the screen.

The new system could easily be used for more tasks according to the user, but the expert thought that the system would then become too big and too expensive. It could be used for estate accounts for other departments as well. But that implied that all users involved would have to agree on the exact meaning of the concepts. It would have meant a new nomenclature for the organisation.

The user in this particular situation could very well imagine the benefit if all concepts were agreed. She would even go so far as to rearrange the declaration form to fit the computer. Currently clerks have to fill it out manually, even with the new system. The form is not fit for a computer printer. This user has left the department for real-estate's declaration. She had shown such good results in the work on requirements specification that the organisation promoted her to the EDP systems planning department.

When it comes to the actual normalisation of the concepts for the computer, the work is done by the DP experts. In this particular case it was, as previously mentioned, undertaken in collaboration, but with only one user who, in turn, was anxious to continue the normalisation. Normalisation here simply denotes further normalisation of concepts already formalised, for certain other purposes, in the declaration form.

In a computer-based information system an expression like 'estate', used in several work contexts for different references, will cause trouble. This illustrates what I mentioned above in the *Theoretical Perspectives* section (starting on page 258) about a specially formalised language for person-to-person communication mediated by a data processing computer system. The work with the requirements specification must have had some impact on the user who was participating in the process as she was given a new position at the systems planning department. Personally, I have noted that those users who are selected for work on requirements specifications tend be those who express a distinctly positive view of computers. If they are proficient in requirements specifications they are likely to spend short lives in non-computer departments and tend to be co-opted by EDP departments.

A Case Study in Malmö Hospital – Computer-Aided Diagnosis System

I encountered a computer-aided diagnosis system when I visited a hospital on other matters and overheard a conversation among doctors, nurses and other staff about computer-aided diagnosis. I introduced myself and told them about my research and asked them if I might ask a few questions. They did not mind my questions but they did not want me to name the clinic nor the persons involved in the interview. This system had been in use for four years and was developed by the EDP department of the hospital. The basis for their systems development was the requirements stated by doctors and the laboratory staff at that time. The staff were all, more or less, of the same opinion of what should be included in the database. They knew what values and limits of variables were valid. For the rest they trusted the DP experts. The doctors and laboratory staff had no notions of the implications of a computer-aided system at that time.

The database was originally developed as a laboratory test-value file which was later integrated with a file of the more common diseases for the current tests. The system should help the doctors to single out the extreme cases for doctors to make a more thorough clinical diagnosis. Comments made by the respondents during the interview:

- The time perspective has changed our opinion on computers. Computers can be very restrictive.
- Many doctors have become computer diagnosticians. You trust the computer data – not your own knowledge.
- There are many more known diseases today, and we are getting more every day, which means that the contents of the database should be more variable.
- We know more today and we would actually want a completely different system.
- The computer could really be a good tool for doctors, if only the systems were correctly designed.
- Should you really let the computer decide whether a person is ill or not?
- The doctors who more often make clinical diagnosis have better results with their treatments.

Many of the original users had either left their jobs or been transferred to other clinics. Most of the current users did not know very much about the procedure of the systems development because they were newcomers, i.e. *second-generation users*. The conceptual model these users had of a computer-aided system did not correspond with their model of their work.

The interview gave me reason to believe that the normalisation of a decision process can have some very powerful effects on language, professional behaviour and culture. The tide of the time had changed the doctors' view of computer aid. They now believe more in the computer diagnosis than in the diagnosis based on personal experience. This could have a great impact on the next generation of doctors.

In summary, the experiences and findings in the exploratory studies implies that prototyping as a technique may be quite useful for both understanding a problematic work situation, communications and active learning, for example:

- prototyping furthers a more process-oriented design, it gives the users a possibility to participate more actively in the design of a new computer-based information system.
- prototyping in the form of *user prototyping* even enables the users to design their own small single-user applications.
- due to implementation of a new computer-based information system people must accept renaming of concepts, used in daily work, since the computer strips off contextual clues and does not accept ambiguity.
- the actual normalisation of the new concepts for the computer is done by DP experts.

- preferably users that favour information technology are selected (by the management) to participate in ISD work.
- second-generation users, who do not know very much about the ISD work for the current system, have a different conceptual model of the system than first-generation users.

The second-generation users learn only to handle the computer and trust the system (cf. Göranzon, 1990, for similar experiences). They become function-oriented, not origin-oriented. Therefore, they are likely to be limited in their views when a change in the system seems appropriate.

Empirical Tests for the Development of the PROTEVS Model

Based on the knowledge acquired from the three cases above I conjectured that the users do not have sufficient knowledge of data processing and hence do not have the language in which they can communicate with the DP experts when stating their requirements. They need some knowledge about both computers and formalised computer language to be able to state their requirements for a new information system. We need working models for this work which enhance learning. The users are often only allowed to participate as *interview respondents* in the more traditional models for ISD.

So where do you let the users in? One solution to the problem could be to let prototype systems act both as a communication tool and as a vehicle for learning in the process of systems development. The best thing would be to let the users themselves develop prototypes (models of small systems), and let a prototype act both as an interpreter and as a limited model for a new computer-aided information system. It could even be used to illustrate the requirements of the new system.

The PROTEVS model was tested and revised in real-life situations. Between 1983–1990 it was tried out in field experiments with skilled-white collar workers based mainly in organisations within the public sector. Since 1989 the model has also been tested in settings with blue-collar workers.

After the initial tests of prototyping with middle management and the economy system described above, I decided to try the prototyping technique with clerks at the personnel department of the town council in Lund. The findings presented in these cases have been presented mostly in Swedish articles and working reports published with the department of Information and Computer Science at Lund University.

The Lund Case – Personnel Department

Together with another researcher and 19 persons from the staff department at the Lund town council, I tried out my conjecture that the future users of IS, both want to and are able to design small prototype systems that could serve as requirements specifications. We met six times, always under the supervision of one of the researchers. During the meetings we introduced requirements specification work common to traditional design, where the DP experts are in charge of the design. As a contrast to traditional design we introduced 'user-controlled design' in accordance with the intentions behind the PROTEVS model.

We did not have the possibility to try out the prototyping on computers. Instead we produced manual prototypes on small wallcharts, and even went as far as to design conceptual relational databases. Both researchers and users wanted to prolong the collaboration in this case, but the offer was turned down by the management. The reason given was lack of resources. The findings and experiences in this case derive from notes documented during group discussions within the project and from working reports by Flensburg (1986, 1987).

Table 3: Problem and possibility identification – Lund case

Problems identified by	*insiders*	*outsiders*	*Possibilities identified by*	*insiders*	*outsiders*
Organization:			*Organization:*		
Lack of time	x	x	Communication	x	x
Discrepancy between the ISD model in use and org. espoused model		x	Conceptual database for the department		
Lack of manpower	x	x			
Project meetings:			*Project meetings:*		
Too short	x		Learning	x	x
No continuation		x	Enthusiasm	x	x
Information technology:			*Information technology:*		
Lack of techn. resources	x	x	–		

The EDP departmental budget did not allow the use of computers for user prototyping experiments. Even when the users wanted to continue to participate in the ISD work it was not possible due to lack of manpower and time. The research project was discontinued after producing the manual prototypes. Once the resources, time, money and aid of experts were used up, the organisation simply had to go on with the ISD work as recommended by the EDP department. The established organisational model for ISD was overruled, a model that comprised active user participation in ISD in accordance with the agreement of co-determination in

Sweden. (When we left the organisation the staff were only allowed to participate in the ISD as interview respondents to the systems analysts from the EDP department. It all resumed its normal pace in a traditional ISD.) To put it in the words of Emery and Emery (1989):

> Come the usual crisis and demands to tighten up, the same managerial prerogatives ... tightens also the progressive belt.

In this case the researchers realised that the art of constructing logical databases of the relational kind is not such a difficult work for the future users, i.e. people who have knowledge the workplace. One of the participants said: 'It was really easy to learn because we know all the data needed for the database'. The identification of problems and possibilities for the Lund case is displayed in Table 3. It was in the Lund case that I first noticed a discrepancy between the established organisational information systems development model and the model in actual use by the EDP department (cf. Argyris & Schoen 1978; Salaway 1987). This phenomenon was also later noticed in the next three cases. The discrepancy is further discussed in Friis (1991a).

The Malmö Case – Social Welfare Department

In this case I was allowed to follow some of the development work for a new mainframe computer-based information system for the administration of the town council in Malmö, as a passive observer. This work was controlled by the communal EDP department. I followed the requirements specification work for the new social security system for elderly and handicapped people for a year and a half. I also had the opportunity to perform some practical experiments with user prototyping. The experiments, performed according to the intentions behind the PROTEVS model, took place during the summer, 1984. All in all, the research project with planning, observations part, field experiments with prototyping, evaluation and documentation lasted from 1983 to 1985. For a more thorough discussion of the process and findings in this case, see Friis (1987, 1988).

Two local welfare district offices were allocated for the field experiments. There were all in all 23 users, out of 51, who participated, 8 from one office and 15 from the other. Some of the users participated only partly, due to summer vacations. Their deputies often replaced them in the project. During this time the users produced two small prototype systems: one 'telephone system' and one 'deputy system'. These prototype systems were later integrated into the overall organisational information system. In this project I again noticed that the future users both wish to and are able to develop small prototype systems that can illustrate their requirements by themselves, provided they have the tools and support from experts for the design work. This is evident in the EDP department's quick acceptance of the user produced prototype systems. These systems were later integrated into the new mainframe IS.

When the users analysed their own work situation in order to produce problem solutions, and requirements specifications, they accepted the responsibility for that solution, and the documentation of it. It was decided by the project that these documents were to be saved and used for educational purposes for the next generation of users.

Table 4: Problem and possibility identification - Malmö case

Problems identified by	*insiders*	*outsiders*	*Possibilities identified by*	*insiders*	*outsiders*
Organization:			*Organization:*		
Lack of manpower	x		Learning	x	
Lack of time	x	x	Communication	x	
Discrepancy between the ISD model in use and org espoused model.		x	Project responsibility		x
Management not sympathetic to project work	x		2 prototype systems	x	
			Document. reqs specs.		x
Project meetings:			*Project meetings:*		
Too short	x		Enthusiasm	x	x
Not all could participate		x	Learning	x	x
Users not aware of knowledge		x	Change in roles	x	x
			Open diaries (evaluations)	x	x
			Communications	x	x
Information technology:			*Information technology:*		
Lack of techn. resources	x	x	Learning	x	x
Tool kit for prototyping	x				

There was an apparent *change in roles*, both of the users and of the participating DP experts. The changing of roles strengthened the belief that there are learning abilities inherent in user-controlled design, as there are in any kind of action taken where one is creative and responsible. Even though the change in roles was apparent to the participating persons, they were not aware of the knowledge they had acquired, not even when I tried to tell them so. They only exclaimed: *'Why, this is really nothing – once we are shown how to do it.'* The participating users complained that they did not have enough time for the design work. If they were really allowed to do the prototype systems design for their requirements specification work, they should be given time for this work. The management did not realise this. They meant that the design work could be done simultaneously with the daily office work. The participating users in the PROTEVS project stated: *'Democracy takes time – so we need time!'*

My logbook was considered beneficial for both openness and communication in the project. The participating users indicated that more than one logbook could be used, enabling us to collect more valuable data. Such logbooks could also systematise data and actions, and the results achieved in the cases needed for the next generation of users. Thus the future logbooks might become instruments of both communication and evaluation.

The discrepancy between the established organisational information systems development model and the model in actual use was noticed in this case as well as in the Lund case. When the resources that enabled user participation in this case were used up, everything went back to normal. The DP experts took over. In the continued system design, performed by the organisation's EDP department, there was no time to involve the users in the design The DP experts who did not participate in the PROTEVS project were the ones who later changed the user-developed prototype systems beyond recognition. This made users antagonistic towards researchers and participating experts. The users felt that we had cheated them in some way, – i.e. helped them for a short time for our own scientific purposes. In user-controlled ISD the future users should have the last word in deciding when all the requirements are comprised in the final system. The problem/possibility identification table for the Malmö case is shown above.

The Falkenberg Case – Wage and Personnel Department

The experiments with user prototyping in Malmö were considered to be rather successful. For the sake of the credibility of the PROTEVS model it was decided that the model should be tested in other settings with clerks and secretaries. I approached the Falkenberg town council and was allowed another try there. The participating department in this case was the 'wage and personnel office'. The research collaboration with Falkenberg town council started in 1986 and was scheduled to continue until the summer of 1990. The findings in this case are discussed further in Friis,

(1991a, 1991b). The findings presented have been discussed and evaluated in group discussions. In this case we worked so closely and affectionately together that we decided, at one point, that we also needed questionnaires to evaluate some of the more problematic project situations. All participants had logbooks in this case: one for each researcher, and one communal for the participants.

The wage and staff department was actually waiting for a new computer-based system from the regional communal EDP department. They had already undertaken a traditional requirements specification in 1984. The users were asked to write their requirements specification but the consultant from the software house, who was to deliver the new target system, did the actual documentation. When we studied this specification later in the project we realised that it was a standard specification which had been used in earlier cases. The department was not satisfied with this specification. The town council hoped that the participation, in a research project concerning user controlled ISD would give the participating departments a possibility to do undertake better requirements specification on their own.

The research project had four phases: a one-year long planning phase, an introductory phase with a two-day scenario outside the workplace, a design phase at the workplace with three restarts due to breakdowns, and a concluding phase.

Two of the breakdowns were caused by lack of time and manpower within the participating department. The communal management and politicians had promised replacements for the participating users, but this promise was never fulfilled. The third breakdown was due to project fatigue with both participants and researchers. This was the most serious breakdown because we started to blame one another for the stoppage of the project. The trust of the participants saved the project and gave us an opportunity to understand this problematic situation. After the first year of planning activities (from autumn 1986 to autumn 1987), the department enquired into the interest of the staff to participate in a research project concerning user controlled ISD. There was a 100% interest. At the start of the project there were 12 participants from the department.

The actual design started in the autumn of 1987 with some introductory meetings where the procedure was very much the same as in my earlier user-prototyping projects. I introduced traditional ISD and user-controlled ISD. In dialogue with the participants, I introduced analysis and description techniques for ISD, and the PROTEVS 'working model'.

Most meetings were held at the 'lunch-table' in the middle of the workplace. This was not a very apt place since the participants were stressed by the telephones and the colleagues who were not attending the project. It was discussed by the participants that we should try to have a longer meeting in a neutral place, i.e. outside the work premises. A kind of full-day scenario was suggested.

The scenario was held in November 1987. For a day and a half the employees tested user-prototyping on computers. I and my colleague Roland Nilsson acted as DP experts. The users were divided in three groups. Each group selected a small current problematic administrative routine to work with: an 'off-duty personnel' system, a 'new employment' system, and a 'sick-leave reporting' system.

We had brought three computers and a printer for that purpose. After a short demonstration of techniques for conceptual modelling, of the Macintosh computers, and the REFLEX software system, the users were left alone with the design work. My colleague and I assisted only on technical questions about the machines and the software. The databases for the prototype systems were designed in third normal form of relational representation. In dialogue with the software the users were told how far to do the normalising. At the end of a day-and-a-half's design work, the users had three small workable prototype systems.

For the users, the scenario was rather successful and they were eager to continue the user prototyping, 'at home', at their workplace in parallel with daily work. They decided to have project meetings once a week with us. Meetings were planned to take place during work hours, every possible Wednesday. We installed eight computers, one extra 40Mb hard-disk and a laser printer at the office in a computer network and with the printer attached. The network served six (sometimes seven) workstations; the software used was mainly REFLEX (Borland), MacWrite and MacPaint. Later on HyperCard was also demonstrated and briefly tried out.

Very soon, after the scenario, we had the first breakdown in the project due to the lack of time of participants from the wage office. They had to rely on their colleagues for servicing the communal employees with, for example, assistance in answering telephones and taking care of visitors. This situation caused stress for the participants, and the project had to be delayed until they caught up with their routine work.

A meeting with the management, politicians, middle management, participating users and the researchers was held and the situation was debated. The management deemed the project useful and educational for the participating department. As a result of this meeting, it was decided to provide additional time and replacements for the staff who participated in the design project. During the next six months we were trying to renew the scenario atmosphere in the project. Between autumn 1987 and March 1988 some of the participants left their jobs and some became tired of project work. The group diminished from 12 to 5 people. After a second breakdown with a restart, two new people joined the project.

The meetings were now held in a separate room and we started the user prototyping in a good spirit. The participants were all working hard during meetings but the work stagnated during the rest of the week. The department was still understaffed and there were no extra personnel (replacements) in sight.

The participants did not feel they could justify user prototyping when we were not present. They even asked for more control from the researchers for activities such as homework between meetings. They actually wanted outside support for the prototyping work. The kind of comments coming from middle management were: *'Do you have time for this kind of play, don't you have any real work to do?'*

The researchers did not support the participants between the meetings. Very soon we had a climate of irritation in the project. This affected both researchers and participants. From the logbooks we read: *'Today there was a noticeable irritation among the researchers.'*

The politicians in the communal service, and the top management had promised extra personnel, i.e. replacements for the participants. This was not fulfilled. The participants said they had no support from the politicians. The politicians together with the communal management were the ones who decided what kind of IS was needed for the different departments. Hence, the participating users felt that it was not possible for them to influence the main design of the future computer-based information system. In reality, the new information system was already decided upon, and the only contact the future users had with the DP experts from the (regional) communal EDP department was tours of study.

I attended one of these educational meetings with the (regional) EDP department, and learned that the participating future users were very alert and inquisitive. They had the experts on their toes with their questions and discussions. The experts admitted that the users had many good ideas for change, but the system was very big and it would be very expensive to change now. The project work was hard for the participants and I noticed project fatigue, which I thought was due to stress and no replacements. I checked the logbooks but found nothing to indicate this. I brought the matter up in a group discussion and the participants began to loosen up and tell about the pressing situation. We also discussed the problem of 'aspect blindness' we had discovered in our group discussions. This was not only due to the 'expert language' used by me but, more to the fact that I had formed the project from my own perspective. I did not listen enough to the others (cf. Nygaard & Sörgaard, 1987).

Very soon we had a third breakdown, this time it was caused by ourselves. I should have noticed this, but I had been too involved in the research results – after all I was the project leader. On this problem we decided it would perhaps be better if one of the participants took over the project leadership. Thus, the participants took over the ownership of the project – otherwise the project would have died.

We decided it was time for a questionnaire so that everybody could have their say on the project, the project climate, knowledge acquired and so on. I volunteered to stay away for a while to let the participants discuss the questionnaire, the project and their own work situation. We decided to correspond by mail if there was anything deemed important.

We sent out 16 questionnaires, with 18 questions in each, to all the users who had attended the project at some time. The questions covered: what the participants experienced as new knowledge acquired in the project; if they experienced the project positively or negatively, if somebody had left the project; and, if they felt they had the support they needed for project work. The last issue gave a surprising result from the current and remaining 8 project participants; they only really experienced support from one another.

Even though the questions of the questionnaire were discussed and agreed upon by the project group before it was mailed, the questions were not experienced as easy to answer by the respondents. I sent two extra questionnaires in letter-form and asked the participants to discuss the questions in groups before sending the answers back to me. The questions were mainly about: why there was project fatigue, was there any reason for continuing the project, how they experienced the collaboration with the researchers, and if they saw any good for research work. The more significant answers were:

- They experienced the researchers as educators/teachers.
- They wanted more planning and control in the research project.
- All people concerned must participate, also management.
- The project was rather successful and they want to continue.
- The prototype examples used by the researchers must be work-related.
- They considered the PROTEVS framework as a user-controlled model.
- They wanted to spread their knowledge of user ISD to other departments.
- They now knew how to specify their requirements.

After mid-September 1988, the users wanted to continue the requirements specification work by themselves. Earlier specification work had been done under some supervision of the researchers. Now the users wanted to work all by themselves. Noticeable results from their work is one complete 'user-developed requirements specification' for a future CBIS of the wage and personnel department. This document also includes an action plan for possible continued user-controlled prototyping. So far this has been approved of by the governing politicians.

The main findings in this case was that there may be many shortcomings in a user-controlled ISD projects due to, for example:

- lack of manpower, hence lack of time for design work;
- not enough planning for user-controlled design work;
- lacked privacy (localities) for design work;
- lack of overall participation in design work.

Table 5: Problem and possibility identification – Falkenberg case

Problems identified by	*insiders*	*outsiders*	*Possibilities identified by*	*insiders*	*outsiders*
Organization:			*Organization:*		
Planning	x				
Lack of manpower	x		Participative ISD	x	
Lack of replacements	x		Scenario	x	
Lack of time	x	x	3 prototype systems	x	
Discrepancy between the ISD model in use and espoused model		x	Learning	x	x
Communication	x	x	Communication	x	x
Management not sympathetic to project work	x	x	Document. reqs specs	x	
Lack of support for project work	x				
Project meetings:			*Project meetings:*		
Too short	x		Learning		x
Expert language (communication)	x		Enthusiasm	x	x
Breakdowns & restarts	x	x	Communication	x	x
Aspect blindness	x	x	Ownership	x	x
All should participate	x	x	Change in roles	x	x
Localities	x	x	Prototyping	x	
Planning		x	PROTEVS	x	
Users not aware of acquired knowledge		x	Logbooks	x	
Information technology:			*Information technology:*		
Expert language	x		Learning	x	x
			Enthusiasm	x	x

It may be of interest to point out that in this case middle management and the DP expert did not want to participate in user-controlled design. We all noticed the changing of roles, which indicated elements of learning in the model. Still the participating users did not believe in their acquired knowledge (cf. Malmö case). When I tried to explicate their acquired knowledge, I referred to the incident with the regional communal EDP department where they disputed the possibilities and limitations of computer-based information systems with the experts. For a short while they recognised their learning. Some of them never really believed in it.

In this case we noticed the same discrepancy between the information systems development model of the organisation and the model used by the DP experts. When the researchers were present the participants felt they had to leave their daily work and concentrate on development work. When we were not there they did not muster any enthusiasm. This may be explained by the fact that they had not had sufficient time for both ordinary work and development work. We also became aware of a certain 'aspect blindness' on the part of the researchers, due to the fact that the research project became more important than the design process.

There is a need for specific localities where the *local designers* can meet and work. This was mentioned several times by the participating users. They needed a place where the designers would not be bothered by telephones and other persons.

We had many discussions of the possibilities and limitations within the PROTEVS framework, e.g., the users also realised the possibility of using the PROTEVS model for evaluating and illustrating desired changes in existing computer-based systems. The problem and possibility identification in this case are shown in Table 5.

The Factory Case – Blue-Collar Workers

The results of the research in Falkenberg gave evidence to the fact that the concept of the PROTEVS model is reliable when applied to white-collar workers. Therefore the model was applied among blue-collar workers. The main problem area here was the creation of a computer-based 'administrative planning system' for the new work organisation of the production line. This was a system which was to be developed, controlled and used by the blue-collar workers. The work organisation of the project meetings consisted of study circles mostly led by the participants.

Some of the findings below are also discussed in Friis (1991b). Additional findings are fetched from the logbooks, and from the HyperCard 'card-box' system design by the workers, and from wallcharts produced during meetings.

The aim of the research was to improve the PROTEVS framework, and to perhaps add a more experimental characteristic to it, and to find out how we may let the future system users take over the responsibility for both the design and use of computer-based information systems.

The design work with the blue-collar workers was carried out in the shop stewards room at the workplace. The shop was thus the meeting place for the shop stewards, the designing workers, management, DP experts and researchers, working in the ongoing project. The experts and researchers were there to support the work when 'called upon' by the workers.

In the shop there was a small computerised workstation. It consisted of two Macintosh SE computers, one 20–40Mb hard disk and one Image-Writer II. For documentation and systems design support we used the HyperTalk programming language within HyperCard and the MacWrite word processor. There was not always room for everyone in the shop-steward's room, and we sometimes borrowed another room from the organisation. This caused no difficulties in general, but it made us and the workers wish we had a special 'design shop' (cf. Järnegren *et al.* 1990), something of the kind described by Emery and Emery (1989), as Participative Design Workshops, and by Jungk & Müllert (1987), as Future Workshops. It can also be described as an 'arena', where workers and management may meet on equal terms when discussing change in the work organisation (cf. Hedberg, 1975; and Gustavsen, 1987).

In this case the participants were not so quick to learn the handling of the computers, mainly because they were not used to office machines such as typewriters. However, as soon as they had mastered the keyboard problem they were quick to learn the software systems we used. After a few weeks they designed small personal systems with buttons and fields designed using HyperCard.

My colleague Hasse Nilsson assisted in the supportive work with the HyperCard design. In this work the two of us noticed the evidence of roles changing again. Once the workers took over the ownership of the design project they acted more confidently. One of the shop-stewards was chosen project leader. The more knowledge concerning ISD and work of change that the workers acquired, the more confidence they showed toward the researchers. These users were not awed by the computers. They even said no to computers but wanted more time for the design of a new work organisation. Other departments in the factory had expressed a wish to participate in the design work. This was deemed a positive involvement by the established 'local design group' and they did not mind supporting the other groups. But, due to financial difficulties and the lack of manpower, this was not possible at that time.

The result so far in the practical experiments comprises both qualitative and quantitative conceptual charts for the work organisation. These have been used by the workers in negotiations with the production management. The experiences so far point toward the need for a specific 'local design shop' where the local designers can work in peace with their design. In a group discussion, about the phenomenon of the change of roles and the new assurance among the workers, we learnt that users need a certain 'know how' of computers and ISD. How else can they talk to computer experts? The workers felt that they needed this 'know how' to be able to even say no to a proposal from an expert and/or specialist.

We also noticed the 'aspect blindness' of the researchers here, as in the Falkenberg case. We also noticed that when the workers took over the ownership of the project - they suffered from the same blindness or 'taken for granted perspective'. There is an 'aspect blindness' with both management and users. In communicating they use the same words, even though these words do not have the same meaning. The problem and possibility identification for the factory case is shown in Table 6.

Table 6: Problem and possibility identification - factory case

Problems identified by	*insiders*	*outsiders*	Possibilities *identified by*	*insiders*	*outsiders*
Organization:			*Organization:*		
Lack of manpower	x		Participative ISD	x	
Lack of time	x	x	Study circles	x	
Communication	x		Learning	x	x
			Project responsibility	x	
			'Card box' systems	x	
Project meetings:			*Project meetings:*		
Too short	x		Enthusiasm	x	x
Planning	x	x	Learning	x	x
Communications	x		Change in roles		x
Aspect blindness	x	x	Ownership	x	x
Taken for granted perspective		x	Knowledge person	x	x
			Logbooks	x	
			Prototyping	x	
			PROTEVS	x	
Information technology:			*Information technology:*		
No knowledge of office machines	x		Learning	x	x
			Users not awed by IT		x

Summary of Findings in Empirical Tests

The problem/possibility identification tables in the four cases discussed above show similarities in character and categories. For clarification, I shall summarise the problems and possibilities identified in all these four cases in one table - a *summary table.* I use numbers to display how many times, in each of the four cases, the same problems/possibilities were identified (Table 7).

Table 7: Summary table of problem and possibility identification in the four cases

Problems identified by	*insiders*	*outsiders*	*Possibilities identified by*	*insiders*	*outsiders*
Organization:			*Organization:*		
Lack of manpower	4	1	Participative ISD	2	
Lack of time	4	4	Learning	3	2
Discrepancy between the ISD model in use and espoused model		3	Scenario	1	
Communication	2	1	Study circles	1	
Management not sympathetic to project work	2	1	Communication	3	2
No support form org.	1		Conceptual database	1	
Planning		1	Project responsibility	2	
Lack of replacements	1		Document. reqs specs	2	
			Prototype systems	2	
			'Card-box' systems	1	
Project meetings:			*Project meetings:*		
Too short	4		Learning	3	4
Planning	2	1	Enthusiasm	4	4
Breakdowns & restarts	1	1	Open diaries /Logbooks	2	2
Communications	2		Communications	2	2
Aspect blindness	2	2	Ownership	2	2
Taken-for-granted perspectives		1	Change in roles	2	3
All should participate	2	1	Knowledge person	1	1
No continuation	1		Prototyping	2	
Localities	1	1	PROTEVS	2	

Problems identified by	*insiders*	*outsiders*	*Possibilities identified by*	*insiders*	*outsiders*
Organization:			*Organization:*		
Users not aware of acquired knowledge	2				
Information technology:			*Information technology:*		
Lack of techn. resources	2	2	Learning	3	3
No knowledge of office machines	1		Enthusiasm	3	3
Toolkit for prototyping		1	Users not awed by IT		1
Expert language	1				

Discussion of Research Findings

In this section I will try to show how problems of UCISD may be resolved, and how we may utilise the possibilities discovered by the participants in the empirical work. This is done from the perspective of information systems development. During the practical experiments all actors became aware of certain major problems/possibilities with user-controlled information systems development. As shown in the 'problem identification tables' above, we grouped these empirical findings under organisational requirements, requirements of the research project, and the information technology environments. This implies that user-controlled ISD should not be studied as an activity apart from the daily organisational work. In the following discussion, I will show the relationships between the problems and the cause and effects of all the problems – seen as a whole.

Some of the problems are similar in character and belong to the same problem area, and I have treated these as one major problem area. There were four major problem areas, and four organisational maps were created on the basis of them.

Lack of Resources

For example, time, manpower, premises, and technical facilities. In this problem area we have mainly discovered economical causes, since management's attitudes were the main obstacles for user-controlled ISD. The idea that ordinary workers can and want to participate in the work of change came from other sources. Still, there is enough literature on how workers' view of a work area is usually more differentiated than management's view of it (see e.g., Katz & Kahn, 1966).

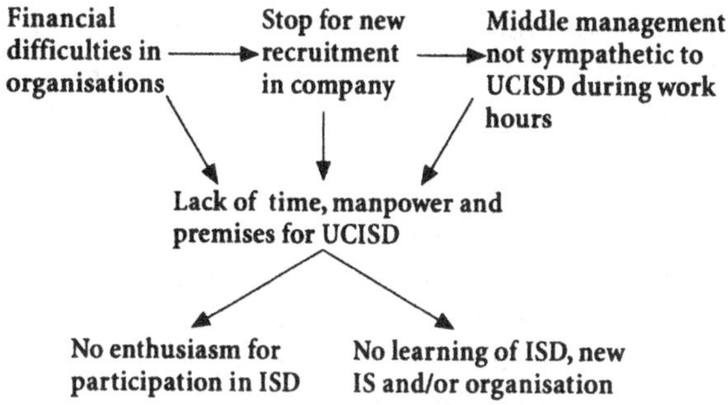

Figure 26: Organisational map of causes and effects for the lack of resources

Financial difficulties in an organisation or company very often put a stop to new employment. When the top management (and politicians in some cases) are in favour of active user-involvement in ISD work they are not always aware of the necessity of replacements for the participants when they do ISD work. Thus we have a negative spiral of no time, no manpower, and of course no extra personnel, i.e. replacements for the participants during project meetings, which also undermines the initial enthusiasm of the users. The effect of lack of resources for active user-participation in ISD is that there will be no possibility for learning ahead of the system and computers and/or of effects on the work for the persons for whom the new system will be applied. The only learning provided to the users, in such situations, is how to operate the new system (Figure 26).

Poor Communication:

- between workers and top/middle management and other decision makers;
- between actors in project groups due to expert language and aspect blindness of some kind;
- between users and EDP departments.

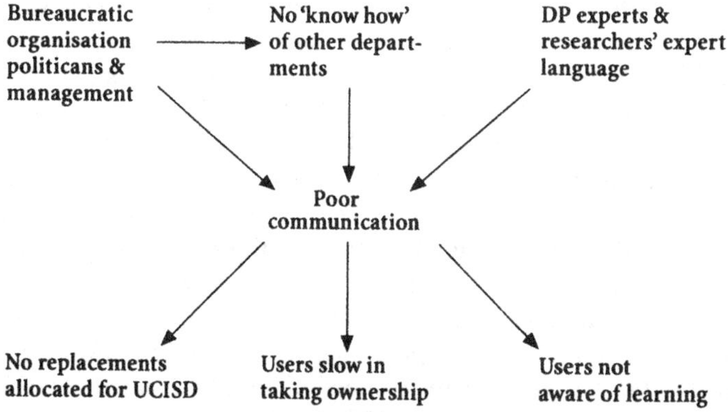

Figure 27: Organisational map of causes and effects for poor communication

We experienced poor communication at more than one level. There was no real communication between the future users (operational staff) and top management (politicians in some cases) due to a hierarchical and bureaucratic organisation. Expressions like: 'They do not listen to us', were common in the projects in the town councils. Due to slow, and sometimes extinct, latitudinal communication within the hierarchy of the organisation, there were no extra personnel allocated for the participating users.

There were also difficulties in the communication between the researchers and DP experts at the beginning of each project, and before the participating users began to master prototyping. The effect of this poor communication was that the users were not able to take over the project.

A lack of communication between departments about their work organisation – no 'know how' of other departments – may also be seen as a cause for poor communication. In a user-controlled ISD one must at least learn of the neighbouring departments' work organisation (Figure 27).

Management

- middle management was not sympathetic to project work for user-controlled ISD;
- middle management did not support the workers who did project work.

We have noticed in more than one case that middle management is reluctant to support project work during work hours. They have no objections if the users do their project work during free time. According to middle management this is an economical problem. When top management allows project work during work hours it does not at the same time give the participating departments financial support for this.

As discussed in the section: *The Research Process* (starting on page 267), and in the Falkenberg case (see page 276), the DP experts from the EDP departments did not really want to give the power of control to the users.

At first, top management was sympathetic to active user participation in ISD, but when they realised how much time and manpower was needed they changed their attitudes. They even saw the research project as nothing but a kind of technical education and it should be sufficient to persuade the users to accept the new (already decided upon) system.

The management's change in attitude, and actions that followed, may be understood in the perspective of Argyris and Schoen (1978), and their discussion of 'espoused theory' and 'theory in use'. The espoused theory may be seen as the initial official attitude of the top management, and the non-supportive actions taken under the pressure of low budgets, can be seen as the theory in use. Some of the effects of the actions were that there was no real allowance for user participation, no user commitment. Thus the users in the research projects needed longer time to take over the project responsibilities (see Figure 28 below).

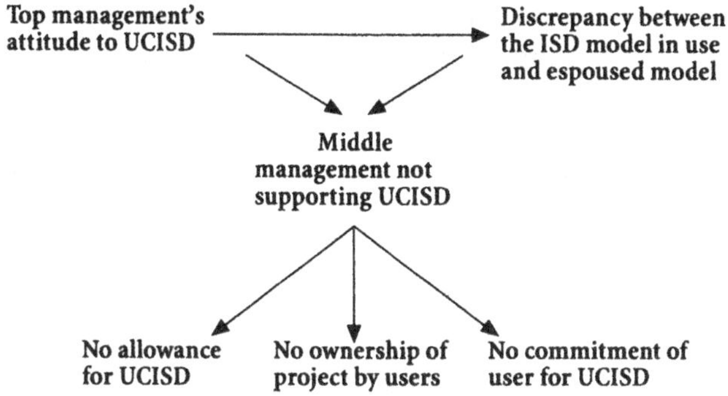

Figure 28: Organisational map of the causes and effects for the management's attitudes to UCISD

Discrepancy of ISD Models

A discrepancy between the ISD model in-use and the ISD model espoused by the organisation. As stated by the users, work democracy takes time. When the allocated time is up, the management entrusts the control of the ISD project to the EDP department, very often with a tighter budget for the work. Thus there may be a two-fold reason why the EDP department does not let go of the reins in the ISD work. In Friis (1991a), I discuss the findings concerning how the management must give over some of the responsibility for the ISD work to the EDP department as a result of low budgets for ISD work. The effect evidently is that there is no active user-involvement in ISD work. The only way for the users to 'influence' the new systems is as interview respondents, and sometimes not even that. Many consultants and EDP departments see top management as the only system owners (see Figure 29 below).

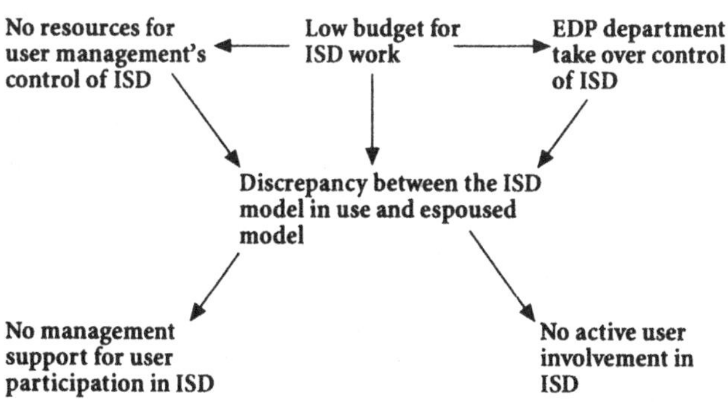

Figure 29: Organisational map for the of causes and effects of the discrepancies between the 'ISD model in-use' and 'espoused ISD model'

When I also consider the findings from the earlier explorative studies (see *Explorative Case Studies of Requirements Specification Work*, starting on page 267) concerning user participation and involvement in ISD work, and add them to the maps presented above I may gain a more inclusive '*global map*' for UCISD problems (see Figure 30 below).

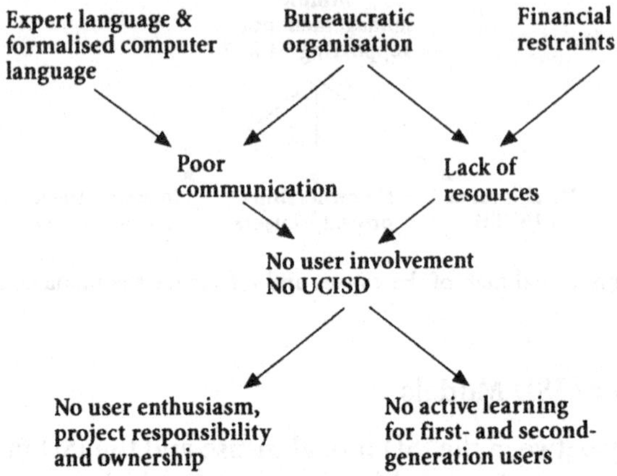

Figure 30: Global organisational map of UCISD problems

The findings that were considered for the global map are: the future-users' enthusiasm for active learning and participation in ISD work, communication problems between users and DP experts, and between users and the computer-based information system, and the problem of the second-generation of users.

Some of the major possibilities displayed in the summary table need a more detailed recapitulation and description to make it clear why they are important for the local theories evolved.

The Changing of Roles

It was evident that the users who participated in the PROTEVS project undertook a change in roles during project work. They became bolder and more innovative than it was common with people in a more traditional user roles. The change of roles was allowed during project time only, because when the project was over, the roles and positions of the participating users went back to 'normal'. In Friis (1988), I discuss the changing of roles in the perspective of 'role theory'. The types of roles the actors took on in, for example, the Malmö case were for the users:

- The *Traditional User Role*. This is a person who has little or no knowledge of the process of systems design and/or computers, and a large uncertainty and unwillingness to participate in the systems development work. The general feeling of this user is: We don't understand the language of the DP experts. We don't want to look like fools. Can we really trust the DP experts? What do they want from us? Are they going to take away our jobs or are they going to change our jobs? No one tells us anything.

- The *Interested User Role*. The unwillingness in the first stage may change into a curiosity, and as the interest awakens within the user group, they start to generate questions like: Could it be that the experts actually want to offer us some assistance with their computers? Could, for example, this tedious problem with the 'Deputy' routine really be solved with the aid of a computer? We would like to know how. But we don't know anything about computers, we are no experts.
- The *Analysing User Role*. This user wants to participate in the systems analysis and to influence the course of change. He/she may recognise the necessity and/or possibility for organisational changes in order to solve a specific problem. For example, a secretary wanted to change the current organisation of work for the deputy routine, in favour of computerisation. Not all secretaries agreed with the proposed solution. But they all wanted to learn more about analysing tools and do further investigations.
- The *Designing User Role*. This user wants to have the last word on whether to computerise or not to computerise in a work situation. This user wants to understand what is taking place behind the 'screen' of the computer, and if given the right tools this user can very well build small prototype systems very well. This user may develop into a 'hacker' – or may not.
- *The Evaluating User Role*. The Evaluator user considers herself/himself as systems owner and finds it appropriate to evaluate it and test the system and perhaps modify it. The user may do this in collaboration with the DP experts.

Some of the DP experts, who participated in the tests, also changed their traditional roles toward a more supportive and 'teacher-like' role. Some of the experts even stated they enjoyed being understood and appreciated by 'ordinary users'. The intervention arising from the experimental systems development that caused the change of roles was not altogether a happy one. The users were quite agreeable with the experiments and said so. But each experiment was going on for a limited period time and when the researchers left, the traditional roles were resumed with one exception. Some of the systems analysts continued to work with parts of the techniques presented in the PROTEVS projects. The result was that the users felt they had been 'let down' by the researchers. One of the secretaries in a PROTEVS project remarked:

> You come here and promise a lot, and you even show us the promised land, and how to get there – then you leave and all is back to 'normal'!

Enid Mumford noticed this situation during the 'Follow Up' phase of the project, 'Designing Secretaries' (1983). When the secretaries were left without the support of an external facilitator they also, like in some PROTEVS projects, lost their influence on further development projects.

In her own words:

> The group apparently lost some power and credibility once I left the scene.

Learning

In the first experiments, I learnt that the users wanted support during both the analysis and design phases in their prototyping. During work analysis they wanted help with not the analysis of their work organisation. which they could do by themselves, but with descriptive techniques, for example, with flowchart and wallchart techniques and graph-drawing techniques. They learned very quickly (cf. Bjørn-Andersen, 1978, 1984) and, once they got the hang of it they even developed their own techniques, for example, for designing documents and future screen images.

In the design phase, the users needed no support with the paper and pencil prototype systems. With the design of the computer-based prototype system, the users needed support to operate the computers, and to construct the databases. In all field experiments we used fourth generation programming languages (4GLs) based on databases of relational representation. I found that even with these techniques the users learned very quickly, and that the tables used in relational representation appeared very natural to clerks.

In the factory case, the users had difficulties handling computers because they were not used to typewriters. When they started to manage the keyboard they were quick to realise the possibilities of the computers. With the software used, HyperCard, they learned of the ways a computer works and that it is rather limited. We all realised that it takes this kind of knowledge for future users to be able to say no to vendors, software houses and DP experts. If you do not know of these matters how can you decide whether to say yes or no? One needs knowledge to decide whether to accept DP expertise.

Ownership and Enthusiasm

When users have acquired the necessary knowledge for user-controlled information systems development they take over the ownership of the project. The ideal situation would be if they owned the projects from the start. When we noticed the project fatigue in the Falkenberg case (described earlier) and the project almost died, we all decided that one of the users should take over the responsibility of the project. As they did, they began to own the project. They even became irritated with the researchers and DP experts and found them meddlesome. They wanted to work on their own. The enthusiasm and the fact that they owned the development project made them independent of the researchers and DP experts. They actually started a user-controlled ISD.

'Aspect Blindness' and 'Taken-for-Granted' Perspectives

These problems were solved within the project groups when it was recommended by the participants to make all assumptions as explicit as possible, and/or that all concerned should participate. Another solution was tried. The persons who possess the 'know how' of the problem area under investigation should act as 'knowledge persons' about that area (cf. Friis 1991b). This illustrates how a 'local theory' may evolve during project work.

When the users are not active enough in their development work we as researchers tend to push them, and worst of all, in the directions of our perspectives – compare the PROTEVS cases. In this way the users are not allowed to own the development (research) projects – we take over. When the users are really active in the analysis and construction work they start to push by themselves. There are some interesting questions concerning 'aspect blindness' and 'taken-for-granted' perspectives; such as:

- How open to all perspectives within a project group must a researcher be?

In our factory case we had come quite far with the user prototyping before we noticed the phenomenon of 'aspect blindness'. In fact, we did not know what it was until we started the discussions and the sharing of knowledge from relevant literature that we found a name for it:

> Lack of ability to see a phenomenon in different aspects, 'aspect blindness', is a common and serious cause of problems.
>
> *Nygaard and Sørgaard, 1987*

We argue that this problem applies to non-expert project leaders in an ISD situation as well as applying to researchers and experts. We learned that not only does 'aspect blindness' affect researchers, but it also affects co-researchers (users) when they are in control of a project. We also found the same for the 'taken-for-granted' perspective. Had we not had such an open and collaborative methodological approach in the case studies, we would perhaps not have noticed these phenomena. The reasons for the project fatigue as due only to over-worked research objects might also have been accepted. This proves that open collaborative research approaches as well as ISD approaches allows for more knowledge about user needs and actions. It both enables and forces a multi-perspective.

Towards Local Theories About User-Controlled Information Systems Development

As a contribution to the knowledge of UCISD in the sense of Elden (1983), I will form some local theories based on the summary table (Table 7) and the logbooks. As previously said above, a local theory is not only a presentation of a solution to a problematic situation, it is also an action plan towards that solution. In various group discussions in the different cases, several local theories about possible solutions and action plans came up. Some of them were only mentioned in the logbooks, some were actually put forward to the management as suggestions for action plans.

The described local theories about the problems in the maps above have been put together by myself, but they derive fundamentally from all participants in the field-research projects. The possibilities added to the local theories are taken from the summary table (Table 7) and from the notes in the respective logbooks. I have displayed a local theory for each problem area described in the organisational mappings above and added the theories of the Falkenberg and Factory cases about premises for user-controlled ISD – the *local design shops* (LDS).

We all learnt that exercising a kind of democracy in user-controlled information systems development is a slow process. The users need extra time for the design work, and they need whole-hearted support from the management. In the projects described above, the users had to do the design work at the same time as they did their daily office work. The solution to this problem is that management should supply enough manpower and replacements to enable the users to have time for design work.

According to the participating users, the 'know how' of the people concerned with a change process such as ISD, is of utmost importance for the organisations if they are to get the information systems they are entitled to (cf. Nissen, 1982). This 'know how' constitutes the very essence of potentialities for an organisation. It should be taken seriously. Therefore we should let all users concerned to take the responsibility for, and the control of future information systems developments (see Figure 31 below).

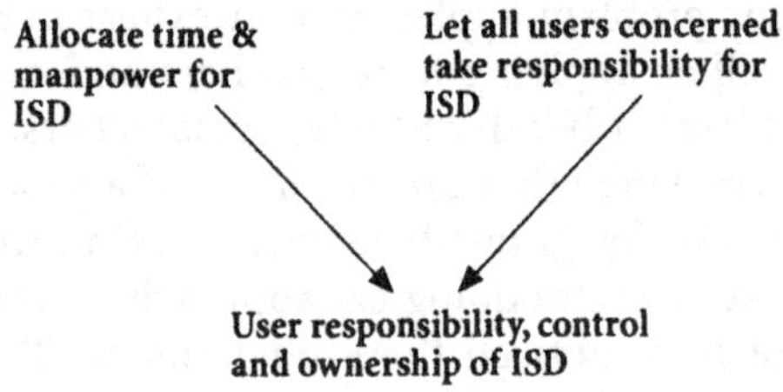

Figure 31: Local theory of user ownership in ISD

To be allowed to do their own development work according to, for example, the PROTEVS model, the users formulated another 'local theory' – the 'theory of premises', with adequate technical facilities, so that they could go there both during working hours and free time if desired. The first test with a kind of design shop where the users took control, was the experience with the 'scenario' in the Falkenberg case.

In the factory case we looked upon the shop-steward's room as a design shop where some of the workers dropped in now and then to talk about, and to give suggestions about the current area of change. They could do this because they knew that all the material for the ISD work was there and because most of the time the project leader (the shop-steward) was also there.

This shop may be conceived in the same way as the 'participative design' discussed by Emery and Emery (1989), or as the 'arena' discussed by Gustavsen (1985), and 'local design shops' discussed by Friis (1991b) (see Figure 32 below).

Figure 32: The local theory of local design shops

We learnt that the communication gap between users and DP experts narrowed considerably when the users were allowed to do prototyping. In the Malmö case we had an example where the DP experts implemented the prototype systems developed by the users in the research project. In the Falkenberg case, the users, during a visit to a software house, were very bold with the DP experts and vendors. In the end the DP experts had to admit that they were not used to such questions from future users. They wanted to stop the interrogation and go on with the demonstrations. Still, they promised to think it all over and come back to the users after a while (see Figure 33 below).

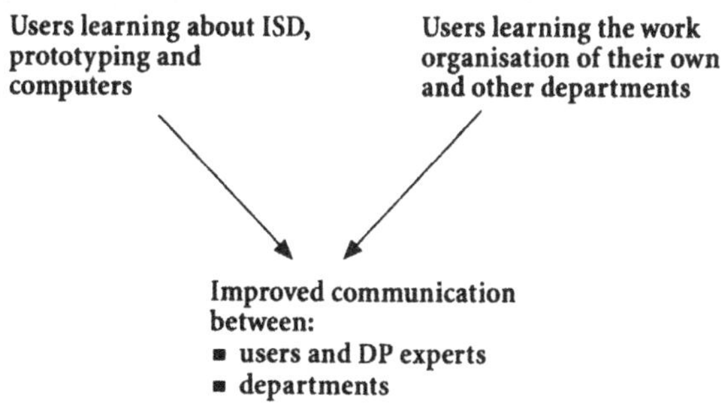

Figure 33: The local theory of how to improve communication of users and experts

I noticed that each organisation declared that they were already practising 'participative systems development', i.e., that the future users were involved in the development work. At least that was the opinion of the top management. It turned out that each organisation had an 'organisational information systems development model' for this work. In reality this was not so, because the EDP departments of the organisations were actually conducting the development in a way they saw fit, and without user participation.

I discovered a discrepancy in participation of users in the two models. In the 'model of the organisation' the participating users were supposed to be elected by their departments, but in the 'ISD model of the EDP department' the few users who actually participated, were selected by the EDP department and the departmental management.

As discussed by Salaway (1987), it is quite common that the EDP departments see the application of the ISD model of the organisation in another way (see Figure 34 below).

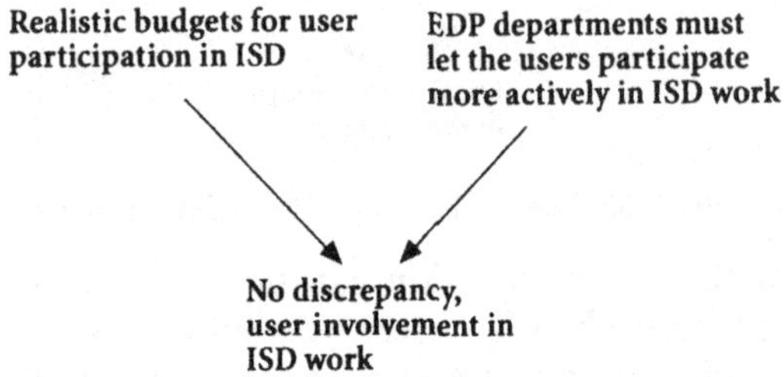

Figure 34: Local theory for active user-involvement in ISD work

In most cases it was agreed that for the sake of a more general solution and action plan to solve contemporary problems of user participation and UCISD we should have:

- budgets for extra personnel that may allow active user-participation in ISD work;
- localities/premises and other resources that will enable the users to actually do ISD work;
- active support from experts and specialists of different kinds when needed and asked for (see Figure 35 below).

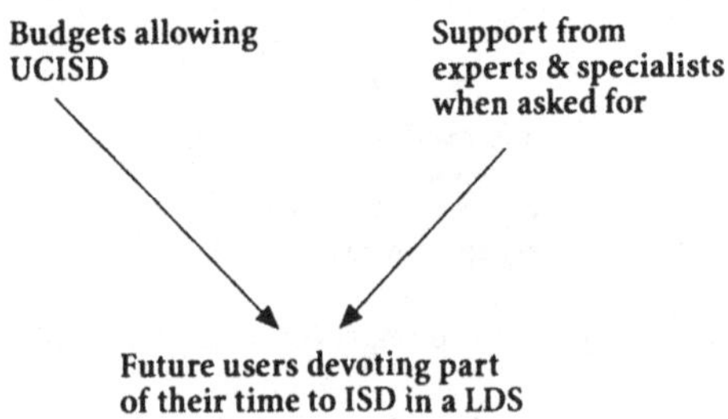

Figure 35: Local theory of UCISD

A development plan, similar to the local theory of UCISD presented above, was actually formed in the Falkenberg case. However, it was never carried out due to financial restraints and lack of manpower within the department.

The Protevs Model

This section is organised in three parts. First, the main features of the model are presented and suitable application domains for the model are described; second, recommended procedures for the users' work with requirements specifications, design, and system evaluations are supplied; third, the environment in which the model may work will be described. The history and the development process of the PROTEVS model is described further in Friis (1984, 1986, 1987).

Alongside the studies of how user-controlled ISD may proceed, I have tried to create a working model for user-controlled ISD. The ISD work done by the future users is mainly work with requirements specification. I claim that the work with requirements specifications and information systems development is so important that it simply cannot be conducted and controlled by DP experts alone. A large part of it should be done by, and with the responsibility of, the future IS users. It is futile to claim that DP experts could acquire all the knowledge needed about different ISS applications. The development of the PROTEVS model has been done in close collaboration with non DP-expert system users. The PROTEVS model is actually named after the prototyping technique. The name is an acronym for 'PROTotyping - an EVolutionary Systems design'. Evolutionary in the sense of continued, ongoing process-oriented ISD.

I claim that the future users should design small prototype systems themselves. The DP experts can provide them with technical instruments for this work. The DP experts' work is to ensure that the users get all possible help in their design work process. Later on the DP experts may integrate the subsystems and make them work.

Main Features

The PROTEVS model is an interactive, process-oriented model similar to action research in contrast to the System Design Life Cycle model which is linear and resembles normal science research. It is context-sensitive in character, as is other types of action research.

So far, the information systems developed using the PROTEVS model are mainly operational and task-oriented systems. The final systems are used to perform certain tasks - defined by the users on both individual and departmental levels (cf. Gunton 1988, Olerup 1989). Gunton defines departmental systems as a combination of end-user systems (EUS), designed with the aid of end-user computing (EUC) for personal use, and operational task-oriented systems, commonly used in various departments of an organisation.

The PROTEVS model should be understood as a framework for how user controlled information systems development, and mainly the work with requirements specifications, and evaluation may be performed. In fact, it is more like a normative approach.

The PROTEVS concept is based on active-learning possibilities for the future users since they participate in the actual design of the ISS. It is also possible for them to influence the appearance of the system, and during the system use to continue the development, as long as the systems exists, i.e. they should be able to change it, or to direct changes they see fit.

As well as being a learning vehicle PROTEVS is also a framework for improved communication with the DP experts. There are too many examples of how difficult it is to establish good communication between the two cultures concerned: the future users of an ISS and the DP experts.

The model is divided in two parts: the upper part is for the requirements specification work, and the lower for the evaluation work of existing systems, and/or for evaluation of ready-made CBIS from the software market. Figure 36 shows a schematic picture of the latest version of the PROTEVS model.

The model consists of five processes indicating the course of events. It is divided between two cultures of actors: the future system users and (in case of ISD) the DP experts. The people working with the procedures in processes 2 and 4 are mainly the users, and the people working in processes 3 and 5 are mainly the DP experts.

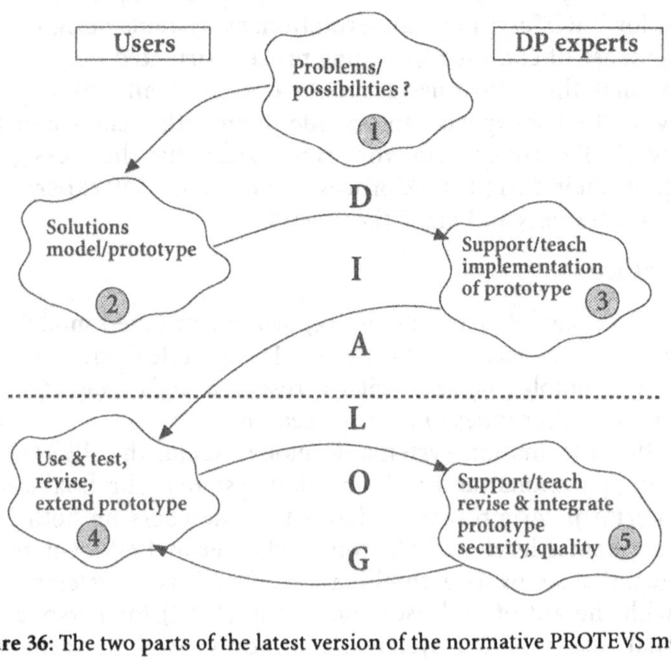

Figure 36: The two parts of the latest version of the normative PROTEVS model

- The users are people whose work situation is affected by the change both directly and indirectly. Thus, all people concerned by the IS application should be allowed to participate.

- The DP experts are the specialists, if needed, who will help to effect the desired solutions (often as prototype systems) – arrived at by the users in their work with the requirements specifications for a new information system.

The first process (1) has been placed between the two cultures to indicate that anywhere in an organisation problems and possibilities may be recognised, and acted upon, i.e., any party in an organisation may be considered users. In the case of the pharmaceutical case study, it was the blue-collar workers who were the initiators of the discussions of change. It is an important issue that the decision to act upon a problem should be taken by people concerned *from all levels* in that problem area.

The first phase of the process is to indicate and create a consciousness of existing problems and possibilities at a workplace, organisation or company. The problems should be handled in some way by the people who work within or in the neighbourhood of the problem area. Process (1) is also placed in the centre of the picture to indicate that any of the actor groups in an organisation may be *problem owners*. In case of DP experts and other kinds of specialists, as users, they are the problem owners.

The basic idea of the model is that the problem owners should themselves be the *problem solvers*. I deem it unjust if the problem owners let other people, like specialists and experts, do this work for them. I question that a person from an entirely different culture is able to solve the problems of a third, fourth or fifth culture. To my mind, only the people who have the true knowledge about the problem area can find a satisfactory problem solution. In an ISD work process, the only people who can possibly come up with the true requirements are the future system users (cf. Checkland, 1981; Emery and Emery, 1989).

The problem solutions arrived at by the users, when working within the PROTEVS model, may take the form of a computer-based prototype system which shows the most significant requirements for a problem solution. This solution or prototype system can also be used as an interpreter when the users are trying to explicate their requirements to the DP experts.

A problem solution may take the form of a manual prototype, i.e., a drawn picture or some kind of flowchart presented on a wall. If there is a large wall available one may very well glue real documents to indicate the direction of data and information transportation. A picture could very clearly show an administrative procedure.

Actually the model is based on the idea that the future system users *can and should* design their own CBIS, at least when the systems are not too complex and/or larger multiple-user systems. This does not mean that the users are taking over all the tasks of the DP experts. They are collaborating with them, in the effort to design the prototype systems, because the users are really creating an interpreting device which is easier to understand than a written document. When working with computer-based prototyping the users acquire 'hands on' learning of the computer and of the ways in which the computer works, i. e., how it reacts to commands of many kinds. Thus they acquire an understanding of the limitations and the possibilities of computers, and of possible CBIS. We learnt from both white-collar and blue-collar workers, during the projects, that:

> One needs 'know how' to be able to say no to a vendor or a DP expert. If one is debating the problems of a specific area without any knowledge of it, one has no arguments.

It is considered hard work for DP experts to understand and implement the problem solutions, i.e., the requirements specifications the users have designed, even when it is illustrated as a computerised prototype system. It may still require quite a lot of collaboration between users and DP experts before the required system is implemented and ready for testing and use. There may still be many instances of misunderstanding. The users must have the possibility to test the system in real-life situations to be able to evaluate if it fits their initial requirements.

In the case of larger development projects which may have been planned by the management, it is still the problem owners who should do the problem solving for each ISS. Problem solving in the case of an ISS is the solution arrived at when working with the requirements specification. Thus, the arrow from process (1) points to the users' side of the model, toward process (2). This also indicates that the users should take over the responsibility, i.e. the ownership of the ISS project. In the case of the evaluation of an existing ISS this work is done in the same fashion, i.e. with the aid of a prototype system showing the desired changes, in process (4). In case of larger systems development, e.g., the implementation of a larger CBIS, the different ISS could be divided between the different departmental user groups.

Recommended Procedures

This is the normative part of the approach in the sense that there are some recommendations of how to perform the prototyping in processes (2) and (4). The procedures are based on the ways the users worked in the empirical cases described above, and on the questions asked by them during prototyping work.

As previously mentioned, the initiation of an ISD work process should be adjusted to the context. The procedures in the model are also context-sensitive. Thus, the start of this kind of ISD project is not necessarily initiated by management or planned in the same way as in more traditional information systems design work. The initiation of a project may quite arbitrarily come from the people concerned in the problem area.

The initiation of the dialogue between users and DP experts is also done by the users of that ISD area. The users as the problem owners and problem solvers may need expert and specialist support. They should be able to call on any kind of expert or specialist to aid them. For the scope of this work, I have chosen to concentrate on one kind of specialist – the DP experts. The model is, though, of such a general character that it may be applied to many kinds of situations of development and change. In fact, one could call the expert culture, on the right side of the model, the culture of *specialists*. The model is thus applicable for the work with, e.g., a dissertation, a research process and/or, for the development of 'local theories'. When the PROTEVS model is introduced to first-generation users it is important for the researcher and/or DP expert to establish a basis of confidence for collaboration. In the PROTEVS model this is done in 3–4 meetings where the 'expert' is the respondent. I stress that these meetings should take the form of a dialogue. In the beginning, the expert may act as a teacher of the activities of ISD (cf. Bjørn-Andersen, 1984), since he/she is acquainted with this work, but as soon as the users become familiar with the process, they should be left to take over. Later on they can also function as teachers and instructors for the next generation of users. A discussion of different forms of co-determination, i.e., work democracy could be the starting point in the initiation of the model. This discussion may take the form advocated by Gustavsen (1985) in his 'democratic dialogue', where he safeguards that all may have their say in a debate of some kind. He prescribes some premises that should be fulfilled to achieve a democratic arena for development work:

- The dialogue is a process of exchange: points and arguments move to and fro between the participants.
- All concerned must have the possibility to participate.
- Possibilities for participation are, however, not enough: everybody should also be active in the discourse.
- As a point of departure, all participants are equal.
- Work experience is the foundation for participation.
- At least some experience which each participant has when he or she enters the dialogue must be considered legitimate.
- It must be possible for everybody to develop an understanding of the issue at stake.
- All arguments which pertain to issues under discussion are – as a point of departure – legitimate.
- The dialogue must continuously produce agreements which can provide a platform for investigation and practical action.

This is a dialogue that may prevent external control of the ISD work. Another possible way to start democratic group work is a 'search conference'. The participants are jointly 'Searching for Common Ground' on issues put forward by the participants. 'In a search conference the ground rules include that rule which insists that even the most far-out, improbable suggestions must be looked at' (Emery and Emery, 1989).

The initial meetings should comprise demonstrations of ISD work at large, and the PROTEVS model in particular. In connection with this, a frame of reference should be explicated since there are many concepts on both sides that need debating. ISD involves CBIS technical concepts, and it is imperative that the users acquire some knowledge of these. This also provides a chance for them to do away with the insecurity of the technical expert language.

This initial state is most important. Of course, this is when the DP expert will learn the language of the users as well. It is our experience that in spite of this introduction, it still may take a while before both parties become secure and drop the politeness towards one another. (cf. Ehn, 1983 where similar experiences from the UTOPIA project have been reported).

The teacher/DP expert should be able to present various analyses and documentation techniques for the users to choose from. Still he/she should be able to encourage the users to design their own techniques for analysis work, because the users should be allowed to invent their own techniques as long as all techniques are agreed on and understood. It was our experience from the earlier research projects that a simple flowchart technique was the most favoured among participating users. This could be just text in boxes that show the chronology of actions in the routines, or it may be just texts/descriptions in a sequential order of actions

The DP expert and/or the project leader should only use work related examples in the initial educational stages. Thus they may enhance the understanding and communication between project participants and DP experts (specialists). Furthermore, the teacher/expert should be able to demonstrate:

- How the work with requirements specifications can and may be done is of importance, even though they may be of different assumptions. The following techniques are suitable: analysis techniques, e.g., flowcharts (Borum, 1977); bubble diagrams (DeMarco, 1978); verbal graphs (Friis, 1986) and/or drawn pictures (cf. Checkland's 'rich pictures', 1981).
- How to do a structural description of a work routine.
- How to make an appropriate documentation of that routine may help the start of an ISD project.
- How to map the necessary information objects needed in a prototype system.
- How to map the necessary data so that the system may answer the questions asked of that system, which means listing the type of questions and retrievals that may be wanted in the future.

It also means displaying how the data should be structured to fit a computer. If there is a long time-interval between meetings the teacher/expert should plan for repetitions. The teacher/expert must be very sensitive and have a good ear for the users' questions.

One way to eliminate this problem is to arrange a 'scenario' of some kind in a place away from the workplace. In our experience a day and a half is sufficient for a scenario. During this scenario the procedure of the work, from problem definition to a ready prototype system, might be run through.

It is imperative that the development process is documented by both teachers/experts and users, for the reason that the participants may be able to evaluate and compare assumptions and conjectures about solutions. These writings will later on form the basis for a more systematised documentation which will comprise the necessary knowledge for continued prototyping for new generations of users. The documentation may also be used at other workplaces with similar problem areas, for educational purposes. The next generations of users should also add to the documentation. In this way 'know how' stays within the organisation and is acquired. The problem selection and/or problem solutions should come from the actual users themselves. If somebody else but the users, e.g., an external expert, delimits the problem area or imposes upon the users which area to select or which solution to decide on it may cause problems, which might be: 'This is not our project, we don't want to waste time on this work, let the experts do it, they have made all the decisions so far'.

If decisions are taken by management without consulting workers, the users do not have the same commitment to the problems, and they do not have the enthusiasm or the drive to achieve something of their own in ISD work. Hence, even the work organisation for an ISD project should be decided upon together with the users. If DP experts make these decisions by themselves, the control is taken away from the users from the start, and may never be returned to the users again.

In case of a departmental IS, or any IS of that size, the actual construction work of the final system is done either in processes (3) and (5) by the DP experts, dependent on how successful and complete the delivered user developed prototype system is. The text in the processes indicates only that there is much work for the DP experts concerning teaching and support for the users, implementation of prototype systems, integration of ISS into the larger organisational system, building security systems and so forth. The actions taken by the experts in this work are not described here. This work is very well described in the literature elsewhere. I leave the choice of methods for this work to the DP experts.

When the users have taken over the ISD project they should start their analysis and design work in the second phase. The work in this phase is concerned with user prototyping, where the prototype system is the illustration of the problem solution, I have seen fit to give a more detailed picture of the procedures recommended (see Figure 37 below).

The problem definitions made by the users should comprise: what the users find problematic in the work routine, what causes the problem and, which change of the work routine they want, and what they mean is good for the organisation. Also, they should specify what objects and requirements they envisage due to the change, including indicators of a successful change.

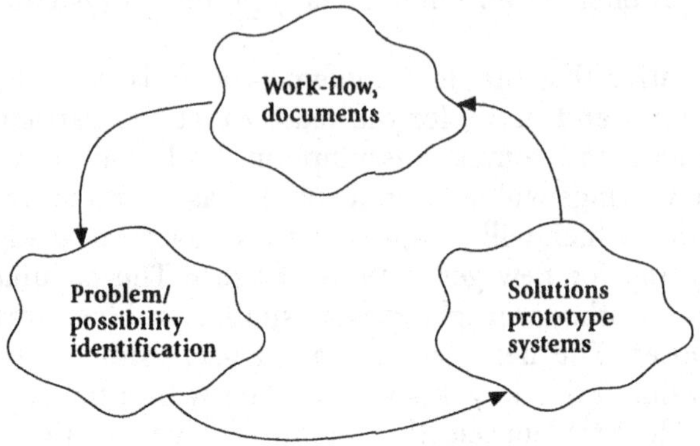

Figure 37: The recommended procedure of PROTEVS development process 2

To be able to find a good solution to the problems/possibilities discovered, one needs to know explicitly how the current work routine is actually functioning. Therefore one should make some sort of graphical schematic picture of the present situation. It is well known, that if one has a good description of a problem one sees half the solution (cf. Elden, 1991). If there is a chaotic situation in the actual routine, you may get a very chaotic picture.

The techniques used in the analysis and the following pictures can vary from workplace-to-workplace. Even 'home-made' techniques are very acceptable. The main point is that the users in the work situation understand the results. Besides, every drawn picture/graph should be accompanied by verbal explanations, so that the next generation of users may be able to learn from the change process.

When the users' and the DP experts' opinion of the prototype system coincides it is time for the experts to implement it for tests in real work situations. This is the job for the DP experts in phase (3).

If, after some time of testing, it is realised that the prototype system is not at all sufficient, the users may reject it or modify it. In the case the system seems adequate, and will be used only as a private system for this particular workplace, it may be put to work at once. When the prototype system is a first solution to a sub-problem, in the form of a requirements specification, it must surely be tested and revised by both users and DP experts before it can be put in a real-life production.

The research collaboration in Falkenberg indicated that the user prototyping technique may also be used for the evaluation of existing IS. The same procedure as in phase (2) is then applicable to phase (4). The prototype system can thus also take the form of a sort of *evaluation prototype system*. This prototype would then be a first solution to a problematic procedure in a current CBIS. In this case as well the prototype system is treated as the requirements specification prototype. The work with testing (and/or extending) and revising, between the users and the experts, in phases (4) and (5) should then continue until the users are satisfied with the system.

In the testing, use and evaluation work performed in activity (4), coordinated with the revision, integration and adaptation of activity (5), the prototype system is adjusted to fit both the requirements of the users, the environment, and the security principles of the organisation – and, it is actually in production all the time.

The meaning behind the lower arrow going from activity (5) to activity (4) is to achieve an evolutionary systems design. The users may now have achieved enough knowledge about information technology to come up with new ideas about changes in the existing productive systems, and to implement these changes as in activity (4). Thus, the users may be in step with, e.g., the changes in the environment, and continue to create prototype solutions as illustrations to new requirements.

There are no specific recommendations for how the DP experts should proceed in activities (3) and (5). There are enough professional writings on this subject in literature. The experts in this branch are mostly so well trained that they have an arsenal of methods, tools and techniques for almost any design occasion. Besides, they belong to a profession that is known for a high degree of responsibility for the end product.

However, recommendations about attitudes and principles are herewith given, which makes the PROTEVS approach different from more traditional systems design approaches. For example, should the DP experts:

- Let the users take over the responsibility for the changes in the work processes in connection with the design of a computer-based information system? In short, they should let the users solve their own problems.
- Work in dialogue with the future users during the entire development process, in the design as well as construction work?
- Broaden their profession to encompass the functions of teacher, supporter and dialogue-partner in the systems design work?

A Suitable Environment for the PROTEVS Model

The procedures in the PROTEVS model and the rules of conduct indicated in the model should be accepted by the whole organisation, i.e., all actors, both workers and management. They should accept the intentions and philosophical attitudes behind the model. This acceptance may be established through conventional negotiations (cf. Ehn, 1988), preceded by a short introduction of the model.

All those concerned with an area of change should be allowed to participate, including the management. This was stressed by the participants in all the field research presented here. In contrast to this, it is recommended by Bødker, Greenbaum and Kyng (1991), that, preferably, only people from the same levels should cooperate in the design.

Nevertheless, a certain flexibility in the frequency of participation for all concerned must be allowed. The responsibility for this kind of ISD work should be placed with those people who are directly concerned with the change. The problem owners should be the problem solvers. The problem owners should take part in defining the objectives of the change.

Support from management at all levels who are willing to cooperate is required. Management and/or leadership is commonly described as either 'employee-centred' or 'production-centred' (Ekvall *et al.*, 1987). In their recent inquiries in the USA and Sweden, Ekvall and Arvonen (1991) speak of a third kind of leadership, a 'change-centred' leadership:

> This third leadership behaviour dimension depicted a supervisor who creates visions, accepts new ideas, makes quick decisions, encourages cooperation, who is not over-cautious and does not stress plans that must be followed.
>
> *Ibid.*

To achieve a process-oriented, user-controlled ISD within an organisation there is a need for design premises – a sort of 'design shop' where the designers may go and do the development work without any disturbance. We learned in the field-work that many good intentions from the management, e.g., allowances for UCISD, failed because there were no localities for the work.

When a group of people work together in an ISD project they need not only knowledge about information technology, they also need space. In the field-work presented here, most of the participants argued for a sort of *design place*. Preferably a 'local design shop', since the project groups often come from small work units. This may not be possible in all organisations. But, at least one design shop could exist in most organisations.

A design shop might well be a temporary institution, lasting only during the lifetime of a large development process. It may afterwards also be used for other purposes, e.g., for second-generations users' educational purposes. The main thing is that a positive attitude to design shops for UCISD, at all levels, is established in an organisation. The ideal situation is of course a *permanent local design shop*.

The idea of a 'local design shop' is based on the principles of 'Participative Design Workshops' (PDW) advocated by Emery and Emery (1989). The actors attending the PDW are:

> ... those whose jobs are under review. It is only from people pooling their various, and usually fragmented, but always detailed, knowledge that a comprehensive and stable design can come.

These people are helped and supervised in their design work by a 'Resource Person' - often a sociologist or an external facilitator (cf. Mumford, 1983). Future Workshops were developed for people without resources who wanted a say in, e.g. environmental issues (Jungk & Müllert 1987). In ISD, the Future Workshops may be used as a technique in combination with 'metaphorical design', as described in Kensing and Halskov Madsen (1991). They are then run by one or two facilitators/DP experts. The Future Workshops and the PDWs are not physical in the sense that 'local design shops' are intended to be. PDWs are more like seminars arranged during redesign of a workplace. Future Workshops may be placed anywhere the design-group meetings are held. My contribution to the Future Workshop and the PDW idea is that I propose:

> In an ISD change situation, there is a need for a physical place for user-controlled information systems development - with the necessary help of experts when needed, as well as the necessary information technology.

For the local design shops we may look to the factory case, the shop-steward's room, for an illustration. The help of experts is meant to include any kind of specialists called for by the users. The approach is partly based on the classical ISD model of Naumann and Jenkins (1982), and partly on the 'dialogue' concept described in the pedagogical method by Freire (1972). The emancipative pedagogy of Freire is one of the basic theoretical fundaments of the PROTEVS approach; to make future-users conscious of their own possibilities and knowledge that make 'work- and user-oriented information systems design' possible.

In Freire's pedagogical model we learn that two entirely different cultures, e.g. teachers and pupils, undergo three activities in a situation of learning: ' ... dialogue, coding and problematisation'.

Freire, and later his followers (Randall and Southgate, 1981), have established communications between different cultures with different languages by drawing pictures of a problem situation and problem process they wanted to learn of. The pictures became the code of communication, with common symbol denominations.

In an ISD approach with a strong element of learning, the two cultures described by Freire, and later Randall and Southgate (1981), may be translated as those of the DP experts and future systems-users (management and operative personnel). In the three activities, dialogue, problematisation and coding, they gain an insight into and understanding of one another in a dialectic process. The experts/specialists may assist in the dialogue and problematisation to establish a communication code within the group.

This cooperation requires knowledge. The operative and practically established knowledge is to be found with the future users. The strategic overall knowledge is mainly to be found with the management. These kinds of knowledge should be coordinated in a dialogue, hence there is a concept of dialogue in two dimensions. One between the DP experts and users and the other between the actors in the user group, i.e. between management and operative personnel (see Figure 38).

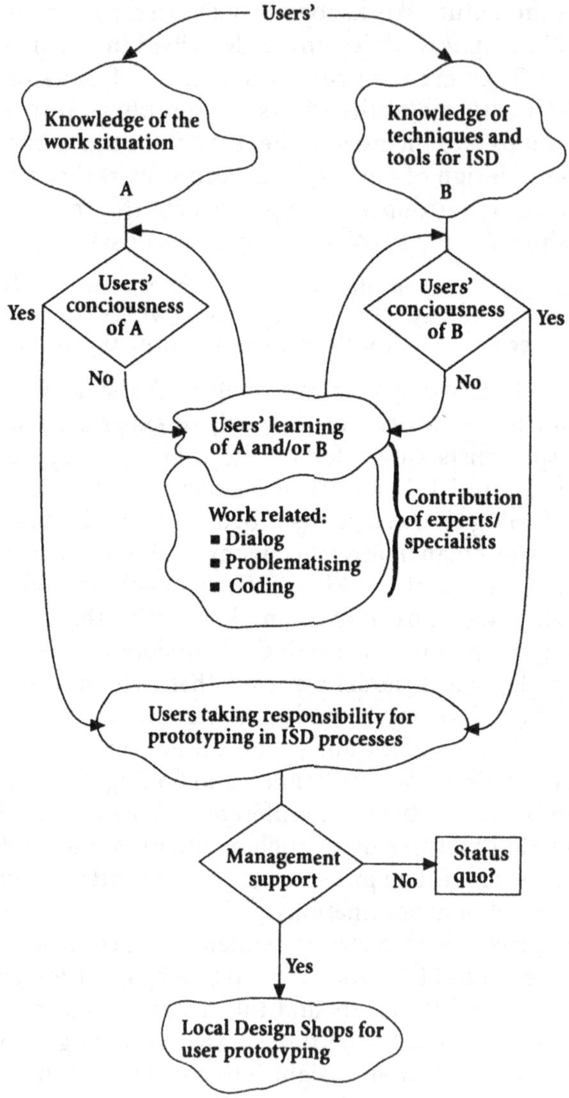

Figure 38: The local theory of the prerequisites for user controlled ISD

Lessons Learned

Most future system-users are *able to* and *willing to* participate in the design of a computer-based IS. This has been corroborated in both the exploratory case studies and in the empirical cases. From these initial developments other future users may learn that designing prototypes on computers is not difficult work if they are provided with time and tools.

The self-confidence evolved, and the new roles of the participating users during user-controlled ISD work, should be taken as an indicator by future users that it is certainly worth while advocating user participation in ISD work. Hands-on learning is acquired during user prototyping which also enhances the possibility of better communication between DP experts and users. Even when the greater part of a new CBIS may be designed by the DP experts, one needs knowledge to say yes or no to further normalisation/automation and other changes in their work situation. They possess the potential to acquire such knowledge. Future users may learn this from the field experiments presented in this work.

They can also learn from the experiences of the participating users in the Falkenberg and factory cases that a specific place for design work is desirable, for example, as in the case of a local design shop.

The implementation of the PROTEVS model in an industry must be preceded by negotiations with both workers and management (cf. Ehn, 1988). The unions have a great share in these negotiations, and the shop-steward's room becomes a natural meeting place. In some cases the shop-steward may take the role of project leader - at least at the start of the project. In the future it is most urgent to further study:

- grounds for elaborating local theories on applying the PROTEVS model for UCISD;
- how the local design shops should be designed in order to support users, and best utilise both experts/specialists and technical facilities;
- how a 'local theory of a suitable environment' for UCISD can become elaborated by considering educational problems as well as other premises of local design in a workplace.

The future research indicated here will be undertaken not only in collaboration with people in the organisations studied, but also with researchers from other areas such as: 'sociology', 'industrial management' and the 'work environment'. This introduces, amongst researchers, some similar problems of a meeting between several cultures as has been reported in this study of users meeting DP experts.

By encountering and having to resolve cultural gaps between themselves, the researchers can gain practical experience of the kind their co-researchers in the organisations have to handle all the time. This should sensitise the researchers to one of the most critical phenomena in their field of study.

References

Alavi, M. (1984). 'An Assessment of the Prototyping Approach to Information Systems Development', *Communications of the ACM*, vol. 27, no. 6.

Argyris, C. and Schoen, D. A. (1978). *Organisational Learning: A Theory of Action Perspective.* Addison-Wesley, New York.

Argyris, C. and Schoen, D. A. (1979). 'Participatory Action Research and Action Science Compared', *American Behavioural Scientist*, vol. 32, no. 5.

Beard, Ch. and Easingwood, Ch. (1989). 'Improving Grounded Theory Methods and Methodology', Working paper 177, Manchester Business School, Manchester, UK.

Berger, P. L. and Luckmann, T. (1967). *The Social Construction of Reality.* Anchor Books, Doubleday &Company, Inc Garden City, NY.

Bjerknes, G., Bratteteig, T. and Espeseth, T. (1991). 'Evolution of Finished Computer Systems: The dilemma of enhancement', *Scandinavian Journal of Information Systems*, vol. 3.

Bjørn-Andersen, N. (1978). 'GONGAR TALVA - Kortleggning og greining av Gongdum i Fyrstingarligum arbejdi', Informationsbolkurin, Forshavn, Farö Islands, Denmark.

Bjørn-Andersen, N. (1984). 'User Driven Systems Design', *Work and People*, vol. 10, no 3.

Boland, Jr R. J. 1978, 'The Process and Product of Systems Design', *Management Science*, vol. 24, no. 9.

Borum, F. (1977). 'Systemkonstruktion og videnskabelig driftsledelse', in EDB, *Arbejdsmiljø og Virksomhedsdemokrati*, red: Borum F, Nyt fra Samfundvidens-kaberne, Denmark.

Bødker, S. Greenbaum, J. and Kyng, M. (1991). 'Setting the Stage for Design as Action' in Greenbaum, J. and Kyng Morten (eds.) *Design At Work.* Lawrence Erlbaum Associates, Publishers, Hillsdale, New Jersey.

Checkland, P. (1981). *Systems Thinking, Systems Practice.* John Wiley & Sons, Chichester.

Cuff, E. C. and Payne, G. C. F. (1979). red, *Perspectives in Sociology*, Korpen, Göteborg.

Davis, G. B. (1984). 'Caution: User Developed Systems Can Be Dangerous to Your Organisazation', MIS Research Center, School of Management, University of Minnesota, Minneapolis, USA, 1984.

de Marco, T. (1978). *Structured Analysis and Systems Specifications.* Yordon Inc, New York, USA.

Ehn, P. (1983). *UTOPIA-projektet (Status rapport (fas I) och arbetsplan (fas II).* Swedish Center for Working Life, Stockholm.

Ehn, P. (1988). *Work-Oriented Design of Computer Artifacts*, Arbetslivscentrum Stockholm, Sweden.

Ekvall, G., Arvonen, J. and Nyström, H. (1987). *Organisation och Innovation.* Studentlitteratur, Lund.

Ekvall, G. and Arvonen, J. (1991). 'Change-Centered Leadership: An Extension Of The Two-Dimensional Model', *Scandinavian Journal of Management*, vol. 7, no 1.

Elden, M. (1979). 'Three generations of work-democracy experiments in Norway' in Cooper, C. and Mumford, E. (eds.) *The Quality of Working Life in Western Europe.* Associated Business Press, London.

Elden, M. (1983). 'Democratization and participative research in developing local theory', *Journal Of Occupational Behaviour*, vol. 4.

Elden, M. and Levin, M. (1991). 'Cogenerative Learning: Bringing Participation into Action Research' in Whyte, W. F. (ed.) *Participatory Action Research.* Sage Publications, Newbury Park, Ca, USA.

Emery, F. E. and Thorsrud, E. (1976). *Democracy at Work.* Leijden: Nijhoff.

Emery, F. and Emery, M. 'Participative design' in *Partipative Design for Participative Democracy*, Emery, M. (ed). Centre for Continuing Education, The Australian National University.

Flensburg, P. (1986). *Personlig databehandling – introduktion, konsekvenser, möjligheter,* Studentlitteratur, Lund, Sverige. In Swedish.

Flensburg, P. (1987). *Systemutveckling med människan i centrum,* Studentlitteratur, Lund. In Swedish.

Floyd, C. (1984). 'A systematic look at prototyping' in Budde *et al.* (eds.) *Approaches To Prototyping,* Springer-Verlag, Berlin-Heidelberg.

Freire, P. (1972). *Pedagogik för förtryckta.* Gummessons Stockholm.

Friis, S. (1984) 'Prototyping and User Developed Requirements Specifications', Seventh Scandinavian Research Seminar on Systemeering, Helsinki, Finland.

Friis, S. (1986). 'Tools for User Prototyping' in Brookes, B. C. (ed.) *Intelligent Information Systems for the Information Society.* Elsevier Science Publ BV (North-Holland).

Friis, S. (1987b). 'User Developed Prototype Systems' in Rasmussen, J and Zunde, P. (eds.) *Empirical Foundations Of Information And Software Science Iii.* Plenum Press, USA.

Friis, S. (1988b). 'Action Research on Systems Development: Case Study of Changing Actor Roles', *ACM computer & society,* vol. 18, no. 1.

Friis, S. (1991a). 'User invovlement in requirements analysis and specifications: in theory and in practice' in Besselaar, Clement, and Järvinen (eds.) *Information System, Work and Organization Design.* North Holland, Amsterdam.

Friis, S. (1991b). 'Dependencies in Research on User Controlled Information Systems Development' in Nurminen, M. I. and Weir, G. R. S. (eds.) *Human Jobs and Computer Interfaces.* Proceedings of IFIP Conference WG 9.1, Tampere, Finland- Forthcoming.

Glaser, B. G. and Strauss, A. L. (1967). *The Discovery of Grounded Theory: strategies for qualitative research.* Aldine Publishing Co, USA.

Greenbaum, J. (1990). 'The Head and the Hart', *ACM computer & society,* vol. 20, no. 2.

Gunton, T. (1988). *End User Focus.* Prentice Hall, Englewood Cliffs, New Jersey.

Gustavsen, B. (1985). 'Workplace Reform and Democratic Dialogue', *Economic and Industrial Democracy,* vol. 6: 461–479.

Gustavsen, B. (1990). *Vägen till bättre arbetsliv. Strategier och arbetsformer i ett lokalt utvecklingsarbete,* Arbetslivscentrum, Stockholm. In Swedish.

Göranzon, B. (1975). *Det praktiska intellektet,* Carlssons Bokförlag, Stockholm. In Swedish.

Hedberg, Bo. (1975). 'Computer Systems to Support Industrial Democracy' in Mumford, E. and Sackman (eds.) *Human Choice and Computers.* North Holland.

Ives, B., Hamilton. S. and Davis, G. B. (1980). 'A Framework For Research In Computer-Based Management Infirmation Systems', *Management Science,* vol. 26, no. 9.

Jenkins, A. M. (1983). 'The prototyping methodology', *Spectrum,* vol. 2, no. 2.

Jenkins, A. M. (1984). 'Research Methodologies and MIS research' in Mumford, E. *et al.* (eds.) *Research methods in information systems.* North Holland.

Jungk, R. and Müllert, N. (1987). *Future Workshops: How to create desireable futures.* London: Institute for Social Inventions.

Järnegren, A. *et al.* (1990). *Ett program för lokalt utvecklingsarbete,* Arbetsmiljöfonden, Stockholm. In Swedish.

Järvinen, P. (1987). 'Notes On Communication And Tool Metaphors Of An Information System' in *The 10th Scandinavian Research Seminar on Systemeering,* Tampere, Finland.

Kalleberg, R. (1989). *Action Research As Constructive Sociology.* Institute for Social Research, University of Oslo.

Katz, D. and Kahn, R. (1966). *The Social Psychology of Organizations.* John Wiley & Sons, USA.

Kensing, F. and Halskov Madsen, K. (1991). 'Generating Visions: Future Workshops and Metaphorical Design' in Greenbaum, J. and Kyng, M. (eds.) *Design At Work.* Lawrence Erlbaum Associates. Publishers, Hillsdale, New Jersey.

Luqi, (1989). 'Software Evolution Through Rapid Prototyping', *Computer*, May.

Mathiassen, L., Nielsen, P-A., and Jepsen, L. O. (1986). 'Back to Thinking Mode', in Sandström, G. and Nissen, H-E. (eds.) *Quality of Work vs Quality of Information Systems*. 9th IRIS Conference, Lund, Sweden.

Mead, G. H. (1930). *George Herbert Mead on Social Psychology*. Chicago Press, USA.

Mead, G. H. (1934). *Mind, Self, and Society*, Morris, C. W. (ed.). Chicago: The University of Chicago Press.

Mumford, E. (1983). *Designing Secretaries*. Manchester Business School, Manchester, England.

Naumann, J. and Jenkins, A. M. (1982). 'Prototyping: The New Paradigm for Systems Development', *MIS Quarterly*, September 1982.

Nissen, H-E. (1976). *On interpreting services rendered by specific computer applications*. Liber Stockholm.

Nissen, H-E. *et al.* (1982). *User Oriented Information Systems – A research program*. Dept of Information and Computer Science, Lund university, Lund Sweden.

Nygaard, K. and Sørgaard, P. (1987). 'The perspective concept in Informatics' in Bjerknes, G., Ehn, P. and Kyng, M. (eds.) *Computers and Democracy – A Scandinavian Challenge*, Avebury, Aldershot.

Olerup, A. (1982). 'A Contextual Framework for Computerized Information System', Nyt Nordisk Forlag Arnold Busck, Copenhagen, Denmark.

Olerup, A. (1989). 'Socio-Technical Design of Computer-Assisted Work: A Discussion of the ETHICS and Tavistock Approaches', *Scandinavian Journal of Information Systems*, vol. 1, August 1989.

Polanyi, M. (1958). *Personal Knowledge*. London, Routledge & Kegan Paul.

Polanyi, M. (1966). *The Tacit Dimension*. Gloucester, Mass, reprint P. Smith, 1983.

Randall, R. and Southgate, J. (1981). 'Doing dialogical research', in Reason, P. and Rowan, J. (eds.) *A Sourcebook of New Paradigm Research*. John Wiley and Sons.

Rzewski, G. (1983). 'Prototypes versus Pilot Systems. Strategies for Evolutionary Information System Development', Working Conference on Prototyping, Namur, Belgium.

Salaway, G. (1987). 'An Organizational Learning Approach to Information Systems Development', *MIS Quarterly*, June 1987.

Sandberg, Å. (1981). Om arbetslivsforskningens metoder och förutsättningar, in Å. Sandberg (ed.), *Forskning för förändring*, Swedish Center for Working Life, Stockholm. In Swedish.

Sandström, G. (1985). *Towards Transparent Data Bases – How to Interpret and Act on Expressions Mediated by Computerized Information Systems* .Studentlitteratur, Lund, Sweden.

Smith, B., 'The Limits Of Correctness', ACM, *Computers and Society*, vol. 15.

Torbert, W. R. (1981). 'Empirical, behavioural, theoretical, and attentinal skills necessary for collaborative inquiry' in Reason, P. and Rowan, J. (eds.) *A Sourcebook of New Paradigm Research*. John Wiley and Sons.

Whyte, W. F., (1991). *Social Action Theory For Action, How individuals and organizations learn to change*. Sage Publications, Newbury Park, Ca, USA.

Zadeh, L. A. (1965). 'Fuzzy Sets', *Information and Control*, pp. 338–353.

Zadeh, L. A. (1973). 'Outline of a New Approach to the Analysis of Complex Systems and Decision Processes, *IEEE Transaction Systems Management & Cybernetics*, vol. 3: 28–44.

Chapter 8

Designing for Knowledge Transfer

Satinder P. Gill

Introduction: Epistemological Basis

What knowledge can be represented? Is it possible to represent practical knowledge? Is it possible to represent personal knowledge? These issues are concerned with design as a process where design is an example of the formation and transfer of knowledge, as well as with two contrasting 'traditions' of knowledge (Gill, 1988; Josefson, 1987, 1988).

One of these traditions is embodied in the idea of a science of mind. This idea has been developed within the field of cognitive science.[1] Here, knowledge has become tangibly linked to a technology. Orthodox cognitivism or computationalism (Fodor, 1976, 1981; Pylyshyn, 1984) holds that cognition can be defined as computations of symbolic representations. The focus is on representation and logic. Knowledge is non-contextual, in that it is time-independent and depersonalised. The other tradition may be termed 'human-centred'[2] (for example, see Cooley, 1987; Gill, 1990a; Rosenbrock, 1988, 1989; Rasmussen, 1989; Smith, 1991), where knowledge is context-based and has a personal and social dimension. Here the interest lies in exploring the interdependence of different aspects of knowledge embedded in knowledge transfer. The premise is that knowledge exists in praxis/experience and has a personal and social dimension.

Non Person-centred

The characteristics of the non person-centred approach to knowledge are propositional knowledge, representation, and the rule. These characteristics will be discussed in turn.

Propositional Knowledge

The propositional knowledge denotes formalised knowledge, wherein knowledge has to be explicitly defined or articulated. It has to be empirically supported or formally proven (Johannessen, 1988a). It is knowledge which does not embody the personal and social dimension and can exist independent of time and, therefore, context. For example, it cannot allow for social and personal values which exist in moral knowledge.

Propositional knowledge has also been termed 'theoretical knowledge'[3] (Göranzon, 1992), 'scientific knowledge'[4] (Woolgar, 1987; Winch, 1958; Rosenbrock, 1988; Collins, 1974, 1975), 'rationalism'[5] (Dreyfus, 1989; Gill, 1984, 1986; Josefson, 1987, 1988; Gill, S. P., 1988), 'positivistic knowledge'[6] (Johannessen, 1988a) and 'universal knowledge'[7] in its various contexts.

Aspects of this notion of knowledge have been traced through the history of ideas to the 'dream of the exact language' of the 17th century in Europe.[8] This 'dream' was an ambition to create a universal language which would allow one to express one's thoughts 'as definitely and exactly as arithmetic expresses numbers or geometrical analysis expresses lines.'[9] This language would embody shared understanding. It would not allow for misunderstanding as it 'would embody and encode all valid modes of argument, so that different people reason together without fear of confusion or error.'[10] This language would be an instrument of reason whereby knowledge is clear and unambiguous.[11] The idea that knowledge must be well-defined is central to cognitive science, where the discipline of programming is regarded as scientifically useful for generating hypotheses about the mind. Underlying this idea of knowledge is the belief that the attempt to express vague concepts helps to clarify them.[12]

The concept of propositional knowledge draws a boundary between that which is knowledge and that which is not. For example, logical positivism[13] (see, for example, Ayer, 1971) argues that the areas of metaphysics, aesthetics, moral inspiration, ethics, lie outside the boundary of knowledge. This is because they are not, in full measure, linguistically articulable and are not scientifically relevant.[14]

The application of propositional knowledge will be discussed in the context of:

1. logical positivism and the idea of an exact language; and,
2. in the context of the traditional, or conventional, model of computational psychology in Artificial Intelligence[15] under the section on the Science of Mind.

Representation

Representation requires that the world be defined in terms of components, each of which mirrors the reality. This leads to a tendency towards reductionism. Ideas about representation of knowledge are tied to ideas about language and thought. The 'dream of a mathematical language'[16] in the 17th and 18th centuries, resulted in the concept of symbolic representation of ideas and the use of logical rules for defining the relations between symbols. Algebra was the model for the ideal language. There was a desire to eliminate all ambiguous or imprecise words and misleading metaphors[17] (Göranzon, 1991).

The Rule

The concept of a rule is linked to the concept of knowledge. Within the context of propositional knowledge, as defined above, a rule lays down the way in which the knowledge is to be interpreted or used.

The concept of the rule in rule-based computer systems embodies the belief in a science of logical thinking which is historically tied to the idea of a scientific law for nature.[18]

Application in Theory and Practice

Logical Positivism and Rationality

The relationship between representation of knowledge and language has been discussed in philosophy in the field of logical positivism[19] where great significance is given to linguistic knowledge. Logical positivism asserted that 'we can only possess knowledge if it can be formulated linguistically and in principle tested on the basis of experience or proved by formal methods'[20] (Johannessen, 1988a). Logical positivism represents knowledge as context independent.[21]

Cognitive Science: the Science of Mind

The characteristics of propositional knowledge, representation and the rule are all essential to cognitive science and artificial intelligence (AI). AI and cognitive science have been described as the last stage of the rationalist tradition in philosophy (Dreyfus, 1989). Cognitive science, also called 'computational psychology' (Boden, 1991), and 'cognitivism' (Varela, 1988),[22] postulates that psychological questions in general, whether they concern belief, problem solving, purpose, choice, language, perception, memory, or even emotion, can be understood as computational questions about mental rules and representations[23] (Johnson-Laird, 1983, 1993).

Within the computer metaphor, the concept of knowledge has been based within the domain of cognitive science and artificial intelligence (AI).[24] AI is the literal construal of cognitive science. AI involves the use of computer programming to study the structure of and function of knowledge.[25] Psychological concerns are with the structure and content of mental representations and the ways in which they can be generated, augmented and transformed. They focus on planning and problem solving (for example, Newell, 1972, 1980).[26] The construction of knowledge-based systems, and in particular expert systems, is the physical representation of AI and cognitive science.

The relationship between representation and language which exists in cognitive science and artificial intelligence is embodied in the idea of cognition. Cognitivists claim that we can only account for intelligence and intentionality if we hypothesize that cognition consists in acting on the basis of representations of concepts which are physically realised in the form of a symbolic code in the brain or a machine. This is a formal syntactic system.[27] In this model of knowledge, it is the form of words which are significant, where they are physically represented as symbols. These say nothing about why 'x' means 'x', that is they do not embody meaning. This particular model may be termed the 'formalist'[28] approach of the Von Neumann machine. Here, information-processing rules are explicitly coded within and accessed by the program. AI programs and computer models are purely formal-syntactic in nature.[29]

Design: the case of Knowledge Engineering

Knowledge engineering is an essential part of the process of designing knowledge-based systems (Kidd and Wellbank, 1984; Hart, 1986). Knowledge engineering here is defined as: the process of selecting relevant information for a system and considering how that information is made available to the system as well as to the user. Traditional knowledge engineering (TKE)[30] claims that all knowledge can be represented in a propositional form. This embodies the idea that knowledge is universal.

A knowledge-based system is a computer-based system which is made up of three parts:

1. the knowledge-base;
2. the inference engine (the programme);
3. the interface.

The combination of (1) and (2) simulates the internal processes of the human mind. Knowledge-based systems have been used in a variety of forms. Some are called 'expert systems' (ES) (Whitaker and Ostberg, 1988).[31] These ES simulate the behaviour of an expert in a particular domain. There have been projects to design medical ES,[32] legal ES (see Susskind, 1989; Leith, 1988), etc. The function of these systems is to provide solutions to problems. There are very few successful applications[33] (SPRU report 1988; Coats, 1988) and those that do succeed are constrained to very clearly definable domains such as configuration where they still need a human support system; for example, XCON, a Digital Computers' system. Today, ES are being combined with neural networks[34] in order to compensate for their inadequacies.[35] This development will not be dealt with in this chapter. Other uses of knowledge-based systems are as decision-making aids (Hayes-Roth, 1984b; Lipscombe, 1989; Mumford, 1987; Wright and Bolger, 1992; Sprague and Watson, 1986), where the system does not make the final decision, but presents useful information. The idea is that the human and the system form an integrated system, whereby the machine is regarded as being more efficient in carrying out some functions of a task (e.g. calculation), and the human is seen to be more efficient at others (using intuition, etc.).

This chapter is interested in one particular part of the design of knowledge-based systems, that of 'knowledge engineering' (KE). The core determinant of knowledge engineering is knowledge 'representation', be it of behaviour or of mind. The idea of knowledge representation, here, is in turn determined by the symbolic representational infrastructure of the computer machine.

In debates on knowledge representation, the knowledge engineering community has seen the inability of experts to express their knowledge and skills in procedures as a human shortcoming (Berry, 1988; Kidd and Wellbank, 1984; Kidd, 1987). In TKE, an emphasis was placed by many of them upon developing techniques to extract the knowledge in a more appropriate form (Gammack, 1987, 1989; Olson and Rueter, 1987). These were initially based within cognitive science. In the late 1980s (Hart, 1986; Smith, 1990), there was a shift towards using methods of communication from other domains such as counselling, social psychology, and anthropology. The objective was to access the practices of the expert and their tacit knowledge. These would still, however, be transformed into appropriate representational/propositional form for the knowledge-base.

Human-centred

Within the human-centred[36] approach, knowledge is context-based and has a personal and social dimension. In this context, the interest of the chapter lies in exploring the interdependence of different aspects of knowledge embedded in knowledge transfer.

The human-centred tradition is discussed within a framework of various human-centred European traditions of both design and application studies on computer-based applications. Much of the literature is primarily concerned with the development and constraints of the concept of scientific knowledge and its use.

Multiplicity of Knowledge Types

The idea of knowledge as consisting in a multiplicity of aspects is in contrast to the idea of knowledge being propositional. Propositional knowledge requires discrete definitions of knowledge which are independent of time, and therefore of context. A multiplicity of aspects of knowledge denotes an interdependence of different dimensions of knowledge, for example between the tacit and the explicit dimension. Propositional knowledge, however, only denotes one dimension of knowledge, that which is explicitly defined. For the former (tacit knowledge), knowledge has meaning in its practice. For the latter (propositional/explicit knowledge), meaning exists in its validity which needs to be either empirically or formally proven.

Challenge to Knowledge as Universal

The class of a multiplicity of aspects of knowledge is opposed to the idea of knowledge as universal. The idea that knowledge is universal is that knowledge is objective and exists outside of time and context. Objective knowledge embodies the notion of rule where a rule lays down the way in which the knowledge is to be interpreted or used. The human-centred approach to knowledge challenges this notion of the rule with the idea of rule-following which shows that there is no particular way in which to enact an idea or a word.[37] Rule-following occurs in the practice of the knowledge. This is discussed further below.

Application in Theory and Practice

Symbiosis of Tacit and Objective Knowledge

Polanyi – Tacit Knowing

The concept 'tacit knowing' was used by Polanyi[38] (1966), in his work on *The Tacit Dimension*. In various discussions on tacit knowledge he is often cited for his expression that: 'we can know more than we can tell'.[39] Polanyi is concerned to show that the aim of exact science to establish a strictly detached, objective knowledge, is fundamentally misleading because tacit thought forms an indispensable part of all knowledge. This idea of tacit knowing is based within a discussion which starts with questions about how we can recognise a face, or any object or phenomenon, and leads into questions about knowledge acquisition. He is interested in how we pass on to others knowledge which we ourselves cannot fully describe. He cites the use of practical classes, and the act of pointing to things to connect words to things, as methods of passing on such knowledge. This act of pointing leaves something out which we cannot tell, and for its meaning to be understood the other person must discover that which we have not been able to communicate. The discussion draws upon Gestalt psychology (Ellis, 1938; Henle, 1961), and the idea of shaping or integrating experience in the process of knowledge acquisition. Here we are introduced to two aspects of knowing: 'knowing how', and 'knowing that'[40] (Ryle, 1949). The former denotes practical knowledge, and the latter theoretical knowledge. The concepts of integration, awareness, interiorization or indwelling form the core of his theory of tacit knowledge and explain the interdependence of the theoretical and practical aspects of knowing. For example, Polanyi argues that mathematical theory can only be learned by practising its application. We can only understand that which is explicit if we can relate to it through practice.

Through the use of various examples, Polanyi identifies four aspects of tacit knowing: the functional, the phenomenal, the semantic and the ontological. The functional aspect is his basic model of the structure of tacit knowing which he uses throughout his argument. He explains it as follows. In recognising a face, we are aware of its features, yet we are unable to identify all those features. The ability to recognise the face occurs through a functional relationship between what he calls 'two terms of tacit knowing'. The first, proximal, term is the features, and the second, distal, term is the face. 'We know the first term only by relying on our awareness of it for attending to the second. It is the proximal term of which we have a knowledge that we may not be able to tell'.[41] He calls this the 'functional structure of tacit knowing'.[42]

Polanyi sees a direct relationship between our bodies and the external world, for example, in the way we see things, in his discussion on perception as an instance of tacit knowing. He cites physiologists who maintained that the way we see an object is determined by our awareness of certain efforts inside our body which we ourselves cannot feel. In this context, Polanyi includes the activity of neural traces in the cortex of the nervous system. His focus on the relationship between our bodily processes and the external world has been of interest to discussions on the nature of craftsmanship skills, such as engineering. He talks of the transformation of the tool into a sentient extension of our body.

He identifies tacit knowing with indwelling. It is not by looking at things, but by dwelling in them, that we understand their joint meaning. This relates back to the point that we can only understand a mathematical theory if we can relate ourselves to it in practising it. We can only understand something if we are within it and understand its practice. We cannot look at the parts and have an understanding of the whole. If we try to focus on the particulars, we can destroy our conception of an entity. For example, if a pianist whilst playing starts focusing on the movements of her fingers, she will be paralysed. However, if she focuses again on her music, she will be able to play again. With this distance the particulars will come back to life and recover their meaning and their comprehensive relationship. Thinking of the parts, rather than of the whole, as the real entity, is a misconception. It can be explained by the fact that the parts seem more tangible. Polanyi does not think, however, that a knowledge of the parts is of no use in the understanding of the whole. For example, he contrasts someone who knows how to use a machine without fully knowing how it works, with an engineer who, he says, has a deeper understanding of its construction and operation. He states, however, that a knowledge of the parts cannot replace its tacit counterpart. For example, the skill of a driver cannot be replaced by the teaching of the theory of the motorcar.

Polanyi talks of the personal dimension of knowing.[43] He considers how a scientist can see a problem. He argues that all knowledge is of the same kind as the knowledge of seeing and working with a problem. To hold such knowledge is to be committed to a conviction that there is something there to be discovered. This experience cannot be represented by any exact theory. It is a personal act of knowing.

To summarise, Polanyi's discussion on how we know something to be something is explained through his model of tacit knowing which has a particular structure of going from particulars to the entity which does not involve an explicit identification of these particulars. His theory of indwelling or interiorization claims that we can only understand that which is explicit if we can relate to it through practice. Knowing has a personal dimension.

Rosenbrock: Model of Symbiosis of Tacit and Explicit Knowledge

The concept of tacit knowledge has been discussed in the domain of engineering as being an essential dimension of expert knowledge. This discussion has taken place in response to an increasing emphasis upon scientific or explicit knowledge at the expense of tacit knowledge, particularly in relation to the development and application of computer-based systems.

This emphasis has raised two fundamental concerns within engineering for the proponents of tacit knowledge such as Rosenbrock (1988, 1989, 1990) and Cooley (1987a, 1987b). One is about the development of engineering theory and practice, and the other is about the construction of purposeful and beneficial systems. An over-emphasis on theory or scientifically based knowledge does not adequately explain the nature of engineering skills[44] (cf. Rosenbrock, 1977). It undervalues the interrelationship between theory and practice. It also ignores the dimension of human purpose in the explanation of skills and in the design of systems. In the case of design, the most extreme representation of the scientific perspective of knowledge is the automated machine[45] (cf. Feigenbaum). This is illustrated in Figure 39 where explicit knowledge denotes theoretical knowledge or scientific knowledge, such as the kind of knowledge found in a textbook. The idea of the automated machine embodies the belief that expert knowledge can be completely explicated (Figure 39(b), cf. Rosenbrock, 1988, p. 318).

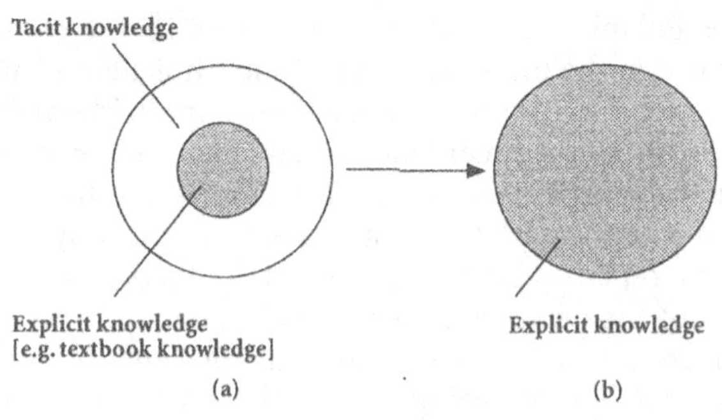

Figure 39: Explication of expert knowledge; an analogy to the idea of the automated machine

However, the concept of autonomous entities, whether they are machines or humans, is fallacious. In the case of machines, some human intervention will always be needed in unexpected situations, such as machines malfunctioning.

The model of symbiosis between tacit and explicit knowledge proposed by Rosenbrock, who is himself an engineer, provides an alternative model of knowledge which meets both of the concerns about the development of engineering theory and practice and the development of systems.

The model of symbiosis between the tacit and explicit aspects of knowledge, (see Figure 40, cf. Rosenbrock, 1988, p. 318) challenges the emphasis on explicit knowledge and the concept of the automated machine (as Figure 39(b) above). The symbiotic interdependence of tacit and explicit aspects of knowledge means that knowledge can never be made completely explicit as it is Figure 39(b). Instead, as knowledge is explicated it gives rise to new tacit knowledge which is required to use this explicit knowledge (Figure 40).

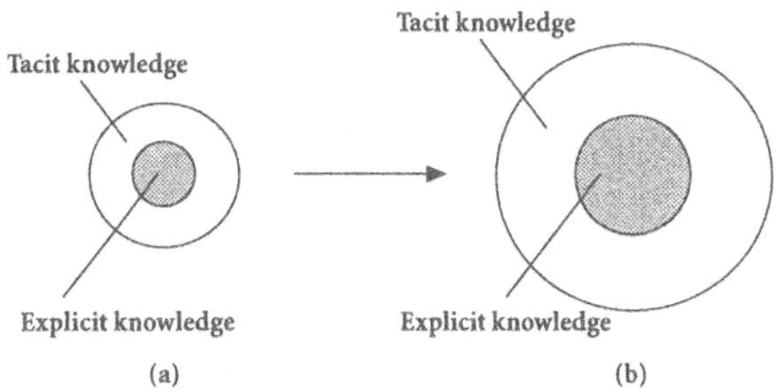

Figure 40: The expansion of explicit knowledge leads to a reciprocal expansion of tacit knowledge required for using the new explicit knowledge

It is in the context of seeing knowledge as a symbiosis of tacit and explicit aspects that Rosenbrock proposes that seeing engineering as an art rather than as a science will provide a better explanation of engineering skill. It is an art because of the tacit knowledge involved in being a skilled practitioner. Although a skilled engineer uses scientific knowledge and mathematical analysis, his/her skill also:

> ... contains elements of experience and judgement, and regard for social considerations and the most effective way of using human labour. These elements partly embody knowledge that has not yet been reduced to an exact mathematical form. They also embody value judgements which are not amenable to the scientific method.
>
> *Rosenbrock, 1988*[46]

These elements make up the tacit knowledge of the skilled engineer. In further describing these elements of experience and judgement, Rosenbrock draws upon Polanyi's (1966) concept of tacit knowing wherein we have knowledge that we are unaware of having. That is, there are certain things we do correctly but which we cannot explain or make explicit in scientific terms. Rosenbrock gives an example of how an engineer performs a task to show the use of tacit knowledge in making judgements. The example focuses on the ability to doubt. An engineer may set up a problem according to an accepted theory and use a computer to process the problem, but he or she may doubt the answer. Further analysis may reveal a mathematical error or the misapplication of the theory thereby confirming the doubt. The doubt involves intuition, experiential knowledge and a personal commitment on the part of the engineer. There is commitment because it is a major responsibility of the engineer to decide when the discrepancy between theory and the real world will lead to error.

Rosenbrock thinks that it is important that engineers recognise the essential element of art and tacit knowledge of their profession. If it is difficult to convince them that this exists, it will be more difficult to persuade them that other professions have this element. This would leave little room for the essential human input when an engineer designs systems in which other people will work. The process of explicating knowledge for systems development will subsequently follow the direction illustrated in Figure 39, and focus on explicit knowledge at the expense of tacit knowledge. This would also ignore the social and personal dimension of working life practice.

The symbiotic model of knowledge, illustrated in Figure 40 above, provides an alternative possibility for the human–machine relationship. Here the essential human input is recognised. The symbiotic model deals with the process of explicating knowledge and the process of the designing of systems. It also provides an explanation of the development of engineering theory and practice.

The necessary interdependence of tacit and explicit knowledge means that as aspects of tacit knowledge are made explicit, through new theories about expertise or in the development of new technologies, there is a reciprocal expansion of art and tacit knowledge. This is required in order to make this new explicit knowledge meaningful and for it to be used. For example, with the development of new technologies, new skills are needed. These could be built upon previous skills. Future technological systems could be designed which accept existing skills. Such systems would enable existing skills to develop through experience with the new system into skills which the changing technology requires. Existing skills need to be seen as relevant and useful, but not static. Technological systems need to be designed so that they cooperate with human skill to make it more productive. This is as opposed to eliminating that skill.

The relationship between tacit and explicit knowledge expressed in Figure 40 provides an explanation of the development of engineering theory and practice. As engineering builds its scientific basis, some of the tacit knowledge embodied in earlier practices becomes redundant. A new body of tacit knowledge emerges with the setting up and interpretation of new methods. The symbiotic model draws attention to human purpose in engineering practice. Rosenbrock's symbiotic model is driven by his wish to see the development of technology and skills as a positive beneficial process whereby purpose is essential (Rosenbrock, 1988, 1992).[47] Rosenbrock proposes that such an approach to skills and systems design would have broader effects on the working life and general standard of living as it would lead to increased productivity. He argues that as productivity increases, unemployment is avoided by a reduction in the working week, by study leave, earlier retirement, better housing and education and other ways in which the benefits of increased productivity can be distributed. Some of the increased productivity can be used to counter negative effects of technology such as pollution and ecological damage. Rosenbrock argues that the model of symbiosis of tacit knowledge/skills and explicit knowledge/technology is needed in order to develop technological systems and working practices together in a human way for the benefit of all.

Cooley: on Common Sense

Cooley (1987a, 1987b), deals with the nature of knowledge and skill of craftsworkers. He believes that knowledge consists in a symbiosis between its objective and subjective parts. These cannot be separated. Cooley attacks three presumptions of knowledge-based computer technology embodied in the description and attitude towards the nature of knowledge and skill. The first is the objectification of subjective knowledge. The second is the reduction of skill acquisition from being a process of knowledge reproduction to being a process of knowledge production. In both these cases, the designers of computer systems have assumed that the subjective can be separated from the objective. The third issue is the damaging effect of the subjective/objective split within human–computer interaction upon the common sense and tacit knowledge of skilled workers.

We will consider these issues in turn. Firstly, Cooley's symbiotic model of knowledge is explained Figure 41 below (cf. Cooley, 1987a, p. 13).

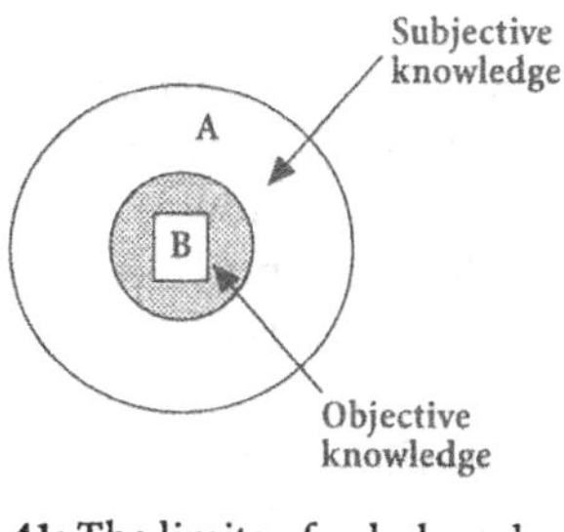

Figure 41: The limits of rule-based systems

In Figure 41, 'A' denotes the knowledge required to be an expert/skilled craftsworker. Within it there is a core of knowledge, 'B', which is the facts of the domain; for example, textbook knowledge. This can be reduced to a rule-based system, i.e. objective knowledge. The annulus AB can be said to represent common sense, tacit knowledge and imagination, heuristics and fuzzy reasoning.

Common sense for Cooley means 'a sense of what is to be done and how it is to be done, held in common by those who will have some form of apprenticeship and practical experience in the area.'[48] Common sense is acquired through learning by doing. Tacit knowledge, for Cooley, is that which is also acquired through learning by doing and by attending to things. The relative levels of the subjective and objective aspects of knowledge which a person utilises vary as one gains expertise. An expert uses more of the subjective aspects and less of the objective aspects of the knowledge in, for example, the use of intuition. An expert has the ability to grasp the situation in front of him/her and make judgements about it. A novice, on the other hand, can only calculate by using explicit rules to make sense of what appears to him/her to be a mass of data. Cooley describes this world of discrete and fragmented data as that of noise. He presents the process of acquiring knowledge as a spectrum going from data to action:

> Data suitably organised and acted upon may become information, and information that is absorbed, understood and applied by people may become knowledge. Knowledge frequently applied in a domain may become wisdom, and wisdom the basis for [normative] positive action.[49]

Common sense and tacit knowledge, which are essential requirements for skill, exist where there is wisdom. We can make calculations with data but we can only make judgements when we have wisdom. Being able to make judgements means that one has acquired expertise.

Cooley questions design philosophies which are based on the data-information end of the spectrum rather than on the wisdom-knowledge end. For example, he rejects the traditional expert systems idea[50] where the ultimate aim is to objectify subjective knowledge (Figure 40). This idea is based upon a belief that the subjective and the objective are separable. For Cooley, it is the interaction between the subjective and the objective that is important.

The second issue which Cooley is concerned with is information technology's focus on the production of knowledge. This focus is based on the notion that knowledge is objective. However, in order to understand the nature of skill, one needs to see knowledge as a reproductive process which is based on the continuous interaction between the subjective and the objective aspects of knowledge. The reproduction of knowledge is part of the process of knowledge and skill acquisition. Knowledge is acquired through learning by doing. Through this process human beings acquire 'intuition' and 'know-how'[51] (Dreyfus and Dreyfus, 1986).

Common sense and tacit knowledge consist of intuition and know-how. Cooley draws upon Dreyfus and Dreyfus' assertion[52] that analytical thinking and intuition are complementary ways of understanding and making judgements. As one becomes increasingly experienced and skilled there is a greater emphasis placed on intuition. Cooley believes that the analysis of working-life tasks and skills needs to account for the holistic balance between the use of analytical thinking and intuition, tacit knowledge and know-how of the worker.

The separation of the objective and the subjective in describing knowledge and skill, and in the subsequent design of computer technology, leads to the third issue. This is that human–computer interaction does not allow for interaction at the subjective level. Instead, it imposes objectivity on working practice. This in turn, Cooley argues, damages the common sense and tacit knowledge of skilled workers. Recognition of the symbiosis of objective and subjective knowledge would however require alternative design philosophies and practices which allow for the subjective aspects in human–machine symbiosis. Such an approach is termed 'human-centred'.

In summary, Cooley argues that it is essential that one recognises the relationship between the subjective and the objective aspects of knowledge. This relationship includes the dynamics of analytical methods and intuition, know-how and tacit knowledge of the skilled craft-worker. The relationship between the subjective and objective aspects of knowledge necessitates a rethinking about the relevance of computer technology and the development of symbiotic systems where the relation between the human and the machine places the human at the centre and not the machine.

Interdependence of Tacit and Propositional Knowledge

A Hermeneutic Approach

During the 1980s the concept of tacit knowledge became the central theme in hermeneutic[53] discussions about professional skills and expertise in Sweden (Johannessen, 1988a; Nordenstam, 1992; Göranzon, 1991; Josefson, 1987, 1988, 1992; Janik, 1988, 1992; Perby, 1988, 1990). The implementation of knowledge-based systems technology in many spheres of Swedish economic life since the 1970s raised concern both about the long-term effects of the technology upon skills and about the nature of design itself (Ehn, 1988). Their studies showed that, in the long-term, experts lose confidence in their ability to judge whether their methods and results are valid in comparison to those of the computer (Gill, S. P., 1995).

Having confidence in one's own judgement and abilities is crucial to being skilled. The case studies[54] and theoretical discussions concentrate on the concepts of praxis, craftsmanship and skill acquisition. Within this context, the discussion on tacit knowledge focuses on the inter-relationship between tacit and propositional knowledge, that is, between practical and theoretical knowledge. This interdependence determines that propositional knowledge only has meaning when we know how to use it, or make sense of it. This entails its practice. This relationship extends from how we relate to the objects in our world, right up to the development of our skills and our language. There are two fundamental aspects of tacit knowledge expressed in the Swedish case studies.[55] These are practical knowledge, and knowledge by familiarity. Practical knowledge is the aspect of performing the knowledge in practice. Knowledge by familiarity covers the aspect of learning a practice through the use of examples of the practice. These may be examples of experiences given by those involved in the practice. Knowledge by familiarity makes for reflection upon one's own practice.[56] It comes close to Polanyi's idea of indwelling (see pp. 11–13).

The hermeneutic discussion draws upon Wittgenstein and Polanyi. Johannessen is a key philosopher in the development of this hermeneutic discussion.[57] Some of the central concepts he has been developing, rule-following and practice, are presented below.

Rule-Following and Practice

Johannessen on Wittgenstein and Polanyi

A discussion on rule-following and practice has been undertaken at the Bergen School of Philosophy (Johannessen, 1988a, 1988b, 1992; Nordenstam, 1987, 1992). They are challenging the emphasis generally placed upon linguistic knowledge or the written word regarded as a superior form of knowing to that of practice.[58] This emphasis upon linguistic knowledge and the written word can be traced back to the assumption that knowledge and language have a mathematical basis. Wittgenstein challenges the concept of a rule as entailed in the concept of a mathematical language, and undergoes a radical shift in his later philosophy in his development of the concept of 'to follow a rule'. This is called his 'pragmatic turn'.[59] In the context of the discussion on tacit knowledge in this chapter, the relationship between rule-following, practice and language is a challenge to idea of propositional knowledge.

Johannessen[60] (1988b), describes Wittgenstein's philosophy as a kind of practice philosophy. To learn to master a language is a matter of mastering human reality in all its complexity. It is a matter of learning to adopt an attitude towards it in established ways, to reflect over it, to investigate it, to gain a foothold in it, and become familiar with it.[61] Wittgenstein includes physical communication such as gestures in his extended concept of language. This is to show that we make use of a variety of means to make ourselves understood. 'A sentence does not say of itself that it is to be taken as, say, an assertion'.[62] Language and human action are intimately interwoven, and so thereby is the relationship between language and reality. Wittgenstein became very interested in the application of the rule and the situation of the user. One and the same rule can be followed in many different ways. What guarantees that a rule is followed in the same way time after time cannot itself be a rule at all. It must depend on our actions and different kinds of spontaneous reactions. Rule-following is a practice. If we do not know the closer details of the current use situation, we will not be able to make up our minds about what is actually said. Therefore, our mastery of a natural language must include a grasp or practical understanding of an enormously large repertoire of situations involving the use of language. In order to understand and respond, we must have situational understanding and judgmental power. The very exercise of an activity might be a constitutive part of the formation of concepts. The content of a concept can be regarded as a function of the established use of its expression. One has mastered a given concept when one is accepted as a competent performer of the series of established practices which incorporates the concept. It is our application or practice which shows how we understand something. 'The practice give words their meaning'.[63] The identity of a rule over time is attained through the exercise of the established set of practices which guarantees that a rule is applied in the same way from one time to another and from one person to another. The rule itself cannot give this guarantee. 'The knowledge built into the mastery of concepts or rules has a partial and non-reducible expression in action. It is not possible to put into words this aspect of action in which the intellectually explicable part of the concept is necessarily embedded.' There is an interdependence between the tacit and the explicit aspects of knowledge.

Design Issues

The non-person centred view of knowledge is that it is propositional. There is a distinction and a separation between that which is subjective and that which is objective. Emphasis is placed on objective knowledge which is independent of time and context. The traditional design perspective embodies the non-person centred view of knowledge. The end product of design is given greater significance than the process of design. The technology is an objective entity whose application has universal predetermined outcomes. The process of design is seen as a series of discrete tasks. For example, traditional knowledge engineering becomes a matter of digging the knowledge out of the heads of experts. It is not about developing an understanding of skill. In this case communication is a functional process which serves the purpose of extracting data according to predetermined constraints. Traditional knowledge engineering separates the designer from the user, and the user from the technology. The chapter takes the view that knowledge is time- and context-dependent and has a social and personal dimension. It is critical of the non person-centred characterisation of knowledge. It is argued that knowledge consists of many aspects which are interdependent. The subjective and objective aspects of knowledge are interdependent and cannot be separated.

For example, intuition and analytical thinking are complementary aspects of knowledge. Given that knowledge consists in interdependent aspects, how can one learn about these aspects of knowledge. For this purpose, knowledge is considered as analogous to skill. In order to contrast the nature of skill with the non-person centred perspective of knowledge, the example of the computer has been taken as a representation of the latter. Humans have the ability to adapt and to innovate. This enables them to deal with uncertainty and breakdown. The computer, however, lacks the ability either to adapt or to innovate. The key to these human abilities is praxis whereby knowledge has meaning in its performance. A computer performs according to predefined functions or rules. Human performance, however, is about rule-following which is a way of doing, whereby the application of a rule is dependent on the situation of the user. Our practice shows how we understand something. One and the same rule can be followed in many different ways. The computer, however, can only follow rules which are defined according to a universal logic. These rules can only be followed in one way. In the event of breakdown, the computer cannot draw upon experience and the situational context to adapt and innovate.

In order to consider how humans can adapt and innovate, the question posed is: How do we acquire knowledge? In the non person-centred perspective, knowledge can be represented in terms of discrete entities. In contrast to this, the person-centred view of knowledge argues that knowledge is a continuous process embodied in dialogue. In dialogue there is no finality about the meaning of a word (i.e. it cannot be represented as a discrete entity). Instead there are shared backgrounds which we tap into when we make an utterance or hear an utterance. It is through, or within, dialogue that we acquire knowledge and skill. Design has been selected as the special case of the application of knowledge and skill. The human-centred perspective, as put forward in this chapter, espouses that in design the particular interrelationship between language, knowledge and skill creates a particular new situation. For example, the outcome of a design process cannot be predetermined. Design is personally and socially situated in praxis and experience. Dialogue is central to the process of design. Dialogue is about participation in a negotiated process. The designer and user collaborate and the user is actively involved in the design of the technology they will use.

The elements of knowledge, skill, dialogue and design identified so far are presented below in the form of a table (Table 8). Although it has been argued above, that knowledge is analogous to skill, they have been placed under separate headings. This is in order to show the different types of knowledge discussed in the various perspectives which are critical of the idea that knowledge is a discrete objective entity. Various aspects of knowledge have been identified. These aspects of knowledge are transferred and acquired without being explicitly described.

Table 8: Concepts used in human-centred literature with respect to knowledge, skill, dialogue and design

Knowledge	*Skill*	*Dialogue*	*Design*
Personal	Judgement	Knowledge transfer	Process
Practical wisdom	Performance	Knowledge acquisition	Participatory
Knowledge by familiarity	Process	Process	Shaping
Common sense	Knowledge acquisition	Tacit dimension	Learning: knowledge and skill acquisition
Intuition and analytical thought	Rule-following	Understanding and misunder-standing	Knowledge transfer
Know-how	Coping with uncertainty	Participatory	
Subjective and objective		Action	
Indwelling			
Interiorisation			

This chapter seeks to investigate the transfer of knowledge through the analysis of dialogue in design.

Summary

It is clear from the discussion above that there is a basis for saying that the non-person centred perspective is inadequate as an explanation of knowledge and skill. It also imposes constraints upon design which have negative implications for human–computer interaction. The person-centred perspective shows the practical and theoretical limitations to representing human knowledge and skill. It provides an alternative framework which provides a better explanation of the nature of knowledge and skill. Here knowledge is situated in personal experience.

Methodologies of Communication and Design

The methodologies described below have been selected from both outside and from within the technological sphere. Those which lie outside it deal with the analysis of discourse. Current techniques being used in this area are ethnomethodology, conversation analysis and discourse analysis.

Certain aspects of these methodologies were drawn upon for the analysis of knowledge transfer in discourse undertaken in a case study. This is briefly described later, herein. Methodologies selected from within the technological sphere are concerned with the study of knowledge and skills, the effects of computer-based technology upon knowledge and skill, and the design process of computer-based technologies. These methodologies embody humanistic design practices and their theoretical bases have humanistic underpinnings.

General Methodological Issues Outside of Technology Design

Ethnomethodology

Ethnomethodology (cf. Heritage, 1984; Garfinkel, 1967, 1986) is based within the domain of sociology. It defines reality as being continuously constructed by people whom it terms 'actors'. Ethnomethodology provides a method for analysing this reality by analysing actors' descriptions and accounts of their experiences using what is termed the 'documentary method of interpretation'.[64] The documentary method is based on the idea that when we make sense of an object, event, or person, we base our interpretation on a 'presupposed underlying pattern'[65] which we have constructed from previous experiences and our general knowledge. Each particular experience can also modify the pattern, as well as being interpreted according to it. The documentary method is termed as such because the particular experience is treated as 'the document of' the 'presupposed underlying pattern'. The method requires the detailed recording of sequences of actions.

The aim of ethnomethodology is to understand how people are able to communicate and coordinate their activities in relation to each other, that is, how people continuously make sense of and construct meaning. Unlike prior approaches to sociology, ethnomethodology acknowledges that people are reflexive. This means, that when people use language they are aware that it will have a certain influence. Language involves motivation, intentionality, and the speaker's assumptions about the listener. In using language, actors make choices which may be based on internalised norms (underlying patterns). Trust is an essential component of communication. A speaker will trust that the listener will share their experience of the situation, and where there is discrepancy in the way they talk, the listener will manipulate his/her background assumptions to make sense of ('accommodate') the expression. Speakers and listeners therefore draw upon their common-sense knowledge and knowledge of contexts in order 'to make definite sense of indefinite descriptive terms'.[66] These actors can sustain, develop or violate some order of activity (some way of doing).

The term ethnomethodology was coined by Garfinkel in the 1950s and arose from his work on jurors (cf. 1974). Jurors, he found, make a systematic use of distinctions using common-sense considerations.[67] Hence the term refers to the practice of common-sense knowledge by people.

Understanding the practice of common-sense knowledge requires direct empirical investigations (case studies) to understand how actors construct and recognise their social activities. Most ethnomethodological research is concerned with institutions and actors' institutional behaviour (see for example, Latour and Woolgar, 1979; Atkinson and Drew, 1992; Lynch, 1985; Wieder, 1974). Institutional knowledge is seen to be socially constituted by the actors and cannot, therefore, be analysed independently of the institutional activity in which it is generated and maintained.

Conversation Analysis

Conversation analysis[68] (CA) is part of sociology. It focuses on analyses of recordings of conversations. The recordings are treated as objective data. This notion of data relies on a belief that sociology must develop scientific methods in order to develop as a field of study (cf. Sacks, 1984). In reaction to the idea of 'ideal' (cf. speech act theory (Austin, 1962; Searle, 1969)) descriptions of language, scientific methods were adopted and the idea of sociology being 'an actual science' was promoted. Data became autonomous from theory: it was now open to interpretation, and to the use of formal methods. Tape-recorded conversations provided the materials to make an initial test for a science of language. They were found to be ideal because they could be studied any number of times by anyone.

Conversation analysis is concerned with the mechanics of language. Interactions amongst people are described as 'products of machinery' (cf. Sacks, 1984. p. 26). It concentrates on ordinary everyday-conversation which, because it is assumed to be non-specific and random, is expected to reveal normative patterns of interaction. The researcher is described as being unmotivated in the analysis, holding no expectations or assumptions about the data. Some subjects which have been covered using this method are studies of conversational organisation, role of gaze and gesture in interaction, and communication in a variety of institutional domains.

Discourse Analysis

Discourse analysis is an umbrella term which covers a broad range of disciplines including sociolinguistics, sociology of language, psycholinguistics, philosophical linguistics and computational linguistics. The common interest is the study of the use of language either in text, conversation, or in other communicative events. As a domain, discourse analysis took form in the 1970s.

Discourse analysis[69] (DA), is partly a reaction to idealisations of language in different disciplines. For example, consider the case of Chomsky's idealisations of speech data (1965). Lyons first criticised Chomsky on this in 1967.[70] Chomsky (cf. 1965), was concerned with producing a generative grammar, i.e. a limited set of rules which generates grammatical sentences. These rules function as representations of people's cognitive systems; of what people actually know. He suggested that this generative grammar is genetic. Chomsky made a distinction between people's ability to produce grammatical sentences and the production of grammatical sentences in actual situations, i.e. between the generative grammar (competence) and performance of language. Performance was marred with speech errors, lapses of memory etc. Chomsky, therefore, worked from idealised sentences and used them to make inferences about the nature of the generative grammar. Such idealisation was criticised for missing the essential features of natural speech and conversation.[71] Lyons (cf. 1967) posed three criticisms of Chomsky's idealised data in comparison to ordinary speech. He described Chomsky's data as regularised, standardised, and decontextualised. Regularised means that any pervasive errors, hesitations, self-corrections, etc., are removed. Standardised means that there is no account of variations, e.g. in pronunciation. Decontextualised means that sentences are separated from any context of their use. It was argued that Chomsky's use of idealised sentences ('proper performance data') to generate a set of grammatical rules rendered such rules incapable of dealing with natural performance data, i.e. ordinary talk. Chomsky, however, held that since each person intuitively knows what is and what is not a proper sentence, intuition is a sufficient criterion for selecting appropriate speech data[72] which is not, therefore, simply idealised. This argument was criticised for its circularity as it required his idea of generative grammars to be true. Chomsky, though, questioned the viability of studying ordinary language performance for the following reason: People are genetically creative, i.e. genetically equipped to produce and acquire language. The majority of each individual's utterances will therefore be unique and each individual is capable of producing an infinite number of sentences. This makes performance of language unpredictable and complex, to the point where it is virtually impossible to derive any consistent theory from a study of it.

Discourse analysts seek to show that language is, contrary to the Chomskian argument, predictable and stereotyped. DA argues that performance data can be systematically handled. DA is concerned with social behaviour. It is less interested in the cognitive dimension, and more interested in how people actually use language in the course of different kinds of interaction.

DA covers a broad range of disciplines. Some of the approaches, namely computational linguistics, and philosophical linguistics/formal linguistics, deal primarily with text. These approaches will not be discussed here. The discussion draws upon the sociology of language approaches, and upon those aspects of psycholinguistics which have been used in the work undertaken on discourse and autobiographical memory (cf. Bekerian and Conway, 1988; Bekerian and Dritschel, 1992; Edwards and Middleton, 1986).[73] Sociology of language analyses the structure of social interaction in conversation. It includes ethnomethodology and conversation analysis which have been briefly described above. Their descriptions emphasise features of social context which can be sociologically classified. They generalise across real instances of language in use, and work with transcribed data. Sociolinguistics[74] is also concerned with the natural forms of language use in the social context. However, it is a development within linguistics, rather than sociology, and is concerned with linguistic forms and grammatical variation in talk. Where the sociologist is more interested in general principles of interaction and discourse, the sociolinguist is more interested in the variations amongst social groups. Within the humanities and social sciences, there is also the analysis of style, rhetoric, argumentation and persuasive communication. Within psychology, in particular the work on autobiographical memory and language, there is a spectrum ranging from quantitative analysis to qualitative analysis.[75] The quantitative approach places an emphasis on the accuracy of recollections, measured against an objective set of data. The particular interest of this thesis is the qualitative approach. This draws upon the idea of 'units of analyses' from psycholinguistics. There is no necessary quantification of a 'unit'. It allows one to work through a dialogue from a general sense of meaning to more clearly identifiable 'units'. These could be phrases, utterances, movements (video analysis), gestures, etc. In qualitative analyses, one can examine the nature of information contained in an account and the manner in which autobiographical memory is recollected. No emphasis is placed upon accuracy or error rates. Qualitative analyses make no assumption about whether information that is recollected actually represents what is in memory. There is an implicit assumption that the form a particular recollection will take depends on any number of factors, such as the purpose of the recall, the nature of social interaction, etc. Thus, a recollection is viewed as reflecting the manner in which information is constructed for the purposes of the conversation, rather than reflecting the contents of memory representations or their organisation.

Applying discourse analyses to autobiographical memory differs from conventional approaches to the study of autobiographical memory in the following ways: definitions based on the structure of memories are constructed; dynamic memory[76] retrieval needs to be considered; content and context in addition to the accuracy of memories is emphasised. The work draws on psycholinguistic theories to develop hypotheses about contextual factors.

Because DA covers such a broad span of approaches, it can acknowledge the subjectivity of the researcher as an influence upon the perception/interpretation of the dialogue and selection of units, which CA, for example, does not. Within social psychology, discourse analysis can be used in 'an open sense to cover all forms of spoken interaction, formal and informal, and written texts of all kinds'. It has been described as a new approach (cf. Potter and Wetherall, 1987) in social psychology which has its roots in more established perspectives in philosophy (in its continental guise), sociology and literary theory.

Methods Within a Technological Domain

Various radical European research traditions in the field of technology question and provide alternatives to mainstream technological developments. Mainstream here denotes market-led developments. These tend to be characterised by a focus on the end product as a measure of success, rather than on the human process of production, and a focus on formalised knowledge.

The radical traditions explore alternatives to such developments in response to certain difficulties which arise from them. One of the difficulties they cite is that the emphasis placed on formalised knowledge undermines the qualitative and subjective human qualities which are part of human skill. The focus on the end-product can lead to deskilling through rationalisation policies, which in turn are based on this emphasis on formalised knowledge. The radical traditions are therefore concerned about the processes of design and use effects of technology upon working life, and their subsequent social and economic consequences. Hence, design methods are being developed to counter the negative effects of technology upon skills and working practices. These methods involve the participation of designers and users in the design of the technology. Each of the groups/approaches discussed below embody a practice of participation in various ways, involving beliefs in the role of the researcher. Some express an explicit political statement of worker participation and require the researcher to be engaged in a process of democratizing the workplace ('proactive approach'). Other approaches believe in worker participation but hold the researcher outside the direct process, calling their methods experimental and their projects, experiments ('user-centred').

A central theme running through all the alternative approaches discussed below is the nature of dialogue. The idea of participation embodies this theme. Dialogue is also discussed in terms of dialectics such as the relationship between theory and practice (researcher/designer and skilled worker/user); tacit knowledge and formal knowledge; the personal dimension/particular experience in relationship to the life-world/general knowledge (this is particular to the German approach of social shaping of technology).

Dialogue is discussed in terms of mediation and negotiation: prototyping as a communication tool (Sweden, user-centred approach) to facilitate dialogue between two cultures (namely of designers and users); the researcher as mediator between designers and users (proactive approach Danish/Scandinavian project on work-oriented design); use of literature and philosophical works as facilitating expression of skill, i.e. the text as a method of mediating knowledge (Sweden: Josefson on practical knowledge).

These particular research traditions and specific projects embody some basic concepts about knowledge, skill and dialogue which are held to be of importance for the investigation of the relationship between tacit and propositional knowledge in knowledge transfer. These projects share in common an interest in the nature of tacit knowledge, the relationship between theoretical and practical knowledge, and the nature of skills.

Praxis - Sweden

The research cited here has been undertaken by a group of researchers at the Swedish Centre for Working Life (SCWL) (Göranzon, 1992; Josefson, 1987, 1988; Perby, 1988, 1990; Gullers, 1988; Karlsen and Oppen, 1988; Schartum, 1988). They are concerned with the nature of professional knowledge and the effects of computer-based technology upon skills. The method they use in their analysis is a case study method which involves participation on the part of the researcher with the environment/people /skills they are concerned with.

The context of the case studies is the mass application of information technology in organisations in Sweden in the 1970s. The Swedes set up a project called 'Education-Work-Technology' (1977–1989) to study the effects of this technology. They were perturbed by the loss of the ability to make judgements on the part of the professional, which takes place over a number of years.

The project involved the undertaking of case studies as a method of gaining deep understanding of professional knowledge in practice. The studies have been described as following a hermeneutic approach which requires the participants to engage in a theoretical reflection upon the practical investigations. The researchers draw upon theoretical ideas from the hermeneutic tradition for their study of skill. In particular, they draw upon philosophical works such as Wittgenstein's later philosophy (1958, 1969, 1977, 1978), Polanyi (1958, 1966), on tacit knowing, and Kuhn's (cf. 1970) work on paradigms. They also draw upon ideas about knowledge, skill and language from the period of the Enlightenment, because they identify this as a critical period in the history of ideas for laying down the basis for modern thinking. The researchers, therefore, cooperated with philosophers[77] in order to arrive at a thematic description of the knowledge embedded in professional skills. Göranzon and the others on the project have adapted the various philosophies and perspectives in the history of ideas into their own hermeneutic frameworks. The fundamental issues are dialogue, theory and practice, reflection and skill, tacit and propositional knowledge.

Ingela Josefson's research (1987, 1988, 1990) is concerned with the expression of practical knowledge in nursing practice. Concern arose about the tendency to shift the education of nurses from teaching hospitals to academic institutions. This gives rise to questions about what constitutes nursing knowledge and what are the most effective methods for passing on this knowledge. The academic sphere tends to lay an emphasis upon scientific knowledge. This is exemplified in the case of doctors who have a scientific training and use a greater amount of formalised knowledge. This is in contrast to a traditional emphasis on practical experience in nursing training. These differences in forms of knowledge reflect a difference in the traditional division of work where doctors deal with cure and nurses deal with care. However, there has been a pressure from within the nursing profession to adopt a scientific approach to their knowledge in order to increase their professional status.

Josefson's aim has been to develop a language and knowledge base of nursing practice whereby the knowledge of nursing is based in its practice. Her work has focused around the concepts of tacit knowing, practical knowledge and knowledge by familiarity (1987, 1988). Tacit knowing is the ability to see a situation and have an intuition about the problem and how to deal with it. Practical knowledge is the performance of skill, of the tacit knowing. Knowledge by familiarity is the ability to make sound judgements by applying one's experience of previous examples at the starting point of interpreting each unique case. It may be described as 'knowing when'.[78]

Josefson's research method is qualitative. Her framework is a hermeneutic one. For each of the two cases (nurses and doctors), Josefson uses different methods to engage the two groups in a dialogue about their practice. With the nurses, she takes the informal approach of a discussion group involving spontaneous dialogue, using forms of discourse from the humanities. These forms are narrative and analytical forms from literature and philosophy. With the doctors, she takes a formal approach of giving lectures to initially engage them.[79] These involve the use of descriptive and analytical forms. These two different approaches reflect the different backgrounds of nurses and doctors, style/structure of presenting knowledge, and expressions of knowledge, recognised as being relevant to their background. Josefson, as the speaker and initiator of the dialogue, seeks to ensure the acceptability of the concepts of tacit knowledge by using forms of expression and styles of speaking which are compatible with those of each group. Lectures were considered to be the most effective form for introducing the doctors to ideas about tacit knowledge and practical knowledge because doctors are familiar and more receptive to this form of expression. Through this method, Josefson was able to find reflexive doctors to undertake a case study approach similar to that undertaken with the nurses. From the humanities, Josefson uses Ancient Greek[80] and other fiction to generate significant discussions amongst the nurses. She discusses philosophical writings such as the Philosophical Investigations (Wittgenstein, 1958) with the nurses. Josefson believes that the humanities and the arts give perspectives and examples, and allow for critical reflection. Plays and other literary forms set people off talking about what they do in anecdotal terms.

Josefson's case study with nurses involves the same group of six nurses over the ten-year period. Upon receiving the interpretations of the various plays and philosophical writings from the nurses, Josefson writes up her experiences of their meetings and gives her interpretation of their discussions together. She becomes their critic and they become hers.[81] Her method is described as a participatory one. Over the years she has created a deep trust and empathy with the nurses. This is crucial to her success in enabling the creation of a language of praxis by the nurses themselves.

Josefson's study does not negate the value of propositional knowledge. Instead, she argues that propositional knowledge forms are an important feature of practical knowledge. It is tested and validated through the experience of unique events (each case). It is assessed by its similarities and disparities with previous examples. Any practice consists in rules and examples which are developed together.

Knowledge by familiarity is essential for the development of skill. Adapting models of knowledge from scientific disciplines cannot improve the conditions for the development of knowledge by familiarity. Josefson's use of methods from the humanities reflects the importance of the use of examples for expressing knowledge and the value of narrative forms. Her use of different methods of discourse to engage doctors and nurses in a discussion on their knowledge, illustrates how different types of knowledge are transferred and acquired through different forms and structures, e.g. the contrast between a lecture and an open dialogue over a play.

Göranzon (1992), aims to develop an epistemology of professional knowledge which is concerned with the maintenance of professional knowledge at the level of the individual, the work group and the community. His case studies[82] focus around the master–apprentice relationship in order to understand the acquisition of skills. Within this context he is interested in what constitutes the ability to make judgements by people who have acquired occupational skills, and in what it means to have mastered one's concepts. He is interested in how occupational skills are transferred.

Action Research and Work-Oriented Design

On the basis of the experiences of the UTOPIA project[83] Ehn (1989) argues for the need for a fundamental understanding of design which concerns an understanding of knowledge in the design process, as well as knowledge in doing design research and knowledge in theories of design. He proposes alternative design philosophies as a way forward, based on a combination of pragmatic interpretations of the philosophies of existential phenomenology (Heidegger, 1962), emancipatory practice (Marx, 1845, 1848; Israel, 1979; Habermas, 1987; Freire, 1972), and the use of ordinary language (Wittgenstein, 1958).[84]

Ehn poses the question of whether it is possible to design useful artefacts without understanding their use within the user's professional domain. Traditional systems design separates the planning stages of a system from it's execution. Systems descriptions and the idea of formal requirements specification are based on this assumption. In the Heideggerian approach, intentions or goals are not prior to action. We do not carry out detached planning in everyday 'being-in-the-world' situations of the use of artefacts, and this is not how expert designers perform. We have intentions but the plan takes form during the activity. The intention determines the activity but is also determined by it. It is only when reflecting back upon a situation that one can separate intention from execution and see them as different phenomena.

Communication is an essential requirement for design. Communication for Heidegger is an act of interpretation. The interpretation the other triggers off in me has meaning to me only within my background of understanding. Only when this background is shared can we understand each other. This is important for considering the communication between designers and users. Language is about action, not description as in the traditional design approach. Following the Heideggerian approach, the design of computer artefacts involves a world of human practice, a practice that is pre-reflective everyday use and the understanding of artefacts.

The Marxist approach[85] to design and use of computer artefacts is an attempt to understand the conditions for social and technical change. Practice is both producing the world and is a product of it. It is the product of historically specific situations, produced by former practice as well as producing the conditions for future practice. In order to develop emancipatory design practices, one needs to relate the design of computer artefacts to three aspects of practice: work, language and morals. In relation to practice as work, design needs to consider the transformation of a given labour process and have user participation in the design process. With respect to practice as language, the design process needs to relate to professional language and the interaction in a given labour process. In relation to practice as morals, design needs to take account of the politics of different power groups involved in a given labour process and in the design process. In contrast to the Heideggerian approach, less is said about individual use of artefacts, but more about the social content of design and use of artefacts, and about the social relations inherent in them.

Ehn considers the application of Wittgenstein's philosophy of the use of ordinary language to the practice of design of artefacts:[86] to understand is to be able to use. To master the professional language, of, say, chair-making, means to be able to act in a correct way together with other people who are familiar with chair-making. To 'know' this does not mean that you explicitly have to know the rule you have learned. To have a concept is to have learned to follow rules as part of a given practice. To use language is to participate in language games, whereby language is a social activity. Professional language games of, say, systems designers, carpenters, typographers, are grounded in everyday language. Because they are socially created, rules of language games can change. The design process is an example of language games which we make up and alter the rules of as we go along.

It is possible, through the mastery of language games, to have creative transcendence of traditional behaviour. With the mastery of language games comes the freedom to follow rules in totally unforeseen but still 'correct' ways. The idea of language games entails an emphasis on how we linguistically discover and construct our world.

To design artefacts which are useful to people, designers have to understand the language games of the use activity, or users have to understand the language game of design, or users have to be able to give complete explicit descriptions of their demands. Traditional systems design approaches focus on the latter. This forces professionals to express their knowledge at a level below the knowledge they possess. The rules they follow are open-ended and many are never made explicit.

Design artefacts, in terms of systems descriptions, may be seen as linguistic artefacts. They can support the interplay of language games between designers and users, and act as mediators of knowledge. For example, in the Utopia project, design-by-doing methods were used such as prototyping, mock-ups, scenarios etc.[87] Design artefacts make for user participation in the language game of design because they give users design activities and family resemblances with the language games they perform in ordinary use situations. They also enable the expression of knowledge at a propositional and practical level. From the designer's perspective, design artefacts allow the designer to express a practical understanding of technical constraints and possibilities by implementing them in mock-ups, prototypes, simulations, experimental situations etc. However, it is important to note that what is learned are correct and incorrect moves in the language games of use and design, not formal descriptions of these practices.

Social Shaping of Technology (Germany)

The social shaping of technology approach (Rauner, Rasmussen and Corbett, 1987), is a holistic conceptual framework for the evaluation, design and development of human-centred technology. The philosophical basis of the framework accounts for the role of personality in working life, and takes the idea of the life-world (cf. Habermas, 1970), and the concepts of praxis[88] and 'dialectic'[89] as its tenet. These are described as follows. People are described as human actors. Life-world is the horizon of meaning which helps the actors to interpret the different situation of an action. It functions partly at the unconscious level as an implicit horizon or context. It also functions as a resource of convictions, etc., which human actors use to establish consensual or conflicting interpretations of specific situations. The communicating actors exist in the contexts of both specific situations (e.g. work) and that of the life-world. The specific situation embodies flexible restrictions whereas the life-world functions more indirectly as both a horizon and resource of interpretation. The life-world can be analysed from three partly interdependent aspects: culture, society and personality. Culture is embodied in mutual understanding, traditional and new knowledge, and meaning. Society is the coordination of actions, norms of legitimacy, social solidarity and integration. Personality is the socialisation of the individual person who has different sorts of competencies, including cognitive, moral, and expressive. The relation between the life-world and the specific situation-based communication may be described as reciprocal action. This relation makes for change of part of the life-world.

The concept of praxis is interrelated with that of the life-world, in the shaping framework. It is used in the sense of the ancient Greek distinction between theory and praxis in that it has a moral dimension to it: praxis means to see human activities as being embedded in situations where the moral question of 'what is to be done' is at issue. In modern society there are different fields of praxis; for example, policy, science, industrial production and individual life. Inside these different fields, it is necessary to reflect the conditions, validity and consequences of the particular praxis. In doing this, the particular praxis is confronted with the overall interrelatedness of culture, society and personality (life-world). Therefore science, just as other fields of praxis, cannot be humane without such reflections.

Communication plays an important role in praxis. Communication through language or through non-verbal forms is seen to be an act of commitment and interpretation (cf. Habermas, 1987). This act of commitment and interpretation has its basis in a common social and historical background.

The concept of praxis described above depends on two interdependencies. One is the mutual interdependence between mind and body at both the conscious and unconscious levels. The other is the relationship between work conditions and the psycho-somatic processes of the workers involved. For the former interdependency the approach draws upon Polanyi (1958, 1966), and Dreyfus (1972, 1986), with respect to knowledge and skill. It sees the relation between mind and body as essential to the acquisition of knowledge and skill. For the latter interdependency, the nature of personality and work are seen to be dialectically related to each other.

The German group cites a lack of respect for traditional skills in the conventional approach to systems development which results in a focus on substitution (of skilled workers) rather than supplementation (of skilled work). Users of a system also have too little influence upon and knowledge of the technology. In order to develop an effective alternative framework, the methods for shaping must meet the requirements of the holistic paradigm described above.

Social Innovation - Britain

The social innovation approach discussed here has been, and is still being developed within the broader societal context. This has been developed in Britain (Gill, K. S., 1986, 1988, 1990b, 1994; Smith, 1991, 1992b). Two projects which follow and develop this approach are described here. The first project, called CAAAT, was developed in the early 1980s. The second project, called MEDICA, was developed in the 1990s. The latter is based upon the philosophy and methodology of the former.

CAAAT Project - UK

The context of this project is a belief that technology must be human-centred. Technology should be used to meet human needs, for example the needs of disadvantaged persons in society (Gill, K. S., 1986), and be constructed through a process of creative participation by users (Gill, K. S., 1984, 1988). Such systems are developmental systems. Technology can be a liberating tool used to enhance people's potential for creative participation in the wider developmental process by enhancing the provision of basic human needs such as education, health and welfare (Gill, K. S., 1988, 1990b, 1994).

The CAAAT project is about the creative participation of handicapped children and their needs. The technology is seen to serve as a mediating tool. Through this mediating tool the handicapped child has the chance to express their world and their culture in a manner that is liberating. In this way, communication between the handicapped and the non-handicapped will not be determined by the expressions of the non-handicapped. The technology functions as both an educational tool and as a physical support. It is a communication tool which could be used to mediate either with a teacher, friend, or family etc. The aim of the project was to give the child a sense of control over their circumstances.

This project uses the metaphor of the theatre where everyone has a role to play. The handicapped person is not to be seen as a passive recipient of technology but an active participant in its construction. The handicapped child's world is different from the majority of people's everyday lives because of institutions, etc., which separate it. Hence, the child needs to be able to express themselves in order that others may participate in their world and that the child may participate in others' worlds. The communication between the handicapped child and others is analogous to the communication between two cultures, and the technology can serve as a mediating tool for achieving this, i.e. the computer as a medium for participatory learning[90] (Gill, K. S., 1984, 1988). The computer, by enabling this communication, could contribute to the self-fulfilment and enhancement of the potential of the learner.

The CAAAT project was based on the idea that effective and appropriate technology can only be constructed through a participatory approach. In this case, the participation took place between the designers and the people in the handicapped child's world, including the child. Dialogue is central to this approach. The CAAAT project required dialogue between the various members of the design team and the child's school teachers, friends, and the parents. Technology does not in itself control the dialogue or transfer the knowledge. It is a tool, functioning as an enabling medium.

MEDICA: A Cognitive Support System for Psychiatry (Smith)

MEDICA is a case study of human-centredness and participatory design (Smith, 1991, 1995).[91] It is based upon the philosophy of the CAAAT project described above. It involves the design, specification, development and partial implementation of 'The Multi-Medical Diagnostic Assistant'. This is a computer-based information system (termed 'tool-set') for use by clinical psychiatrists to assist them in making diagnoses.

The MEDICA tool-set has been developed using a method of rapid prototyping. The prototype acts as a mediation tool in the design process. The MEDICA project contrasts itself to the ergonomic approach to systems design which predefines the user's role. The design of MEDICA recognised a plurality of roles. The experts/users involved in the design were a group of highly-skilled clinical psychiatrists who were consultants. Their knowledge represented the complexity of the domain. The experts demonstrated that it is not possible to provide a rational description of what people do as there is no one normal process of medical communication. The designers were interested in the stories that psychiatrists use when speaking to each other, and when speaking about how the computer system could be used. The psychiatrists had ideas about what a device could do, but had no concrete images. There was a lot of negotiation that took place amongst the experts and the designers about the device. This involved a modification of their stories. Psychiatric information is mainly verbal. Psychiatrists do not use a great deal of quantitative techniques. It became important that, for the effective design of the artefact, the original verbal structures of the experts were maintained.

Requirements for the system came through the process of dialogue between the designers and experts. They were not determined by the designers. The problem for the designers was one of how to construct a system which meets these requirements. The designers ended up constructing fragments of systems (prototypes) for the experts to explore with. The experts used these examples for getting reflections. In one case, an expert asked for a write-up of the tool-set for a seminar he wanted to give to get other experts thinking about the future.

The MEDICA project drew on the CAAAT work in the following ways. CAAAT was very reflexive about the process of design which it believed was about ordinary people. It tried to examine the roles of those involved in the design process as a complex social process. There were no formal requirements for the computer tools until the end of their design. In other words, the requirements evolved out of the process of constructing prototypes and people's experiences.

Résumé

The discussion on methodology and methods from both the non-technological and technological domains is summarised in three tables below. Table 9 is a résumé of the non-technological domain, Table 10 and Table 11 are a résumé of the technological method domain.

Table 9: Methods of examining discourse: consideration of knowledge and dialogue

Methods of examining discourse	*Knowledge*	*Dialogue*
Discourse analysis	Performance	Natural speech: *predictable and stereotyped* Holistic: *language exists in social context* Atomistic: *units of speech*
Ethnomethodology	Knowledge in action Socially constructed Common-sense knowledge	Language as a social construct, not a linguistic construct Trust Reflexivity Internalised norms
Conversion analysis	Mundane knowledge	Predictable Description of structure and patterns e.g. turns

Significant aspects of these methodologies are now drawn out for the model of knowledge transfer developed in the author's research. The methodologies from the non-technological domain are all concerned specifically with human communication. They have rejected the emphasis placed upon a theory of language as this has resulted in idealisms of reality which do not reflect actual communication.

These methodologies, namely ethnomethodology, conversation analysis, and discourse analysis, differ from each other as they have arisen from different disciplines which have their own particular discussions. Ethnomethodology studies how people are able to communicate and coordinate their activities in relation to each other. It acknowledges that people are reflexive: when people communicate they have motivations, intentionality, and any speaker has assumptions about their listener. Trust is an essential component for communication. For example, when there are discrepancies in speech, a speaker will assume that their listener will accommodate these differences. Speakers and listeners draw on their common-sense knowledge and knowledge of other contexts (previous experiences) when trying to make sense of each other. In the study of institutional behaviour, institutional knowledge is seen to be socially constituted and cannot be analysed independently of institutional activity.

Table 10: Features of the different design traditions with respect to knowledge, skill and dialogue

	Knowledge	*Skill*	*Dialogue*
Praxis	Practical Tacit Knowledge by familiarity Reflection Propositional	Judgement Acquisition Reflection	Participation Language: based in practice
Social innovation	Social Personal Tacit	Judgement Acquisition	Participation Social interaction Cultural understanding Practice
Democratic participation	Personal Subjective	Acquisition: apprenticeship	Participation Language games Communication: an act of commitment

Table 11: Feature of the different design traditions with respect to design methodology, analysis and evaluation.

	Design	*Specific methods*	*Analysis of results*	*Evaluation approaches*
Praxis	Design involves different concerns from that of use	Case studies Hermeneutics Work circles Theoretical reflection	No formal analysis of data	Participatory
Social innovation	User participation Learning process Social and technical change Social participation Emancipatory	Case studies Social projects Prototyping	No formal analysis of data: continuous analysis during design of systems with participants	Effectiveness or appropriateness of technology
Democratic participation	Shaping Responsibility User control Conditions of use Learning Politics	Case studies Experiments Prototyping Action-oriented	Work-oriented Involvement of participants Organisational mapping and local theories	Suggestions from participants for follow-up work

Conversation analysis is a derivative of ethnomethodology. It focuses on structure and the mechanics of conversations. The researcher is unmotivated in the analysis and the data (recorded conversations) is objective.[92]

Within discourse analysis, this chapter focuses on work on autobiographical memory and psycholinguistics. The work on autobiographical memory is based on an implicit assumption that the form a recollection takes depends on a number of factors such as the purpose of recall, nature of social interaction, etc. Recollection reflects how information is constituted for the purpose of the conversation, rather than reflecting the contents of memory representations and their organisation.

The methodologies from the technological domains are radical research traditions which question the focus on the end-product and on formalised knowledge, a focus which leads to deskilling. They focus instead on processes of design and effects of technology use, and social and economic consequences. These traditions are praxis (Sweden), democratic participation (Scandinavia, Germany), and social innovation (Britain). Central to all these methodologies is the importance given to dialogue. Dialogue is a participatory process where the participants' interests have to be accounted for. It inevitably involves some kind of mediation as participants come from various backgrounds. Dialogue allows for the formation and sharing of knowledge which traditional design methods of eliciting knowledge (in order to capture expert knowledge, identify systems requirements, etc.) fail to do.

The radical traditions place an importance on the tacit dimension of the knowledge of people in, say, a particular profession, as being necessary for understanding the formal knowledge of that profession. In the design methodologies of these various traditions, there is a continuous relating of theoretical knowledge from particular research fields (research teams) to the knowledge based in the practical domains of professions and other users from social domains (as opposed to working life domains) such as those inhabited by handicapped children. Some specific concepts from these design methodologies have been significant for the model of knowledge transfer, developed by the author in her research,[93] and have been adapted by her for the analysis of the relationship between tacit and propositional knowledge in discourse.

These are knowledge by familiarity, practical knowledge, propositional knowledge, mediation, personality and personal commitment, purpose, and learning.

Knowledge Transfer – a Framework

Knowledge transfer is defined as the communication and acquisition of knowledge between participants. A case study was undertaken by the author (Gill, S. P., 1995) to study the influence of propositional and tacit knowledge upon the nature of knowledge transfer.

The case study involves an analysis of spontaneous discourse. The concept of a unit is adopted from discourse analysis (DA). This is used for breaking up the dimensions of the discourse for analysis. Units allow one to discern aspects of the discourse and to then build up a picture of the discourse. The descriptive notation of discourse developed in conversation analysis (CA) is used for representing certain dynamics of the discourse, e.g. intonation and body language. The data consists of transcripts, videotapes and notes. The case study involves participatory observation which, in this case, means that the research was based within an institution and that the data was collected during the process of design by being present at design discussion meetings. The analysis follows a qualitative human-centred approach to knowledge and draws upon the framework described below (see Figure 42).

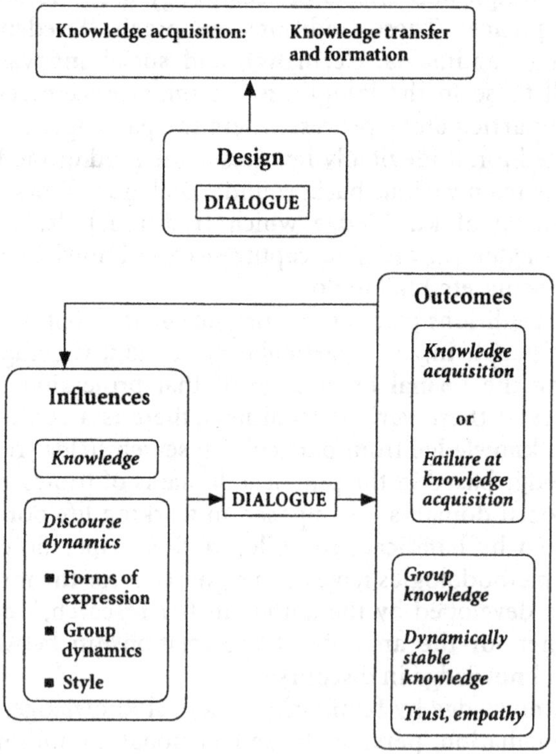

Figure 42: Framework of knowledge transfer

The case study undertaken by the author (Gill, S. P., 1995) was concerned with the discourse of a design team of a multimedia communications computer system. The technology that is being designed is an audio-visual communications computer-based system. The data analysed was from one of the design meetings that took place during its development and implementation. The technology itself has been termed RAVE by the organisation. RAVE is an acronym for the Ravenscroft Audio-Visual Environment. The design team is known as the RAVE team. The team was set up in February 1990, after a considerable period of time had been spent in the organisation discussing how such a technology should be set up and its implications.[94]

Case Study Analysis

The analysis is based on the hypothesis that design involves the communication of knowledge and the transfer of this knowledge. In order to study this process of knowledge transfer, the analysis identifies the moments where a discrepancy in knowledge arises and where it has been resolved, and investigates the problem solving paths. According to the model of knowledge transfer, the processes involved in problem solving and decision making cannot be explained solely in terms of the application of formal knowledge. Instead, these processes require different types of knowledge, expressed through various forms of communication, which are both formal and tacit.

Concluding Remarks

This chapter takes as its basic premise that design is about the communication of knowledge. The traditional approach to design places a significance on the communication of propositional or formal knowledge.[95] The human-centred approach taken here is that the effective communication of knowledge has both propositional and tacit dimensions, where the tacit dimension of knowledge is inseparable from the propositional dimension. The tacit dimension of knowledge consists of practical knowledge, personal knowledge and experiential knowledge; involving, for example, assumptions and values. In all cases (propositional and tacit knowledge), the goals, assumptions, knowledge, and the compatibility of these between speakers and listeners will determine the effectiveness of knowledge transfer. Also significant for the effective communication of knowledge is what is termed 'discourse dynamics'[96] which includes the use of examples to transfer knowledge by familiarity, and the influence of empathy, personality, and power. These aspects of knowledge are considered to be critical in design practice. The chapter has taken as its case study a design team constructing a computer-based communications technology. Their combined expertise represents a highly technical domain of expert knowledge. The study has focused on the breakdowns in communication which result from discrepancies in knowledge amongst the participants. It has traced how knowledge is successfully transferred in order to help overcome these discrepancies and achieve a working agreement on design questions. Discrepancies are defined as differences in levels of knowledge, assumptions and goals. They are indicated by misunderstandings or breakdowns in communication. Both speakers and listeners are described as 'problem solvers' of these discrepancies, which are resolved when speakers/listeners judge that they have adequately transferred and acquired knowledge from each other such that the design process can move on. The way in which people speak has been undermined, for example, in knowledge engineering which focuses on content. The work of Josefson, Gilligan, and Bruner shows that ways of speaking reflect ways of being. It is not therefore surprising that personal narratives and hypothetical stories in the case study (cf. Gill, S. P., 1995), were found to be significant ways of telling one's knowledge, although they were not always the most appropriate and effective way of communicating knowledge. Formalised expressions are effective, and sometimes necessary for knowledge transfer, where one has propositional knowledge of the domain. In this case study, propositional knowledge was technical or design knowledge. The case study has also shown that propositional knowledge is not only expressed in formalised terms, and that forms of expressing propositional knowledge are bounded by the particular discrepancy in knowledge. Practical knowledge is an intangible category and also highly complex. Unlike the case studies upon professional skills, reviewed here, the research presented in this chapter has taken dialogue as the focus for a study of knowledge and skill. It was found to be impossible to describe the practical knowledge of the speakers/listeners. This is tacit knowledge. It is practice. The framework for knowledge transfer developed in this chapter provides a basis for the study of practical knowledge and the investigation of differences in knowledge in different domains.

References

Anscombe, E. and Geach, P. T. (1979). *Descartes, Philosophical Writings.* Introduction by A. Koyré. Middlesex: Nelson University Paperbacks. First published in 1954 in the Nelson Philosophical Texts Series.

Atkinson, J. M., and Heritage, J. (eds.), (1984). *Structures of Social Action: Studies in Conversation Analysis.* Cambridge University Press, Cambridge.

Austin, J. L. (1962). *How to do things with words.* Clarendon Press, Oxford.

Ayer, A. J. (1971). *Language, Truth and Logic.* Pelican, London.

Barr, A. and Feigenbaum, E. A. (eds.), (1981). *The Handbook of Artificial Intelligence.* Addison-Wesley, MA.

Barsalou, L. W. (1988). 'The Content and Organization of Autobiographical Memories' in Neisser, U. and Winograd, E. (eds.) *Remembering Reconsidered: Ecological and Traditional Approaches to the Study of Memory,* pp. 193–243. Cambridge University Press, New York.

Bartlett, F. C. (1932). *Remembering: A Study in Experimental and Social Psychology.* Cambridge University Press, Cambridge.

Bekerian, D. A. and Dritschel, B. H. (1992). 'Autobiographical remembering: An integrative approach' in Conway, M. *et al.* (eds.) *Theoretical Perspectives on autobiographical memory,* pp. 135-150. Dordrecht: Kluwer Press.

Bekerian, D. A. and Conway, M. A. (1988). 'Everyday Contexts' in Davies, G. M. and Thomson, D. M. (eds.) *Memory in context: Context in memory*, pp. 305–318. Chichester: John Wiley & Sons.

Bench-Capon, T. J. M. (1990). *Knowledge Representation: an approach to AI.*

Berger, P. L. and Luckmann. (1971). *The Social Construction of Reality: A treatise in the sociology of knowledge.* Penguin.

Berry, D. (1987). 'The problem of Implicit Knowledge', *Expert Systems*, 4: no. 3.

Berry, D., Hart, A. (1990). *Expert Systems: human issues.* Chapman and Hall.

Bjerknes, G., Ehn, P. and Kyng, M. (eds.), (1987). *Computers and Democracy: A Scandinavian Challenge.* Avebury, Aldershot.

Bloomfield, B. P. (1988). 'Expert Systems and Human Knowledge: A view from the sociology of science', *AI & Society*, 2: 17–30.

Boden, M. (1988). *Computer Models of Mind: myths and mechanisms.* Cambridge University Press, Cambridge.

Boden, M. (1991). *Artificial Intelligence in Psychology: Introductory essays.* MIT Press, Cambridge, Massachusetts.

Bødker, S. (1987). 'A Utopian Experience' in Bjerknes, G., Ehn, P. and Kyng, M. (eds.) *Computers and Democracy - Scandinavian Challenge.* Avebury.

Bramer, M. (1988). 'Expert Systems in Business: A British Perspective', *Expert Systems*, 5.2.

Brewer, W. F. (1986). 'What is autobiographical memory?' in Rubin, D. C., (ed.) *Autobiographical Memory.* Cambridge University Press, Cambridge.

Brown , P. and Levinson, S. C. (1987). *Politeness: Some universals in Language Usage.*

Bruner, J. (1990). *Acts of Meaning.* Harvard University Press, Cambridge, Massachusetts.

Carnap, R. (1956). *Meaning and Necessity.* Phoenix Books, London.

Carnap, R. (1959). *Introduction to Semantics and Formalisation of Logic.* Cambridge, Massachusetts.

Chomsky, (1965). *Aspects of the Theory of Syntax.* MIT Press, Cambridge, Massachusetts.

Coats, P. K. (1988). 'Why Expert Systems Fail', *Financial Management*, 17: 77–86.

Cooley, M. J. E. (1987a). *Architect or Bee? The Human Price of technology.* Hogarth, London.

Cooley, M. J. E. (1987b). 'Human Centred Systems: An Urgent Problem for Systems Designers', *AI & Society* , 1: 37-46.

Cooley, M. J. E. (1991). *European Competitiveness in the 21st Century. Integration of Work, Culture and Technology.* Brussels: FAST, Commission of the European Communities.

Collins, H. M. (1974). 'The TEA Set: Tacit Knowledge and Scientific Networks', *Science Studies*, 4: 165–86.

Collins, H. M., Green, R. H., Draper, R. C. (1986). 'Where's the expertise?: Expert Systems as a Medium for Knowledge Transfer' in Merry, M. (ed.) *Expert Systems '85.* Cambridge University Press, Cambridge.

Collins, H. M 'Expert Systems and the Science of Knowledge' in Bijker, W. E., Hughes, T. P. and Pinch T. J. (eds.) *The Social Construction of Technological Systems: New Directions in the Sociology and History of Technology*, pp. 229–348. MIT Press, Cambridge, Mass: .

Collins, H. M. (1992). *Artificial Experts: Social Knowledge and Intelligent Machines.* MIT press. Cambridge, Massachusetts.

Conway, M. A. (1990). *Autobiographical Memory: An introduction.* Oxford University Press, Milton Keynes.

Coulthard, R. M. (1977). *An Introduction to Discourse Analysis.* Longman, London.

Diaper, D. (ed.), (1989). *Knowledge Elicitation: principles, techniques and applications.* Ellis Horwood, Chichester.

van Dijk, T. A. (ed.), (1985). *Handbook of Discourse Analysis*, vols. 1, 2, 3, 4. Academic Press, London.

Drew, A. and Heritage, J. (eds.), (1992). *Talk at Work: Interaction in institutional settings.* Cambridge University Press, Cambridge

Dreyfus, H. L. (1972). *What Computers Can't Do: A critique of artificial reason.* Harper & Row, New York.

Dreyfus, H. L. (1989). 'Is Socrates to Blame for Cognitivism?' in Göranzon, B. and Florin, M. (eds.) *Artificial Intelligence, Culture and Language: On Education and Work.* pp. 219–228. Springer-Verlag, London.

Dreyfus, H. L. and Dreyfus, S. (1986). *Mind over Machine: the power of human intuition and expertise in the era of the computer.* The Free Press, New York.

Edwards, D. and Middleton, D. (1987). 'Conversation and Remembering: Bartlett Revisited', *Applied Cognitive Psychology*, 1: 77–92.

Ehn, P. (1988). *Work Oriented Design of Computer Artifacts.* Arbetslivscentrum, Stockholm.

Ellis, W. D. (1938). *A Source Book of Gestalt Psychology* (includes an introduction by K. Koffka). Routledge & Kegan Paul, London.

Fann, K. T. (1969). *Wittgenstein's Conception of Philosophy.* Basil Blackwell, Oxford.

Feigenbaum, E. A. and McCorduck, P. (1984). *The Fifth Generation.* Pan Books.

Fodor, J. A. (1976). *The Language of Thought.* The Harvester Press, Sussex.

Fodor, J. A. (1981). *Representations: Philosophical essays on the foundations of cognitive science.* Harvester Press, Brighton.

Fodor, J. A. and Pylyshyn, Z. W. (1988). 'Connectionism and Cognitive Architecture: A critical Analysis', *Cognition*, 28: 3–71.

Friis, S. (1991). 'User Controlled Information Systems Development', Lund*: Lund Studies in Information and Computer Science*, no. 6.

Gammack, J. (1987). 'Modelling Expert Knowledge Structures using Cognitively Compatible Structures' in *Proceedings of the Third International Expert Systems Conference, London 2-4*, pp. 191–200. Learned Information (Europe) Ltd, Oxford.

Gammack, J. (1988). *Eliciting Expert Conceptual Structure using Converging Techniques.* PhD dissertation. University of Cambridge.

Garfinkel, H. (1967). *Studies in Ethnomethodology.* Polity Press, Cambridge.

Garfinkel, H. (1986). *Ethnomethodological Studies of Work.* Routledge, Kegan and Paul, London.

Gilbert and Mulkay. (1984). *Opening Pandora's Box: A Sociological Analysis of Scientists' Discourse.* Cambridge University, Cambridge.

Gill, K. S. (ed.), (1986). *Artificial Intelligence for Society.* Wiley, Chichester.

Gill, K. S. (1986). 'The Knowledge-Based Machine: Issues of Knowledge Transfer' in Gill, K. S. (ed.) *Artificial Intelligence for Society*, pp. 7–17. John Wiley & Sons, Chichester.

Gill, K. S. (1988). 'Artificial Intelligence and Social Action: Education and Training' in Göranzon, B. and Josefson, I. (eds.) *Knowledge, Skill and Artificial Intelligence*, pp. 77–90. Springer-Verlag, London.

Gill, K. S. (1990a). *Summary of Human-Centred Systems Research in Europe.* SEAKE Centre, Brighton University.

Gill, K. S. (1990b). 'Cultures, Languages, Mediation' in Göranzon, B. and Florin, M. (eds.) *Artificial Intelligence, Culture and Language: On Education and Work.* Springer-Verlag, London.

Gill, K. S. (1994). 'Human-Centred Shaping of Social Innovation' in Schmid, F. *et al.* (eds.) *Computer Integrated Production Systems and Organisations: The human-centred approach*, pp. 1–28. NATO ASI Series. Springer-Verlag, Berlin.
Gill, S. P. (1988). 'On Two AI Traditions', *AI & Society*, 2: 321–340. Springer-Verlag, London. A version of this [Gill, S. P. (1988). 'Knowledge and Skill Transfer through Expert Systems: British and Scandinavian Traditions'] is published in Kelly, B. and Rector A. (eds.) *Research and Development in Expert Systems V: Proceedings of Expert Systems' 88*. Cambridge University Press, Cambridge.
Gill, S. P. (1990). 'A Dialogical Framework for Participatory KBS Design' *in Proceedings Tenth European meeting on Cybernetics and Systems Research*. Vienna, April 1990.
Gill, S. P. (1992). 'Dialogue and Design of Computer-Based Technology' in Y. Masuda (ed.) *Human-Centred Systems in the Global Economy*, pp. 217–228. Springer-Verlag, London.
Goffman, E. (1959). *The Presentation of Self in Everyday Life*. (Reprint, 1990). Penguin Books, London.
Goffman, E. (1981). *Forms of Talk*. Blackwell, Oxford.
Goodwin, C. (1981). *Conversational Organisation: Interaction between speakers and Hearers*. Academic press, New York.
Goody, E. (ed.), (1978). *Questions and Politeness: Strategies in Social Interaction*. Cambridge University Press, Cambridge.
Göranzon, B. (1988). 'The Practice of the Use of Computers. A Paradoxical Encounter between Different Traditions of Knowledge' in Göranzon' B. and Josefson, I. (eds.) *Knowledge, Skill and Artificial Intelligence*, pp. 9–17. Springer-Verlag. London.
Göranzon, B. (1992). *The Practical Intellect: Computer and Skills*. Springer-Verlag, London.
Göranzon, B and Josefson, I. (eds.), (1988). *Knowledge, Skill and Artificial Intelligence*. Springer-Verlag, London.
Greenwell, M. (ed.), (1988). *Knowledge Engineering for Expert Systems*. Ellis Horwood.
Greenwood, L. H. G. *Aristotle, Nichomachean Ethics Book VI*. CUP.
Grice, H. P. (1975). 'Logic and Conversation' in Cole P. and Morgan J. L. (eds.) *Syntax and Semantics III*: Speech Acts. Academic Press, New York.
Habermas, J. (1970). 'A Theory of Communicative Distortion', *Inquiry*, 13: 360–375.
Habermas, J. (1987). 'Intermediate Reflections: Social Action, Purposive Activity, and Communication' in J. Habermas *The Theory of Communicative Action: Life world and system: a critique of functionalist reason*, (translation) T. McCarthy. Polity Press, Cambridge.
Hart, A. (1988). 'Knowledge Acquisition for Expert Systems' in Göranzon, B. and Josefson I. (eds.) *Knowledge, Skill and Artificial Intelligence*, pp. 103-111. London: Springer-Verlag.
Haugeland, J. (1985). *Artificial Intelligence: The Very Idea*. MIT Press, Cambridge, Massachusetts.
Hayes-Roth, F. (1984). The Machine as a Partner of the New Professional, *IEEE Spectrum*, 21 (6): 28–31.
Hayes-Roth, F. (1985). 'Knowledge-Based Expert Systems – the state of the art in the US', *Knowledge Engineering Review*, 1: 18–27.
Heidegger, M. (1962). *Being and Time*. Harper & Row, New York.
Henle, M. (1961). *Documents of Gestalt Psychology*. Berkeley & Los Angeles: University of California Press, Cambridge University Press.
Heritage, J. (1984). *Garfinkel and Ethnomethodology*. Polity Press, Cambridge.

Israel, J. (1979). *The Language of Dialectics and the Dialectics of Language.* Harvester press, London.

Janik, A and Toulmin, S. (1973). *Wittgenstein's Vienna.* Simon and Schuster, New York.

Janik, A. (1988). 'Tacit Knowledge, Working Life and Scientific Method' in Göranzon, B. and Josefson, I. (eds.) *Knowledge, Skill and Artificial Intelligence,* pp. 53–65. Springer-Verlag, London.

Janik, A. (1990). 'Tacit Knowledge, Rule-Following and Learning' in Göranzon, B. and Florin, M. (eds.) *Artificial Intelligence, Culture and Language: On Education and Work,* pp. 45-56. Springer-Verlag, London.

Janik, A. (1992). 'Why is Wittgenstein Important?' in Göranzon, B. and Florin, M. (eds.) *Skill and Education,* pp. 33–40. Springer-Verlag, London.

Jefferson, G. (1984). 'On the organisation of laughter in talk about troubles' in Atkinson, J. M. and Heritage, J. (eds.) *Structures of Social Action,* pp. 346–369. Cambridge University Press, Cambridge.

Johannessen, K. and Nordenstam, T. (eds.), (1981). *Wittgenstein – Aesthetics and Transcendental Philosophy.* Holder-Pichler-Temsky, Vienna.

Johannessen, K. S. (1988a). 'Rule-Following and Tacit Knowledge', *AI & Society,* 2: 287–302. Springer-Verlag, London.

Johannessen, K. S. (1988b). 'The Concept of Practice in Wittgenstein's Later Philosophy', *Inquiry,* 31: 357–370.

Johannessen, K. (1992). 'Rule-Following, Intransitive Understanding and Tacit Knowledge: An Investigation of the Wittgenstenian Concept of Practice as Regards Tacit Knowing' in Göranzon, B. and Florin. M. (eds.) *Skill and Education,* pp. 41–63. Springer-Verlag, London.

Johnson-Laird, P. N. (1983). *Mental Models. Towards a Cognitive Science of Language, Inference, and Consciousness.* Cambridge University Press, Cambridge.

Johnson-Laird, P. N. (1993). *The Computer and the Mind: an introduction to cognitive science.* 2nd. ed. Fontana, London.

Josefson, I. (1987). 'The Nurse as an Engineer', *AI & Society,* 1: 115–126.

Josefson, I. (1988). 'The Nurse as Engineer - The Theory of Knowledge in Research in the Care Sector' in Göranzon, B. and Josefson, I. (eds.) *Knowledge, Skill and Artificial Intelligence,* pp. 19–30. Springer-Verlag, London.

Josefson, I. (1990). 'Language and Experience' in Göranzon, B. and Florin, M. (eds.) *Artificial Intelligence, Culture and Language: On Education and Work,* pp. 71–76. Springer-Verlag, London.

Karlsen, T. K. and Oppen, M. (1988). 'Professional Knowledge and the Limits of Automation in Administrations' in Göranzon, B. and Josefson, I. (eds.) *Knowledge, Skill and Artificial Intelligence,* pp. 139–147. Springer-Verlag, London.

Kidd, A. (ed.), (1987). *Knowledge Acquisition for Expert Systems.* Plenum Press, London.

Kidd, A. and Wellbank, M. (1984). 'Knowledge Acquisition' in Fox, J. (ed.) *InfoTech State of the Art Report on Expert Systems.* InfoTech: Pergamon.

Kolodner, J. L. and Riesbeck, C. R. (1986). *Experience, Memory and Reasoning.* Erlbaum, Hillsdale, N. J.

Kowalski, R. (1974). 'Predicate Logic as a programming language' in *Proceedings of IFIP '74.*

Kowalski, R. (1982). 'Logic as Computer Language' in Clark and Tarnland (eds.) *Logic Programming,* pp. 2–16. Academic Press, London.

Kuhn, T. S. (1970). *The Structure of Scientific Revolutions* (2nd edition). University of Chicago Press, Chicago.

Laessoe, J. and Rasmussen, L. (1989). *Human-Centred Methods.* Reproset Press, Copenhagen.

Leibnitz, G. W. (1666). *Dissertatio de Arte Combinatoria.* G. Phil. University of Leipzig.

Leith, P. (1988). 'The Application of AI to Law', *AI & Society*, 2: 31–46.

Levinson, S. (1983). *Pragmatics.* Cambridge University Press, Cambridge.

Lipscombe, B. (1989). 'Expert Systems and Computer Controlled Decision-Making in Medicine', *AI & Society*, 3: 184–197. Springer-Verlag, London.

Lyons, J. (1970). *Chomsky.* Viking, New York.

Marx, K. and Engels, F. (1971). 'Thesis on Feuerbach' in Bottomore, J. B. (ed.) *Karl Marx - Selected Writings in Sociology and Social Philosophy.* (First published in 1845). Penguin, Harmondsworth.

Marx, K. and Engels, F. (1980). *The Communist Manifesto.* (First published in 1848). Penguin, Harmondsworth.

Michie, D and Johnston, R. (1984). *The Creative Computer.* Penguin, Middlesex.

Minsky, M. (1963). 'Steps towards AI' in Feigenbaum, E. A. and Feldman, J. (eds.) *Computers and Thought*, pp. 95–128. McGraw-Hill, New York.

Mumford, E. 'Sociotechnical Systems Design - Evolving Theory and Practice' in Bjerknes, G., Ehn, P., Kyng, M. (eds.) *Computers and Democracy - Scandinavian Challenge.* Avebury.

Needham, J. (1927). *Man as Machine.* Kegan Paul.

Negrotti, M. (1991). *Understanding the Artificial. On the future shaping of Artificial Intelligence.* Springer-Verlag, London.

Newell, A. and Simon, H. A. (1972). *Human Problem Solving.* Prentice-Hall, Englewood Cliffs, N. J.

Newell, A. (1980). 'Physical Symbol Systems', *Cognitive Science*, 4: 135–183.

Newell, A. (1981). 'The Heuristic of George Polya and its Relation to Artificial Intelligence' in Groner, R. Groner, M. and Bischoof W. F. (eds.) *Methods of Heuristics.* Lawrence Erlbaum Press, Hillsdale, N. J.

Nordenstam, T. (1979). 'Ethical Competence and Development) in Barmarck, J. (ed) *Perspectives in Metascience.* Goteborg.

Nordenstam, T. (1987). 'Moral Rules and Paradigms' in *Archivo Di Filosophia.*

Nordenstam, T. (1992) 'Language and Action' in Göranzon, B. and Florin, M. (eds.) *Artificial Intelligence, Culture and Language: On Education and Work.* Springer-Verlag, London.

Olson, J. R. and Rueter, H. H. (1987). 'Extracting Expertise form Experts: Methods for Knowledge Acquisition', *Expert Systems*, 4: 3.

Ostberg, O. and Whitaker, R. (1988). 'Channelling Knowledge: Expert Systems as Communications Media', *AI & Society*, 2: 197–208. Springer-Verlag, London.

Ostberg, O., Whitaker, R. and Amick, B. (1988). 'The Automated Expert: Technical, Human, and Organisational Considerations' in *Expert Systems Applications.* Teldok, Sweden.

Partridge, D. (1987). 'Human Decision Making and the Symbolic Search Space Paradigm in AI', *AI & Society*, 1: 103–114.

Perby, M. L. (1988). 'Computerisation and Skill in Local Weather Forecasting' in Göranzon, B. and Josefson, I. (eds.) *Knowledge, Skill and Artificial Intelligence,* pp. 39–51. Springer-Verlag, London.

Plato, (1981). *The Republic.* Translation by Gruber, G. M. A. Pan, London. *c.* 1974.

Polanyi, M. (1966). *The Tacit Dimension.* Doubleday, New York.

Polanyi, M. (1958). *Personal Knowledge: Towards a Post-Critical Philosophy.* Routledge, Kegan and Paul, London.

Polya, G. (1945). *How to Solve it.* Princeton University Press, Princeton, N.J.

Potter, T. and Wetherall, M. (1987). *Discourse and Social Psychology.* Sage, London.

Price, C. J. (1990) *Knowledge Engineering Toolkits.* Ellis Horwood, New York.

Pylyshyn, Z. W. (1984). *Computation and Cognition. Toward a Foundation for Cognitive Science.* Cambridge MA: MIT Press, Bradford Books.

Rasmussen, L. B., Laessøe, and Tøttrup, P. (eds.) *The Electronic Sketch Pad: Prototype Observation and Organisational Context.* Copenhagen: Reproset.

Rasmussen, L. B. (1994). *Human Centred Methods of Social and Technical Design.* To be published in book series (ed.), AI and Society. Springer-Verlag, London.

Rauner, F., Rasmussen, L. B. and Corbett, M. (1987). *The Social Shaping of Technology and Work.* Institut Technik & Bildung: Universitat Bremen.

Rauner, F., Rasmussen, L. B. and Corbett, M. (1987). 'The Social Shaping of Technology and Work: Human-Centred Systems', *AI & Society*, 2: 47–62.

Rosenbrock, H. H. (1988). 'Engineering as an Art', *AI & Society*, 2: 315–320. Springer-Verlag, London.

Rosenbrock, H. H. (ed.), (1989). *Designing Human-Centred Technology: A Cross Disciplinary Project in Computer-aided Manufacturing.* Springer-Verlag, London.

Rosenbrock, H. H. (ed.), (1990). *Machines with a Purpose.* Oxford University Press, Oxford.

Rosenbrock, H. H. (1992). 'Purpose and automatic control', *Computing & Control Engineering*, March: 88–90.

Rumelhart, D. E. and Norman, D. A. (eds.), (1986). 'Parallel Distributed Processing: Explorations in the Microstructure of Cognition', vol. 1: *Foundations*; vol. 2: *Psychological and Biological models.* MIT Press, Cambridge, Massachusetts.

Ryle, G. (1949). *The Concept of Mind.* Hutchinson.

Sacks, H., Schegloff, E. A., and Jefferson, G. (1974). 'A simplest systematics for the organisation of turn-taking in conversation', *Language*, 50: 4: 696–735.

Sacks, H. (1984). 'Notes on Methodology' in Atkinson, J. M. and Heritage, J. (eds.) *Structures of Social Action: Studies in Conversation Analysis.* Cambridge University Press, Cambridge

Sandberg, Å. (ed.), (1979). *Computers Dividing Man and Work – Recent Scandinavian Research on Planning and Computers from a Trade Union Perspective.* Swedish Centre for Working Life, Malmo.

Searle, J. R. (1969). *Speech Acts: An Essay in the Philosophy of Language.* Cambridge University Press, Cambridge.

Schank, R. C. (1982). *Dynamic Memory.* Cambridge University Press, New York.

Schank, R. C. (1990). *Tell Me a Story: A new look at real and artificial memory.* Charles Scribner, New York.

Smith, D. (1991). 'Project MEDICA: A Cognitive Support System for Psychiatry: A Case Study of Human-Centredness and Participatory Design in Action', Working Paper No.6, SEAKE Centre, Brighton University.

Smith, D. (1992). 'The Psychology of Apprenticeship: A Discussion Paper' in Göranzon B. and Florin M. (eds.) *Skill and Education*, pp. 83–100. Springer-Verlag, London.

Smith, D. (1992). 'Can the Family Survive the New Communication Media? 'Human-Centredness' and the Challenge of Consumer Technologies' in Masuda, Y. (ed.) *Human-Centred Systems in the Global Economy*, pp. 59–70. Springer-Verlag, London.

Smith, S. (1990). 'A Counselling Approach to Participatory Design' in *Proceedings HCS '90.* Brighton University, Brighton.

Sprague, R. H. and Watson, H. J. (eds.), (1986). *Decision Support Systems.* Prentice-Hall, Simon & Schuster, Englewood Cliffs, New Jersey.

Susskind, R. E. (1989). *Expert Systems in Law: a jurisprudential inquiry.* Clarendon Press, Oxford.
Tilghman, B. R. (1988). 'Seeing and Seeing-As', *AI & Society,* 2: 303–314. Springer-Verlag, London.
Toulmin, S. (1991). 'The Dream of an Exact Language' in Göranzon, B. and Florin, M. (eds.) *Dialogue and Technology,* pp. 33–43. Springer-Verlag, London.
Tulving, E. (1972). 'Episodic and semantic memory' in Tulving, S. and Donaldson, W. (eds.) *Organisation of memory.* Academic Press, New York.
Varela, F. J. (1988). *Cognitive Science: A Cartography of Current Ideas.* Paris: Editions du Seuil.
Varela, F. J., Thompson, E., Rosch. (1991). *The Embodied Mind: Cognitive Science and Human Experience.* MIT Press, Cambridge, Massachusetts.
Wardaugh, R. (1985). *How conversation works.* Basil Blackwell, Oxford.
Weizenbaum, J. (1984). *Computer Power and Human Reason.* Pelican, London.
Wiener, N. (1949). *Cybernetics, or Control and Communication in the Animal and the Machine.* Wiley, New York.
Winch, P. (1958). *The Idea of a Social Science and its Relation to Philosophy.* Routledge, Kegan and Paul, London.
Winograd, T. and Flores, C. F. (1986). *Understanding Computers and Cognition. A new foundation for design.* Ablex Press, Norwood, N. J.
Wittgenstein, L. (1958). *Philosophical Investigations.* Blackwell, Oxford.
Wittgenstein, L. (1969). *On Certainty.* Blackwell, Oxford.
Wittgenstein, L. (1977). in Anscombe, G.E.M (ed). *Remarks on Colour.* Blackwell, Oxford.
Woolgar, S. (1988). *Science, The very Idea.* Ellis Horwood, Chichester.
Woolgar, S. and Latour, B. (1979). *Laboratory Life: The Social Construction of Scientific Facts.* Beverley Hills; London: Sage Publications.
von Wright, G. H. (1972). 'Wittgenstein on Certainty' in von Wright, G. H. (ed.) *Problems in the Theory of Knowledge.* Haag.

Notes

1 In talking of cognitive science, this chapter is specifically referring to orthodox cognitivism. It is not referring to Bruner's vision, for example, of the cognitive revolution (1990), and interpretivism of folk psychology.

2 This is a generic term for approaches to the study of issues pertinent to the design and application of computer-based technologies which differ from what is termed a machine-centred approach. The human-centred approach is best explained in a comprehensive report by Dr. K .S. Gill; ref: *Summary of Human-Centred Systems Research in Europe.* SEAKE Centre, Brighton University, 1990.

3 'theoretical knowledge', in this context, is analogous to 'knowing that', as distinct from 'knowing how'; cf. Ryle (1949). It has been described as mathematically-structured knowledge (cf. Göranzon, 1991).

4 Some of these authors (particularly those from the domain of the sociology of scientific knowledge (SSK)) critique the descriptions/formal language of science, contrasting them to its practice. They do this by either undertaking apprenticeship in the specific scientific area, and/or by recording its everyday discourse. For further literature, in addition to Woolgar (1987), and Collins (1974, 1975), see for example, Gilbert & Mulkay (1984), Woolgar and Latour (1979). Rosenbrock (1988) is critical of the application of the formalist notion of science (as in propositional knowledge) to the

domain of engineering at the expense of sensibility and human purpose - discussed later, herein.

5 Rationalism denotes the emphasis on objectivity in terms of the mind–body split (cf. Descartes). Toulmin (1991), discusses the "seventeenth century Cartesian ideal of intellectual exactitude, ... it's idolization of geometrical proof", citing the "rationalist procedures of Newtonian mathematics" (*ibid.* p. 41) and compares this ideal to that of the 16th-century humanists (for example, Bacon), who believed in a 'humanly fruitful' science. The contrast lies between an emphasis on theory and on praxis.

6 Johannessen (1988a) discusses the higher status in society placed upon analytical and empirical validity for one's having knowledge, versus the doing of knowledge, i.e. its practice. He traces this back to logical positivism, cf. the Vienna circle (Carnap, 1956), chapters II and III.

7 As in scientific law for nature, for example, Galileo: 'The book of nature is written in mathematical symbols'; Newtonian physics: 'a vote in favour of theoretical cosmology, not for practical human dividends' (cf. Toulmin, 1991). Leibniz's (1666) *characteristica universalis* (universal system of characters) is the dream for a universal unambiguous, context independent, language, and thereby knowledge.

8 For reference to an outline of the development of the dream of exact language see Toulmin, *op. cit.* He discusses Leibniz's (*op. cit.*) project to develop a language based on mathematical symbolism. Johannessen also cites this period and the idea of an exact language in a discussion on the rule and the positivist conception of knowledge, in Johannessen, K. (1988), 'Rule-Following and Tacit Knowledge', *AI & Society* Journal, vol. 2.4. Springer-Verlag, London.

9 Leibnitz, *op. cit.*

10 Toulmin, *op. cit.* p.36.

11 Göranzon, *op. cit.* p.1.

12 Boden, *op. cit.* p. 32.

13 Ayer, A. J. (1971) *Language truth and logic.* Pelican.

14 Johannessen, *op. cit.* p.290.

15 See Boden (1991), p. 10. This is in contrast to developments in parallel processing and connectionism (see for example, Rumelhart and Norman, 1986) which have been seen to be a challenge to the conventional model based on the von-Neumann machine (cf. Fodor and Pylyshyn, 1988). However, all these various approaches of AI and computational psychology have originated from the same mid-century ideas about the brain's logical-computational potential. See also Boden (1988), for a comprehensive overview of differing perspectives in AI and cognitive science.

16 Cf. Liebnitz, *op. cit.*; see Toulmin, *op. cit.*

17 Göranzon, *op. cit.* p. 43.

18 See note 8.

19 See Johannessen *op. cit.*; see Ayer *op. cit.*

20 Johannessen, *op. cit.* p.288.

21 It is important to note that the logical positivistic view of knowledge has been traced as far back as Plato's dialogues about expertise and skill and his conception of rationality. Although a historical discussion cannot be developed in this paper, it is important to note the historical context for understanding the current thinking. Cf. Dreyfus, H. L. (1989) *Is Socrates to Blame for Cognitivism?* p. 219; Johannessen (1988a) *op. cit.*; Plato, *The Republic.* Penguin, 2nd ed. Translation, D. Lee; Toulmin, *op. cit.*

22 Varela, F. J. (1988) *Cognitive Science: A Cartography of Current Ideas.* Editions du Seuil, Paris. See also Varela, F. J. *et al.* (1991). Fodor and Pylyshyn (1988) describe

'classical cognitive science' as an extended attempt to apply the methods of proof theory to the modelling of thought; Pylyshyn (1984) describes cognition as a type of computation.

23 Boden, *op. cit.* p.30.

24 See for example, the classical AI paradigm works of Fodor (1976), Newell (1980), Newell and Simon (1972), Pylyshyn (1984), Michie and Johnston (1984), Minsky (1963), Feigenbaum and McCorduck (1984). Also see *The Handbook of AI*, (eds.), Barr and Feigenbaum (1981), and Haugeland (1985).

25 Classic AI programming languages such as Prolog (Kowalski, 1974, 1982) are based upon systems of logic. Prolog is based on predicate logic.

26 NB. Newell (1972), discusses the significance of Polya's (1945, 1962, 1965) idea of heuristics which greatly influenced the work on problem solving in AI. Polya, himself, was a mathematician, not directly involved in AI or cognitive science. AI drew upon his heuristics but omitted those aspects of his theory which were 'ambiguous' or intangible; for example, his emphasis upon personal motivation or commitment which he considered to be essential for effective problem solving.

27 Varela, *op. cit.* pp. 18–21.

28 Boden, *op. cit.* p. 20.

29 It is noted that Boden (1991) has questioned whether this model is purely formal-syntactic when she criticises Searle's assumption of this in his argument about translation and the Chinese box. Searle (1980), argues that a computer cannot understand. Understanding entails intentionality. A computer cannot have intentionality because it is a formal syntactic system. He therefore argues that AI ideas or computational theories based on them cannot describe or explain mental processes. Boden argues, however, that any simple program has some semantic properties and that computational theories are not essentially incapable of explaining meaning. This is because explanation involves assimilation of something to something else which is analogous to it but not identical with it. Searle also argues that the physical nature of the brain produces intentionality and that on intuitive grounds, the metal and silicon cannot. Boden argues that the brain's ability to generate intentionality is intelligible only through its information processing capability. The computer has such a capability.

30 This is distinct from current approaches to knowledge engineering which embody a humanistic/interdisciplinary approach to design. TKE is based on the ideas of Feigenbaum (1983), and 'traditional expert systems design'; see Gammack (1987, 1989), Feigenbaum, E. A. and McCorduck, P. (1983) *The Fifth Generation*. Addison-Wesley, MA.

31 For example, see cf. Bramer (1988), Ostberg (1988), Hayes-Roth (1984a). For a critique of ES, see, for example, Bloomfield (1988), Collins *et al.* (1986), and Collins (1992).

32 Lipscombe (1989), provides a critique of the development of expert systems in medicine and suggests that approaching the design/purpose of these systems to facilitate problem analysis rather than problem solving will provide for more effective tools. See also Lipscombe (1991), Alvey (1983).

33 Partridge (1987), puts this down to the explanation problem in ES technology which is inherently limited to the relatively static and relatively context-free domain of abstract technical expertise. In this article, he strongly recommended ES designers to concentrate on 'low road' applications and stay away form complex 'high-road' ones, otherwise there will be a software crisis.

34 For a comprehensive introduction to neural networks/connectionism, see the 'Special Issue on Connectionism', *AI & Society*, 4:1.

35 See *Computing*, issue no. 1992; see for example, Medsker (1991), Jones and Hiemstra (1994).
36 See note 2.
37 Cf. Wittgenstein (1958). See Johannessen (1988a, 1988b, 1992).
38 Polanyi (1966) *The Tacit Dimension*. Doubleday, New York.
39 *Ibid.*, p. 4.
40 Ryle, G. (1949) *The Concept of Mind.*
41 Polanyi, *op. cit.* p. 10.
42 *Ibid.* p. 10.
43 Polanyi has developed his theory of personal knowing in his work, *Personal Knowledge: Towards a Post Critical Philosophy* (1958).
44 As in the case of control systems design.
45 Rosenbrock is concerned that the notion of the automated machine allows for the idea of a workerless factory. It rejects the tacit dimension of worker's knowledge and skill, seeing them as automata, reducing their tasks to ever simpler and more closely definable fragments (cf. Fordism). He cites Needham's view in 1927 that, in science, man is a machine; or if he is not, then he is nothing at all.
46 Rosenbrock, H. (1988) 'Engineering as an Art', *AI & Society*, vol. 2, no. 4.
47 Rosenbrock has been involved in a European venture upon Human-Centred CIM Systems. (CIM – Computer Integrated Manufacturing). This project attempted to apply the symbiotic model in the construction of a human-centred CIM system.
48 Cf. Cooley, 1987a, p. 12.
49 Cooley, M. (1987a), *Architect or Bee?* Hogarth Press (new edition), p.11. Cf. cybernetics, see also Cooley's chapter herein; see Wiener (1949).
50 Feigenbaum, *op. cit.*
51 The terms, 'holistic similarity recognition', 'intuition', or 'know-how', have been adopted from Dreyfus and Dreyfus (1986). They use them synonymously to denote: 'the understanding that effortlessly occurs upon seeing similarities with previous experiences ... intuition is the product of deep situational involvement and recognition of similarity.' They, in turn, have drawn upon Polanyi (1966).
52 Dreyfus and Dreyfus (1986), present this balance between analytical thinking and intuition in their theory of skill acquisition. Here, there are five stages of development from being a novice to gaining expertise (*op. cit.* chapter 1). These are 1. novice; 2. advanced beginner; 3. competent; 4. proficient; and 5. expert. A novice follows context-free rules. The relevant components of a situation are defined to enable the novice to recognise them. The novice lacks any coherent sense of the whole task and can judge his/her performance only in terms of how well the learnt rules have been followed. The advanced beginner, through practical experience in concrete situations, learns to recognise situational elements, which cannot be defined as objective context-free features. Recognition occurs through discerning similarities to prior examples. The competent learner can make plans and choices in order to achieve certain outcomes. Being proficient means that one has acquired an intuitive ability to use patterns without reducing them to features. The competent performer still relies on analytical thinking in combination with intuition. An expert, however, can discern whole scenes without decomposing them into elements. Because of their experience they can use their intuition to make decisions and cope with uncertainties and critical situations.
53 The hermeneutics of Göranzon and Josefson are based on interpretations of Wittgenstein, Polanyi, and Kuhn; whereas the hermeneutics of Ehn covers interpretations of Heidegger and Marx as well.

54 These case studies and discussions are reviewed further below.

55 Two additional aspects of tacit knowledge, based on an interpretation of the Scandinavian, (cf. Gill, S. P., 1990, 1992) are 'experiential knowledge' and 'personal knowledge'. Gill, S. P., in her PhD dissertation (Gill, S. P., 1995) defines 'personal knowledge' as that which is specific to the individual personality, gained from our personal life experiences (like family culture, school, friends), and expressed in our social values, beliefs etc.; 'experiential knowledge' is that of direct experience which can cover specific experiences, such as within the context of the workplace, e.g. knowledge gained from interaction with work colleagues, group culture, organisational culture etc.

56 See also Schön (1983), for another significant discussion on practical knowledge in, *The Reflective Practitioner: How Professionals Think in Action.* Basic Books.

57 See Janik (1988), for an overview of the hermeneutic approach adopted in particular by the Scandinavian researchers based at the Swedish Centre for Working Life.

58 They cover the fields of ethics, morality, and aesthetics; cf. Nordenstam (1979, 1987), Nordenstam and Johannessen (1981), Johannessen (1981). See also Tilghman (1988).

59 See Fann (1969), for a discussion of this shift.

60 Johannessen, K. (1988b) 'The Concept of Practice in Wittgenstein's Later Philosophy' in *Inquiry,* 31: 357–370.

61 Johannessen, K. (1992) *Rule-Following, Intransitive Understanding and Tacit Knowledge.* pp. 41–63.

62 Wittgenstein, L. (1958) *Philosophical Investigations.* §199, p. 81.

63 Wittgenstein, L. (ed), (1977) *Remarks on Colour.* G.E.M. Anscombe. §317.

64 Cf. Garfinkel (1967) *Studies in Ethnomethodology.* Cambridge: Polity Press.

65 Cf. Garfinkel, *op. cit.*

66 Heritage, p. 144

67 'a person is 95 per cent juror before he comes near the court'; cf. 1967b. p. 110.

68 See for example, Atkinson and Heritage (1984), Sacks, Schegloff, and Jefferson (1984), Jefferson (1979, 1984, (on the analysis of laughter)), Goodwin (1981), Grice (1975).

69 Cf. Van Dijk (1985) *Handbook of Discourse Analysis,* vols. 1, 2, 3, 4.

70 Lyons (1967).

71 Potter and Wetherall (1987).

72 *Ibid.*

73 On autobiographical memory, see also Conway (1990), Rubin (1986), Barsalou (1988), Brewer (1986), Kolodner (1986), Schanks (1982, 1990). The emphasis upon dialogue and autobiographical memory; cf. in Bekerian *et al,* draws upon the work of Bartlett (cf. 1932).

74 Wardaugh (1985, 1986), Brown and Levinson (1987), Goffman (1959, 1981), Goody (1978).

75 Cf. Bekerian (1992).

76 Cf. Shank (1982).

77 With reference to their use of Wittgenstein, Polanyi and Kuhn, they work with Johannessen (cf. 1988a, 1988b, 1992), Nordenstam (1979, 1987, 1992), Janik (1988, 1990, 1992), Janik and Toulmin (1973). With reference to the period of the Enlightenment they work with historian's of ideas such as Toulmin (cf. 1991, 1992).

78 This emerged from a discussion with Josefson, April 1995, London.

79 Cf. Dialogue Group discussion, Darwin College, February, 1993.

80 For example, the tragedy of Antigone. Here, the nurses see Creon as representing the institution, the decision-makers who have no understanding for their everyday practices. They see the doctors as the inflexible ones and see Antigone as representing the

voice of the people. From the discussion with the doctors, they saw the nurses as the inflexible Creon. This is from a discussion in the Dialogue Group, Darwin College, Cambridge, February 1993. Josefson wished to illustrate how the use of a play can enable the expression of experiences and raise self-awareness.

81 Ref. Dialogue Group, Darwin College, Cambridge, meeting in February, 1993.

82 See Göranzon (1992), for a discussion upon these studies.

83 UTOPIA is an acronym for 'Training, Technology and Products from the Quality of Working Life Perspective'. In the 1970s in Scandinavia there was a democratic movement involving a collaboration between academia and the trade unions in reaction to the negative effects of new computer-based technology upon workers' skills and work environments. This movement created two major projects, of which the largest was UTOPIA. See also Bodker (1987), Ehn (1988), Bjerknes *et al* (1987), Sandberg (1979).

84 Ehn, *op. cit.*; see also Gill, S. P. (1995: 44–46), for a brief resume of these.

85 *Ibid.* Chapter 3, 'Emancipatory Practice', pp. 82–102.; Part III Designing for Democracy at Work, pp. 247–364.

86 *Ibid.* Chapter 4: 'Language-games - A Wittgenstenian Alternative'.

87 These are discussed by Gill, K. S., in Chapter 1 of this book.

88 As in the sense of ancient Greek philosophical distinction between theory and praxis. See, for example, Aristotle's *Nichomachean ethics.*

89 as in the 'dialectical relationship between nature and technology', cf. Rauner, *op. cit.* 1987, p. 6.

90 This draws upon Paolo Freire's pedagogy, cf. 1971

91 Project MEDICA: A Cognitive Support System for Psychiatry. Working Paper, 1991, SEAKE Centre. MEDICA (1989–1990) was a component project of the 'Exploratory Action' phase of the AIM Programme (under the aegis of the Directorate General XIII of the EC Commission); NB. For Smith (1995), see his chapter in this book.

92 The ideals of the objectivity of data and the objectivity of the researcher are not held by the author.

93 cf. Gill, SP (1995), Dialogue and Tacit Knowledge for Knowledge Transfer, PhD Dissertation, University of Cambridge.

94 See Appendix A in Gill, S. P. (1995 – PhD dissertation) for description of the RAVE system.

95 i.e. the traditional approach to knowledge engineering.

96 Cf. Framework of knowledge transfer: Figure 42, page 346.

Chapter 9

Designing Practice-Based Learning Environments

Felix Schmid

Introduction: Industrial Revolution and the Use of Computers

The 1980s and 1990s have been characterised by a move from self-contained manufacturing and service operations to large organisations (not necessarily under a single command or ownership) whose activities are integrated through the use of electronic communications and computers. These developments have not occurred overnight but are the result of a much longer evolution of the manufacturing and service industries.

The changes brought about by the emergence of powerful computer technologies are often compared to the changes which resulted from the industrial revolution: many authors have described this transition from craft-based work to factory production, the move from individual skill to the division of labour with its emphasis on dexterity and speed rather than an awareness of the process as a whole. Bowers (1984), wrote:

> The sense of time and human activities became intertwined in such a way that change came to be seen as the expression of progress. The rapid development of technological innovation during the latter part of the eighteenth century and the industrial revolution of the nineteenth century strengthened the idea of unlimited abundance.

Similar feelings were at the heart of the economic optimism of the 1960s, e.g., the 'Deutsches Wirtschaftswunder'. The 'first' Industrial Revolution progressed in two stages: a beginning in the last quarter of the 18th century with the mechanisation of the cotton and woollen industries of Lancashire, Central Scotland and West Yorkshire and the continuation with the mechanisation of the iron and steel industries from circa 1830 (Crystal, 1990). The transformation meant that instead of small quantities of expensive goods, industry could supply standard products of repeatable quality, with vastly improved productivity (Rembold, 1993). While these changes led to entirely different working conditions in the production of artefacts, the creation of markets for mass-produced goods also meant that industry had to recruit workers whose background lacked any prior involvement with manufacturing. This development was very welcome in a period where: 'The continual rise in the population made it indeed impossible to provide work for everyone in the English village. Agriculture had absorbed all the hands it required. Great national industries, like cloth, were migrating back out of the country districts to which they had moved in the later Middle Ages ...' (Trevelyan, 1942, p. 486). However, at the same time a great many craft-based businesses, such as hand loom weavers, disappeared, with their skilled workers ending in abject poverty. But:

> An important group had even accepted, indeed welcomed, industry, science and progress (though not capitalism). These were the 'artisans' or 'mechanics', the men of skill, expertise, independence and education, who saw no great distinction between themselves and those of similar social standing who chose to become entrepreneurs...
>
> *Hobsbawm, 1968, p. 90*

Although this group of 'engineers' was not very large it created its own myths, including that of the superiority of machines over people. Very often the men (gender-specific labels intended) became successful businessmen in their own right. In general, however, the situation in the cities was barely better than that in the countryside. Between 1800 and 1840 entrepreneurs had to employ workers on subsistence wages because much of the value added was not reinvested in the system by their backers, a phenomenon which is familiar in the 1990s. The new markets were still in the course of being established and workers were not yet perceived as potential customers. Naturally, people changed a great deal as a result of all the pressures and experiences, as observed by Robert Owen:

> The general diffusion of manufactures throughout a country generates a new character in its inhabitants; it will produce the most lamentable and permanent evils. The manufacturing system has already so far extended its influence over the British Empire as to effect an essential change in the general character of the mass of the people.
>
> *Owen, 1815*

Not only Owen, but also a number of other 'modern' industrialists, such as Titus Salt in Bradford, developed models of industrial communities which showed the way towards creating a better environment for workers, at the workplace and at home. 'Some time in the 1840s all this began to change, and to change rapidly, the readiness to accept legal supervision of working conditions – as by the admirable Factory Inspectors – increased. British industrialists now felt rich and confident enough to be able to afford such changes' (Hobsbawm, 1968: 123–4). For a while, Britain had become, to use a cliché, the 'workshop of the world'. New hierarchies and new opportunities for training, education and personal development for the masses emerged. Working hours became regulated and salaries rose to levels where most people could live without constant fear of the future – apart from the Great Depression of the 1930s.

Recent Developments in Industrial Production

The industrial society whose social patterns persisted in Europe until around 1985, the type of society to which many 'threshold countries' aspire, existed only since the late 19th century. Major cracks in this 'Western' system had already appeared around the time of the first oil crisis of 1973/74. However, the seeds for a new revolution had been sown before and during World War II: the new levels of mass production necessary for the war machine led to the development of mechanically programmable machine tools. After the war, rising standards of living meant that the increased productivity could be absorbed without leading to overcapacity (Finniston, 1989). The advent of NC (numerically controlled) machines, invented at the MIT in 1947, initially had an impact only on the feasibility of more and more complex components. Once the new manufacturing processes and computer-controlled systems of the late-1970s and early 1980s had arrived, productivity and quality of output could grow at rates never imagined before, thus providing the foundation on which 'mega-corporations' could be built with scant regard to overhead costs but still employing relatively large workforces.

By the mid-1980s though, the process of industrial concentration and the move to ever larger and more complex manufacturing facilities ceased as low cost producers emerged in the 'third world'. Changing markets threatened industrial stability and with it the social fabric. European industry had to change. As stated earlier, this process was and is compared to the industrial revolution of the 19th century:

- the introduction of new technologies with advanced capabilities;
- the modern understanding of systems behaviour can be compared to the early understanding of the behaviour of rigid bodies;
- the replacement of human organising power by that of computers (similar to the substitution of water and steam power for human muscle);
- the introduction of new materials to all sectors of the industrial economy;
- the development in data communications with new levels of speed and accuracy parallels that of the emergence of steel ships and of railways in the first industrial revolution;
- new markets are developing in China and the Pacific Rim in general.

However, there are also key differences between the two 'revolutions' (see, e.g., Finniston, 1989):

- the capabilities of modern technology are far more advanced, by orders of magnitude, than those which replaced craft-based industries;
- the cost involved in acquiring and maintaining advanced technology is very high and investors are far more risk-averse than in the past;

- today, the needs of a large population can often be satisfied by a proportionally very small workforce;
- first-world markets are generally not expanding in volume terms but qualitatively, with increased product diversity in some sectors and very much reduced variety in others;
- improved communications and better transport links have rendered competition worldwide and therefore unpredictable;
- the emerging new markets are often poor, volatile and very price-sensitive. Productivity in most sectors of the economy has risen dramatically and the productive capacity far outstrips demand;

People, displaced by technological or economic developments in one industry, can no longer expect to be re-employed quickly elsewhere. The skill levels required to operate and manage the new technologies in their pure forms are very different from those available amongst existing workforces. Computer integration is only one of many such applications of advanced technologies to manufacturing and service organisations. It has received more attention than many others though, not just because of the fascination with its 'pseudo-intelligent behaviour', but because of its scope for displacing human workers at all levels of an organisation, rather than simply replacing manual workers. It is also believed to have further long-term implications which cannot yet be predicted. Technological pessimists would proclaim that any people displaced because of mismatched skills or productivity increases will simply be marginalised. Technological optimists firmly believe that the expansion of the economy as a result of greater productivity and the availability of new technologies will provide full employment.

Technological and Computing Euphoria

Around 1985, microprocessor based systems started to emerge, offering automation possibilities beyond engineers' and managers' wildest dreams. High-speed data networks became available, together with the necessary central information storage devices. The concept of the 'lightless factory', often described as a Japanese invention, was the result of these technological advances, a concept promising high productivity, flexibility, top quality and very low labour costs, due to the almost automated design of products and the development of the corresponding automatically controlled means of production, the automatic handling of goods at all stages of manufacture and the reduction in labour costs to nearly zero.

This scenario did not materialise, however, since it soon became apparent that the investments required to achieve such performances were beyond the resources of even the biggest and most powerful manufacturing organisations. A prominent victim was General Motors of the USA, where the company management spent large funds, a sum equivalent to the price of acquiring one of its Japanese competitors, on developing a fully automated car factory, an enterprise resulting in a failure which brought the company close to collapse (Gannon, 1988).

The relatively indiscriminate use of computer based systems was aimed at displacing the 'lower' levels of industrial workforces but, at the same time, created many new support and management functions elsewhere in the companies. Rather than being able to establish a lightless factory run by a few experts from home, industrialists were faced with top-heavy organisations with high wage bills and under-qualified machine minders and cleaners on the shop floor. This was clearly an undesirable outcome and other routes towards higher industrial effectiveness were sought.

What went wrong? Some critics maintain to this day that the systems used were simply not powerful enough, that software and hardware were still flawed. As technological optimists they are convinced that better systems will fulfil the early promises. Technological pessimists though argue that any system whose complexity exceeds a level at which it could still be operated by a human being will ultimately be doomed to failure. Both groups tend to overlook one of the main factors in the equation even though productivity of the industrial workforce had risen to very high levels in the 1960s and 1970s, direct labour costs, expressed as a proportion of product cost, remained more or less constant, partly due to the higher skill levels required by new technology (Boer, 1994). Its complete removal could therefore not be achieved in a cost-effective manner.

Finding an Alternative

While it would be easy to side simply with either the technological optimists or the pessimists, choosing one or other of the extreme positions, we would run the risk of denying manufacturing the use of powerful tools which can enhance its effectiveness and thus profitability. The data and information handling capability of computer systems must be used if manufacturing is to continue to provide a contribution to Gross National Product (GNP) and to offer challenging and reasonably paid work to a proportion of a country's population. The contribution of people, individuals and groups, can also have a major impact.

New systems must be designed by engineers and scientists who have a thorough understanding of the needs of users and who are willing to adapt the concepts to the size of manufacturing entity for which they are intended. Companies must choose technologies which fit in with the status of technological work in the respective society, the relative status of the technologies evaluated, the level of training of personnel and the professional backgrounds of the future users (d'Iribarne, 1986). It is thus necessary to develop new paradigms for the computer integration of manufacturing, paradigms based on people and computers as partners.

People, Organisations and Technology

A new vision has been promoted for the manufacturing industry under the heading of POT (people, organisation and technology), as part of the Manufacturing, Organisation, People and Systems (MOPS) programme, supported by the UK Department of Trade and Industry and the European Community (Evans, 1994). This vision is founded on the principle of using organisation, people and technology 'as equals' to develop a manufacturing strategy that will enable companies to respond to the dynamic markets of the 1990s. It emerged as one of the products of the human-centred systems debate of the 1980s and is equally applicable to service industries. In this section we provide a background and reflect on some aspects of the debate.

The Meaning of Human-centredness

The concept of 'human-centredness' and the ideas that underlie the approaches were first developed and formulated by Howard Rosenbrock in the early 1980s (Rosenbrock, 1983), but are to some extent enhancements of theories first defined by the more progressive members of the ergonomics community, such as Branton in (Oborne, 1993). The concept of human-centred design goes along with and includes many principles well known in the field of 'human work design': personality promotion, skill-based manufacturing and complementary design of man–machine systems (Schmid, 1993). Support from the European Community, worried about the decline of European competitiveness, and from the research funds of a number of countries led to the 'invention' of HCCIM or Human Centred Computer Integrated Manufacturing.

In general, a human-centred technology is one which extends human skill and its application to real-life situations. The technology itself must be designed so as to optimise the synergy between human skill and computer power. The work within the factory must be organised such that in all areas people are able to apply a substantial range of their skills rather than just a small 'useful' part thereof. Individual skill and competence should be increased through a balanced combination of learning by doing and formal training and education. An example of such an approach is given by Rosenbrock (1989) who developed an exemplary model of an alternative machine-tool programming system, combining several different approaches. His team designed a prototype CNC lathe controller as a demonstration project to show that the removal of human purpose from the system and its replacement with a mechanistic process results in a more constrained and therefore less powerful solution than the approach which combines human skill and machine capability. In his approach, the operator is presented with a CAD-like image of the component to be made and can then programme the lathe with the number of cuts, speeds and feeds either himself or by accepting suggestions made by the computer, derived from algorithms or earlier inputs by the operator for similar tasks. In general, the operator's inputs override those of the computer although the latter checks for dangerous situations. A short quote from the preface to *Person Centred Ergonomics* ... (Oborne, *op. cit.*), may serve to highlight some of the reasons for retaining human beings in systems:

> ... far from being sources of error in a system, people at work bring unique characteristics to the system which no machine would ever hope to match. They have a sense of responsibility in their work, a quality of flexibility and adaptability and the ability to predict events and courses of action for the future.

Human-centred Systems Design

In a human-centred system, traditional design and manufacturing skills are used to their fullest extent. The range of activities for which each operator is responsible is maximised, to include, for example, production work as well as quality and planning related duties. As a result, each person will develop a general knowledge of the whole production process and must be given the freedom and opportunity to comment on any aspect of it. Cherns (1975) developed ten principles of socio-technical design, four of which are at the heart of human-centred systems design:

- *compatibility*: the way in which a job or system is designed should be compatible with the objectives of the design, e.g., a system requiring participation for its operation must be designed in a participative manner;
- *minimal critical specification*: the specification of a job or system should identify and specify no more than is absolutely essential;
- *information flow*: information for action should be directed first to those whose task it is to act;
- *power and authority*: those who need resources to carry out their responsibilities should have the power and authority to command those resources.

The concept of human-centred manufacturing relies on skill enhancement rather than skill replacement. The German Industriegewerkschaft Metall (IG Metall, the German trade union) realised the need for this in 1976 and lobbied for new approaches for the training of workers and a change in attitude in industrial relations in the 'old industries' (steel, textiles etc.). The Lucas shop stewards committee took up the theme in their campaign to replace arms production with socially useful products (Cooley, 1987).

There are many ways of increasing human-centred working within as well as outside manufacturing industry through the application of computers. True interactive computer-based systems are needed and are being designed for a variety of human work situations, perhaps more so in areas such as banking and air-traffic control than in manufacture. The reason for these positive developments is that the introduction of the computer has forced people to think deeply, perhaps for the first time, about the fundamentals of processes, both qualitatively and quantitatively. This need for reflection on the key issues in manufacturing is not limited to management; the integration of computers involves everyone engaged in the production process.

The Person-centred Ergonomics Perspective

Human-computer integration in manufacturing and elsewhere is, in essence, no more than the application of the fundamental concept of ergonomics to computer-supported systems, that is:

> designing the 'machine' (or any aspect of the environment with which the person has to interact) to match the operator's abilities
>
> *Oborne, op. cit.*

The operator is part of one or several feedback loops maintaining a system in the desired state or near to it. This traditional view of the ergonomic approach has been expanded in recent years to include the person's story in the analysis, where 'story' refers to an individual's social and cultural background. Person-centred ergonomics views the interaction between human and machine as one which must be controlled and dominated by the operator(s) in the system.

Human beings have special abilities and dimensions of flexibility which must be taken into account when designing a system. A primary feature which humans can bring to systems is a sense of purpose. Since this differs from person-to-person the system must be designed to cope with the general thrust of the intentions or purposes. This purposeful behaviour is not predetermined but evolves from continuous 'computation' of information provided by the environment. According to Branton (in *ibid.*), the performing of such computations, that is, the existence of 'unselfconscious intentions' may help explain skilled activity. I would contend that this is one of the factors which affects the design of experiential-learning environments where predictions of motivation, learning and leadership performance are difficult and unreliable.

Operators, people, bring to a system a collection of inherent strengths and weaknesses which interact with the system to change it. In Branton's opinion (Oborne, *ibid.*), operators in any system invariably turn it from being a closed-loop entity, in which information flows between components with (theoretically) maximum efficiency for correcting deviations, to an open-loop system. The deviation corrections are effected by the operator on the basis of his or her mental model of the system and its operation. Thus system design must be defined by the abilities and backgrounds of the operators who inject human factors into their work such as autonomy, responsibility and creativity.

People Interacting with Computers

Computer systems, if they are to be subservient to human beings, must be designed in a human-centred way, to paraphrase a statement by Rosenbrock quoted below. Design efforts must not be limited, as a matter of course, to 'interface improvements' such as the use of mice, windows and graphics in the case of computers for office use. The efforts must be aimed at sharing meaningful tasks between person and machine, for example, allowing a human operator to decide on the rate of injection of plastic into a mould, based on her experience, while providing, as a backup, a number of proven time profiles stored in the computer. Since computer-based technology is still relatively new, there is a chance for cross-disciplinary teams to work together not only in establishing the new technology but also in getting this technology accepted by the users. Even though changes in working procedures may be radical, properly managed process innovation and skill development can open opportunities for new, more satisfying and thus more human-centred ways of working.

In the past, and to some extent even today, computers have been used often to de-skill jobs. A typical example of such an approach were the computer-controlled human workplaces for parts insertion into printed circuit boards (PCBs) which were developed in the early 1980s: parts were automatically presented in a carousel to the operator while a light-point indicated the insertion position. The only contribution demanded from the human being was the dexterity involved in pushing two or three 'legs' of a component through a set of holes!

The development of human-centred, computer-based tools provides people with assistance in the management of complex situations. The combination of human skills, such as fast and adaptive decision-making, with computer capabilities, such as quick storing, processing and retrieval of information, offers a good approach to dealing with highly dynamic problems. The computer-based tools either simply provide information to the user in a structured form or they help by offering options for approaching the solution of problems. Thus they can be viewed as elements of a decision-support system. Algorithms for decision making can either exist in the system or can be added by users as and when required. It is important that the user can check the method of operation of the algorithm at any moment. In some instances it may be necessary to limit the level of automation provided by the computer systems so as to leave sufficient scope for the work of the human.

Various projects have been set up to design alternative methods of implementing technology in the form of 'flexible manufacturing systems in which the operator is not subservient to the machine' (Rosenbrock, 1983). The European Community project on 'Human Centred CIM Systems', described by Ainger (1990), was one of the key projects concerned with human-centredness in manufacturing. The objective of the project was to produce human-centred CIM components, in which human skill and its application are optimised, in harmony with leading-edge computerised manufacturing technology.

A Cautionary Remark

It is to be expected, unfortunately, that the human-centred approaches postulated will only be applicable in some areas of manufacturing, particularly in 'leading-edge' production, where competition is not purely on cost but includes factors such as timeliness. While it would be theoretically interesting to eliminate competitive pressures as threats to human-oriented work designs by means of trade barriers, McKinsey & Co (1993) state in a recent study that global competition is essential if productivity, measured as value-added per hour worked, is to increase and thus create conditions ensuring company survival. It must be accepted that this free international trade will militate against human-oriented work designs as long as there are no internationally binding agreements or consumer pressures which guarantee comparable levels of industrial employees' welfare in all markets. Such a levelling of the 'competitive playing field', although desirable from the human work angle, will also result in a levelling of living standards between newly industrialised and developed nations. There is, clearly, a built-in conflict between different goals. However, this must not be used as an excuse not to teach the skills required for the successful development and implementation of human-centred systems.

Transforming Companies and Organisations

To remain competitive in world markets, manufacturing enterprises must take full advantage of the potential for increases in productivity, better quality and greater flexibility provided by more responsive organisational structures, better utilisation of human skills and experience and the power of modern computer-aided technologies. To achieve these goals, a skilled, cooperative and motivated workforce is required. Traditional management hierarchies and forms of control must be modernised. More participation by employees is needed at all levels, in planning, designing and implementing new technologies and systems. Technical systems must be designed not just to meet economic and technical goals, but also to satisfy organisational and human requirements.

Experience suggests that manufacturing systems and other complex systems do not operate effectively and do not yield the expected benefits unless attention is also paid to the organisation and people issues during design and implementation. As discussed at the beginning of the section *People, Organisations and Technology* (starting on page 366), there is now an increased recognition that organisation and people issues are important, although most companies adopt a fairly narrow approach as to how these issues should be addressed. Managements attempt to find ways of overcoming the resistance to change and the reluctance of people to accept new technology but fail to provide an entirely new environment and philosophy. Lang (1990), discusses this reluctance of companies to move from interfacing to integration because it requires more change than quick-and-easy projects.

Some companies have started to change the way they use technology and have begun to consider organisation and people issues in some depth. The changes needed are, at times, quite simple but must be structured well. Organisation and people issues should be considered at all stages, from business strategy right through to manufacturing systems design and implementation. Instead of looking for the technological fix, companies need a broader and more balanced approach that addresses manufacturing organisation, people and technological systems as an integrated whole. It is necessary to overcome the belief that technology provides the answers to all problems. Managers must develop a new and broader vision of the 'Factory of the Future'.

Organisations and Strategy

The oft-stated demand for flexibility, adaptability and improved responsiveness to customers' requirements, combined with the need to motivate people and to make use of their skills, judgement and experience, mandate change in organisations, work practices and technology. These facets of an enterprise must be developed in a way that will allow highly-trained people, at all levels of the company, to adapt their work strategies to the variety of situations which they will have to face. There is a need, therefore, to move away from an organisational model characterised by:

- large number of hierarchical strata;
- centralised and cumbersome decision-taking;
- under-qualified, single-skilled operatives with little competence or authority;
- complex information-handling and management processes;
- static structures and reactive handling of markets.

Such a company structure and its order management process are discussed by Creux (1993), based on a real company analysed as part of a UK Department of Trade and Industry (DTI) supported Manufacturing Organisation, People and Systems (MOPS) project. Creux advocates the replacement of multi-layered organisation trees, still common despite being encumbered by complex and time-consuming decision processes, with organisational models based on:

- flat organisational hierarchies;
- decentralisation of decision making and control;
- increased competence and authority for shop-floor staff;
- multi-skilled employees and teamworking;

- skill-supporting technologies, simplified handling and use of information;
- continuous improvement involving all employees.

Benefits of such a balanced manufacturing strategy are reported to include: shorter throughput times, reduced inventory, improved product quality, streamlined material flows, easier production planning and control. Put simply, more economic operating conditions and improved job satisfaction which can lead to better motivated people (e.g., Skinner, 1992; Hill, 1989; Slack, 1991).

The appropriate starting point for the development of a human-centred, computer-integrated manufacturing system is thus the consideration of strategy. Two strategic components must be considered here:

- the corporate strategy and, associated with this, how computers can help realise business objectives;
- the method and priorities adopted by the organisation in the development of its computer systems.

Although a number of methodologies exist and others are being developed to reflect new approaches to systems development, evidence seems to suggest that no one methodology can guarantee successful systems development on its own. As a general rule it can be stated that usable methodologies must be flexible and adaptable to almost any given situation. It may be better to have a set of guiding principles which cover the key issues and to allow system developers to use these as tools in a manner which is appropriate to a particular organisation rather than prescribing a flawed all-purpose methodology.

Users are often disappointed in a system once it is delivered. This statement is particularly true of computer-based systems. One reason given for this is an inaccurate specification of users' requirements. It is often wrong to imply that this is all the users' own fault. It is frequently the case that the people for whom the system is intended are left out at critical stages of the development process and, even when they are asked to participate, they are often not adequately briefed or guided when stating their requirements. This problem is described in Rachel's study of the design of a new customer-oriented software system in a water company. Users were theoretically involved in the process of designing the system but, in fact, they were ignored (Rachel, 1994). This situation could potentially be improved through the adoption of changed patterns of training and education for young managers.

The Process of Change, Education and Training

The change process involved in moving from an existing mix of organisational methods and technical tools needs to be planned in a way which is appropriate to humans. Possible patterns for changing business environments and company structures are shown in Figure 43 (adapted from

Ainger, 1990). For the example, it has been assumed that the company needs to be transformed to reach a new stable *modus operandi* by changing both its organisation and the technologies used to run its systems. The main part of the diagram is used to illustrate three possible migration paths from one structure to another. The two structures differ in terms of the level and complexity of the organisational methods and technological tools applied. The two subsidiary figures indicate the respective rates of change as a function of time, For example, path 3 refers to a transformation dominated by organisation-oriented approaches.

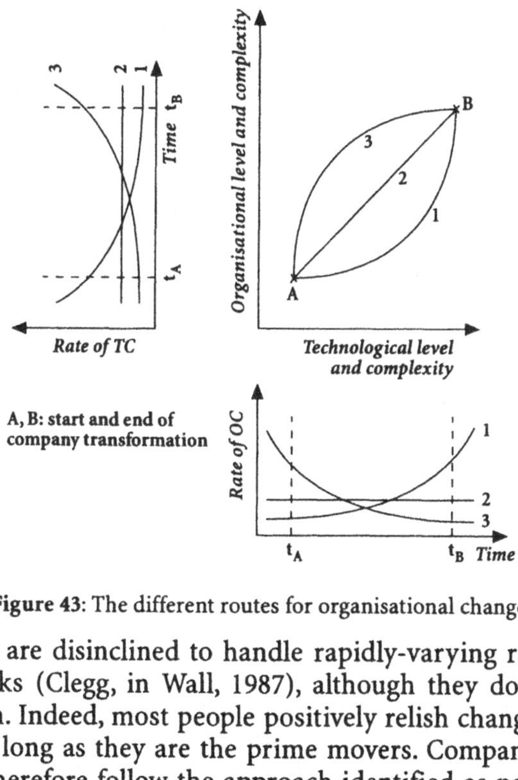

Figure 43: The different routes for organisational change

Most people are disinclined to handle rapidly-varying rates of change with high peaks (Clegg, in Wall, 1987), although they do not object to change as such. Indeed, most people positively relish change in their private lives - as long as they are the prime movers. Company transformations should therefore follow the approach identified as path 2 in Figure 43 and must involve the people concerned with the outcomes.

The diagram illustrates the need for a balanced process of organisational and technological transformation where humans are able to adapt in an incremental fashion to a changing work environment.

The third axis of the main graph would show the rate of change in training and attitude required of the humans involved in the process. This third axis forms the context for this chapter. It is quite feasible, and indeed very traditional, to teach engineers, designers and applied scientists in a manner which is represented by a path very similar to (1) in Figure 43. Although this will probably result in trained people capable of handling and solving the problems which are current today, it is unlikely that the people taught will adopt the ability to change and the associated ability to manage change. People also have to be taught how to specify their own needs and requirements in terms of systems and organisation. A discussion of the specialist education and training needs of manufacturing engineers should provide some insights into the more general problem of education for change.

Educational Needs in Advanced Technology Environments

Many managers are promoted from the ranks of engineers who often lack some of the skills and qualifications necessary to perform well as managers. Some authors suggest that the majority of engineers become managers at an early stage of their careers. Other authors use the terms engineer and manager interchangeably in the context of modern manufacturing (e.g., Skinner, 1985). Between 1985 and 1995 awareness of these issues has grown, leading to the trialling of continuous professional development (CPD) in the shape of voluntary schemes administered by the Engineering Institutions. The Engineering Council (1995) has addressed the issues in its guidelines for the professional development of engineers which includes clear recommendations for postgraduate education of all engineers wishing to be granted chartered status.

Since managers play such key roles in operating complex systems of the type encountered in advanced manufacturing and, in particular, in CIM installations, it would appear appropriate to ensure the best possible preparation of engineers, the future managers. In the past, the vast majority of engineering courses have been characterised by the transfer, from teachers to students, of great quantities of information, from the fields of mathematics, the natural sciences, technology and applied engineering. Unfortunately, this approach is not inline with the requirements of either the individuals or industry.

Manufacturing and Education for Manufacturing Engineering

'Historically, companies met the need for manufacturing engineers by promotions from the ranks of machine operators' (MSB, 1985), who would have possessed some of the key skills required for the new role as a result of their shop-floor experience. By 1980 in the USA, for example, there were fewer than half a dozen universities offering specific Manufacturing Engineering programmes (MSB, 1985).

The situation in the United Kingdom in the early 1980s was similar to that in the United States, if not worse, in the sense that the political environment was not supportive of manufacturing industry as a main contributor to GNP and that industry was no longer seen as a creator of wealth. But while the UK government promoted the growth of service industries, there were other views such as that of Sir Kenneth Corfield, Chairman of STC, who wrote in 1983: 'Increasingly it is being seen that Britain has the knowledge and inventiveness to spur innovation but that this is not translated into production. ... Inadequate attention is given to the huge problem of funding and resourcing the resultant product for production' (Corfield, 1983). Later, in 1989, Finniston looked back and declared:

> Britain's main wealth creating activity is manufacturing. Although a vigorous service sector is very important, only 18% of its activity is tradeable internationally compared with 80% in manufacturing. ... It is necessary ... to increase the speed of response of ... advanced technology manufacturing
>
> *Finniston et al., 1989*

By that time Parnaby (1987), at Lucas had already derived from a similar realisation the idea of creating a new discipline, Manufacturing Systems Engineering or MSE, to encourage the development of a new breed of engineer capable of specifying, designing, implementing and managing production operations. Although this term of Manufacturing Systems is now part of the industrial vocabulary it is a term which requires definition, best perhaps in Parnaby's own words: a manufacturing system is:

> an integrated combination of processes, machine systems, people, organizational structures, information flows, control systems and computers whose purpose is to achieve economic product manufacture and internationally competitive performance.
>
> *Ibid.*

Parnaby had first introduced the concept in Lucas owned businesses from where it spread, partly as a result of the quality of his key dissemination tool, the Lucas Mini-Guides. The provision of a near scientific base for the design, introduction and management of production operations was one reason for the pervasiveness and great success of the MSE approach.

Commenting on the rate of technological change, Finniston *et al.* (1989) state that education and training must become a continuous process, that engineering courses must be broader based, more flexible and more responsive to the changing technologies. They draw the conclusion that the formation of more and better manufacturing systems engineers is essential. The necessary skills have to be made available to students learning something about the manufacturing world in which they live. They should develop and extend their awareness of engineering through both specialised (technology-based) and integrated (broad-based) engineering applications. To design and operate the modern manufacturing or process system requires, in their view, engineers with multidisciplinary skills:

> The design of the artefact must be related to its production, ultimate performance and use - and this desirable goal cannot be achieved without people who are educated, trained and have the attitude to adopt change at the right time and in the right way ... Graduate engineers are needed who have received an appropriate education to enable them to appreciate and work

> across traditional engineering disciplines, combining and integrating appropriate engineering concepts, systems designs and the business of engineering.
>
> *Ibid.*

Unfortunately though, some of the features which had militated against engineering at the time of writing of the Finniston report in 1980, status and pay, still obtained in 1989 (and, to some extent, still do so today). However, Finniston *et al.* also conceded that engineering courses had changed a great deal and had broadened their scope (Finniston, 1989). Naturally, this development process has continued until today, perhaps tempered by the requirement for mass higher-education which requires a more careful use of resources. From much of the available literature on the subject, and based on the author's own experience, it would appear that one of the key difficulties in forming manufacturing engineers is the issue of defining a 'body of knowledge' which encompasses all an engineer needs to design, implement and run complex manufacturing systems (see Shea, 1985; Brummett, 1985). In most cases the manufacturing systems engineer is described in terms of the objectives he or she has to achieve rather than in terms of skills and knowledge. For example:

> A Manufacturing Systems Engineer should be capable of coordinating endeavour on a broad front, whilst maintaining a clear perspective of ultimate business goals and managing technology in the appropriate fields.
>
> *Finniston, 1989*

This move from a content-defined to a process- and objective-defined 'job description' is a phenomenon more and more common in the professional world, since much of the factual knowledge which was important in the past can now be retrieved from computer databases and knowledge-based systems whenever necessary. Since 1989, the profession of manufacturing systems engineer has become well established, despite the problematic definition, and systems engineers trained with a manufacturing focus have made a substantial impact on manufacturing industry. Discussion has now moved to issues concerning engineering education in general (EPC1-6), in particular as a consequence of the impact of mass higher-education.

Summarising the Problem and Attempting a Solution

For convenience, and so as to reflect common industrial practice, it may be acceptable to use the expressions 'engineer and manager for advanced manufacturing technologies and systems', 'engineer for CIM' and 'manufacturing systems engineer' as approximate descriptions even though different companies and institutions might wish to make clear distinctions. The problem facing an educator wishing to prepare engineers for a career in manufacturing systems engineering and, by dint of argument, in manufacturing management, is multi-faceted: there are a number of valid definitions for the scope of the professional involvement but there is no clearly defined body of knowledge, although there is a demand for a broad base in classical engineering subjects; in addition, there is a long list of skills and personal characteristics which such an engineer should possess but there are few, if any, guidelines covering which of these are important and how they can be imparted.

Conventional classroom teaching, even if enlivened by video presentations and industrial guest-lectures, may be well suited to the transfer of factual knowledge and method-based approaches, e.g., systems analysis using Laplace transform, but it should not be applied to the teaching of management and social interaction skills. Most of these can only be enhanced through personal experience of situations where such skills can be applied.

It has become the conviction of this author that the skills needed by future manufacturing engineers can be taught and developed by using project work of a novel nature as a teaching platform. While this project work cannot replace more than a small proportion of the conventional lecture, laboratory and tutorial-based teaching on an engineering course, it provides a space where students can experiment and learn about their own strengths and weaknesses in a near professional work situation and team environment. Moreover, the project approach presents the opportunity for integrating the knowledge and skills, gained from studying individual subjects over the whole duration of an academic programme, into a cohesive framework tailored to the tasks facing the professional manufacturing engineer.

The need for a course in Computer Integrated Manufacturing (CIM) at Brunel University arose from the conversion of an existing Production Technology degree programme into the Brunel Manufacturing Programme (BME) (*ibid.*). CIM was perceived as the key to future success in manufacturing industry and a team of academics was assembled to develop the course. Very soon though, the members of the team realised that the major issues in improving the competitiveness of manufacturing industry related not so much to the implementation of new technological approaches but to the development of skills which would facilitate the process. A decision was therefore taken to develop a new educational approach, which would open up a range of opportunities for students to develop their skill and knowledge in a team-based situation, working on a practical task rather than being given information in lectures and tutorials. Within an individual module, rather than on the scale of a complete programme of study, the author and his team created a course offering such an integrated approach, which follows the precept on the need for integration stated by Finniston: 'integrated engineering, which derives from the ubiquity of the computer, is industrial treatment of the artefact as a whole and not just the final assembly of its parts. It is not CAD or

CAM or CIM ... the final artefact must be designed with final economy not just of cost, but ease of production, assembly, maintenance, final quality.' (*ibid.*). This practice-based and strongly people-oriented learning environment was labelled the CIM approach - because this was the first topic taught in this manner (Schmid, 1995).

An existing laboratory area was cleared and the members of the team defined an overall task, the development of a computer-integrated manufacturing facility, as well as a range of sub-tasks related to computer hardware and software, robotics, Direct Numerical Control (DNC) of machine tools, Computer-Aided Design (CAD) and manufacturing cell design. Up-to-date computing equipment, donated by IBM, was made available to the students who had to form an organisational structure modelled on a business enterprise. The students were given a free hand and a small budget to manage the development process themselves. At first, it was expected that each iteration of the course, lasting eight months, would be self-contained. However, once the magnitude of the overall task became apparent, a decision was taken to develop the manufacturing facility and all its support function over a number of years, mirroring the process of continuous improvement which should characterise any well run manufacturing operation. Each year, the team responsible for the course, together with the students involved, deliberated ways in which the complexity of both the manufacturing plant and the equipment used could be developed to mirror ever more exactly an industrial situation. Hence, 'practice-based learning' was born.

The Engineer and the Design Tradition

Before attempting some modern definitions of the term engineer, it may be useful to undertake a historic excursion, concentrating on Britain, where engineering found its most vital expression, and France, where it found its codification and strongest recognition.

Work we would describe today as engineering existed from the earliest of times, but it was invariably ascribed to craftspeople or builders, from the invention of the throwing wheel to Archimedes, from the Roman builders of aqueducts to shipwrights, to the master-builders of the great cathedrals. Filippo Brunelleschi (1377–1446) and Michelangelo Buonarotti (1475–1564) deserve a mention, the former because he scandalised his contemporaries by building the giant cupola of the Duomo in Florence without a supporting framework, made of wood, as was the convention at the time. On the basis of his erudition and wide range of interests Brunelleschi must have been the renaissance engineer incarnate (see also Cooley's chapter in this volume). Leonardo da Vinci, as a matter of course, should be mentioned too, but his achievements (and his plagiarising of the ideas of the 12th Century) would fill a book... While the great architects had a high profile, there was little awareness of the future importance of the role of people like the eminent Gutenberg, 'the prototype of the entrepreneurial and innovating engineer' (Noblet, 1984), who appeared on the scene as early as the 15th century.

The job title, 'engineer' is old. In the 2nd century AD the word 'ingenieurs' was used by Pliny the Younger (book 10, letter 40), although he referred to it as a vernacular expression. In the 11th and 12th centuries it appeared in the form 'engegneres' and referred to the designers of war machines. The first 'officially designated' engineer was English, 'Waldivius Ingeniator', King William I's Royal consulting engineer, advising on the state of the fortifications. The first managing engineer, 'Magister Ingeniator', was English too, a monk who was overseeing a group of what must have been the first 'technician engineers', each in charge of a specific aspect of a big building project. The French, however, were not far behind with the creation of the 'Corps du génie' in the late 13th and early 14th century (Isnard, 1984). For some time to come, the title of Engineer was to remain in the preserve of the military. Nevertheless, the knowledge required was clearly spelt out and included mathematics and geometry at an advanced level. An early definition could have been phrased thus:

> The engineer is a person who designs and operates war machines, a task for which he or she uses knowledge of mathematics and geometry, combining scientific methods with a natural curiosity and ingenuity.

Clearly, this is not the positive image of the engineer which today's professional engineering groups would like to foster although many of the great advances in engineering of today have been spawned by the demands of the 'defence' sector.

Modern Times and Recognition

In France, the meaning of the word 'ingénieur' started to change around the year 1600 (Vérin, 1984) with the establishment of the 'Corps des Ingénieurs des Ponts et Chaussées' such that Furetiére could say in 1727 'Se dit aussi par rapport Ö l'architecture civile, d'unàhomme, qui par les machines qu'il invente augmente les forces mouvantes, autant pour traéner et enlever les fardeaux, que pour conduire et élever les eaux'. Diderot (1772) could refer to a third (non-military) class of engineers who 'improved the great roads, built bridges, beautified streets and repaired the canals'. Diderot and Furetiére both had observed, at the beginning of the 18th century, the emergence of the criteria for judging the modern engineer which applied to France as well as England: timeliness, soundness and economic production (Vérin, 1984). In as far as its formal recognition was concerned, the transformation from the Royal Engineers, Ingénieur de l'Armement and the Ingénieur-Architecte to the civil(ian) engineer had to wait until the late 18th century, the time of the invention of the steam pumping engine in Britain and the establishment of the Corps des Mines in France, an elite body of, originally, 57 engineers charged with supervising the mining industry. Interestingly, in England the early professional engineers were described as 'mechanics' and were usually recruited from the ranks of millwrights. This was not meant as a derogatory term but was praise in that it signified that they understood the complexity of mechanical systems (the word mécanicien in French was even modelled on mathématicien, that is, a true scientist). Of course, schools of engineering existed only in a moral sense: most 'graduates' were practitioners who had learnt their trade from their employer, see Figure 44. Only the two Brunels were exceptions to the rule – but Brunel Senior was of French origin and had trained in France (Poitou, 1984).

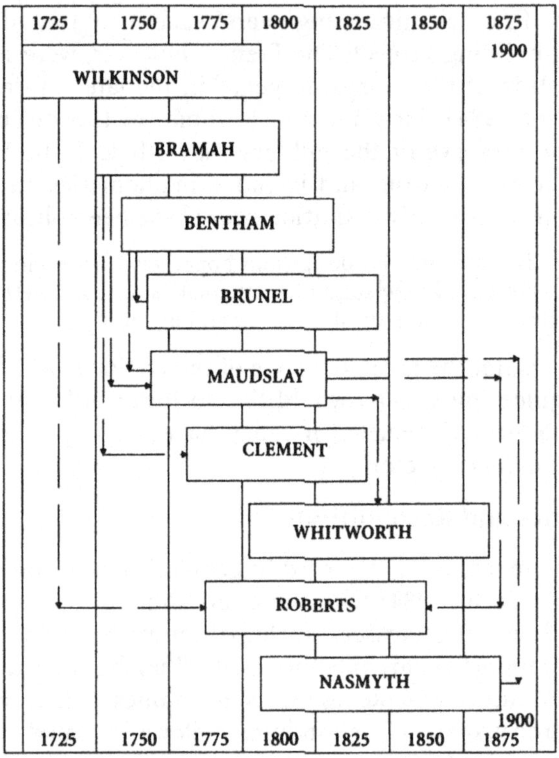

Figure 44: Who's who: the 'school' of the great English engineers

In France, the large-scale breakthrough for the profession of ingénieur came as late as the beginning of the 19th century when the first purpose-educated engineers shaped the second phase of the industrial revolution. This process was started by Napoleon with the creation of the école Polytechnique in 1794 which was allowed to recruit a small number of students destined for non-military careers (Noblet *et al.*, 1984).

In Britain the establishment of engineering as a codified profession with its own high-calibre training had to wait even longer, until the end of the 19th century (Guagnini, 1984). At first, until about 1880, professional engineers in England were trained at nightschool, e.g., the Manchester

Technical School (now UMIST), whereas scientists were educated at University – even though Owens College in Manchester (now Manchester Victoria University) had as its mission the support of local industry.

Factory owners and managers were at that time still suspicious of highly educated engineers and preferred to train well-qualified skilled people so that they could take on more complex tasks, such as designing, improving and maintaining machines, in particular the big steam engines whose failure would paralyse the whole mill. The attitude towards engineers in Britain is to this day influenced by this early mistrust of professionalism.

Outside Britain engineers were the heroes of the 19th and of the 20th centuries even though authors like Alain and Phillippe d'Iribarne (d'Iribarne, 1987), have detected great differences between the status and background of engineers in France, Germany, the USA and Japan. Names like Eifel, Edison, Riggenbach, Picard, Giesl, Siemens, Tupolev and Marconi spring to mind. This was true right up to the early 1970s when the prestige of engineers came to be questioned as a result of the more and more apparent collapse of the world's ecology, for which they were held to be partly responsible. A 'heroic' definition in this context would have been:

> The engineer is the person who can solve any technical problem, no matter what it is and no matter how difficult it might be.

English literature does not feature many engineer heroes, with the exception of hagiographies of the 19th century engineers and of authors like Neville Shute (1954), who related his own experience as a site engineer involved in airship development.

Since the beginning of the 20th century 'engineers' have thus been accepted as equal partners of the scientist and, subsequently, their role has expanded, to include areas such as cybernetics and ecotechnology. However, nothing changes and genetic engineering (carried out by biochemists and biologists) attracts some of the opprobrium which was attached to the word ingenium in medieval times.

A Modern 'European' Definition

For the purpose of this chapter the definition of the engineer will be crafted as a composite of the British Chartered Engineer, the French 'ingénieur diplômé' and the German 'Diplomingenieur'. One of the published definitions we find is 'An ingénieur is defined as an engineer who uses creative ability over a wide range of problem solving situations, particularly at a conceptual level' (Macleod, 1992(1)). This definition may well be applicable in some circumstances. In its generality though it does not help us distinguish the 'modern' engineer from the 'historic' engineer, especially so since it uses the word 'engineer' as part of the definition! It is, as a matter of fact, a carbon copy of the definition of the French engineer, as put forward by the d'Iribarnes (1987), a conceptual designer who sets out to make the most 'beautiful technical object', regardless of cost or function. Three terms might be of assistance in finding a more up-to-date definition, all originating from the quality debate of the late 1980s and going beyond the 'timeliness, soundness and economic production'. They are: predictability, reliability and repeatability. The difference between craft engineers who were, in essence, experimenting on the basis of experience, and what we might call professional engineers can be defined by the results of their activities:

> A modern engineering product, designed and created as the result of an engineering project or as the output of an industrial process must be predictable in terms of performance, time of delivery and cost. It must be reliable and safe. If it is one of many products to an identical specification then it also needs to be repeatable in terms of functionality, performance and appearance.

In order to achieve such an end-result the engineer must use a scientific approach. Since her or his work is normally undertaken in a commercial environment there is also a need for an economic perspective. A tentative formal definition for the purpose of this discussion could thus be:

> An engineer is a professional person who uses knowledge, conceptual and creative ability to solve, subject to time constraints, the problems involved in developing, designing, producing and maintaining goods or services which perform predictably and reliably in line with the specification agreed with a customer while working within clear financial limits.

The earliest engineers satisfying this definition, Watt, Stephenson and Whitworth (inventor of the standard screw thread), were practical people who could identify problems which were amenable to solution with the available technologies or further developments thereof. An example of this was Brunel's realisation that future marine transport needs could only be satisfied by the construction of steel-hulled ships. In the view of one of the most important British historians the engineers:

> ... were better paid than their fellow workmen, they were on the average more intelligent, and they took the lead in educational movements. They were in the forefront of progress and invention.
>
> *Trevelyan, 1942*

The Shaping of Learning Environments for Good Engineers

> Adult education received its first impetus from the Industrial Revolution in the desire of mechanics for general scientific knowledge ... From 1823 onwards mechanics' institutes, begun in Scotland by Dr. Birkbeck, spread through industrial England ... from 800 to 900 clean respectable looking mechanics paying most marked attention to a lecture on chemistry.
>
> *Ibid.*

> Technical education emerged from the middle of the 19th century as a revolt against the use of genesis without analysis.
>
> *Mcleod, 1992(1)*

Environments for the successful acquisition of engineering knowledge and skill, the formation of good engineers, can be created in many different ways. They range from the early engineering apprenticeships of Watt's times to computer based learning systems. However, depending on the aspect of the engineer's role which is to be developed, e.g., analytical ability or synthetic thought, one environment may prove more suitable than another. Each environment though requires its own range of methods for teaching and forming the aspiring engineer. Apprenticeships would use learning by imitation and correction, supported by much practice, while the most modern systems might use virtual reality to let the learner 'experience' and resolve problem situations.

A key aspect of designing learning environments has always been the question of reality, of providing a situation for the advanced learner which is as close to real life as possible while the actions of the learner must endanger neither people nor systems, directly or indirectly. There is usually a continuum between the two extremes of this dichotomy but the choice of solution is also strongly influenced by cost considerations.

The discussion here focuses on the academic route to the profession of engineer. However, it may be useful to illustrate some aspects of the wider debate on the best approach to the university education of engineers which has been current since the 1970s. In doing this we should be mindful of the words of Corfield in which he paints a very positive image of the rounded engineer:

> The treatment of engineering as a liberal profession and the claim that engineering offers, with the humanities, an excellent basis for living one's life as contrasted with earning one's living may fall less easily on the ear, lie less easily on the mind ...
>
> ... engineering as a way of life and whose values ... are contributing to the creation of wealth from which, in turn, the other values, so highly prized in older societies, are themselves derived ... let me, therefore, make quite clear that education is the key to future wealth creation and that the world belongs to those who understand this and act upon it.
>
> *Corfield, 1983*

Knowledge and Theories of Learning

> The process as well as the context of formal education should be designed to prepare students to live mature, effective, adult lives.
>
> *Alschuler, 1970*

Only through poor usage has the word 'teaching' come to mean those things teachers do while getting paid for being in a classroom (after

Pittenger and Gooding, 1971). This conventional image of teaching is not very old, dating back to the renaissance for children of the middle class and to the early 19th century for all children – the 'ragged schools' leading to universal instruction (Crystal, 1990). Rather than being a 'broadcaster', the traditional view, a teacher must be a designer of effective learning situations taking account of students cognitive abilities and developmental stages. Wherever possible teachers should be 'narrow casters', using the teaching and learning approach most suited to a particular situation and student. Tradition though can be very strong, especially so in the case of schooling, since the pervasiveness of the experience in our society favours a virtually unchanging transfer of dogma from one generation to the next, often in cycles of fashion. While educational reformers battle to introduce more appropriate forms of interaction, economic pressures continue to favour classroom-based learning and schooling, in effect, the broadcasting of knowledge. At the most basic level, university teachers of engineering could simply follow the EPC's approach which is described in detail in their publications (EPC 1, 3, 5, 6). Unfortunately though, there are no simple and conflict-free definitions in the field of teaching and learning, even the word 'knowledge' is used at different levels of the discussion in different ways, depending on the context. The discussion below starts first with this issue of knowledge from a philosophical point of view and then provides a framework of qualities, faculties and knowledge in which an engineer's, or indeed any human being's problem-solving ability can be analysed. This is followed by a section on learning theories and issues of practical implementation.

The Traditional Concept of Knowledge

Human activity is based on knowledge and its application to particular situations and objectives. Epistemologists have found many ways of describing and classifying knowledge and there are different opinions as to the way knowledge has developed throughout the history of mankind and as to what is its current state. In philosophy, there is the distinction between 'Wissen' and 'Erkenntnis' where the former can be translated as all-embracing knowledge while the latter could be described as rational knowledge (a dictionary translation does not exist). For the purpose of this chapter, 'cognis' is defined as circumscribing the world of rational knowledge and 'gnosis' the spiritual realm while 'knowledge' contains both. Education in general and universities in particular, can be said to be concerned predominantly with the transfer and appreciation of 'cognis' which, unlike the realm of 'gnosis', can be codified.

In the context of engineering Rosenbrock (1989(2)), distinguishes two domains of 'knowledge', on the one hand that of 'explicit knowledge' and on the other that of skill and 'tacit knowledge'. People in general, not just engineers, can function as professionals in their environments by accessing both. Rosenbrock postulates that, in future, both domains must continue to expand while remaining approximately in proportion to each other. Such a development is necessary to maintain a good quality of working life, according to Rosenbrock.

J. S. Brown (1989), uses the term 'implicit knowledge' for the domain of skill and tacit knowledge and distinguishes between two types of learning: *formal training* and *informal learning*. It may be pertinent to add here that J. S. Brown's distinction between learning and training is probably polemic rather than necessary. The writer will therefore be using the terms formal and informal learning. J. S. Brown asserts that explicit knowledge can be codified and transmitted in a highly structured manner, formal learning; while tacit knowledge and skill tend to be adopted as the result of practice and experience, informal learning. Although this clear separation of domains of knowledge and associated types of learning may be valid at a very elementary level, it does not take into account that 'explicit knowledge' can be acquired through informal learning, followed by a phase of reflection. Conversely, in the opinion of the author, 'implicit knowledge' can be formally learnt and then internalised through practice. Inferring from the contexts in which the authors discuss knowledge, it can be assumed that they deal exclusively with the realm of cognis although some of their approaches could be extended to the realm of gnosis, opening up interesting opportunities for a debate on the way in which gnosis is adopted. This debate could also include issues such as tangibility, for which there is no space here.

Other Philosophical Approaches to Knowledge

Beyond the classifications of knowledge offered above, there are many others, suffice it to mention, as an illustration, Macleod's, 1992(1): operational knowledge is required to carry out vocational tasks, broad knowledge goes beyond what is directly relevant to vocational tasks while deep knowledge concerns basic assumptions and fundamental principles that can only be used through understanding. Branton argues that:

> Much 'non-conceptual mental work' must be involved in skillful pursuit of purpose. Consciousness is something which some of our colleagues in that most recent branch, cognitive ergonomics, seem to by-pass. They concern themselves mainly with human functions that are verbalisable and computational, overlooking vast areas of non-linguistic, non-conceptual, mental activity ... while the problems of explaining such basic functions as selective attention or directed awareness still remain with us. The largest part of our knowledge is still 'tacit knowledge' ...
>
> quoted in *Oborne, 1993*

Bowers (1984) attacks the dominance of codified theories and abstract thought and argues for a teaching approach integrating common-sense experiences and tacit forms of knowledge (accepting that the teacher has acquired these in a specific cultural context). He argues that the oft demanded freedom from cultural values is bought at the price of a new dependency on technicisms. He deplores the exclusive recognition of measurable reality. Mapping Bowers' views on the taxonomy of capability the writer interprets his views as favouring complex skills over measurable skills and understanding over recall and know-how.

Development of Knowledge and Ways of Learning

While it is generally accepted that the cognis (or rational knowledge) available to societies and individuals has expanded, even exploded, throughout the development of humankind and its history (Bowers, 1984; Rosenbrock; 1990, p. ix) the distribution between the explicit and implicit domains has varied, as have the opportunities for formal and informal learning. In Figure 45 the author presents an attempt to amalgamate J. S. Brown's and Rosenbrock's findings into a single diagram which indicates qualitatively how the relationship between the realms and domains of knowledge and the approaches to learning have changed in historic times, that is, the period for which written records exist. The horizontal and vertical axes respectively refer to learning opportunities and types of knowledge while the oblique axis refers to time and 'progress'.

The following should be kept in mind with respect to the content and purpose of the diagram: implicit rational knowledge is shown as part of 'cognis' even though it could equally well be shown as part of gnosis – the difference is in the application, that is, rational knowledge tends to be used for practical purposes rather than philosophical or academic discussion. The relative sizes of the rectangles of 'cognis' and 'gnosis' indicate the significance of the two domains over the whole of a society at different times in history. The gain of 'cognis' indicates the increasing respect for rational thinking. The position of the rectangle of cognis with respect to the diagonal (and the axes) is used to represent theoretically possible developmental patterns in a given social context in terms of learning and knowledge. Learning and knowledge patterns of 'typical' individuals are shown as feint dotted lines. The patterns for individuals viewed as eccentric or avant-garde would lie partly outside the 'standard' pattern for a period. The information presented is purely qualitative in nature and must not be viewed as anything but approximate.

The information in the diagram should not be over-interpreted. It is applicable only to Western European society and is simply intended to show how the learning and knowledge patterns both for whole societies and for individuals within a society may change over time. It cannot show in detail the dynamic processes involved in learning, although an attempt to do this is inherent in the way in which the knowledge and learning patterns of individuals have been drawn: the individual pattern given for the late 20th century would be that of a conventionally educated engineer with little practical background, having formally learnt most of the very substantial 'explicit knowledge'. Informally learnt 'explicit knowledge' has been rationalised (arrow 1), while formally learnt engineering practice

(EP1) elements have been internalised (arrow 2), that is, they have become 'implicit knowledge'. The pattern for the 18th century could be that of Watt, an apprenticed engineer who managed to make 'implicit knowledge' explicit (arrow 3) but still used much informal learning of skills.

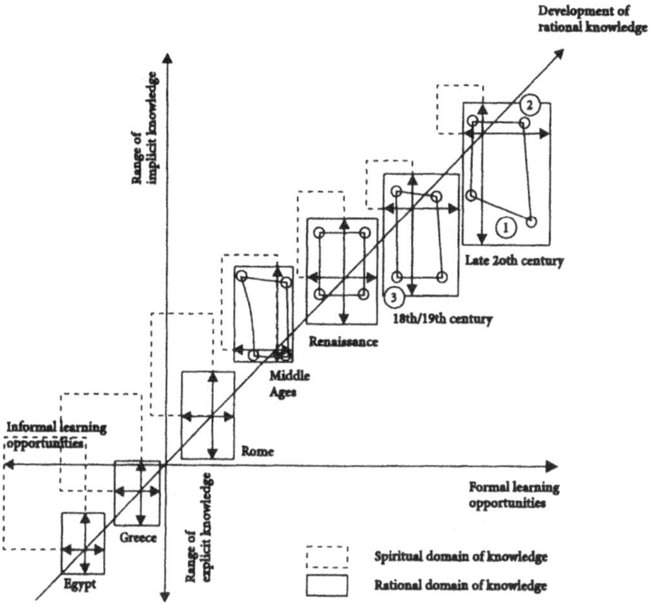

Figure 45: Development of knowledge and learning through history

The pattern shown in Figure 45 for the Renaissance period portrays a 'balanced' person using the available knowledge and learning styles to best effect. The range of 'explicit knowledge' is no longer as limited as in the Middle Ages but informally learnt 'implicit knowledge' still predominates. This Renaissance person is, of course, a theoretical construct but she conforms nicely to the images of future engineers painted by Cooley and J. S. Brown in their concept of a 'cognitive apprenticeship'. In this the term 'apprenticeship' stands for affective or informal learning of both explicit and implicit 'knowledge' while they apply the qualifier 'cognitive' for a formalised context for this learning. Cooley (1993), refers to the result as a state of 'wisdom'.

It may well be a reflection of the writer's optimistic view of the world of engineering that he feels that there are already many engineers who come close to this ideal image, perhaps not so much as the result of intelligent course design but because of their background, curiosity and personal motivation. Beyond the teaching though, a university period should also have afforded opportunities for working in an open and equal manner with experienced engineers and academics, the tutors they encountered at the university.

Bloom *et al.*'s Taxonomy of Learning

A group of American educationalists published in 1956 a 'Taxonomy of Educational Objectives' with the aim of helping teachers better to understand learning processes and thus to improve course design (Bloom, 1956). This was not intended to be a Dewey-style classification based on an analysis of terminology but a taxonomy aimed at identifying the inner logic of learning situations. Learning objectives were defined in behavioural terms, that is, they were used to circumscribe the intended scope of the modification of the learners' behaviour. Bloom *et al.* divided the set of potential learning objectives into the cognitive, affective and psychomotor domains of learning. They did not attempt to analyse the psychomotor domain, concerned with manipulative and motor skills, since it is relevant to apprenticeships rather than to classroom teaching. In all domains complex behaviour is built on simpler behaviour and the awareness of the individual increases with the complexity of the behaviour. Bloom *et al.* contrast simple objectives, such as rote recall or remembering of knowledge, with complex objectives, such as understanding, insight, really knowing and true knowledge. In practice, a complete separation of the cognitive and affective aspects of learning is impossible. Indeed, where appropriate affective behaviours, exist the cognitive learning of subject matter presents few problems (cf. Weiner in Dienstbier, 1991), this is particular true of situations which allow for peer reinforcement.

Krathwohl, Bloom *et al.*, in the second volume of the taxonomy published in 1964, write: 'we find that learning experiences which are highly organised and interrelated may produce major changes in behaviour related to complex objectives in both the cognitive and affective domain.' An important element of success is, in their view, isolation of the learner from the 'normal environment'.

Cognitive Domain of Learning

'The cognitive domain... includes those objectives which deal with the recall or recognition of knowledge and the development of intellectual abilities and skills.' (Bloom, 1956). These objectives thus include the behaviours of remembering, reasoning, problem solving, concept formation and (limited) creative thinking. The success of cognitive learning is determined by the situation, the interest and the learner's liking for the subject – clear pointers to the significance of affective factors. Bloom *et al.* include knowledge and intellectual abilities and skills into their taxonomy of cognitive objectives:

1. *Knowledge*: the recall of specifics and universals, methods and processes, patterns, structures and settings. The knowledge objectives emphasise the psychological processes of remembering.
2. *Comprehension*: the lowest level of understanding, that is, grasping what is being communicated.

3. *Application*: the use of abstraction in particular and concrete situations.

4. *Analysis*: the breakdown of a communication into its parts such that the relative hierarchy of ideas and their interrelationships are made explicit.

5. *Synthesis*: the process of working with pieces, parts and elements to create a pattern or structure not clearly there before.

6. *Evaluation*: judgements about the value of material and methods based on internal evidence and external criteria.

Each set of objectives is divided into a number of sub-goals whose exact descriptions can be found in Bloom (1956). The objectives should not be considered as independent of each other but as part of a continuum.

Affective Domain of Learning

The affective domain of learning 'includes objectives which describe changes in interest, attitudes, and values, and the development of appreciation and adequate adjustment.' (Bloom, 1956) The development of the cognitive domain is, unfortunately, not a sufficient condition for the achievement of objectives in the affective domain even though Krathwohl and Bloom acknowledge the possibility of using cognitive objectives for facilitating affective goals. The continuum of goals in the objective domain consists of:

1. *Receiving or Attending*: the learner must become sensitised to certain phenomena and stimuli.

2. *Responding*: the learner is not just willing to attend but is actively attending, displaying 'interest' and gaining satisfaction.

3. *Valuing*: the learners show through their behaviour that they attach 'worth' to the thing or phenomenon being addressed and that they are committed.

4. *Organisation*: the learner encounters situations for which more than one value is relevant and has to build up a value system.

5. *Characterisation by a value or value complex*: the individual acts consistently in accordance with the values internalised through having attained lower level goals.

The highest achievement in the affective domain is the development of an individual Weltanschauung or world view, a value system having as its object the whole of what is known or knowable. It is possible to establish relations between the categories of the two domains.

Table 12: Affective continuum

Classification methods >>			Affective Components (Krathwohl's Range)					EPC and Schmid Categories							
								Qualities		Faculties				Content	
Domain	Category	Sub-category	Receiving	Adjustment	Values & attitudes	Appreciation	Interest	Aptitudes	Attitudes	Recall	Skill	Knowledge	Understanding	Atoms 2 and 1	Relations & links
Psychomotor															
Cognitive	1 Knowledge							x		x				x	
	2 Comprehension		x					x		x				x	
	3 Application		x					x			x				x
	4 Analysis		x								x	x			x
	5 Synthesis		x									x			x
	6 Evaluation								x			x			x
Affective	1 Receiving	1.1 Awareness					x		x			x	x	x	
		1.2 Willingness to receive					x		x				x	x	
		1.3 Controlled or selected attention				x	x		x				x		
	2 Responding	2.1 Acquiescence in responding			x	x	x		x				x		x
		2.2 Willingness to respond		x	x	x	x		x				x		x
		2.3 Satisfaction in response		x	x	x	x		x				x		x
	3 Valuing	3.1 Acceptance of a value		x	x	x	x		x				x		x
		3.2 Preference for a value		x	x	x	x		x				x		x
		3.3 Commitment to a value		x	x				x				x		
	4 Organisation	4.1 Conceptualisation of a value		x	x								x		
		4.2 Organisation of a value system		x	x								x		
	5 Characterisation with a value complex	5.1 Generalised set		x									x		
		5.2 Characterisation											x		

In such an analysis Knowledge corresponds to Receiving and Attending, Comprehension to Responding, Application to Valuing, Analysis and Synthesis to Conceptualisation and Evaluation to Organisation. Krathwohl *et al.* give the following example for Receiving: 'listens to music with some discrimination as to its mood and meaning and with some recognition of the contributions of various musical elements and instruments to the total effect.' The corresponding objectives in the cognitive domain are Comprehension, Application and Analysis.

Table 12, based on Krathwohl (1964), but adapted and expanded significantly by the writer, is included here to demonstrate the accuracy of the taxonomy when it is compared with the everyday usage of terms which refer to behaviour forming part of the affective continuum. The writer has mapped the terms concerned with the affective domain onto the cognitive domain, as suggested by Krathwohl *et al.* The resulting diagram is also used to illustrate the relationship between the writer's categorisation of capabilities and Bloom's as well as Krathwohl's analysis.

Nearly all cognitive objectives have an affective component. The interest objective, for example, can be the affective component of all or most cognitive objectives of a learning institution. As Krathwohl *et al.* (1964) observed though, many course managers start out with objectives rooted in the affective domain of learning but are content for these to move into the cognitive domain in order to facilitate testing. They describe this phenomenon as 'shift of intent'. For students' development to remain holistic, embracing both the affective and cognitive aspects, this shift must be resisted.

A Taxonomy of Capability

In the introduction to their taxonomy of learning, Bloom *et al.* express the fear 'that the availability of the taxonomy might tend to abort the thinking and planning of teachers with regard to curriculum, particularly if teachers merely selected what they believed to be desirable objectives.' This caution is very germane to the following section since many of the recommendations made by the EPC and by the author can be viewed as substantially prescriptive and may therefore be taken on board unthinkingly. Bloom *et al.* also state that 'each theory of learning accounts for some phenomena very well but is less adequate in accounting for others'.

Critique of the EPC Taxonomy

'Tools' selected for teaching purposes should always be assessed in terms of their contribution to the aims of the education process. In engineering (and elsewhere) the development of a professional capability is probably the most important aim.

When the writer embarked on his background studies to provide a theoretical framework for the assembly of learning methods forming the CIM approach (Schmid, *op. cit.*), research very quickly led him to the Engineering Professors' Conference (EPC, expanded from 1 January 1994 into the Engineering Professors' Council) and its working groups. Over a period of five years the members of this body carried out an in-depth investigation of best practice in engineering education, putting forward a number of new approaches to curriculum design. Their conclusions and recommendations must be rated very highly in an international comparison in that they were prepared to start almost with a fresh sheet of paper. At first the writer was hopeful that he would be able to adapt the results of the EPC's work on teaching methods and their classification without any modification. However, he soon found that he disagreed in several important respects with the findings of the sub-committees of the EPC. Even though these disagreements are mostly concerned with terminology they lead to a different structure and some new definitions.

In their publications the working groups of the EPC addressed the issues of content, quality and assessment in first degree engineering courses with a view to improving the capability of future engineers. They undertook an analysis of the current situation and then embarked on a very thorough process of definition of the terminology used in teaching and learning and the approaches advocated by experts in the field. In their first contribution on the issues they wrote (EPC 1, 1989):

> ... distinguishing between (i) skills (either manual or intellectual) which are learnt mainly through practice; (ii) knowledge, which is simply information committed to memory, and (iii) the deeper learning variously described as understanding, conceptual learning, or meaningful learning.

Based on the very valuable efforts of the EPC the writer has developed his own taxonomy of learning which, he hopes, is slightly more robust and rigorous than their approach. As shown below he defines three categories within the domain of capability and learning: qualities, faculties and knowledge. All of these can pertain to individuals, teams and groups (cf. the concepts of the learning organisation and learning society). The taxonomy differs most markedly in its replacement of the term 'knowledge' by the term 'recall', which he uses to describe a memory-based faculty and in using the word 'knowledge', for the content of learning, instead. This is, in fact, in line with one of the dictionary definitions of knowledge: 'the sum of what is known'. The writer uses the term 'capability' in a meaning which is even wider than that postulated by the EPC Working Group on Quality in Engineering Education (EPC 1, 1989; EPC 6, 1993), in that he also includes in his requirements personal qualities, in addition to faculties and knowledge (or content).

Components of Capability

Engineers can be described as capable with respect to the execution of their duties if they have adopted and acquired the qualities and faculties needed to gain and use knowledge and if they possess the knowledge requisite for solving engineering problems, management tasks and for attending to related duties. The writer is of the opinion that there is a relevant difference between 'acquire' and 'adopt' which are both used, in general parlance quite indiscriminately, to describe the behaviour of the student during learning. He applies the term 'adopt' to more active learning in formal situations while he uses 'acquire' for relatively passive learning in substantially informal situations. These usages are in line with the dictionary definitions.

Knowledge (Content)

As discussed in the section *The Traditional Concept of Knowledge* (starting on page 384), in relative detail at a philosophical level, cognis, or rational knowledge, exists in two forms, explicit and implicit. Here these two terms are used to structure 'classical' descriptions or terminologies of knowledge (see the section *The Shaping of Learning Environments for Good Engineers* – starting on page 382). The term 'content' will at times be used interchangeably with cognis when discussing qualities and faculties. Table 13 provides an overview of the definitions and attributions. Links indicated as '(x)' are perceived as weaker than others.

An example of 'data' as a basis for inference is 'knowledge' about the pin allocations of different RS232 connectors. It is only useful in conjunction with facts. Facts are here taken to be meaningful associations of data while information is defined as the result of processing data and facts, possibly attaching a 'quality rating'. An example of 'schemata' would be the learnt moves a football player can access in milliseconds.

Table 13: The definitions and attributions of knowledge

Terminology of Knowledge	*explicit*	*implicit*
atoms of knowledge—		
data	x	(x)
fact, formulae	x	x
information	x	x
fiction	x	x
sources[a]	x	
ideas	(x)	
relations and links -		
schemata	x	x
rules	x	x
principles	x	(x)
concepts		x
intuitions		x

Note: (a) the awareness of resources which can help overcome a lack of memorised information

Knowledge is transmitted through teaching and training, it is adapted to a person's needs through reflection and unconscious modification and retained either directly or as a consequence of practice. Together, knowledge and the faculty of recall help pass exams, but not much more. In professional contexts, knowledge is only of use in combination with faculties and qualities which permit its application to real problems. Students and other learners must therefore not only be guided in the acquisition of knowledge but must also be helped to enhance their faculties and to develop their set of personal qualities.

Content is adopted and acquired by learning activity in both the cognitive and affective domains, the latter being dominated by acquisition.

Qualities

For the purpose of this chapter human qualities are classified, in a relatively arbitrary manner, as follows:

- *aptitudes*: predispositions facilitating the acquisition of manual and mental faculties, of the natural or acquired kind; e.g. dexterity with numbers;
- *attitudes*: settled behaviour; e.g. confidence, enthusiasm, motivation (EPC 1), ethical behaviour, reliability (Macleod, 1992(2)), openness for new ideas;
- *values*: principles, standards; e.g. truthfulness (Macleod, *ibid.*), honesty, fairness;
- *emotions*: instinctive feelings; e.g. likes and dislikes.

Qualities (sometimes quoted as natural, innate, intrinsic) cannot be taught formally, they are acquired or modified through processes in the affective domain of learning. However, this must not be used to abrogate teachers' responsibility for developing qualities.

Faculties

In the following enumeration no differentiation is attempted between so-called 'manual' and 'mental' faculties. Both require the intervention of intellect in their acquisition and application.

- *recall*: ability of memorising, storing and retrieving data and information accurately; e.g. knowing the pin-out of an RS232 interface connector. Recall can be directly useful in passing exams but is only a prerequisite for success in acquiring and using other faculties;
- *skill*: ability to do specific things without necessarily being able to understand the processes by which one does them - such as speaking, writing, designing, playing tennis etc. Skills can be further subdivided into manual and mental skills and both of these again into measurable and complex skills. The correct execution of skill dependent actions is controlled mainly by automatic processors, on the basis of unconscious schemata;
- *know-how*: more general acquired and adopted by accumulating experience of successful problem-solving in a well-defined field, which may require substantial recall effort and the use of a number of skills; e.g. the activities involved in repairing a television set. It can be

adopted through the classic apprenticeship. The correct execution of know-how based task activity is controlled mainly by stored rules;

- *understanding*: intellect, reasoning power. The ability of grasping concepts and *using* them creatively, the basis of (logical) thinking. Essential to cope with novel situations requiring planning and conscious analytical processes, such as designing new artefacts or explaining not yet encountered phenomena. The power to work with intuitions;
- *Analysis and Synthesis*: two key elements of the engineering approach, require a mix of all the faculties listed, although analysis is perhaps biased more towards the first three while synthesis relies more on skill, know-how and understanding.

Table 14 is included here to show a possible set of relationships between knowledge, qualities and faculties. It is intended to illustrate the complexity of the issues and their classification rather than to present a definitive picture of the component parts of the different faculties. The structure is evolutive in that it moves from the simple to the complex (Bloom, 1956).

Faculties can be taught using different methods belonging to processes in both the cognitive and the affective domains of learning (see the section *Affective Domain of Learning* – starting on page 389), depending on their complexity. The components of 'qualities' and 'faculties' are, at times, taken together under the joint heading of 'powers' by the writer.

Table 14: Relationship between knowledge, qualities and faculties

faculties → *are functions of ↓*	*recall*	*skill*	*know-how*	*understanding*
aptitudes	x	x	x	
attitudes	x	x	x	x
values		x	x	x
emotions				x
data		x		
fact, formulae		x	x	x
information		x	x	x
fiction		x	x	x
sources		x	x	x
ideas				
schemata		x	x	x
rules		x	x	x
principals			x	x
concepts			x	x
intuitions				x

Analytical and Synthetic Faculties Versus Knowledge Acquisition

Traditional engineering education stresses the importance of an in-depth knowledge of the science and technology base of engineering. Students can adopt factual and procedural knowledge in lectures and tutorials, by recording information and applying paradigmata to examples. Only very rarely though will this force students to question assumptions and preconceptions. They tend to learn factual content and methods, for the reason given above, an approach which can appear, on the surface, very successful, especially so since it is resource efficient. It pre-supposes, however, that on graduation engineers will not immediately take up positions of responsibility since most first have to develop a 'feel' and understanding for engineering.

The acquisition of both analytical and synthetic faculties is essential for students who are expected to undertake placements in industry at an early stage of their careers as, for example, on a thin sandwich-course (where academic studies alternate with industrial experience). The ability to analyse a problem and to develop reasonable hypotheses about its essence and genesis cannot be taught in an abstract manner, students have to experience the underlying gnostic processes. While such an experience may be provided in carefully designed, game-based environments, these would not reflect the reality of engineering where uncertainty is high both as regards the process of problem solving and the kind of solution which may be achieved. The intellectual development of the student should never be measured in terms of the correct application of well rehearsed methods but in her or his ability to try a number of different approaches and to assess their relative performances.

Teaching and learning is thus no longer a simple question of 'talk and chalk' but rather a complex process of formation and long-term motivation. The issue of learning is therefore addressed in more detail in the following sections.

Cognition and Problem Solving

Careful preparation of individuals for their future tasks is essential since:

> The capacity of the human mind for formulating and solving complex problems is very small compared with the size of the problems whose solution is required for objectively rational behaviour in the real world – or even for a reasonable approximation of such objective rationality.
>
> *Simon, 1957*

The formation of engineers must have as one of its stated objectives the education of people able to optimise solutions within the available resources, that is, to cope with problems subjected to a very large number of constraints. As discussed earlier, successful interaction with the environment, be this at work or in any other living situation, requires both analytic and synthetic mental activity to arrive at the correct actions or reactions. The term 'cognition' as used in this dissertation subsumes both reasoned and intuitive approaches in human behaviour, that is, activities which satisfy both cognitive and affective objectives. As used by the author cognition refers to decisions and responses based on rational knowledge (cognis), but not emotive and wilful actions which run counter to conventional logic. Such behaviour is not considered here because it does not fall within the scope of university teaching.

Capability and Problem Solving

James Reason provides an excellent overview of human problem solving in *Human Error* (1990). The following comments are therefore largely a summary of his analysis, adapted to fall in-line with the definitions used by the writer. Reason defines a problem as:

> a situation that requires a revision of the currently instantiated programme of action. The schematic mode of control can only operate satisfactorily when the current state of the world conforms to the regularities of the past. The departures from routine demanded by these situations can range from relatively minor contingencies, swiftly dealt with by pre-established corrective procedures, to entirely novel circumstances, requiring new plans and strategies to be derived from first principles.

Human beings will try to reduce cognitive strain in problem solving and concept attaining by applying pre-packaged solutions to recurring problems, that is, a form of persistence forecasting. Although perhaps not the most efficient or elegant strategy, in many cases problem solving on the basis of cues that have proved useful in the past achieves better results with far less effort than most other forms of problem solving. This human propensity to use 'shortcuts' has been identified by Rasmussen and Jensen (1974) and their visualisation is reproduced here in Figure 46. The classic step ladder though has been adapted by the author to reflect the faculties described in the section *Faculties* (page 394) and is shown to require a foundation of experience stored in the shape of schemata, rules, concepts etc. Rouse (1981), states 'human beings, if given the choice, would prefer to act as context specific pattern recognisers rather than to calculate or optimise' and defines two types of rules, symptom rules and topographic rules. The former work on the basis of a full match between problem cues and situation specific stored rules and will only be useful if the problem situation is accurately mapped. The latter rules rely on an actual or mental map of the system and a logical consideration of structural and functional relationships. For any rule to be selected it must satisfy four criteria to some degree:

1. the rule must be recallable (available);
2. it must be applicable to the current situation;
3. it must have some expected utility;
4. the rule must be simple.

The rules, also labelled 'production systems' by Rouse, suggest a link between a stored pattern of information relating to a given problem situation and a set of motor programs appropriate for corrective actions. Loosely interpreting, it could be said that the patterns correspond to the schemata formed about past experiences mentioned above, under the heading of skill. However, Rouse accepts that the human being will adapt the problem-solving strategy to the situation, similar to Reason's view. Rouse's comments also go some way towards providing a basis for the discussion of the differences between serialist and holist learners.

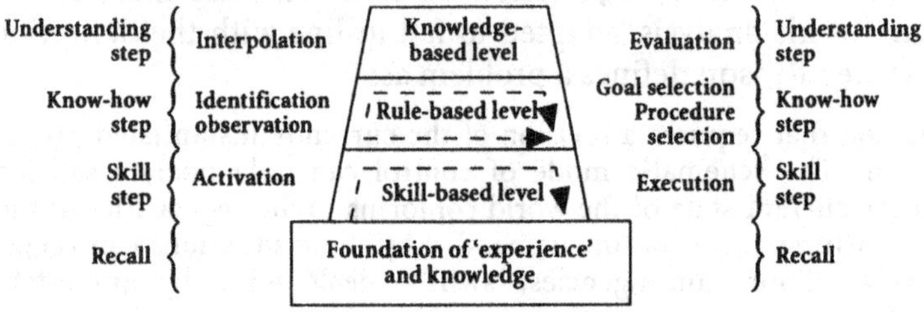

Figure 46: Rasmussen's Decision-Making Stages

In contrast to Reason's beliefs, Rouse's position is fairly close to behavioural psychology which concentrates on explaining problem-solving activity purely in terms of situation-response patterns. This world view has influenced the development of many of current approaches to training. This rigid mechanistic view is not universally shared though: Branton (in Oborne, *op. cit.*), is quoted as stating 'that past experience is used to predict and anticipate. It should, however, not be inferred that anticipation means automatic, cue-triggered (stimulus response) behaviour.' From his work with railway drivers he concludes that 'If correct anticipatory behaviour, once acquired, were rigidly automatic, such incidents (passing of signals at danger) could not occur.' To control a system (of any kind) efficiently, people need ways of predicting the outcome of their actions. Another important observation of Branton's refers to interest and boredom: 'what is known about ... modern man in particular: there is an irrepressible, spontaneous need for sensory inputs.'

So far, the discussion has addressed learning as a general issue rather than an activity taking place in an institution, e.g., a school or university. Since the discussion here is concerned with the professional education of engineers, the characteristics of institutional learning must be studied in some detail.

Learning in Institutional Contexts

At a practical rather than theoretical level, perhaps best described as phenomenological, we find three methods for classifying human learning which may help in the design of learning situations. In contrast to the preceding discussion there is no overlap between these classifications, in fact, they can be used as the three axes of a coordinate system to describe differences in learning situations. The system as shown in Table 15 represents learning based on the dimensions of students' natural propensities. The first axis, depth of learning, deals with the quality of the process, the second axis with the timing or sequence of the process while the third axis concerns the mode of instruction. The teaching tools which may be used have been selected more or less at random as examples to indicate how the classification could be applied.

Table 15: Classification of learning situations

Quality and Sequence →	Surface Learning		Deep Learning	
Medium ↓	Serialist	Holistic	Serialist	Holist
Verbalise	Lecture	Taught example session or class	School class	Non-tutored learning
Visualise	Video and diagrams	Descriptive textbook	Programmed learning	Analytical textbook
Do	Standard laboratory session	'Tasks' type activity	Open-ended laboratory work	Open-ended project work

Surface Learning and Deep Learning

Marton and Säljö (1976), identified different levels of processing of information among Swedish university students. Their research was directed at what was being learnt, rather than at how much. The authors classified the different conceptions of the content of learning tasks into a number of categories or, in their words, according to levels of outcome. They discovered that the associated learning activities could be described as surface-level processing and deep-level processing (or conceptual learning). The deep approach is described as being based on the intention to understand, to question, to compare statements with experience etc. (to grasp the intentional content of the learning material), while the surface approach reflects an inclination to memorise without question, to practise skills as instructed, not to look for meanings etc (to learn the text or 'sign' itself). However, this classification of the learning processes and outcomes must not be viewed as binary - both belong to continua. Deep learning goes beyond the transmission of factual knowledge but helps students acquire relations and links. A most important discovery made was that some students may never be able to adopt the deep approach in some subjects or even in all subjects. From his own experience the writer would be happy to concur with this view since he often had to advise students on particular strategies for the choice of final-year options! Marton and Säljö (*ibid.*) also discovered that retention of learning was good; but that changes in outcome, when testing was repeated after an interval, tended to favour less-deep levels. Some students would adapt their learning strategy to the type of testing which they expected: it may be useful to let students know in advance that an exam will not simply test the amount of content learnt but also understanding.

The surface approach is adequate for many elements of a standard university engineering curriculum where a good memory and well practised skills are sufficient to pass exams with flying colours. This type of learning behaviour should not be rejected though: deep learning cannot take place without the prior learning of some skills and the acquisition of knowledge. This is a precondition which had been uppermost in the minds of the CIM course's designers at the time of its conception. It was decided at the outset that an integrating course could only be offered in the fourth year, after three academic periods, where students accumulated skills and knowledge and three industrial periods, where they were able to garner experience and develop their personal and interpersonal skills. Through several elements of the programmes in years one to three students would already have been able to practise conceptual learning.

Holist and Serialist Learners

Beyond the distinction between deep- and surface-learning there is also some benefit in differentiating between holist and serialist learners (Pask, 1976). The former type of learner gains a broad overview before filling in the detail, while the latter need to follow a well structured and logical path. Pask stresses the importance of understanding in learning a topic. He defines understanding as learning for which the learner has been given an explanation and derivation. Holist students who have many goals and working topics often entertain correct beliefs (adopted content and views) about topics outside their direct area of work. A serialist learner only moves to a new topic when she or he is completely certain about the one currently being studied. Students are often unable to change a learning strategy once adopted, even if it has proved to be unsuccessful. According to Pask, teachers should therefore try to provide a range of learning opportunities adapted to the different learning styles although he also states that all students should be introduced to holist strategies. He warns of the institutional bias towards serial-learning situations and serial recall for testing. Lectures and traditional distance-learning materials tend to favour serial learners whereas well-designed laboratory work, student-centred tutorials, discussion oriented use of video support and nonconventional learning situations are more appropriate for holist learners. Pask also makes a distinction between comprehension learning and operation learning and stresses that both are necessary for creating understanding.

Verbalise, Visualise, Do

A similar dimension can be established for the choice of the mode and media of instruction most suited to individual students's needs, depending on whether they are verbalisers, visualisers or doers (EPC 1, 1989). Verbalisers learn more easily if information is provided in written or spoken form whereas visualisers respond better to diagrams and pictures. Doers prefer learning from practical activities regardless of whether they are dealing with information or concepts. The type of material and its structure are of particular significance in designing distance learning materials - which must serve a 'serialist doer' equally as well as a 'holist verbaliser'! An ideal course would, however, not offer the option of using either deep- or surface-learning, it would encourage students to develop both aptitudes. The design of the situation must be tied closely to the component of capability being taught.

Insightful Learning

Insightful learning is a particular form of deep-learning which is discussed by Hergenhahn (1988). He lists the following characteristics of insightful learning: the transition from the pre-solution to the solution is sudden and complete; performance based on a solution gained by insight is usually smooth and error-free; a solution to a problem gained by insight is usually retained for a considerable length of time; a principle gained by insight is easily applied to other problems. The approach has clear links to the Gestalt theory of learning.

The Need for a Rich Learning-Environment

A rich learning-environment offers scope for both deep- and surface-learning and gives serialist as well as holist learners the opportunities they need. Rich learning-environments are also advocated by Hebb (from Hergenhahn, 1988), who found that organisms reared in an enriched environment learn faster than those reared in relative isolation. Theories and background relating to rich learning-environments are described in some detail in the first of the bulletins of the Engineering Professors' Council on educational issues in engineering (EPC 1, 1989). Although Table 15 appears to indicate that traditional lectures fulfil a very minor role, it is important to realise that they often play an important role in kindling students' interest when presented in an enthusiastic and clear manner. A fairly standard HCCIM lecture, for example, can move about half the students present to explore the subject by themselves and can lead to the formation of student networks which continue after graduation. Staff in the Department of M&ES at Brunel University (Schmid, *op. cit.*), have always been aware of this need for rich learning-environments which help maintain students' interest. The original concept of SEP (the Special Engineering Programme) even has its own word for the course elements which provide opportunities for open-ended work: 'Technik' circumscribes the elements of the programme which introduce the main (student) action components: action provides for the reinforcement of learning and the development of important engineering skills. The term Technik was borrowed from the German language where it can express the application of methods as well as a technical content: on SEP it stands for an ethos in which the design, innovation and problem-solving aspects of engineering are emphasised (after Clark, 1985(1)). Even the most recent discussions concerning the learning of mature people stress learning which arises from:

> being thrown into a problem from which you can extract concepts and theory, rather than by first teaching concepts and theory and trying to apply it to the problem
>
> *Abell, 1994*

The creation of a rich learning-environment on its own, however, is not a guarantee for successful learning. Abell (*ibid.*), stresses this point in a discussion of case-based teaching: learning should be an iterative process between cases (the study of practical situations) and concepts. Not only is the structure of the learning of great importance but the learning processes which are engendered must be relevant and supported by tutors and mentors. One approach used successfully elsewhere is that of experiential learning, first proposed by Kolb *et al.* (1974). Practice-based learning (PBL), is a slightly modified, but still recognisably 'Kolbian' form of instruction.

Designing Contexts for Engineering Education

The education process is aimed at improving students' performance in both simple and complex tasks through learning. It is an important requirement that this successful performance must not be restricted to tasks occurring at the university but must be transferable to later professional activity. As a general observation, good performance is related to achieving a satisfactory standard in two domains of learning, the cognitive domain and the affective domain (EPC, 1992). Depending on the educational or industrial situation where the results of learning are to be applied, one or other domain will be more relevant. In general, the capability for learning will be domain specific and learners may therefore require guidance about their strengths and weaknesses so that they can choose a career path which optimises the use of their potential or of their intellectual and emotional resources.

Engineers must function well at all levels of Rasmussen's 'step-ladder' (cf: Figure 46 above). In particular, they must be able to decide, subconsciously or otherwise, when they may take a short-cut. Their education should therefore be designed to impart all the components identified above, as well as the experience to handle the processes involved.

It is important to ensure that the 'schemata', and to some extent the 'rules', being imprinted correspond as closely as possible to the present and future 'real' environment in which the engineer will be working. These tools should be taught or supplied in a structured format which allows absorption in a manner which will later not result in situation-inappropriate recall. The facts and information presented should be of relevance to allow the students easy and motivating access.

Shaping Teaching and Learning

Brandt (1984), as a scientist and academic teacher who bases his argumentation on research conducted at the Centre for University Didactics (HDZ) of the Technical University of Aachen, expects universities to equip students with:

1. domain-specific factual knowledge;
2. problem-solving strategies;
3. a. the ability to work in an independent and self-critical fashion on scientific questions;

   b. the curiosity and the courage to question scientific methods and authorities;

   c. the desire to communicate their experiences and achievements to others – including the general population;

   d. abilities such that they feel happy to cooperate with many different groups; and,

   e. attitudes which allow them to transcend received standards and wisdom.

While objectives (1) and (2) can be achieved more or less wholly in the cognitive domain of learning, the aims listed under (3) require a combination of the cognitive abilities of understanding, analysis, synthesis and judgement and the affective resolve to operate within one's own value system. Brandt's list thus has more than a passing resemblance to some of Rogers' principles – listed on page 406. If the universities are to satisfy the demands of students and society, they must go beyond the realms of cognitive teaching and learning processes. Brandt quotes a number of studies dating from the mid-1970s to support his thesis that university teachers must adopt different teaching methods, particularly those which support experience-based or experiential learning. He feels that the university teachers must themselves develop an understanding of the complexity of the situation.

Branton (quoted in Oborne, 1993), forces teachers to face three important responsibilities:

1. we must not rely on 'programming' students, we must work at all levels: affective, cognitive and subconscious;

2. in setting up learning situations we must ensure that students acquire the skills needed to predict outcomes based on cues which may be fuzzy or conflicting;
3. the working environment must be stimulating and provide a wide range of sensory inputs.

For satisfactory performance in their chosen field of activity the future engineering graduates must acquire not only a level of competence appropriate to engineers setting out on a professional career but also the ability to build on this foundation, a claim which is echoed in the EPC's demand for 'preparation for continuing education and training (CET)', (EPC 3, 1991).

Problem-solving at all levels, from conceiving a product through to handling the environmental issues of its production, operation and ultimate disposal, is the most relevant dimension of engineering work. Rasmussen's view that most human activity is related to problem-solving of some kind is therefore highly pertinent. Most theories of learning can claim to handle a particular aspect of the process of formation better than others, however, in reality only a combination of approaches will succeed in preparing people to handle complex situations requiring all facets of capability. The writer feels most at ease with the theories advocated by Thorndike, Guthrie and the Gestaltists but has sympathies with the advocates of transaction-based learning. The key performance criterion for any approach must be whether or not it teaches students to learn for themselves.

Applying Theories of Learning and Motivation to Education

Research has identified a number of requirements for an education which could be expected to result in the formation of 'good engineers', most notably a call for reality-oriented learning. Particularly so in the UK, two rationales are given for this move away from theory-oriented to practice based learning:

- industry's perceived need for immediately useful engineers;
- academics' realisation of graduates' difficulties when confronted with 'real' problems.

Although the two objectives appear to be similar at first sight, there are two very different underlying concerns in that industrial managers' needs are frequently stated in terms of directly relevant knowledge, combined with experience of applications, while the academic aims can be paraphrased as learning to learn (see below) and coping with the often ill-defined problems occurring in real life. The industrial objective can be satisfied to a substantial extent by ensuring that up-to-the minute and relevant knowledge is transmitted and that engineering paradigmata are learnt, in part, through practical experience (laboratories and applied examples). The academic objectives, however, require more sophisticated approaches, that is, completely new learning environments.

A brief review of conventional teaching approaches will be followed by an extended discussion of experiential learning and its derivatives.

Conventional Teaching Methods

Traditionally, university education has relied on a mixture of lectures, tutorials, seminars, laboratory sessions and individual project work to prepare students for their future tasks (cf. EPC1-5). Lectures are a very useful tool for the transfer of factual information and for the presentation of intellectual concepts. Tutorials run by experienced teachers can serve to further the understanding of concepts while seminars provide opportunities for students to demonstrate that they are capable of structuring and transmitting information themselves. Laboratory sessions are usually intended to train students in narrowly-defined skills or to allow them to experience the practical manifestation of theoretical concepts. Although these methods are suited to preparing engineers and managers for their technology-oriented tasks, they are rather less useful in generating an understanding of people and business-related issues. These aspects of engineering education were traditionally imparted during the first two or three years of postgraduate industrial experience. However, the cost of such induction periods is no longer acceptable to many businesses. As a result, universities and other institutions of higher education have been forced to develop alternative approaches.

The traditional teaching and learning methods should not be dismissed out of hand. Most of engineering education should still be based on traditional approaches, because the vast quantities of factual and conceptual information which must be transmitted cannot be taught in any other manner. Many of the skills required by engineers (and managers) are of a mechanistic nature and do not necessarily benefit from open-ended learning situations.

The Case Study Approach

The Master of Business Administration (MBA) movement has led to the development of the case study teaching method which is based on real-life problem situations. In a formal case study, of the type created at the Harvard Business School, the background to the problem is described in some detail, a great deal of emphasis being placed on one particular angle, e.g., the issue of technological change. The information given would have been vetted and approved by the company which provided the case study situation. Students are invited to analyse the situation and the company's approach to generating a solution. The analysis should lead, wherever possible, to the definition of a generic problem and a widely applicable solution method which should allow students to identify similar situations once they have moved into professional positions. Alternative response possibilities are also discussed, so as to allow paradigm based solutions. This is both the strength and the weakness of the case study approach: a traditional case study does not necessarily develop engineers' and managers' problem solving abilities, it is based on pattern recognition and relatively automatic responses. The approach does, however, teach students to work fast and to develop a methodology for solving case studies. But reality is rarely presented in the form of a background paper with questions! Learning which goes beyond case study must therefore be based on experience which is founded on real life.

Experiential Learning

Carl Rogers' Principles of Learning

In 1969, Carl Rogers formulated ten principles of learning, listed below, perhaps one of the first comprehensive programmatic statements on learning based on a real environment and relying on students' interest in the subject studied, coupled to participation and furthering of independent work. The principles are really about freedom, choice and the need for an education where people learn to learn such that they are able to enhance their own performance. Principles 1, 2, 6, 7, 8 are non-controversial and derive more or less directly from theories of education and motivation. Principle 10 is no longer challenged even though it may have been viewed as very progressive in 1969! The need for life-long learning and openness to change is now recognised fully. Principles 3 and 4 deal with obstacles to the process of learning; however, they are not meant to prevent teachers from introducing 'threatening learning' and 'change'. They simply advise caution in introducing such elements. Principle 5, in Rogers' own view, is more or less an extension to the preceding principle. He emphasises that threats to the organism - which would also include the rational mind - are different from those to the ego or self, as represented by the person's perception of himself or herself. The experience of the CIM course at Brunel University (Schmid, *op. cit.*) bears out this observation to a substantial degree. Principle 9 is, in the writer's opinion, far too idealistic. The 'real world of work' cannot rely on basic self-criticism and self-evaluation. Evaluation by others is unavoidable and students, and all other people, must learn to handle this type of threat.

10 Principles of Learning **[Rogers, 1969]**

1. Human beings have a natural potential for learning.
2. Significant learning takes place when the subject matter is perceived by the student as having relevance for his/her own purposes.
3. Learning which involves a change in self-organisation - in the perception of oneself - is threatening and tends to be resisted.

4. Those learnings which are threatening to the self are more easily perceived and assimilated when external threats are at a minimum.
5. When threat to the self is low, experience can be perceived in differentiated fashion and learning can proceed.
6. Much significant learning is acquired by doing.
7. Learning is facilitated when the student participates sensibly in the learning process.
8. Self-initiated learning, which involves the whole person of the learner (feelings as well as intellect), is the most lasting and pervasive.
9. Independence, creativity and self-reliance are facilitated when self-criticism and self-evaluation are basic and evaluation by others is of secondary importance.
10. The most socially useful learning in the modern world is the learning of the process of learning, a continuous openness to experience and incorporation into oneself of the process of change.

Models of Experiential Learning

John Dewey (1929), and other pragmatists believed that experience and what is drawn from it are the raw data out of which real learning grows. Based on this early work, Kolb, Rubin and McIntyre (1974) developed a model of learning which relies on a structured use of experience and which they labelled 'experiential learning' (EL). The restrictive adjective 'structured' is important since it stands for the realisation that experiences on their own do not provide a learning opportunity. Kolb *et al.*'s approach was based on a cyclical process, the experiential-learning cycle, as shown in Figure 47.

Most recent literature sources (Boud, 1978; Kolb, 1984; Lederman, 1984 and 1992; Hammel, 1986; Thatcher, 1986; Gibbs, 1993), continue to refer to this type of approach as experiential-learning, a term which has unfortunately been devalued through overuse. An alternative term, heuristic methods, is perhaps not ideal since it has the connotation of a trial-and-error approach which is not appropriate. In an extension of this model, postulated by the writer, the learning of complex concepts and subjects can be likened to a spiral process leading to understanding at ever-increasing levels, as shown in Figure 48. If experiential learning is to happen with some degree of success it must take place in four stages:

1. concrete experiences (do, praxis). This is followed by:
2. reflective observation, that is, assessing the outcome of the practical experience, which then leads to:

3. abstract conceptualisation (the forming of abstract concepts and generalisations, that is, theory building, hypothesising), and which requires:
4. the testing of the implications in new situations, that is, active experimentation, allowing the student to resume reflection in a new cycle.

Experience without reflection and generalisation, without the development of hypotheses, results in ineffective, non-retained learning. Kolb *et al.* strongly refute the allegation that the cycle corresponds to trial-and-error learning. It is the reflection stage which differentiates experiential learning from an approach which is said to be that of a child. In terms of general learning theory this part of the cycle, as well as the generalisation phase, are clearly cognitive elements while both application and experience are more related to the affective domain of learning.

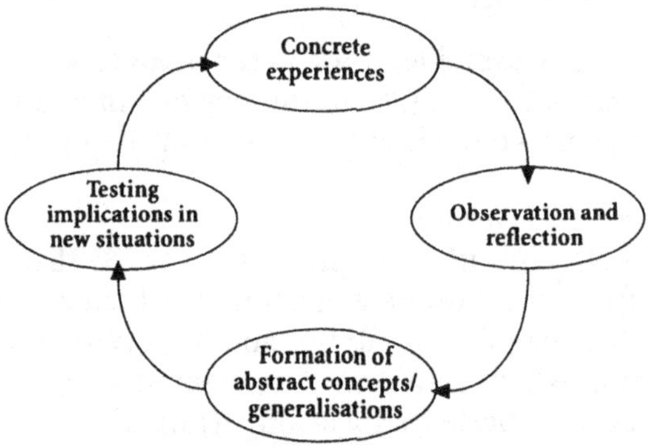

Figure 47: Kolb *et al.*'s experiential-learning cycle

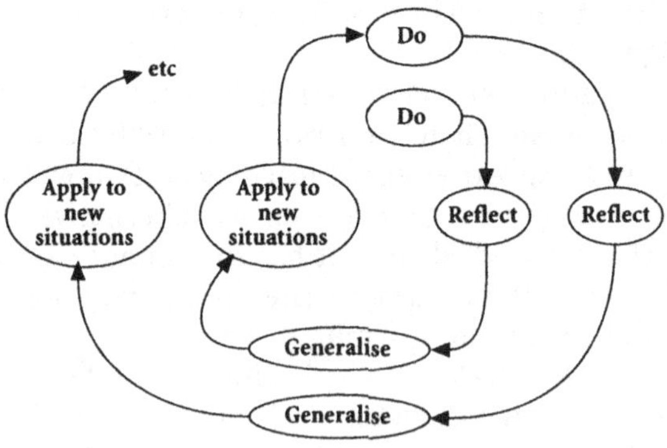

Figure 48: Spiral model of experiential learning (developed from Kolb)

The cycle is an idealised representation rather than a reality which holds true for all individuals. As found by Newland (1987), a majority of people exhibit a bias for one or a combination of the phases (or dimensions) of the experiential-learning process, that is, they will spend more time and effort while they are engaged on one or two phases of the process, even though they must eventually complete the cycle. Teachers who wish students to use the complete cycle must design learning situations amenable to this. They must insist, in particular, on the reflection stage without which the process cannot be deemed successful.

Handy (1985), in his introduction to *Understanding Organisations*, offers his own learning cycle, reproduced in Figure 49, which differs from Kolb *et al.*'s model in replacing Reflection by Questioning and in that he introduces an Experimentation stage rather than Kolb *et al.*'s Application to New Situations (or Testing). In Handy's model, the student is expected to question either the situation or the teacher's input, to conceptualise by moving from the particular to the universal, to experiment for better understanding and prediction and to consolidate by internalising. Gibbs (1993) has re-interpreted Kolb *et al.*'s cycle slightly although, beyond adding example activities as shown in Figure 50, he has not made significant changes.

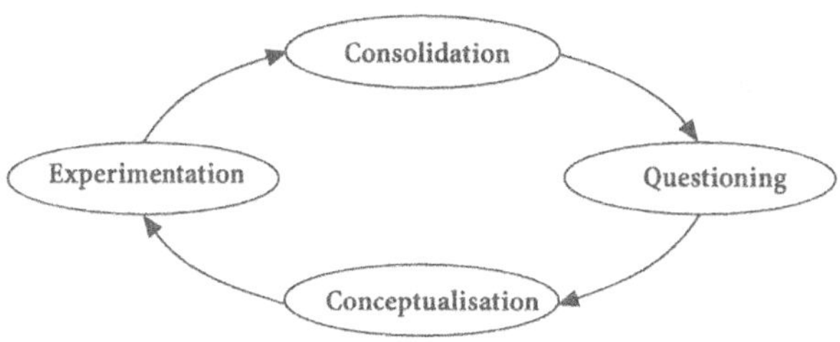

Figure 49: Handy's model of learning

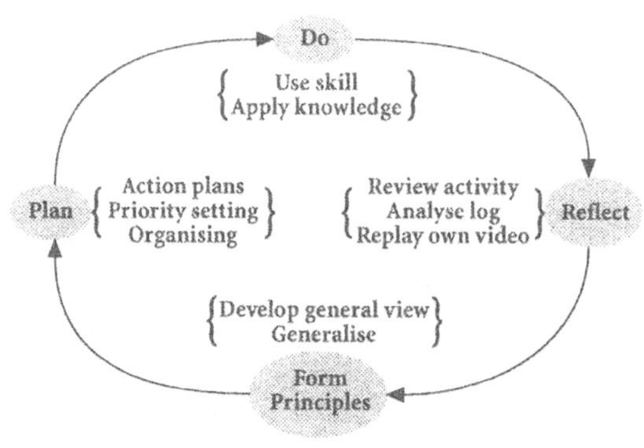

Figure 50: Gibb's interpretation of the experiential-learning cycle

It may be of interest to compare the traditional experiential-learning cycle with that of the five-phase model of problem solving in project engineering, familiar from the quality discussion, shown here in Figure 51. The main differences are the focus on an individual task or problem rather than a learning activity, and the division of the plan or experimentation phase into a development stage and an implementation stage for the solution. The monitoring of the solution can be equated directly to the Do phase of Kolb *et al.*'s model since it involves the experiencing of reality. The two cycles are very similar, particularly in that they both highlight the issues of change and continuous improvement: while a cycle is being completed, the environment will change so much that a new improvement cycle must be started immediately, problem solving is thus a fairly exact image of the learning situation.

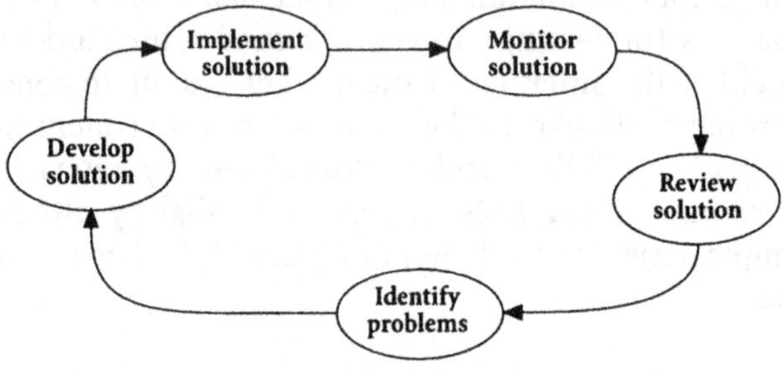

Figure 51: The traditional problem-solving cycle

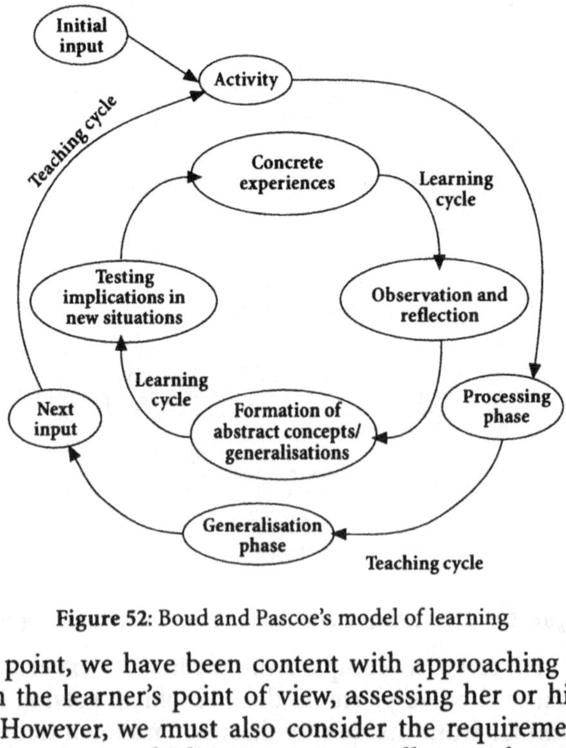

Figure 52: Boud and Pascoe's model of learning

Up to this point, we have been content with approaching the learning process from the learner's point of view, assessing her or his needs and approaches. However, we must also consider the requirements imposed on the learning process which is necessary to allow teachers to fulfil their role. This role is defined, at least partly, by the demands of the customer, society, which relate to the need for measurable achievement and useable outputs. The next paragraphs will be concerned with these issues.

The model proposed by Kolb *et al.* fails to interpret the teacher's role. It is therefore necessary to adopt the enhancement of the theory, provided by Boud and Pascoe (1978), based on Kolb, Bach and Miller and represented diagrammatically in Figure 52. This includes the influence of the teacher on the learning process of the student: during the Input Phase the teacher structures the learning process by providing content and time-oriented boundaries, but does not prescribe what is to be experienced as learning. During the Activity Phase, teachers may offer input but most of their role will be limited to monitoring and advising the learner who will work within the boundaries defined earlier. During the Processing Phase the teacher has to counsel and advise the learner to optimise reflection and to open up areas which may have been overlooked. The role of the teacher in the Generalisation Phase is concerned with the creation of a learning situation where the students can exchange results freely with other members of a group and where the teacher ensures that the learners go beyond simply restating 'facts' with insufficient analysis. The New Input Phase is concerned with deciding, in cooperation with the learners, which of a number of hypotheses should be tested further.

Often, experiential learning is implemented as team-based learning, involving groups of peers. For such teamwork to become effective, a sufficient time period must be allocated for the group to develop a suitable working relationship and, where necessary, to overcome initial problems connected to unfavourable group dynamics. Although some group-based experiential learning may be established specifically so as to provide training in conflict resolution and associated skills, this will not be the general pattern. Familiarisation and group-working skills may be acquired through an activity such as the SCANCO case described in Evans (1994), which can be used as an introduction to team-oriented experiential learning. Ever since Kolb *et al.* published their paper in 1974, the concepts and the terminology of experiential learning have provided a recognised framework to describe and advance a method for transmitting knowledge content, faculties and, to a limited extent, qualities.

Experiential Learning and the Elements of Skill Development

According to Gibbs' (1993) interpretation, implementing effective skill development involves four elements, which are perceived to be related to the experiential-learning cycle (ELC) of Figure 50 and to the problem solving cycle shown in Figure 51. The labels though, Training, Demand, Monitoring and Assessment (as shown in Figure 53) are completely different. The starting element, Training or teaching, would have to be equated to the phase of the ELC which Gibbs describes as Plan (whereas a more Kolbian author would refer to 'testing in new situations'). This is, in fact, the stage of the ELC where the learner receives an input in Boud and Pascoe's model. A new Training element is triggered if the Assessment has shown that learning or internalisation has not taken place to a sufficient degree. Demand is closely allied to the all-important Do phase of the ELC in that students are given time and targets (demands) to practice the skill(s). All learners, not just students, require well-designed learning activities to practice the skills as part of this demand element. The use of the term 'demand' is not in conflict with the third of Rogers' ten principles of learning even though it might appear to be so at first sight. Changes in self-organisation cannot be avoided in learning situations. It is essential that the whole environment presents a minimum of external threats. Once practice has taken place, or while it is still taking place, the students monitor their performance, e.g., they reflect on their work using checklists. Monitoring is not a teacher-led activity in this view.

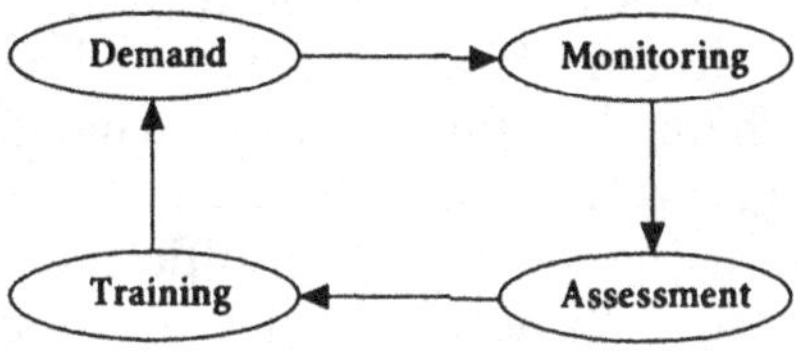

Figure 53: Gibb's skill-development model

The Assessment element is essential since it gives students both feedback and a reward-based target towards which they can orientate their work. Tutorial comments at this stage must address both the content of the work submitted and the quality of the process which led to its completion. The Assessment phase is, perhaps, the part of the cycle which differs most from the ELC although it could be argued that assessment can only be useful if it addresses learners' ability to Generalise.

Simulation and Gaming

In recent years, the concept of experiential learning has become somewhat devalued as a consequence of its indiscriminate adoption by education policy makers and teachers, some of whom fail to understand the need for implementing all the stages of the experiential-learning cycle or who are normally not prepared to invest the effort necessary for successful performance in all phases of the experiential-learning cycle and, especially, the reflection stage. A second and perhaps even more significant concern is students' perception of the concrete or practical experience, doing, as a simulation or game. In simulation and game-based teaching the construction of the activity (the creation of a simulated reality), its content and the debriefing phase are all-important. The discussion of Guthrie's work on learning (1952) helps explain why the design of the learning situation and the provision of appropriate sets of stimuli are so difficult but also important: since complete learning can be assumed to take place in just one trial, that is, since a particular set of stimuli can lead to the learning of an appropriate response without a need for repetition of the same situation, the situation must be free of conflicting sets of stimuli and should not present the same sets several times over. In a review of recent literature (from 1963 to 1991), on the topic of the effectiveness of games and simulations for educational purposes, Randel *et al.* (1992), compared their impact with that of conventional instruction. The investigation covered social sciences, maths, language, arts, physics, biology and logic, that is, a representative spread. (Computer) games proved very successful in mathematics and also in the more atomic aspects of language learning, while there was no significant benefit in the social sciences. Here only naive beginners performed better in an economic simulation than in a lecture-cum-discussion situation. Having found that out of 60 studies 22 favour games while 38 show no difference between the two methods, Randel *et al.* conclude that gaming can show beneficial educational effects if the approach is used for subject areas where very specific content can be targeted and where objectives can be defined precisely.

The research yielded the consistent finding that games and simulations are more interesting than traditional classroom instruction. The greater interest holds true even when controls for initial novelty (Hawthorne effect, see Handy (1985)) were used. Randel *et al.* advise the use of gaming or simulation situations where classes have motivation problems. They also suggest that research into the effectiveness of games and simulations is not yet sufficiently advanced to give clear guidelines for successful implementation.

Gosenpud and Miesing (1992), studied the influence of the environment - in terms of a large variety of factors - on the performance of small student groups in a business simulation. They found that two motivational and two interest variables had a significant influence on the outcome while ability, confidence, cohesion and organisational formality did not. The motivational factors were a 'desire to play the game' and 'easy to work with' team-mates while the interest variables were the students' specialisation and plans for the future. Organisational formality had some influence on performance but was not shown to be significantly correlated. The authors of the study explain the lack of significance of cohesion and formal organisation with the small proportion of the duration of the simulation during which the students were working in teams (five out of twelve weeks).

Stewart (1992), differentiates between post-experimental and post-experiential debriefing. While the former is mostly encountered in the context of psychological experiments, the latter is germane to the type of simulation involved in experiential learning which often has a behavioural aspect. It must be designed to achieve positive change and not just as a method to remove negative consequences. The debriefing is a phase separate from the experience being processed. Amongst other goals of the debriefing process, Stewart cites 'recognising ethical issues' and 'developing analytical skills' as important. 'Participants should leave with a clearly defined view of themselves as manager and team member' (Bailey, 1990, quoted by Stewart *ibid.*). This is perhaps the most relevant element of the assessment and appraisal processes which take place on the CIM course (Schmid, *op. cit.*).

Experiential Learning in Relation to Simulation and Gaming

Experiential learning can be said to contain, as a subset, simulation and game-based learning and is often assumed to be virtually equivalent to these approaches, an insidious simplification. In any learning cycle where the 'concrete experience' is intended to approximate a (near) real-life situation, distinct from a game or simulation, effective management becomes very important. Course organisation and monitoring must ensure that the learners do not interpret the semi-artificial situation created in terms of a simulation or game. Such an interpretation would make it more difficult for students to translate the lessons learnt into practical decisions at a later stage of their studies and in professional life. The resource implications of providing the appropriate teaching inputs for (near real) experiential learning also differ substantially from the inputs required for running games. However, it is not possible to draw up a definitive resource comparison between, e.g., a simulation and a lecture with associated discussion. While the former will demand a great deal of resource in the case of a social work study, there would be no staff input in the case of a computerised simulation, at least not at the point of delivery. A lecture and tutorial, conventionally seen as a very economical form of instruction, will appear resource intensive by comparison.

Even though 20 years have elapsed since the concept of experiential learning was first mooted, it is still accepted as a powerful approach to managing learning of all types. While it was originally lauded as a panacea, today it is clear that other forms of learning are also of importance. Indeed, in some situations other forms of learning must precede experiential learning in order to provide a starting point for the cycle. Problems arise in experiential learning as the result of inadequate involvement of teachers and wrong perceptions of learners.

Branton (in Oborne, *op. cit.*), is of the opinion that the constant performing of computations, that is, the existence of 'unself conscious intentions' may help explain skilled activity. This is one of the factors which affect the design of experiential-learning environments where predictions of motivation, learning and leadership performance are difficult and unreliable because people's models and the use they make of the models are changing over time, a cornerstone of the transaction based philosophy of learning.

Practice-Based Learning

The model of learning, reproduced in Figure 54, is a combination of the cycles proposed by Kolb *et al.* and Handy. The writer has inserted Handy's *Question* phase into Kolb's cycle because only a questioning attitude can lead to useful reflection. Learners will only be motivated to reflect on new learning if they perceive a difference between previous or remembered experience and the latest contact with reality. Depending on the situation and the strength of disagreement, this discovery and thus the questioning of the relationship between reality and own perception will take place spontaneously or it will need to be prompted by a teacher, tutor or other person external to the process.

> The process of articulating one's thinking is a vital part of the process of converting experience into learning or of using one's conceptual apparatus in a concrete experience. The process can be very much assisted by promoting discussion of an experience by raising to the surface the thoughts and embryonic ideas of the individuals in the learning situation.
>
> *Thatcher, 1986*

In essence, this process can be likened to post-experiential debriefing.

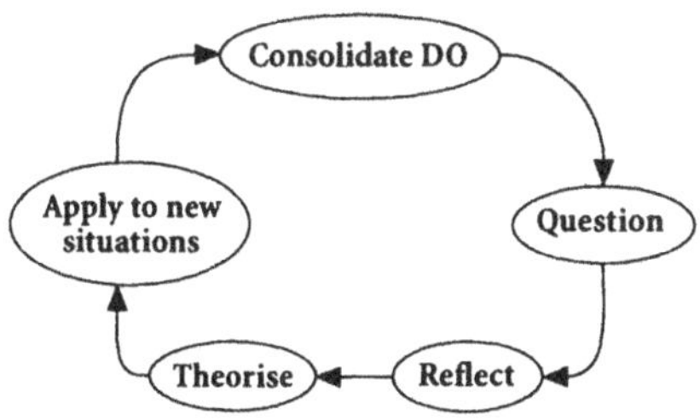

Figure 54: Combination of Handy's and Kolb's models of learning

Lederman (1992), offers a three-phase model of the debriefing process which follows on from the concrete experience and which ends with the generalisation phase of Kolb *et al.*'s model. This debriefing sequence is shown in Table 16. It thus substitutes two stages for the reflection phase. The first stage involves the introduction of the learners to a systematic process of self-reflection, that is, the development of a questioning attitude. The second stage can be equated, more or less, to the reflection phase in the experiential-learning model. This provides further justification for the modified cycle presented in Figure 54. In the paper quoted, Lederman provides a detailed discussion of the three-phase model, together with a list of relevant literature.

The whole practice-based learning cycle can be viewed as an opportunity for the consolidation of new learning which is internalised as the result of the cognitive stages of Questioning, Reflection and Generalisation and the affective processes of Testing on New Situations and Use in the real environment. Day-to-day use of the new skills may lead to the discovery of new discrepancies between reality and experience or unexpected ramifications of having internalised previous learning.

Table 16: Lederman's 3 phases of the debriefing process

Phase	*Purpose*	*Description*
Phase 1	Systemic reflection and analysis	Phase 1 is an introduction of the participants to a systemic self-reflective process about the experience through which they have just come
Phase 2	Intensification and personalisation	Phase 2 is the refocusing of participants' reflections onto their own individual experiences and the meanings they have for them
Phase 3	Generalisation and application	Phase 3 is the exploration that takes participants from their own individual experience to the broader applications and implications of that experience

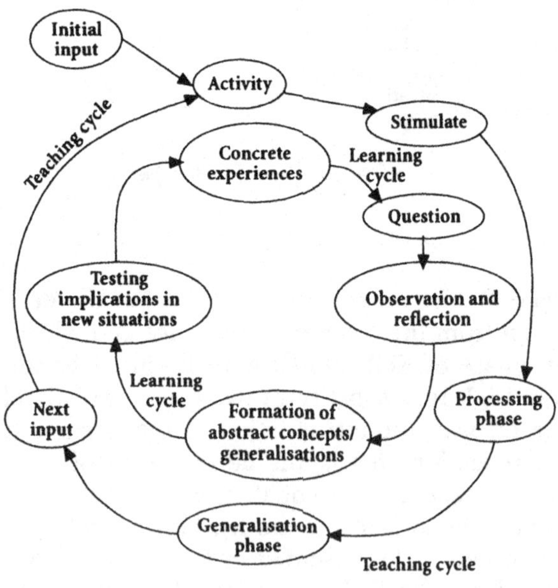

Figure 55: Expanded version of Boud and Pascoe's learning model

Where the practice-based learning cycle is used to improve the performance in a particular activity or set of activities, rather than to learn new skills, the question or enquiry phase becomes an assessment stage during which the effectiveness of the previous learning is measured or questioned. The reflection phase must then result in either an acceptance or revision of the applicability of the approach learnt. The suggested model highlights the demands on the teacher since he/she must create the right attitudes on top of supporting students in the task-oriented phases. In Figure 55, the author interprets Boud and Pascoe's approach of combining the teacher's role with the student's learning cycle presented in Figure 54. The teacher's role in stimulating and directing students' enquiry and questions is highlighted. Unfortunately, it is near impossible to find a completely meaningful relationship between the elements of the practice-based learning approach and Gibbs' cycle for moderating learning.

The Role of the Teacher

It could be argued that the role of the teacher in practice-based learning is very much reduced, that the students learn of their own accord, that they learn by interacting with each other. However, the role of the teacher in coordinating, managing and developing a course of the type represented by the learning activity is important, especially in terms of the adoption or modification of qualities. This can be seen as an expression of Bowers' views on the duties of teachers. Bowers (1984), is opposed to one of the tenets of liberalism in education, because for him: 'the phrase *autonomous individual* reflects the myth of being free of the past'. Instead, he stresses the shaping influence of the past on individual identity. The process of socialisation reproduces in the individual the tacit historical knowledge of a culture. He observes though, that the traditional authority arising from social practices, conventions and patterns of belief is being de-legitimised and must be replaced by, in his words, 'new definitions'. In his view teachers are responsible for giving people the tool for negotiating these definitions, that is, communicative competence. He declares that this is made difficult by the Taylorisation of the teaching process (subject-specific teaching). Teachers must allow students to make informed choices and must insist on scrutiny of taken-for-granted beliefs.

The Role of the Learner

Practice-based learning demands that the learners and groups or teams of learners adopt a professional attitude to the learning situation, that they consider it as a real situation and accept the demands and pressures which follow from this. They must be prepared to work under their own control according to externally defined guidelines and must question their experience at all stages of the process. Learners and tutors must strive together to maintain motivation, by providing feedback on performance and by establishing a meaningful discussion on course design, course content and course management. Practice-based learning is a concept which has been derived from Kolb's and Handy's models of learning. Its key characteristics listed below:

- Real-life objectives.
- Realistic deliverables.
- Intermediate targets.
- Structured learning process.
- Experiential-learning cycle.
- Questioning phase.
- Debriefing and appraisal.
- Maintaining motivation.
- Teamwork orientation.
- Problem related duration.
- Strong teacher support.

The concept has been developed in an attempt to enhance students' learning by formalising a transition in the experiential-learning cycle which is ill-defined in the original concept. The modification of Boud and Pascoe's model highlights the more significant involvement of the teacher necessary during the questioning phase if real-world oriented learning (a term including both experiential- and practice-based learning) is to become fully effective.

The features listed above reflect the experience gained in operating a part of an engineering course at eight institutions involving about 300 students.

Project-Based Learning: an Exemplar From Engineering Praxis

Universities and Polytechnic Institutions in Germany, Austria and Switzerland have always required students to complete a substantial piece of independent project work at the end of their studies. In England, final-year projects are a relatively recent phenomenon on engineering courses; for the materials science course at Cambridge, for example, this requirement was introduced as recently as the 1960s. The objectives though in asking a student to carry out a substantial piece of research or design work on his or her own have not changed since the late 19th century, that is, the early years of formalised academic engineering education. The French Grandes Écoles have not yet introduced final-year projects. The assessment of a graduand's readiness for work in industry is based on a long industrial placement which must be documented in a major report.

In setting a final-year project, the first objective is to 'prove' that a student is ready to take on a task and complete it successfully within a deadline, working independently from beginning to end. A second objective is to provide an opportunity for assessment which goes beyond the testing of content which can be learnt by rote. A third objective is often the solution of a technical problem in which the supervisor or an industrial sponsor has a strong interest but where 'a pair of hands' comes in useful... Perhaps the most significant objective is the selection of those students, from amongst the best in a year, who are most likely to succeed as researchers – where the definition of research is perhaps best described as 'old-fashioned'.

The image of engineering work which stands behind these final tests of graduands' ability is not realistic. It is the concept of the engineering designer or engineering researcher who is given the specification of a product or process and then goes away to her or his drawing board or laboratory to work quietly on their own and to return with the finished design when required. A quote from Finniston (1985) may serve to illustrate this view: 'Engineers ... are problem solvers.' The image is, in other words, the myth of the *Great Engineer*, the Stephenson, Brunel or Thorn. A myth because, in general, the great engineers worked not by themselves but with apprentices and skilled craftspeople who would contribute their ideas and experience.

Purpose of Project Work

Project work on engineering courses, be this laboratory-based work or design tasks, is intended to equip students with faculties which are deemed to be essential in professional work. Dowdeswell and Harris (1979), classify the aims of project work at university as follows:

Student Aims

1. adopt an active approach to learning in a real context;
2. assume greater responsibility for own learning;
3. greater depth of knowledge in limited area based on interest;
4. integrate existing skills and develop new ones;
5. optionally: work with other students to develop team ability;
6. optionally: work in an interdisciplinary context.

Project Aims

1. atmosphere in which student can feel maximum involvement;
2. atmosphere of reality/research;
3. reasonable chance of success;
4. close relationship between student and supervisor(s);
5. atmosphere of minimum constraints;
6. discourage passive assimilation;
7. discourage uncritical approaches;

8. encourage students to use lateral thinking approaches.

As a matter of course, the project aims must include the encouragement of attitudes listed under the heading of student aims! There are also a number of implied aims (implicit aims) which must be satisfied if project work is to be successful:

1. project work must be sufficiently adaptable to satisfy the needs of many different students and situations;
2. project work must fit in with and be complementary to other forms of teaching;
3. it must be designed to constitute an integral part of the curriculum;
4. the demands of the project must not exceed the available time, manpower and resources;
5. the process and its end products must be assessable.

The assessment procedure must be acceptable to students and staff. Dowdeswell and Harris' analysis is the most thorough of the reviews of project work encountered by the writer, but some other results are also of interest. In the eyes of the members of a working party of the Engineering Professors' Conference, project work is the key tool for developing skills and understanding on engineering courses. Through conventional project work students are expected to acquire the skills of (adapted from EPC 6, 1993):

1. analysing a problem using a structured approach;
2. researching potential solutions through the selection and study of relevant literature;
3. finding and implementing a solution to the problem;
4. gathering data to validate the approach used;
5. reflecting on the solution process; and,
6. communicating the results verbally and in writing.

These are very much the skills of the classical engineer who is conducting research and development as an individual, a working situation which has become less and less pervasive of the growth in scale and complexity of engineering and scientific projects where many people from different backgrounds and with a variety of skills must cooperate in multi-disciplinary and multi-functional teams.

Campbell (1990), goes further than this in setting the following objectives for a project based element of the final year of a sandwich type engineering degree course:

> To expose the student in an educational environment to an industrially relevant engineering problem ... to develop in the student the critical

> faculties, appreciation of practical problems and communication skills required by the practising engineer in industry ...

He advocates close links with industry for the projects which are undertaken on an individual basis and closely supervised by a university tutor. On most engineering programmes accredited by a major engineering institution, students are expected to complete several different pieces of project work. Where such work is undertaken before the final year it is usually of a straight forward nature, starting with very basic and well-defined tasks at the beginning of a course. It tends to be rather more substantial and complex in the case of final-year projects. But even at this stage the tasks must be clearly delimited so as to allow a student to reach a conclusion within a reasonable time and with limited support from the supervisor. Instances are rare where original results are achieved without clear input from the supervisor. Depending on the ethos of the institution and of the supervisor, a final-year project can be purely an assessment opportunity, a test of mental strength or a learning experience.

Usefulness of Final-Year Student Projects

Project work in the final year of an engineering course should be designed to allow students to apply the powers they have always possessed or acquired during the course. At the same time it should allow students to experience, in a controlled environment, the type of work situations which they will face in later life. At least in the latter respect the conventional (final-year) project work fails all but those students who will later be working as individual researchers. Depending on the choice of project and the academic or industrial support it may succeed in providing an opportunity to apply faculties and knowledge learnt during the course to a real problem. In the author's experience this is not a common outcome in the case of manufacturing engineering graduands. This is confirmed in a further quotation from Finniston (1985):

> Students are given guidance in many ways on how to go about achieving solutions, e.g. by case studies, by project involvement at university, vacation experience etc. These are not of themselves sufficient to measure the capability of individuals ...

One of the important realisations, and perhaps the author's main objection to the exclusive use of conventional student project work, is linked to Gibbs' observation (1993), that any skill which has once been taught must also be used, monitored and assessed if it is to be internalised by the student. If any of the later elements in the chain are missing then there can be no feedback to the student. On the SEP course at Brunel, staff introduced a third-year project, based on group tuition, specifically to ensure that students could obtain feedback on their performance in the key stages of a design project before undertaking the individual fourth (final) year project (Ellis, 1985).

While final-year projects reflect one range of views of engineering work rather well, namely that of research, industrial design, product design or development work, they singularly fail to show whether the prospective engineers have the skills required for the successful integration in the ever larger and ever more multidisciplinary teams becoming the norm for project work in industry:

> They will usually be integrated into an experienced team at the beginning of their professional careers and assigned concrete tasks.
>
> *Hernaut, 1993*

Individual Project Work Versus Project Work in Teams

Corfield (1983), was very clear on this subject of the management of engineering projects:

> … we are perhaps only on the fringe of learning the techniques by which large teams of engineers can be brought together to work not in series but in parallel to reduce the time-span of complex projects.

and

> The more progressive educators are asking industry to spell out its requirements and have adopted a more flexible approach…

In most institutions the students are directed to complete the final-year project and the associated thesis on their own or, occasionally, in teams of two although a few institutions will accept larger teams. Even where group-work is permitted the project must be undertaken in a way which allows an individual assessment of the students' contributions thereby fostering a degree of competition within the mini-team. This is a requirement imposed by the engineering institutions and can be very useful in stimulating the generation of new ideas but not for learning about co-operative work. Teams are of particular importance for new projects in AMEs (Advanced Manufacturing Environments) and for their management. The author defines AMEs as manufacturing situations where advanced manufacturing technologies and methods are being used. In fact, it is very probable that it is the teamworking skills which are essential and differentiating qualities of the manufacturing engineer, although she or he must also have an excellent grasp of traditional engineering disciplines. Conventional project work cannot provide the necessary learning opportunities since they are too limited in both scope and content. The practice-based learning approach discussed in the section *Practice-Based Learning* (starting on page 415) and tested with the CIM course described by Schmid (1995), may provide an alternative route for acquiring skills and understanding needed to work effectively in AMEs and in advanced technology situations in general.

Designing a Project-Based Course for Future Advanced Manufacturing Engineering Professionals

Advanced Manufacturing Environments (AMEs) are characterised by the use of advanced technologies (CAD, CAM, CAQ etc.) and advanced management approaches (just in time, lean manufacturing etc.). They are aimed at combining the benefits traditionally associated with mass production with those normally associated with job shops and batch manufacturing: short lead-times, customisation and economies of scale. The powers required for successful practice as a manufacturing engineer in a team in an advanced manufacturing environment can be divided into two categories, namely, technical and scientific components on one hand and people and systems oriented components on the other:

- Technical and scientific faculties and knowledge.
- Software design methods.
- Interface design and use.
- Methods of computer communication.
- Robotic technologies.
- Workpiece handling and interfacing.
- Database design.
- Machine tool operation.
- Measurement, quality, control.
- People and systems-oriented qualities and faculties.
- Systems-oriented thinking.
- Strategic thinking.
- Interpersonal communication.
- Motivation.
- Team leadership.
- Ability to initiate and promote change.
- Ability to manage change.
- Activity planning.

Some of the above are equally applicable to working as an individual while others are only relevant in team-oriented work situations. In addition to the generic faculties and powers listed above:

> Ingenieurs need additional in-depth fundamental knowledge of their specialized fields, general knowledge of problem-solving methods in engineering and special application knowledge in accordance with workplace demands.
>
> *Hernaut, 1993*

Most courses aimed at developing competence for working in computer-integrated manufacturing follow the approach used at Cambridge

University on their MEng course where groups of students are given a quite narrowly defined 'integration task', for example, the control of an automatic assembly system by the computer. While the task can be difficult it is limited both in time and scope: the duration is three to four weeks full- time and both the equipment to be used and the outline solution are fixed by the course management.

As an alternative to university-based learning Züst *et al.* (1991), describe an industry-based teaching environment for CIM which is designed to give students an appreciation of the problems of integrating different technologies with the help of computers. However, the approach chosen does not involve the young engineers in the process itself – they experience neither the problems of building networked solutions nor those inherent in successful teamwork. Students simply watch designers and production operatives carry out their tasks using computer systems.

The writer has identified a range of paradigmata which cannot be described here, for reasons of space and time. These relate to the teaching of advanced engineering skills, with a particular emphasis on manufacturing industry. However, he has included some of the more interesting novel teaching approaches in his thesis (Schmid, 1995).

Concluding Remarks: Looking Ahead

The education of people and the development of methodologies and techniques are only part of a larger consideration of the worlds of work and learning: the question of tangible returns must be resolved if industry is to continue to put effort into supporting the design of computer integrated systems, in particular those involving a human-centred approach. Developers, users and, perhaps more importantly, stakeholders must be able to derive clear benefits from the major investments required.

The criteria for determining benefit is usually based on financial and functional issues and may not include consideration of what is best for the people involved. We have to assume that the economic imperative can be satisfied by solutions which go beyond the technology dominated patterns being advocated today. Extended studies of the philosophical, societal and ergonomic issues surrounding human work, 'tools' and automation should point the way towards a more optimal integration of people, systems and computers. In this pursuit researchers must be mindful of all aspects which characterise a soft system: customer(s), actor(s), transformation(s) world image, environment, purpose and social context, but particularly those referring to human needs and economic prerogatives. Theses features of soft systems are discussed by Checkland (1981) and Ainger (1994). At some future time, tasks involving the cooperation of people and computers may well be viewed as 'jobs of work' rather than complex sociotechnical problems, in the tradition of skill and craft-based occupations so prized in times past.

Acknowledgements

The author would like to thank Chris Ellis, Regula Schmid and Tania Hancke for supporting him with feedback, advice and encouragement while he was working on his doctoral thesis. He would like to thank Karamjit Gill for his help in transforming part of the thesis into a chapter for the present volume.

References

Abell, D. (1994). 'The IMD Approach to Case Studies' in an interview with J. Gray in *ECCHO, Newsletter of the European Case Clearing House.* Milton Keynes, Spring 1994.

Ainger, A. (1995). 'The Helical Systems Design Process', A. Ainger with F. Schmid, *Computer Integrated Manufacturing Systems Journal,* April 1995.

Ainger, A. (1994). 'Synopsis and Discussion on the Helical Approach to Systems Design and Build' in the proceedings of the NATO Advanced Study Institute *PEOPLE and COMPUTERS - Applying an Anthropocentric Approach to Integrated Production Systems and Organisations.* NATO ASI Series,

Ainger, A. (1990). 'Human Centred CIM Systems' in *PGC5, Proceedings of an IEE Colloquium.* Savoy Place, London, 1990.

Alschuler, A .S. (1970). 'Teaching Achievement Motivation' in *Theory and Practice in Psychological Education,* A.S. Alschuler, D. Tabor and J. McIntyre. Middletown, Conn., USA.

Bloom, B. S. (1956). 'Taxonomy of Educational Objectives' in Bloom, B. S., Engelhart, M. D., Frust, E. J., Hill, W. H. and Krathwohl, D. R. *The Classification of Educational Goals, Handbook 1, Cognitive Domain.* Longmans, London.

Boer, G. (1994). 'Five Modern Myths', *Accountancy,* June 1994: 82–83.

Boud, D. (1978). 'Experiential Learning' in D. Boud, D. and Pascoe, J. *Australian Consortium on Experiential Education.* Sydney, 1978.

Bowers, C. A. (1984). *The Promise of Theory, Education and the Politics of Cultural Change.* John Dewey Lecture, New York, 1984.

Brown, J.S. (1989). Source of Figure quoted in Cooley, 1993.

Brummett, F. D. (1985). 'The U.S. Manufacturing Engineer: Practice, Profile, and Needs' in *Proceedings of The Symposium on Education for the Manufacturing World of the Future.* Series on Technology and Social Priorities of the National Academy of Engineering, National Academy Press, Washington, 1985.

Campbell, D. R. (1990). 'An Undergraduate Student Project on Digital Filter Design with Industrial Cooperation' *International Journal of Electrical Engineering Education,* vol. 27: 141–151.

Checkland, P. (1981). *Systems Thinking, Systems Practice.* John Wiley, Chichester, UK.

Cherns, A. (1975). 'The Quality of Working Life', vol.1: *Problems, Prospects and the State of the Art.* Collier Macmillan, London, and Free Press, New York, 1975.

Clark, C. (1985:1). 'A New Enhanced Engineering Programme for Manufacturing Industries'; Clark, C., Medland, A. J., Rakowski, R.T., and Wild, R., *International Journal of Applied Engineering Education,* vol. 1, no.1. Pergamon Press, Oxford.

Corfield, Sir. K (1983). 'The Making of an Engineer', *Royal Society of Arts Journal,* June 1983, London.

Cooley, M. (1987). *Architect or Bee?: human price of technology.* Hograth Press, London.

Cooley, M. (1993). 'Skill and Competence for the 21st Century' in *Proceedings of the IITD 24th National Conference.* Galway, 1–3 April, 1993.

Creux, S. (1993). 'Business Process Redesign: The Toolbox Matrix'. MSc AMS Dissertation, Brunel University, September 1993.

Crystal, D. (ed.), (1990). *The Cambridge Encyclopaedia.* Cambridge University Press, UK.

Dewey, J. (1929). *Experience and Education.* Macmillan, New York.

Diderot, D. (1751–1772). *EncyclopÉdie ou Dictionnaire raisonnÉ des sciences, des arts et des mÉtiers.* D. Diderot (ed), d'Alembert, Voltaire *et al.*, Paris, 1772.

Dowdeswell, W.H. (1979). *Project Work in University Science.*

Dowdeswell and Harris, N. D. C. in *Learning Strategies in University Science*, pp. 203–239, University of Wales Press, Cardiff, 1979.

Ellis, C. D. (1985). Design Project Case Study: 'Capacitor Design for High Temperature Applications'; Ellis, C. D., Rakowski, R. T., Marsh P. T. C., *International Journal of Applied Engineering Education*, vol.1, no. 5, pp. 349–353.

EPC, Engineering Professors' Council (from March 1993), publisher of the *Bulletin of the EPC.*

EPC 1, Engineering Professors' Conference, 1989. 'Quality in Engineering Education' in *The Report of the EPC Working Group on Quality in Engineering Education, No. 1 of the Occasional Papers of the EPC.* Liverpool, July 1989.

EPC 3, Engineering Professors' Conference, 1991. 'The Future Pattern of 1st Degree Courses in Engineering' *a Discussion Document Prepared by the EPC Working Party, No. 3 of the Occasional Papers of the EPC.* Liverpool, February 1991.

EPC 5, Engineering Professors' Conference, 1992. 'Assessment Methods in Engineering Degree Courses' a *Discussion Document Prepared by the EPC Working Party on Assessment, No. 5 of the Occasional Papers of the EPC.* Liverpool, December 1992.

EPC 6, Engineering Professors' Conference, 1993. 'Developments in First Degree Courses in Engineering' *a Discussion Document Prepared by the EPC Working Party on First Degree Courses, No. 6 of the Occasional Papers of the EPC.* Liverpool, May 1993.

Evans, S. (1994). 'Developing an Anthropocentric View of Modern Manufacturing, A Case Study Approach – The SCANCO Example' in *Computer Integrated Production Systems and Organizations: The Human-Centred Approach, Proceedings of the NATO Advanced Study Institute 'PEOPLE and COMPUTERS – Applying an Anthropocentric Approach to Integrated Production Systems and Organisations.* NATO ASI Series, Series F: Computer and Systems Sciences, vol. 134, pp. 255–262, Springer, Heidelberg, 1994.

Finniston, Sir M. (1985). 'Engineering and the Future', *International Journal of Applied Engineering Education*, vol.1, no.1. Pergamon Press, Oxford, 1985.

Finniston, Sir M. (1989). 'Integrated Engineering and its Influence on the Future of Engineering Education in the U.K.' in Sir Monty Finniston, Duggan, T.V. and Bement, J.M. *International Journal of Applied Engineering Education*, vol. 5, no. 2. Pergamon Press, Oxford, 1989.

Gannon, P. (1988). 'The Human Factor Returns' in *The Independent* (Newspaper), 10 October 1988, p. 18, London, 1988.

Gibbs, G. (1993). *Developing Students' Transferable Skills.* Lecture Text, Oxford Centre for Staff Development, Oxford Brookes University, Oxford, 1993.

Gosepund, J. and Miesing, P. (1992). 'The Relative Influence of Several Factors on Simulation Performance', *Simulation and Gaming*, vol. 23, no. 3, pp. 311–325, September 1992.

Guagnini, A. (1984). *La formation des ingÉnieurs en Grande-Bretagne Ö la fin du XIXe siäcle*; see Noblet, 1984.

Hammel, H. (1986). 'How to design a debriefing session', *Journal of Experiential Learning*, vol. 9, no. 3, 1986.

Handy, C. (1985). *Understanding Organisations.* 3rd Edition, Penguin, London

Handy, C., (1993). *Understanding Organisations.* 4th Edition, Penguin, London,

Hernaut, K. (1993). 'Engineering Education in Germany, An Industry Perspective', *International Journal of Engineering Education*, vol.9, no.1. Tempus Publications.

Hobsbawm, E.J. (1968). *Industry and Empire*, London, 1968. Volume 3 of *the Pelican History of Britain*, Penguin, 1975.

d'Iribarne, A. (1987). Nouvelles technologies et culture franáaise. Le mariage du noble et du vil, Alain and Philippe d'Iribarne, *Revue Franáaise de gestion*, no. 64, Sept–Oct 1987.

d'Iribarne, A. (1990). 'Advanced technological forms which reflect the cultural, economic, educational, geographical and social reality in Europe', Appendix Part 1, FAST Report, *European Competitiveness in the 21st Century: the Integration of Work, Culture and Technology*, Brussels, 1990.

Isnard, M. (1984). Le Genie Militaire, in M. Isnard and Y. Cabon, *Encyclopedie Universalis*. Paris.

Kolb, D. A. (1984). *Experiential Learning: Experience as the source of learning*. Prentice-Hall, Englewood Cliffs, N.J., USA.

Kolbe, D. A., Rubin I.M. and McIntyre, J.M. (1974). *Organisational Psychology. An Experiential Approach*. Prentice Hall, New Jersey.

Krathwohl, D. R., Bloom, B. S. and Masia, B. B. (1964). in Taxonomy of Educational Objectives; The Classification of Educational Goals. *Handbook II; Affective Domain*; Longmans, London, 1964.

Lang, R.G. (1990). 'The Path of Least Resistance and the Path of Highest Return in the Pursuit of a Computer Integrated Enterprise' in *Proceedings of Rensselaer's Second International Conference*, Rensselaer Polytechnic Institute, Troy, New York. IEEE Computer Society Press, Los Alamitos, California, 1990.

Lederman, L.C. (1984). 'Debriefing: A critical re-examination of the post-experience analytic process with implications for its effective use', *Simulation and Games*, vol. 15, 415–431, 1984.

Lederman, C. (1992). 'Debriefing: Toward a Systematic Assessment of Theory and Practice', *Simulation & Gaming*, vol. 23, No. 2, June 1992.

Macleod, I. A. (1992:1). 'The Competence of an Ingenieur', *European Journal of Engineering Education*, vol.17, no.4, 1992.

Macleod, I. A. (1992:2). 'Towards a Core Curriculum for Structural Engineers' in McKay (ed.) *Education for Structural Engineers*, Sheffield.

Marton, F. and Säljö, R. (1976). 'On Qualitative Differences in Learning: Outcome and Process', *The British Journal of Educational Psychology*, vol. 46, pp. 4–11.

Newland, R. A., Powell, J. A. and Creed, C. (1987). 'Understanding architectural designers' selective information handling', *Design Studies*, vol. 8, no. 1, Butterworth.

Parnaby, J. (1987). 'Education and Training in Manufacturing Systems Engineering' in Parnaby J. and Donovan, J.R., IEEE Proceedings, vol. 134, part A, no. 10: 816–824, December 1987.

Pask, G. (1976). 'Styles and Strategies of Learning', *The British Journal of Educational Psychology*, vol. 46: 128–148.

de Noblet , J. (ed) *et al.* (1984). 'Les ingÉnieurs', *Culture Technique*, no. 12, Centre de Recherche sur la Culture Technique, Paris, March 1984.

Oborne, D. J. *et al.* (1993). *Person Centred Ergonomics: A Brantonian View of Human Factors*, [including original articles by P. Branton]. Taylor & Francis, London, 1993.

Owen, R. (1815). *Observations on the Effect of the Manufacturing System*, 1815, quoted from Hobsbawm (*op. cit.*).

Pittenger, O. E. and Gooding, C. T. (1971). *Learning Theories in Educational Practice – An Integration of Psychological Theory and Educational Philosophy*. J. Wiley and Sons, New York, 1971.

Poitou, J. P. (1984). *Dessin technique et division du travail*, see Noblet (1984).
Rachel, J. M. (1994). 'Getting the Human to the Centre of the System' in Schmid, F. *et al.* (eds.) *Computer Integrated Production Systems and Organisations*. Springer-Verlag.
Randel, J. M. *et al.* (1992). 'The Effectiveness of Games for Educational Purposes: A Review of Recent Research', *Simulation & Gaming*, vol. 23, no. 3, September 1992.
Rasmussen J. and Jensen, A. (1974). 'Mental procedures in real-life tasks: a case study of electronic trouble shooting', *Ergonomics*, vol.17, 1974.
Reason, J. (1990). *Human Error*. Cambridge University Press, Cambridge, 1990.
Rogers, C. (1969). *Freedom to Learn: A View of What Education Might Become*, Charles E. Merrill Publishing Co. Ltd., Columbus, Ohio, 1969.
Rosenbrock, H. H. (1983). *Flexible Manufacturing Systems in which the Operator is not Subservient to the Machine*, Research Project Mimeo, UMIST, Manchester, 1983.
Rosenbrock, H. H. (1989:2). *Designing Human-Centred Technology*, Springer Verlag, London, 1989.
Rosenbrock, H. H. (1990). *Machines with a Purpose*, Oxford University Press, Oxford.
Rouse, W. B. (1981). 'Models of human problem solving: Detection, diagnosis and compensation for system failures'. Preprint for *Proceedings of IFAC Conference on Analysis, Design and Evaluation of Man-Machine Systems*, Baden-Baden, FRG.
Schmid, F. (1992:1). 'Product and Artefact Study: Essential Tools for the Engineering Education Process' in Schmid, F. and Ellis, C. D. *Proceedings of the World Conference on Engineering Education*, Portsmouth. Published by *Computational Mechanics*, Sept. 1992.
Schmid, F. *et al.* (1993). 'The Human-Centred Approach to the Implementation of Computer Integrated Systems' in *Control and Dynamic Systems*, vol. 61; *Computer Aided Manufacturing: Computer Integrated Manufacturing* (CAM/CIM). Academic Press, San Diego, 1993.
Schmid, F. (1995). *Practice Based Learning for Engineering Students*, PhD thesis, Brunel University.
Shea, J.F. (1985). 'The Changing Face of U.S. Manufacturing' in *Proceedings of The Symposium on Education for the Manufacturing World of the Future*, Series on Technology and Social Priorities of the National Academy of Engineering, National Academy Press, Washington.
Shute, N. (1954). *Slide Rule*. Heinemann, London
Simon, H.A. (1957). *Models of Man*. Wiley, New York.
Skinner, W. (1985). *Manufacturing, the formidable competitive weapon*. John Wiley, New York.
Skinner, W. (1992). 'Missing the links in manufacturing strategy' in C. A. Voss (ed.) *Manufacturing Strategy – Process and Content*, pp. 13–51. Chapman & Hall, London.
Slack, N. C. D. (1991). *The Manufacturing Advantage*. Mercury Books Ltd, London.
Stewart, L. P. (1992). 'Ethical Issues in Post-experimental and Post-experiential Debriefing', *Simulation & Gaming*, vol. 23, no. 2, June 1992.
Thatcher, D. (1986). 'Promoting learning through games and simulations', *Simulation/Games for Learning*, vol. 16, no. 4.
Trevelyan, G. M. (1942). *English Social History*, London, 1942. Pelican edition 1979.
Vérin, H. (1984). *Le mot: ingÉnieur*, see Noblet, 1984.
Wall, T. D. (1987). *The Human Side of Advanced Manufacturing Technology*, Wall, T. D. *et al.* John Wiley & Sons Ltd, London.
Züst, R. (1991). Wie wird CIM-Ausbildung attraktiv?, in R. Züst and F. BÅchting, *io Management Zeitschrift*, 60(7/8), 43–47, Verlag Industrielle Organisation, ZÅrich, 1991.

Chapter 10

Workplace Innovations: the Making of a Human-centred Industrial Culture

Francesco Garibaldo

Introduction

Workplace innovations in the form of group-work, team-working, life-long learning are part of broader social and cultural horizons for the making of a human-centred industrial culture. These innovations go further than the socio-technical working life innovations such as human factor oriented or user-centred approaches. They are societal in the sense that they affect, and are affected by, all societal levels of the industrial culture, micro (the workplace level), meso- (the contextual level, that is the local societal context), and macro- (that is the wide social level such as the country and /or the regional level) levels. At the micro-level, the predominant tendency is the integration of roles and functions. The meso-level refers to the leading relationship between each organisation and its environment; it is a process of mutual adaptation. It is at the macro-level where public and collective policies succeed or fail in creating the public vision and the general knowledge for supporting both the processes of innovation and adaptation, i.e. the process of change at both micro- and meso-levels.

Innovations of group-work and 'team-working' stem from two historical periods. The first period is that of the rationalisation of the old Tayloristic -Fordist organisations at the beginning of the 1980s. The second period is concerned with the Japanese experience of 'quality circles' and 'just-in-time' (JIT) innovations. The driving force of change at the organisational level is the increasing integration of roles and functions of work life on a wider scale than before. As to 'group-work', this change leads to a new paradigm of industrial culture. There is a growing awareness both on the management side and on the union side, that the competitive edge of a single firm will depend more and more on its ability to utilise and develop its human resources because this is the most effective way of dynamic adaptation to a changing environment. In this perspective, group-work deals less and less with classical workforce flexibility and more and more with the *capability* and the *willingness* of each person to be committed to the best performance of the organisation. In other words the organisation will depend more and more on its employees' *expertise* and *willingness* to attain this capability. Indeed, new possibilities arise. However, the problem that the West has to cope with is: *how to gain employees' personal commitment? Is it possible to gain this commitment within the framework of the Western country and locally-based systems of social and institutional regulation?* Without this commitment any change, however smart the design, will revert back to the spirit of the Taylorist–Fordist model of work.

According to this group-work perspective, the old 'metric system' to measure work and productivity becomes quite inconsistent with the new concepts, where new problems and issues arise on how to reward work. The competitive edge of the organisation strategy will depend on the employee's commitment but this, in turn, depends upon an effective integration of functions and roles, that is, the performance of teams and of clusters of teams. Such a radical change at the organisational level requires a general cultural change. This means setting up an integrated agenda suitable for both management and labour. Furthermore, the agenda has to cope with broad economically and socially important issues both at the meso and macro-levels. This process is problematic because social actors have to cope with difficult alternatives and dilemmas; they have to choose what is relevant because 'catch all' solutions do not exist.

These alternatives and dilemmas as part of the actors' strategies, the creation of a self-sustaining process which requires the fulfilment of two social conditions: *trust* and *commitment*; trust between social partners and the commitment of top-level management. These conditions are, however, dependent on both the social environment and the vision of the organisation which are being shaped by a new industrial revolution of workplace and management culture, education policies and processes of knowledge transfer. There is increasing recognition of placing the analysis of the social environment and organisational vision within a broader industrial cultural framework of work life, living conditions and learning society at the micro, meso and macro-levels.

From this societal perspective, the overall process of societal change could be considered both as a learning process and a political process. Hence, workplace innovation is concerned with both analytical and action oriented activities from a human-centred perspective.

Social Trends: the Relations Among Different Levels of Analysis and Actions

New Trends

One of the major issues facing the information society is the changing nature of the workplace, the nature of the shift from skill-based production to competence-based production, from expert-centred innovation to knowledge-based innovation, and from hierarchical organisations to more distributed and networked organisations. We believe that these shifts are part of the wider societal changes in economic and social spheres. Just as societal development is not linear, so workplace shifts are non-linear. People have to cope with alternatives, if not true dilemmas. The overall process of societal change could be considered both as a learning process and a political process (Garibaldo, 1994; Reich, 1992), selecting alternatives and highlighting impacts are both analytical and action-oriented activities.

The analysis of the societal developments means an understanding of different levels of analysis, mainly three levels: the micro-level (the workplace level), the contextual level (the local societal context or the meso-level) (van Beinum, 1993), and the macro-level (the wide social level such as the country and/or the regional level).

The Micro-Level

At the micro-level, there is a growing shift from the regulatory mode of Taylorist–Fordist model towards the integral model of the market: integration of roles and functions. It is important to stress that such new philosophy of integration provides an explicit alternative to the regulatory model of the mass market and the Taylorist–Fordist organisation which has dominated so far. We can briefly summarise these characteristics (see Table 17) and compare them using 'functional alternatives'. The comparison shows that the previous equilibrium is out-dated, and that changing market conditions require a new form of production organisation.

The market is now the driving force of the change process. This process of transformation has partly been beyond the strategic control of the management. It began with the Bretton Woods agreement, and was impacted by the oil crisis in the 1970s. There was also a transition from the prevailing price-based competition to the quality-oriented competition, as well as the impact of the Japanese competition. In spite of more than a decade of intense restructuring of systems and organisations, this transition is far from being completed.

The European debate on the future of industry illustrates this ongoing transition. It is dominated by the view of the so-called 'lean production', promoted by the MIT report entitled *The Machine That Changed the World* (Womack *et al.*, 1990). In this perspective the process of globalisation is a strong determinant of homogenisation without offering any other alternative. The same idea underlies the 'Ohmae view' of globalisation. There is a 'new best way' to productivity and competitiveness: the Japanese way; what people have to do is to implement this best way, or adopt this model, for managing industrial firms.

We strongly disagree with this viewpoint. In our perspective, as Jürgens puts it, the lean-production debate is largely sustained by the 'Japanese myth' with many arguments drawing their persuasive force from here ...

there – Womack *et al.* – description of Japanese reality is pallid and often stereotyped (Garibaldo, 1993: 19).

Table 17: The Taylorist–Fordist regulatory mode and functional alternatives. *From*: Jürgens, U., Malsch, T. and Dohse, K. (1993) *Breaking From Taylorism*, Cambridge University Press, p. 5

Elements of the Taylorist–Fordist regulatory mode	*Functional alternatives*
Standard product	Product variety
Assembly line	Module production / production island
Single-purpose mechanisation	Flexible mechanisation
Unqualified mass worker	Qualified (skilled) worker
Low work motivation (indifference)	High work motivation (identification)
Conflictive labour relations	Cooperative labour relation
Hierarchical management	Participatory management
Vertical division of labour	Vertical job integration
Horizontal division of labour	Horizontal job integration
Workers tied to specific jobs	Job rotation
Machine/ assembly line determined work pace	Work pace independent from production cycle
Individual work	Group-work
External control of time and motion	Self-control of time and motion

A stage of strategic learning at the workplace has already begun, it can be described as revolutionary, because it breaks both the cultural and practical continuity with previous stages. What still remains to be seen is whether a balance or rather a 'virtuous' circle between market and production could be easily established, with well-defined rules, as was the case for most of this century. In other words, will there be a new equilibrium or will change be the only constant in the future of the industrial culture? No doubt the Japanese model of workplace innovation (e.g. quality circles, JIT) provides a basis of this new trend. New solutions have been proposed which are now grouped together under the general label of lean production as an attempt to respond to the Japanese challenge and understand their experience. This is more than a new management philosophy. The growing emphasis on the concept of total quality concentrates mainly on the structure of company management and control. It modified the model of organisational control. Merli (1988), claims that in this model, management is making a crucial shift from top-down hierarchy to bottom-to-top participation, accompanied by the 'philosophy' of empowerment, participation and involvement of the workforce, in contrast to previous models based on centrally-managed mechanical systems.

If this description reflects a broad sense of the industrial transformation that is currently taking place and likely to continue from now to the end of the century, we cannot ignore asking and answering some important political questions. The first question concerns power relations among the social subjects within the company.

According to the traditional view, company bureaucracy, or hierarchy, is organised into a line of command consisting of more than one hierarchical level. For each level a combination of technical and managerial responsibilities is defined. The line of command proceeds in a single direction, from top-to-bottom. In ideal conditions, the execution of a job is predefined by the top-level management. In short, just like a mechanical machine, the company is totally predictable and its functions and performance are programmed. Since the post-war period this unrealistic model has been refined with increasing acceleration, for example, by identifying the functional necessity of lines of communication from bottom to top, as well as horizontally. This revision has led to increasingly complex models with respect to the original distinction between 'line' and 'staff'. Mechanical analogies have been transformed into systemic biological analogies.

Many basic ideas of 'scientific management' still prevailed until the end of the 1980s. Even the increasingly refined cultural and organisational elaboration of a new industrial culture has not stopped a robust phase of rationalisation which occurred in key sectors of industry throughout the world during the 1980s. For example, even a partial control of management won by trade unions during the 1960s and 1970s was lost.

A relaunching of 'scientific management' occurred both at the level of traditional tools and at the level of intermediate bureaucracy of 'line' management (the department heads). What is the role of intermediate bureaucracy? The traditional view assigns it a dual role: a vertical and active role, and a horizontal and passive one. The first consists of observing a group of people and machines, and guaranteeing that the company achieves the programmed performances, on the basis of a power delegated to intermediate bureaucracy and its own responsibilities. The second role provides a connection between the superior hierarchy and the activity of the organic and functional company with respect to strategic duties which intermediate bureaucracy is responsible for. Finally, a third and decisive role of intermediate bureaucracy must be noted: its political function. Bureaucracy is directly exposed to trade-union pressure in order to regulate and exclude labour performance from the unilateral exercise of the company. So intermediate bureaucracy must constantly achieve a balance between the exercise of discipline and the organisation of agreement.

It is no coincidence that, all over the world, the battle for rationalisation was fought at the level of the intermediate bureaucracy. During the 1980s, the key issue for top management, supported by the ownership, was to gain back the total control of the area of manufacturing transformation at any price. This happened to be the case everywhere, with differences related to the nature of pre-existing industrial relations.

What happens when the reorganisation process of world industry goes well beyond the rationalisation of the early 1980s? A certain breadth of self-regulation is extended to all levels of the company. Accordingly, intermediate bureaucracy is caught between two alternatives: on the one hand there is a process of verticalisation (that of command of production and conditioning of demand), which aims at achieving the systems efficiency through information systems. Intermediate bureaucracy is thus deprived of much of its technical and functional tasks. On the other hand (that of manufacturing transformation), the duties of regulation, control and discipline are questioned in direct proportion to the degree of extension and depth of the forms of self-regulation.

However, if the technical and functional contents disappear and the tasks of regulation, control and discipline are reduced, what is then left of industrial bureaucracies? It is mainly left with a communication and political function. Of what type and size should this bureaucracy be? Is there a way to transform this bureaucracy from the current state into a new one without damaging it?

The changing role of the industrial engineer illustrates this issue. For example, the roles and functions of the new quality control industrial technician is not dissimilar from that of industrial engineers in the quality control department. The shift from a notion of quality as a subsequent check to that of a first time quality, has thus led to a breakdown of the parallel bureaucracies of the quality inspectors.

In the manufacturing area, which is still of great employment significance for many industrial sectors, we have reached a crucial watershed: *either* the magnitude of the problem is ignored and only marginal adjustments are made, thereby continuing to concentrate on the classic rationalisation of costs; *or* the view introduced by total quality is assumed, thereby facing the problems connected with reducing bureaucracy within the company. It is not just a matter of reducing the number of middle-level managers and high-level managers, but rather that of redesigning the general functioning of manufacturing activities, transferring responsibilities and powers, and reorganising hierarchical levels and duties.

New workplace innovations so far seems to be oriented towards teamworking (here too the reference point is the Japanese experience). Considerable problems arise even at this level since two opposing solutions might be taken: the first solution attempts at saving as much of the old structure as possible. Accordingly, techniques of by-passing the intermediate hierarchy are used, in order to accumulate the advantage of the old system with more fluid forms of indication and adjustment of the bureaucratic structures. The insistence on 'quality circles' comes from this viewpoint. By definition, the 'circles' are parallel to the old structure and constitute an attempt at revitalisation.

The second solution aims at reorganising the company structure at a manufacturing level, by eliminating a considerable layer of intermediate bureaucracy and shifting a group of operational functions to collective and self-regulated responsibilities of the group. The groups formed in the area of manufacturing represent a new structure and not a parallel organisational form of support .

Complex organisations which are submitted to this difficult transition seem to imply:

1. a partial watering-down of the 'welfare state' at a company level, which was one of the hinge elements of the legitimisation of top management; and,
2. a revision of the extent and depth of the powers and functions of the bureaucracy for making a shift into the area of self-regulation.

This change must face the active, and perhaps politically represented, resistance from the hierarchies in question, as well as withstand all the possible passive resistance. It must also actively involve the workers, which leads to theoretical and practical implications that are not difficult to imagine.

It is not simply a difficult passage but a mandatory one: in fact sooner or later problems will have to be faced involving the distribution of power and the determination of the motivations necessary for participation.

Such a complex process cannot be governed in a purely autocratic way by a group of enlightened managers, supported by the ownership, who bring about the change with the attitude typical of a surgeon, who has to act without remorse.

The determination of the top management and the ownership can contribute in eliminating at least, in part, the old bureaucracy which resists. However, the problem remains as to how a new hierarchy will be produced and an active and stable approval by the workers will be gained, thus guaranteeing the efficiency of self-determination.

The North-European experiences during the 1960s and 1970s is evidence of the repeated failures of attempts at participation directed by the management with pure techniques of staff management.

A process of verticalisation can take place in various forms, not necessarily strongly hierarchical. The deepest significance of the linearisation process lies in the integration of activities for the purpose of strategic company management, which were once distributed between 'line' and 'staff', and attributed at different hierarchical company levels (R&D, design, publicity, distribution, etc.). These stages and functions are verticalised and integrated, with an accentuation of the elements of cooperation between the company functions (see Figure 56). The overall functioning of the company thus becomes more transparent and the political nature of the strategic choices more evident. Also, the responsibilities of the top management in any single stage of accomplishment of a strategy are more evident. If the control of some key variables in achieving company performance has been dematerialised and has passed into the verticalised area (i.e., into strategic management), the regulated functioning of a company can no longer be prescribed, at least to a great extent, and must depend on intelligent adjustments, not predefined by operators in the manufacturing sector.

It must be noted, however, that it is extremely difficult for creative behaviours to cohabit in strictly totalitarian social climates. Such a problem affects the Japanese companies too. It is certainly possible to construct a community linked by non-rational restrictions, whose ideal is the clan (Ouchi, 1982), and with very strong elements of identification between individuals and the community. Something similar has already happened in the countries of 'real socialism'. In such a situation, behaviours can be intensely collaborative without necessarily being commanded, though they will basically be of a gregarious and thus conservative kind. The community will be based on principles that prevent changes occurring.

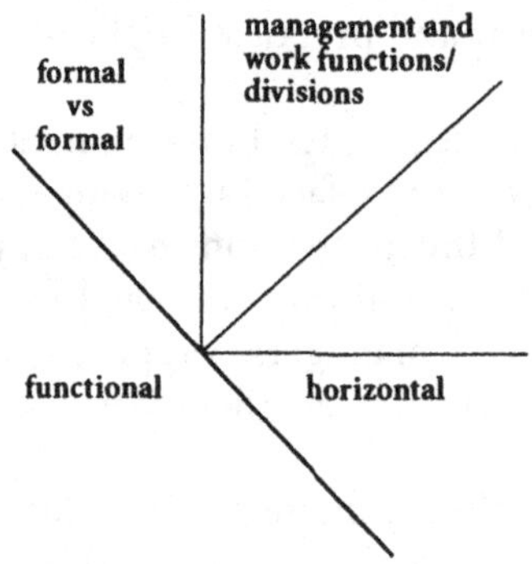

Figure 56: Five divisions/integration areas: horizontal, functional, vertical, power and control between management, employees, formal and informal

The dilemma for the top management is very intricate. The solution so far proposed on the international scene can be divided into two areas: the first is based on the self-sufficiency of the top management and of the ownership; the second is centred on the awareness that a revolution of this type cannot be brought about without defining new power relations between the ownership and top management on the one side, and organised workers on the other.

The self-sufficiency solutions attempted by managers to reproduce the Japanese experience in the European context can be described as follows:

1. a simplification and reduction of the intermediate hierarchy;

2. a decentralisation of internal management responsibility;
3. the introduction of forms of group-work in the strategic part of the manufacturing area, complementing traditional forms;
4. an improvement in company communications;
5. a relaunching of the forms of motivation which best symbolise a philosophy of participation, such as shares linked to company performance;
6. little is done with respect to the very meaning of participation in the clan characteristics of the Japanese solution, such as life employment, etc.

This is an attempt to divide the workforce into various layers with differential access to the company structure. For most part, it is a form of a 'mystical communion' which is dependent on the nature of refined communication techniques. For a minority it is matter of power attribution within strictly functional limits that does not reduce the social control of the company. Such a strategy is obviously the most reassuring one, being the least antagonistic towards traditional managerial culture. Throughout Europe and in the United States, there has already been strong resistance, if not actual open opposition, by representative sectors of the trade union movement. The defence of a traditional social control system leads to costs which can sometimes completely cancel out the expected benefits (see Table 18, below).

Table 18: The main sociological dimensions in an organisational change – based on the three-dimensions scheme by Linhart (*AI & Society*, vol. 8.3)

Internal functioning	*Philosophy of social relations*	*Work organisation*
The means of articulating and making coherent the different functions of the firm according to clearly defined criteria, in terms of references and objectives	The different means of workforce ideological mobilisation	The formal determination of work operations and of skills, as well as the condition of their realisation
Is it a synchronous process?		
Very fast and powerful?	Very fast and powerful?	When?

So, according to this table we need to know the three dimensions in order to have an integrated assessment of the 'team' phenomenon.

On an international scale, the most significant case concerns flexible technological applications, both in the manufacturing area (FMS) and in technical offices (CAD) or applications of an integrative nature like logistics. Fiat factories provide a good example of how high these costs of a self-sufficiency solution by managers can be. Of course, the proponents of these directions of change minimise the impact of structural changes of the organisation, and instead focus only on the division of tasks in the managerial area.

However, the trend towards this 'self sufficiency' seems to be significant throughout the West, especially in Northern-European countries. Similar changes are also taking place in Denmark, Holland, Germany and Italy, as well as in the United States and Canada. Management has abandoned the old model of social control with a broad and collectively negotiated transfer of self-regulation for the workers. To some extent this has restored a situation of non-transparency of the group-work methods for company management, the 'black box' of the old craft-based labour. This type of 'self-regulation' can be highly competitive, if strict coercive mechanisms of the Taylorist model are not replaced by a participatory model which enables negotiation and democratic means of agreement between autonomous and responsible subjects.

The main European industrial trade unions are gradually becoming aware of the strategic nature of total quality. The point at issue is not how to defend the employees from these processes, but how trade unions can make positive use of them, and also vice versa: how can enterprises involved in innovation processes benefit from the shaping of competencies of the workforce and their organisations? This means that the participation of workers and trade unions should not just be considered a necessary concession in order to guarantee workers' consensus but also a potential resource for innovation. Consideration of the alternative shaping concepts may be useful in order to implement radical and long-term processes of modernisation which are oriented at both the economic interests of the enterprise and the social interest of the workforce. If we consider workgroups as a possible clue in assessing the state of progress of total quality, we can see significant innovations at a European level. In the Federal Republic of Germany, for instance, the metalworkers' union has gradually overcome the traditional hostility and mistrust towards forms of group-work. For example, Ig Metall, formally defined workgroups as a fundamental principle for labour organisation of the future (IG Metall, 1991; Roth, 1992).

But what do managers and trade unionists mean by 'workgroups'? Some European trade unions consider workgroups positively insofar as they are as uniform as possible, in order for the workers to achieve the highest levels of qualification. On the managerial side, there is a preference for forms of cooperation between workers with different levels of qualification, in order to emphasise the elements of flexibility of worker performance. In Italy, CGIL, the Italian trade union, decided to support self-regulated forms of labour within the limits of a new system of industrial relations founded on a new system of industrial relations which is based on the principle of 'codetermination'.

So far this orientation of 'group-work' has achieved positive results in innovative small and medium-sized companies in the North of Italy which were the first to introduce total quality. Generally speaking, these Italian workgroups admit marked professional differences within the group, in direct proportion to the breadth of the work areas covered and to the degree of self-regulation possible within the group. An agreement between social partners at the workplace seems possible only if an equilibrium is established between the two sides of the balance: on the one side, flexibility and an intelligent cooperation towards the creative and dynamic solution of unexpected problems, and on the other side, the formal recognition of a collective power of self-regulation of aspects concerning the methods of labour, including those once split in terms of control and supervision. Such a balance is meaningful only within a system of industrial relations which rejects a pure, minute-by-minute, assessment of the relations between the social partners. It partly requires a theory of the company as a democratic space where a conflict between the management and the ownership, on the one hand, and the workers' delegation on the other, can occur and be seen as a positive ' thrust' for change.

Only in a democratic context of this type do the conditions for the debureaucratisation of the managerial structure exist without resorting to autocratic solutions which would start off the evil spiral again: authoritarian and despotic management - alienation of content and meaning of labour - reduction in responsibility, demotivation - authoritarian escalation - increase of an organism of control. Without a reorganisation of powers to this extent and the formal definition of a 'constitutional pact', only oligarchical forms are possible, which are increasingly segmented in order to create a compromise between social control and structures of flexibility.

The Contextual or Meso-Level

Our interpretation of the contextual level is strongly related to the change both in the way the theorists consider the relationship between the organisation and its environment, and the new interdependency between each organisation and its environment.

As to the theoretical perspective most of the scientists agree on the statement that there is an active relationship between change within each organisation and change in the related environment. As van Beinum points out:

> Organisational change is therefore rooted in the interdependencies between the organisation and its environment. Consequently, the unit of analysis and the unit of organisational change is not the organisation as such, but it is the organisation and its environment.
>
> *van Beinum, op. cit.: 181*

The main relationship between each organisation and its environment is a process of mutual adaptation. It is a very fragile process because of its multi-dimensional structure, and this is especially true in a fast-changing environment. The process of adaptation is the actual way of metabolising an exogenous process of change whether the exogenous process is a major change in the practice of managing an organisation (i.e. starting from within organisations); or an overall institutional change, i.e. starting from the social level, either a country or a region such as the EU (North, 1990). The processes of metabolisation are culturally and historically based: thus it is quite impossible to plan a process of change in the contextual environment, in a self-sufficient way. The contextual level is both the unit of analysis and the unit of change, accordingly:

> the issues involved are too extensive and too many-sided (meta problems) to be coped with by any single organisation ... The response capability required to deal with these meta problems is inter- and multi-organisational
>
> *van Beinum, op. cit.: 182*

But the difficulties stemming from the inter and multi-organisational features of these issues are only one side of the coin. The other is what Fernand Braudel calls 'civilisations'. 'Put briefly,' he says:

> the measuring stick by which the confused mass of words is judged and classified (in order of importance), and which is not any less confused than are men themselves, is the time it takes them to disappear from the face of the earth ...

He continues, noting that:

> by 'structure', the commentators of social reality mean an organisation, a coherence of fairly stable relationships between social reality and the masses ... Some structures, lasting a long time, become stable elements for an infinity of generations: they hamper, hinder and determine the course of history ... Like obstacles, they are characterised as limits, in a mathematical sense, from which human beings with their experiences cannot in any way free themselves ... Even our mental processes are long-term prisons
>
> *Brandel, 1973*

In other words, a successful process of change needs some kind of personal involvement in the overall goal of the process of change itself. In turn, it implies some kind of permeability of the process of change to the value and views of the people involved. The bilateral trait of the process of change makes the process of adaptation very difficult. When this process fails or falls short, the overall process of change is jeopardised both at the organisational level and at the contextual level.

As to the changes which have occurred in the last ten years, the major innovation is a new kind of integration between the organisation and what the socio-ecological approach considers the transactional level (van

Beinum, *op. cit.*). This is the environment the organisation interacts with in carrying out its primary task. The integration at this level is much higher than in the past due to the 'just-in-time' techniques and to the so-called information society (Bangemann Report, 1994), and the related so called interconnectivity paradigm (OTA, 1994). It is at this level that the transaction costs play a central role and:

> These kinds of transaction costs are on the rise in today's global, knowledge-based economy comprised of many more players and fewer standardised, mass-produced products.
>
> *Ibid.: 31*

The networked economy, either electronically linked or not, will be tested at this level. The new experiences of integration – the so-called 'comakership' – among producers and suppliers will be tested at this level, etc. According to one of the conclusion of the FAST research project on 'The future of Industry in Europe' (FINE), the spatial context and proximity of firms will become more important in a global environment in order to favour processes of innovation (van Banemer, 1994). On the basis of the proximity of firms it is easier to link the knowledge of different firms and other actors with one another. Therefore, the formation of production clusters may favour innovation processes. They constitute a favourable framework for mutual learning and the formation of the workforce.

However, crucial to this perspective of innovation is the wider socio-economic environment, which provides space for enacting enabling public policies such as those by local, national and regional institutions and/or collective policies by union and employer representatives. In this wider space, general policies could test the capability of the environment to affect the change in the daily life of individuals, of organisations and of the so called 'meso-society'.

The Macro-Level

At this level the significant tendencies are:

1. the process of globalisation of the market and the rise of a customer-oriented process of production;
2. the convergence of manufacturing, telecommunications and information technology, forming the basis of the information society;
3. the birth of the knowledge-based society.

It should be noted that these tendencies are not a linear progression but a social process (Belussi and Garibaldo, 1994). In summary, the integration of the international economy:

> ... has given rise to a global economy in which patterns of international trade primarily reflect patterns of international production. Specialisation takes place on the basis of parts and specialised components, rather than on the

> exchange of finished products as in the past... Patterns of direct investments abroad also highlight this trend... Multinational corporations are also driving the trend toward globalisation. To compete in today's global economy, companies must integrate their activities on a world-wide basis, allocating activities among a number of countries to gain the greatest advantage... When not fully integrated into multinational corporations, these firms are networking their activities across global boundaries.
>
> *OTA, op. cit.: 12; Dankbar, 1993; Garibaldo, 1993*

On the grounds of this international integration the competitive advantage rests more and more on the capability to meet customer demands, as well as on the time-to-market through the innovation and production processes as a whole and the quality of products and services. As David Whitwam of Whirlpool CEO puts it:

> Our strategy is based on the premise that world class cost and quality are merely the ante – the price of being in the game at all. We have to provide a compelling reason other than the price for consumers to buy Whirlpool - built products. We can do that only by understanding the consumer better than any one else does and then translating our understanding into clearly superior product designs, features and after-sales support.
>
> *Whitwam, 1994*

The new communication technologies strongly support this perspective allowing the creation of a networked economy (Bradley *et al.*, 1993), but what is necessary is a shift from a Taylorist–Fordist paradigm of productivity to a new one that should be built on the basis of both a widespread availability of the general knowledge congruent with these trends. This is a matter of public policies, and of universal rights to continuous education as a social condition of productivity, a matter of democracy as well as that of industrial relations. This shift implies a totally different concept of the relationship between market and public policies as well as of labour.

The transition from the old economy to the networked and knowledge-based economy (Reich, *op. cit.*), looks like the setting up of a railway system in the first industrial revolution. It is something too expensive, too complex and dependent upon too many social and institutional preconditions, destined to be the natural outcome of the unleashed market forces. It is argued that a global strategy of innovation should be defined by public institutions as well as by social factors. This statement raises many questions and among them the most important is: what kind of relationship is there between the meso- and the macro-levels?

It is important to emphasise the relevance of the meso-level for enacting strategies and policies of innovation, and the complexity of planning a course of action in a democratic and participatory way. According to some scientists this could lead to a completely process-oriented concept of change; this means a radical indirect process approach: the substantive issues necessary to the process will be generated in a spontaneous way through the dialogue of the participants. It looks like a *tabula rasa* setting. It is, however, worth pointing out that the concept of a democratic decision-making process and participatory design doesn't imply *a laissez-faire* idea. On the contrary, as Emery pointed out, the idea seems, in some way, an antagonist to a democratic decision-making process (Emery, 1993). It is our belief that it is the macro-level at which public and collective policies succeed or fail in creating the public good and the general knowledge which feed and support both the process of innovation and adaptation, i.e. the process of change, that will take place and will be shaped at the meso-level.

Alternative Paradigms of Team-working and Their Social Implications

Groups Versus Teams

From an historical perspective there are two different periods in the experience of working in groups within organisations. The first period is that of the rationalisation of the old Tayloristic–Fordist organisations at the beginning of the 1980's. Referring to the automobile sector Jürgens, Malsch and Dohse point out:

> The main goals for production groups were an increased deployment flexibility on similar work stations and the reduction of cycle loss time on the lines
>
> *Jürgens et al., op. cit.: 335*

In this case, it is very difficult to state that the Japanese group principle – 'using informal aspects of group relations for the goals of productivity and social integration' (*ibid.*: 48) can be operationalised. It seems more realistic to consider this as a logical development of the efforts aimed at some kind of reform of the old paradigm. It is a matter of loosening the degree of horizontal and vertical fragmentation of tasks and of enhancing the possibility of group formation. Bearing in mind the five dimensions of division/integration of tasks and roles, it is clear that the other three dimensions, i.e. the idea of some kind of de-bureaucratisation of the organisation as a whole, is out of the question. The central issue for unions and management is the scope of self-regulation and its implications such as a certain degree of decoupling between performing tasks and the pace of the assembly line and/or the possibility of obtaining qualification alignment (Qvale, 1995).

The second period is clearly marked by a benchmarking process with reference to the Japanese experience. The Japanese myth was stereotyped; but it was not only a matter of propaganda by management to maintain the previous situation, things are actually changing (Jürgens *et al., op. cit.*:

376). The driving force of change at the organisational level is the integration on a wider scale than before; it seems realistic to say:

> we can observe a growing de-coupling of the system elements of the Taylorist–Fordist mode as we presented them in Table 17, above. According to our observations, this structural connection began to break up in the 1980s. This situation offered considerable scope for design and a corresponding mood of change could be observed at many sites
>
> *Ibid.*

As to group-work this change leads to a new paradigm. There is a growing awareness both on the management side and on the union side of the very fact that the competitive edge of a single firm will depend more and more on its ability to utilise and develop its human resources as the cheaper and most effective way of a dynamic adaptation to a changing environment. In this perspective, group-work deals less and less with classical workforce flexibility and more and more with the capability and the willingness of each person to be committed to the best performance of the organisation; in other words, the organisation will depend more and more on its employees' expertise and willingness to attain this capability. Indeed, new possibilities arise.

Capability here means the expertise to cope with a complex and turbulent environment, and expertise, adaptive expertise (Meghnagi, 1993a), which depends on a very complex interaction between the individual's general knowledge, on-the-job learning processes, entitlements for job organisation and job design. This process, in turn, will interact with the dynamics at the meso- and macro-level.

Willingness here means the availability of people to be committed to a brand new way of performing jobs based on autonomy, responsibility, pro-active style in job performance. This availability will depend, as to its scope and the capability to be long lasting, on the meso- and macro-level, too. But the major question is: why should people be ready to confront a new demanding way of working in order to accomplish their jobs? This issue is strongly related to what Jürgens *et al.* call the 'group principle' (Jürgens *et al.*, 1987: 92-110). This was the leading and central innovation in the Japanese experience:

> The group's spectrum of functions for the individual is almost all-encompassing: it is a substitute family and a social network, an educational authority (for late entrants for example), a place for learning (on-the-job and in its function as quality circle), an organiser of leisure time, a unit for performance regulation (time allocation and efficiency control do not take place in respect to the individual job, but rather in respect to the work area of the entire group) and quality regulation (the production group is responsible for the quality of its area).
>
> *Jürgens et al., 1993*

The relevant feature of the Japanese experience seems to be a social experience between community and society. What about the West? The idea of acting as Japanese and considering the Japanese experience as a whole – a model – and as a recipe book (*ibid.*; Linhart, 1995), is loosing its impetus among the scientific community as well as among managers and policy makers (Dunlop, 1994). There is a growing belief that the answer for the West is not a strategy which aims at catching up with a stereotyped Japanese model or to implement the paraphernalia of so-called 'lean production'. To reach this awareness, however, has taken a tremendous waste of time and money. Paradoxically, this new belief puts more emphasis on the core group of innovations embedded in the Japanese experience, the process of removing bureaucracy (Garibaldo, 1995), and among these the major one is: the personal commitment of each person to cope with a proactive way of accomplishing his/her job; i.e., it is out of the question:

> The fact that there is a new major breakthrough in the theory and experience of managing industrial firms and/or organisations; the breakthrough comes first of all from the Toyota experience; that experience was the point of departure of a new theory... The strategy is largely based on the specific Japanese social context but, as a theory, it contains a core set of concepts that can be the conceptual point of departure of a specific country/local process of interpretation/metabolisation, enabling the creation of a functional equivalent of Japanese industrial performances.
>
> *Garibaldo, 1993*

The Japanese way of personal commitment lies in the community-like paradigm of the group principle; the community-like paradigm is rather unsuitable to the Western culture, which is based on the division and balance of powers. A strong drive in that direction will imply a severe breakdown of shared democratic values. Thus, if it is not the case to copy Japan, the problem that the West has to cope with is: how to gain employee's personal commitment? Is personal commitment in the communal sense possible within the framework of the Western social and institutional regulations which are rooted in nationally- and locally-based systems? Without this commitment, any change, however smart the design, will fall short. Hence it is possible to sum up these considerations by sketching two alternative heuristic and normative concepts of working in a group: groups vs. teams.

Groups are:

1. a form of cooperative relationship mainly aimed at reaching the classical workforce flexibility that is the personal capability of affording different tasks in a wider area in order to cope with different levels of production goals;
2. consequently, groups are homogeneous as to the member's capability of accomplishing many tasks and generally speaking, they are located

inside a specific function and/or department of the organisation; basically these experiences are located at the shop-floor level;

3. the knowledge requirements are primarily based on-the-job training;
4. a form of compromise between management and labour, based on a trade-off between flexibility, on the management side, and a mixture of qualification alignment and more functional autonomy on job performance, on the labour side;
5. a form of work reorganisation between the old paradigm and the newly developing one; the degree of innovation is dependent on the scope and the nature of the autonomy. In the 1980s it was possible to find a very wide range of experiences dependent upon the social actors' policies, the critical variables stressed in each organisation, the leading tendencies both on a meso- and macro-level.

Teams are:

1. a form of cooperative relationship aimed at creating a social environment for both the integration of tasks, functions, roles and the employee's personal commitment to the newly developing international standard of competitiveness; the key issues are quality and customerisation, self-regulation of the integration of personal activity within the team and of the team with other teams upstream and downstream as well as sidestream;
2. consequently, these experiences are located both within each function and department and among different functions and departments; generally, they will affect each level of the organisation;
3. teams are primarily based on a good background of general knowledge and continuous education;
4. at present, management and labour are wondering whether or not it is possible to create a new form of social compromise on these grounds. They are considering the 'whys' and the 'wherefores' of such a compromise. International evidence makes it clear that a new social compromise could not be built only on the basis of the improvement of work conditions; the nature of management demand on the workforce leads to the very nature of the basic social relations of employment. It seems more and more inadequate to utilise the concept of workforce in the classical Marxian version of a general and abstract capability of producing wealth; a knowledge-based economy implies a different kind of employment relation based on a different structure of entitlements for employees. The old 'metric system' to measure work and productivity becomes quite inconsistent with the new concepts, where new problems and issues are raised on how to reward work. The competitive edge of the organisation strategy will depend on the employee's commitment, but, this in turn depends

upon an effective integration of functions and roles, that is on the performance of teams and of clusters of teams;

5. a form of work organisation dependent on a different paradigm of organisation both from the viewpoint of organisational engineering and from that of the role of organisations at the meso- and macro-level. The radicality of change at the organisational level requires a general change where the unit of analysis and the unit of organisational change is at the meso-level.

Strategies, Alternatives, Dilemmas

The polarised and normative way in which the two paradigms are sketched above emphasises the necessity of a cultural and political shift for management, labour and society. As Qvale (1995) points out:

> One tentative conclusion to draw may be that a jointly and strongly felt need for change, may be a more important condition for large scale change in worklife than peaceful and trusting industrial relations in preceding periods with stable growth.

We hold that economic and social pressures are the main driving forces behind the ideas of teamwork and their operation at each societal level. This requires two major requirements for change: the adaptive and the proactive ones. The former could be described as the understanding of a need for change in order to remain competitive on the market; the latter as a vision of management and/or unions and/or work councils in order to anticipate emerging trends and avoid a state of crisis or of inadequacy vis-a-vis the global competition. In many European countries it means overcoming the old vision of industrial relations between Labour and Capital based on antagonistic competition, and reach the joint awareness of a need for change. The problem then is to find a common ground and to list the key points of this process as well as to map out the main steps of a strategy for development. What seems very important on the basis of the experience of the last thirty years, is to avoid a separation/divergence between the so-called human factors and the economic and technological factors. What is required a more holistic framework for reaching a common ground (*ibid.*).

The search for common ground and the process of creation of the awareness of issues of common interest between Labour and Capital and of general social interest could start from the core concept of the new standard for competition: quality and customer-driven economy. For instance, all the experiences of reshaping work organisation at the moment in small- and medium-sized industrial firms in North-East Italy started from the problem of reaching the ISO 9000 standard but the successful ones are those which were able to define a common agenda with labour representatives at the plant level (Garibaldo *op. cit.*).

What is very interesting about these experiences is that all of them, the successful and unsuccessful ones, had to deal with team-working. This process is very difficult, indeed, because social actors have to cope with true dilemmas; they have to choose what is relevant because a 'catch all' solution doesn't exist.

It seems useful to go into three leading issues in the creation of a self-sustaining process (Qvale, *op. cit.*):

a. trust and commitment;

b. the nature of social cooperation in team-working and its social implication;

c. the role of the meso-level and the setting up of a new specific space for action.

Trust and Commitment

According to literature and experience two critical factors for a successful process of change, especially for team-working are (Ehn, 1988):

1. trust between social partners; and,
2. the commitment of top management.

The idea of trust is illustrated here with two case studies, one with a positive trend and the second with a negative end; the first concerns an Italian food company, Barilla, and the second concerns a small Italian firm. In the Barilla case study the group-work consisted of workers from the marketing sector, product development sector and plant design sector. The investigation focused on the question: 'What, in your opinion, facilitates or impedes creativity in your job?' The findings of the case study is: 'Several times we have underlined the presence of trust in the group and, in our opinion, trust represents the basic factor of creativity.'

Even if it is not possible to explain the single creative act, it is legitimate to formulate hypotheses of the environment that can encourage it. Trust, as observed by many, derives from a rich environment and, in turn, allows for the emergence of thought for the primary process, the premise for creativity. From another perspective, that of Winnicot (1974), we could say that trust allows for the manifestation of the *True Self* which possesses the creative potential; whereas the lack of confidence, the obscurity of the decisions, their 'trickling down from above', all these factors stimulate the defences of the complacent *False Self*, a defence that can lead to behaviour characterised by obedience and diligence, but certainly does not engender a spirit of adventure, i.e. creativeness. Indeed, the defensive environment leads to a passive adaptation. Such an environment hampers the creative potential, because it eliminates or flattens out the mental disposition that underpins it. But also because what is 'new' is unconsciously considered a transgression from the established hierarchical order, a transgression that generates guilt and thus, as a defence, inhibits the intellectual function that presides over the creative processes. It could also be said that in this environment the creative act is considered to be too strong a risk, (IRES, 1994). According to the Barilla case study:

> ... the presence of trust in the group that reproduces its presence in the work environment, allows us to state that the latter is organised in such a way as not to undermine, or undermine significantly, the trust that each individual has placed in it
>
> *Ibid.*

and the participants express this feeling in a clear way:

> ... a person expresses himself at his best, freely... if he doesn't always feel judged, touched by what is said... if he feels he is in the realm of purposeful freedom... which means that if someone says something that is not shared by the others he doesn't feel that he has his back against the wall... even if he says something stupid... the atmosphere here is reassuring, so it enables people to express themselves better... Besides, we can also focus on trivial things, because until something has actually been shown to be trivial, it just isn't...
>
> *Ibid.*

Indeed, from this case study it is legitimate to suppose that Barilla has encouraged creativity of participants.

The Nature of Social Cooperation

The kind of social cooperation within a group of people carrying out a common goal can vary considerably. Social analysis is not able to assess both the reason for the success or failure of team-working and the nature of group dynamics within teams. It is necessary therefore to turn to a psychological perspective. First of all, this perspective highlights the nature of the interaction:

> There is a sociological and a psychological definition of the group... which is very different. In the first instance, we have a description of those activities and actions that are accomplished at the group level. Vice versa, the psychological perspective is more oriented towards the description of what happens in the group at the mental level. It not only takes account of the actions that take place in it but also of all the mental parts that are activated by being in the group, whether they are of a rational or emotional kind.
>
> *Rebecchi, 1995*

> At the first stage of group working, there is inevitably a series of resistances that hamper the work itself. These resistances have been catalogued and described in various ways; some are well-known (for example, dependence, pairing up, attack and fleeing). Instances of these occur in all the work groups... In short, it may be said that when defences are triggered in a group, the collective activity is compromised; the group functions as a group work only when the defences are absent.
>
> *Ibid.*

From a psychological perspective, it is possible to describe the following factors for group-working:

a. the existence of a definite theme, an exact goal
It seems obvious but 'each one of us brings everything of himself to the meeting he/she is taking part in his/her personal and family problems etc.' (*ibid.*). There needs to be an authority to se the agenda and ensure it is achieved. Regretfully this means that the group-work is not in itself as democratic as someone might think it is.

b. there must be a leader
Once the above two conditions are achieved, this does not necessarily mean that the workgroup will become effective. It depends on many other conditions, which are dependent on both the social environment and the vision of the organisation. Two primary issues arise:

1. the possibility of transforming the context by the group; in many cases, however, the environment can impede the activities and the goal of the group;
2. group-work is basically a functional activity, illustrated by the following:

> ...the group's work is essentially heterogeneous as it combines different experiences of a more complex knowledge... The group work questions Taylorism because it not only establishes that the duties broken down by Taylorism must be reassembled (the so-called reparative aspect), but suggests that group work modalities must be utilised in order to realise creative and complex jobs.
>
> *Ibid.*

It is possible to see a strong convergence between the sociological /organisational perspective according to our view point, and the psychological perspective, according to the English school (Bion, 1961). Table 19 sums up the key factors of this convergence. The empirical evidence of the Barilla study case supports this kind of close relationship between the scope of creativity and self regulation, as being the main reason for team-working. It also determines the degree to which the advanced paradigm of working in group can be implemented.

The analytical digression on the nature of team-working has been influenced by our bias against the very simplistic concept of 'team-working' which is prevalent among social actors and policy makers. The basic concept seems to be that team-working is only a matter of social and organisational engineering: what in the old organisational model was designed as being divided, in the new model it has to be integrated. According to the analysis above, the setting up of teams in the socio-organisational perspective means to jeopardise the old structure as a whole and this is very difficult to attain and requires a large coalition of forces within each organisation involved in the process of change. This process will take a very long time because it will change the daily life of the organisation which is the solid base of the organisation. It will also interact with all social features – visions, values, culture, vested interests, and will involve all levels of analysis and action – micro-, meso- and macro-. It means that setting up teams cannot be taken for granted and defined *a priori*.

Table 19: Table of convergence developed by Francesco Garibaldo and Emilio Rebechhi

Issues	*Sociological perspective*	*Psycological perspective*
Type of working in group	Grouping vs. teaming	Group driven by basic emotion vs. workgroup
Cooperative relation in the two basic types	Mutual equivalence of each other of different knowledge	Missing vs. integration of knowledge
Goals	Quantitative flexibility of the workforce vs. integration of the team's tasks in the environment and self-regulation of the feedback clues	Defence of the group's members against strong emotions vs. cooperation to achieve a common task
Nature of the relation among the members at the best of the range of possibility	Defence against the environment demands vs. some degree of control of the environment	Emotionally-based vs. a shared mental space felt as necessary to accomplish common goal
Inside organisation	Missing vs. division of labour and leadership functional to the control of the boundaries of the group activity	Charismatic head vs. leadership functional to the stability of the setting
Driving force	External and out of control: the motivation and the driving force is totally exogenous vs. a combination of external and internal motivations and forces	Basic emotion rooted inside group: self-referent vs. a texture of emotion and rational tasks: reality referred

If the actual process of change is a multidimensional process, the setting up of teams is the core of the meta-nature of the process itself. Such costly process of change cannot be undertaken by single organisations. What is required a diffusion perspective and this implies social transparency and availability of a course of action based on the following virtuous spiral:

- knowledge dissemination;
- on-the-field experiences;
- evaluation/feedback process;

- knowledge store;
- knowledge dissemination.

From a psychological perspective, the very idea of the setting up of workgroups as the natural outcome of a decision by the board of directors or of some social reformer initiative is meaningless. In this perspective, the unit of analysis and action could not just be the single workgroup or the organisation in itself; to work in a group is a very complex outcome of a change of perspective of society as a whole. It is necessary to initiate this process at the primary school level in order to teach people how to work in a group. According to literature. This in turn, means organising the whole teaching process in the same way.

We therefore regard the inter-relationship between the micro, the meso- and the macro-level, as central to the formation of workgroups or teams.

A New Space Encompassing the Micro, Meso- and Macro-Levels

The connection between enterprises and schools seems quite evident; what is not so evident is the nature of the relationship. According to the previous analysis it can be described as a bilateral relationship for two main reasons:

1. the close connection between the social structure within the enterprise or the organisation and the nature of the learning process;
2. a strong demand for continuous education.

These reasons, in turn, are interwoven. In fact the birth of a network and knowledge-based economy and society (Cook, 1994; Bradley *et al.*, 1993), means that 'the level of innovation within a cluster (production clusters) depends on the existence of a high level of common knowledge and an advanced system of collective norms' (Telljohann, 1993), and the way to foster innovation is the setting up of 'collective activities of enterprises, public authorities and research institutions that support mutual learning' (*ibid.*). On the other hand, the labour market could evolve in two opposite directions depending on the choices social actors and public authorities will make for the future. It is quite self-evident that the nature of these choices are strongly related to the nature of the organisational and social change within organisations. In short, there are different patterns of consistent frameworks connecting the levels of the organisation (the micro) with the meso- and the macro-levels. These different patterns and alternative frameworks are a matter of the interaction of the policies by all the social, public, political, cultural actors.

For a heuristic reason, it might be useful to sketch a polarised model of two internally consistent and alternative frameworks. The discriminant factor is the way of coping with the new criteria for competition (Garibaldo, 1993):

1. The full utilisation of the whole potential of human labour as the strategic factor. This means designing some kind of redundancy at the micro-level as well as at the other two levels. Redundant and widespread knowledge capability as well as a new set of rights for employees are not a waste of resources but issues of competitive advantageous.
2. The segmentation of the labour market into a core part and other secondary parts is becoming increasingly contingent as to the strategic management of the economic process. The core part is made up of permanent employees, i.e. those who are able to match the new skill and competency requirements of international competition. In this perspective, the investments in knowledge and rights have to be restricted to a few and to a minimum.

Having made this choice, we consider that the remainder of the workforce has to be either coherent or ineffective. In essence, the innovation at meso- level is at the heart of the matter of a successful change.

Concluding Remarks: Work and Managerial Culture, Education Policies and Transfer of Knowledge

We are witnessing the emergence of new relationships between innovation and the preservation of the patrimony of knowledge of work and management culture. The idea of globalisation is wrongly believed to be a progressive and linear realisation of a predefined set of ideas and innovative practices. We believe this idea is too simplistic and incongruous with the historical analysis. It is maintained that when the innovative process does not deal with incremental innovations but instead with a break in the paradigm (Kuhn, 1970), it is a process of metabolisation from a specific culture of forces which cannot be ignored or encapsulated (Jacobs, 1985). The process of metabolisation is neither natural nor guaranteed. There are and there have been regions, cities (Landes, 1969), nations, firms and institutes which were not able to realise a process of metabolisation when faced with an external radical change in their control. The process of metabolisation has a greater chance of success than a process of deliberated and programmed learning.

Two historical processes can be cited: one related to a nation and the other to a city, namely Prussia and Bologna. In both cases, it (the process) entailed confronting, in diverse historical periods, the birth of a new industrial reality; in the first case – Prussia (*ibid.*) – the birth of industry, in the second case – Bologna (Meghnagi, 1993b) – the passage from the silk industry to the mechanical industry. In each case, the obligatory reference point was England, and in both cases the fundamental choice was the setting up of schools which would directly teach the operators – workers and managers – the new culture and also how to establish a close relationship with the working process which on the whole signifies techno-professional instruction. This created a virtuous cycle between the school as a diffusion agent and the formalisation of a new industrial culture, the creation of new industrial firms and the availability of the labour-force.

The importance of education processes for innovation is well recognised. The differences are apparent when you examine the nature of the education process and the concrete meaning of the transmission and elaboration of knowledge. The concept which emerges is that of competency. The interest in analysing competence emerges from the need to understand the:

> ...transmission and elaboration of knowledge, taking account of the different contexts of knowledge, culture and power and forms in which these are socially defined, distributed and evaluated ... It is inadequate, for such a purpose, to refer exclusively to technical qualifications or to the promotion of professional knowledge. However, it is necessary to place at the centre of analysis, the recognition of the connection between working conditions and living standards, the comprehension of the global and procedural character of the transformations, the recognition of its own time and context, the assumption of an historical dimension which are linked in order to interpret and tackle them to social dynamics.
>
> Cf. *Marx*, 1857–1858; Reich, *op. cit.*

There is therefore the need to realise an education process similar to that in the XVIII century under altered historical conditions. This means going beyond the classical limits of an instructive process which is only scholarly and formalised. What is the core of this instructive process? The core of this new education process is the creation of a culture of group-work which requires a modern form of working cooperation. This cooperation needs to be capable of creating a bridge between general knowledge (Polanyi, 1967), on which the working process is based, and tacit knowledge (Ehn, *op. cit.*), (non-formalised or 'non-formalisable' knowledge) which constitutes the element of value – the expert competence – which can support proactive behaviour (Meghnagi, *op. cit.*). There is a need for the notion of competency because it:

> ...consents the overcoming of the alternative, gaps in the education cycle, between the value of theory and that of practice, since every theoretical acquisition has practical implications in an inadequately significant measure. Emphasis is placed, concerning the interdisciplinary character of knowledge, on many aspects which are difficult to understand and interpret so as to allow the decoding of the reality.
>
> *Reich, op. cit.*

Group-work cannot be taught theoretically as it requires personal experiences of working with others and the realisation of contributions of others for one's own work. This implies either a radical reform of schools or their internal structure, methods of teaching, as well as a different integration between formal teaching activities and the working context. This requires a real intertwining between education and working processes, a close liaison between skills and the upgrading of the workforce's skills. The introduction of new technology increases the quantity and quality of information. Professional competencies for controlling and managing complexities deriving from working situations are increasingly changing. Consequently, the workers must acquire theoretical, practical and specialised instruction, connected for example, to managing resources, comprehension of production processes, the relationship between product/ market, and firm policies.

Therefore, the necessary education cannot be aimed at defining specific professional profiles or transmitting only operative abilities concerning new machinery, but instead it must deal with the extension and the diffusion of competencies at various levels of the organisation. The limits of division of competencies disappear in favour of polyvalency and flexibility.

It is for this reason that the acquisition of theoretical, practical and specialised knowledge complicates such a framework, because instruction must be continuous so as to accompany and support the processes of change. This demands continuous training as an instrument for accelerating the adaptation of the workers to industrial and technological changes (Meghnagi, 1992).

> In the model which is proposed for the description of work within the organisation, competencies become a category of analysis which includes:
>
> *declaratory* competencies, founded on concepts and contents; *procedural* competencies which include knowledge of procedures and rules for solving a problem and consents to a more widespread control of the problem-solving process; *contextual* competencies relative to the possible forms of decisions and interventions in given situations and *strategies* with respect to other unexpected situations.
>
> *IRES, 1995*

Such a hypothesis involves a huge shift in public and private resources in favour of the education process (in the wide sense of the term). All this would be unrealistic if there was not the contemporary development of a different concept of the labour market and the relationship between the labour market and social change. The problems which emerge at this point, are:

1. the actual availability of innovative knowledge;
2. the process of storing and diffusing practical knowledge which begins from organisational training;
3. how to cope with the apparent dilemma in 'modern' planning:

> ... how does the expert make a contribution of planning without alienating people? This almost has the making of a paradox for social planners: the

> more knowledge experts accumulate, the greater the gap in understanding between them and the people and the less likely they are to go along with their plans for implementation ...

or, in other words,

> ...how can we expect to improve our planning in the face of relatively decreasing knowledge?
>
> *Cook, op. cit.*

4. what kind of relationship can be established between innovation as an exogenous factor and an innovative endogenous capacity.

Acknowledgements

This chapter is based on the report of an exploratory study carried out by a European network called 'WORKPLACE EUROPE' funded by the European Commission. The network was organised by the Istituto Ricerche Economiche e Sociali (IRES). The Study Team consisted of the following scientists: Palle Banke, Danish Technological Institute (DK); Francesco Garibaldo, Istituto Ricerche Economiche E Sociali (I); Karamjit S. Gill, SEAKE Centre, University of Brighton (UK); Ulrich Jürgens, Wzb Wissenschaftszentrum Berlin (D); Oscar Marchisio, Studio Giano (I); Piero Mussio, Università Di Roma (I); Thoralf Qvale, Work Research Institute (No); Emilio Rebecchi, Università Di Bologna (I); Ed Van Sluijs, Merit (NL); Gyorgy Szell, University of Osnabruk (D); Volker Telljohann, University of Osnabruk (D). Other contributors to the Study were Giuseppe Sciortino (I) and Volker Telljohann, University of Osnabruk (D).

References

Bangeman Report (1994). *Europe and the Global Information Society.* Recommendation to the European Council, DGXIII, European Commission, Brussels.

Belussi, F. and Garibaldo, F. (1994). 'Waiting for the Future: a possible modellisation of the post-Fordist society: why the 'old times' are not yet finished and the 'new times' are still so far?' Paper presented at *New Visions of the Post-Industrial Society*, Brighton, July 1994.

Bion, W. (1961). *Experiences in Groups and other Papers.* Tavistock publications, London.

Bradley, S. P., Hausman, J. A. and Nolan, R. L. (1993). *Globalisation, Technology and Competition: The Fusion of Computers and Telecommunications in the 1990s.* Harvard Business School Press, Boston.

Bradley, S. P., Hausman, J. A. and Nolan, R. L. (1993). *Globalisation, Technology and Competition: the fusion of computers and telecommunications in the 1990s.* Harvard Business School Press, Boston: 5, 11.

Braudel, F. (1973), *Scritti Sulla Storia.* Mondadori, Milano.

Cooke. P (1994). 'Building a 21st Century Regional Economy in Emilia Romagna'. International Workshop, *Industrial Districts and Local Economies Development In Italy. Challenges And Policy Perspectives*, Bologna, 1994, May 2–3, (cf. Telljohann, *op. cit.*).

Dankbar, B. (1993). *Economic Crisis and Institutional Change.* Maastricht Universitaire Press,

Dunlop, J. T. (1994). *Commission on The Future of Worker Management Relations.* Fact Finding Report, U. S. Department of Labour and U. S. Department of Commerce Washington, May 1994. It is the so called Dunlop Commission, by the name of the chairperson: Prof. John T. Dunlop.

Emery. M. (ed.), (1993). *Participative Design for Participative Democracy.* Centre for Continuing education, The Australian National University, Canberra.

Ehn, P. (1988). *Work Oriented Design of Computer Artifacts.* Arbetslivscentrum, Stockholm.

Garibaldo, F. (1993). *Processi Di Globalizzazione* – Spazio impresa.

Garibaldo, F. (1994). *Organisation and Framework for Managing the Social Dimensions of Change.* IRES Materials, 10, Roma.

Garibaldo, F. (1995). La Qualita' Totale Come Dilemma Politico – Riflessioni Dall' Italia – forthcoming in German translation in ARBEIT, IRES, Rome.

IG Metall, (1991). Tarifreform 2000 – ein gestaltungsrahmen fur die Industriearbeit der Zukunft. Frankfurt a.M. ('Qualified teamwork' is the core structure of an 'intelligent system of production' proposed by the IG Metall. The concept of qualified teamwork connects a high as well as a broad level of technical and social qualification with elevated opportunities of self-regulation and a corresponding level of employee motivation).

IRES, (1994). *The Barilla Case Study Reports* – IRES, Rome.

IRES, (1995). The Force project, *Technological Innovation and Evolution of the Professioanl Profiles: Needs Analysis and Vocational Up-dating Strategies.* IRES, Roma.

Jacobs, J. (1985). *Cities and the Wealth Of Nations.* First Vintage Books Edition, New York.

Jürgens, U., Dohse, K., Malsch, T. and Stromel, H. P. (1987). 'The Communication Structure between Management and Shop Floor: a Comparison of a Japanese and a German, in Jürgens, U., Malsch, T. and Dohse, K. (1993). *Breaking From Taylorism.* Cambridge University Press.

Kuhn, T. S. (1970). *The Structure of Scientific Revolutions,* 2nd Edition. Chicago University Press.

Landes, D. S. (1969). *The Unbound Prometeus,* Cambridge University Press.

Linhart, D. (1985). 'Managerial Innovation', in *AI & Society,* vol. 8.3.

Marx, K (1857-1858), *Grundisse,* Harmodndsworth, Penguin Books, 1973

Meghnagi, S. (1993a). Acquisizione Del Sapere E Competenza Negli Studi Di Matrice Cognitivista, in state-of-the-art papers in connection with Garibaldo, F., *op. cit.*

Meghnagi, S. (1993b). La Formazione Delle Competenze Professionali Nei Processi
Riorganizzativi – in *AA. VV – Auto E Lavoro,* Ediesse, Roma: 69–70.

Merli, G.(1988). 'IL Total Manufacturing Management' in *La Fabbrica Automatica*: 297.

North, D. C. (1990). *Institutions, Institutional Change, And Economic Performance.* Cambridge University Press.

OTA (1994). *Electronic Enterprises. Looking to the Future.* OTA–TCT–600, Washington, DC: U. S. Government Printing Office, May 1994. Chapter 2 and 3.

Ouchi, W. G (1982). *Theory Z.* Avon Books, NY.

Plant, in Trevor, M. (ed.), *The Internationalization of Japanese Business. European and Japanese Perspectives.* Campus/Westview, Frankfurt: 92–110.

Polanyi, M. (1967). *The Tacit Dimension.* Anchor Books, Doubleday & Company, New York.

Qvale .T. U. (1995). 'The Role Of Research For The Social Shaping Of New Technologies: Designing A Research Strategy', *AI & Society,* vol. 8.3.

Rebecchi, E. (1995). Difficulties and Potentialities of Group Work, *AI & Society,* vol. 8.3.

Reich, R. B. (1992). *The Work of Nations.* Vintage Books, NY.

Roth, S. (1992). 'Japanization or Going Our Own Way?' in Hans-Bockler-Stitfung (ed.) *Japanization or Going Our Own Way? Internationalization and Interest Representation.* Dusseldorf

Telljohann, V. (1995). 'Labour Oriented Research', *AI & Society*, vol. 8.3

van Beinum, H. (1993). 'The Kaleidoscope of Work Place Reform' – in Naschold, F. *et al.*, *Constructing the New Industrial Society*. Van Gorcum, Assen/ Maastricht.

von Banemer, S. (1994). 'Firms and Regional Contextual Factors of Industrial Development, Synthesis of the FINE regional network' in *Future of Industry Paper Series*, FOP 373/Vol. 18 Brussels, CEC DG XII FAST programme, Brussels.

Whitwam, D. (1994). interview with *The Right Way to Go Global*, Harvard Business Review, March–April 1994: 143.

Chapter 11

Meanwhile, Out in the *Real* World: Developing a Commercial Human-centred Software Application

David Smith

Health Warning

This chapter is not a high-power academic treatise. It will not attempt to break new ground in philosophy; it will not attempt to moderate the use of every other word by a full bibliographic reference or the '™' symbol; it will not draw on abstruse psychology, sociology or, indeed, any other kind of '-ology'; it will not mention Wittgenstein (well, not again, anyway!). Furthermore, it will not attempt to justify or explain these omissions. So if you require all or any of these things, this is *not* your chapter. Pass by quickly and leave the rest of us in peace!

What I *will* try to do is to present a case-study based on an attempt to design, develop and bring to market, a software application[1] embodying some of the principles of 'human-centredness'. The chapter will lay out and explain some of the challenges, problems and compromises involved in bringing a design concept out of quasi-academic research and into a situation where fairly large numbers of people are willing to use it in their professional work and, even more significantly, to pay for it.

Project MEDICA

AMIGOS grew out of MEDICA: a small project (total budget 300 000 ECU) in the 'Exploratory Action' of the first AIM (Advanced Informatics in Medicine) Programme, which ran under the aegis of Directorate General XIII of the European Commission between 1989 and 1991. Directed by Julian Hilton at the University of East Anglia in, UK, Project MEDICA involved collaboration between partners from the UK, the Netherlands, Spain and Sweden. This multinational team of clinicians, academics and industrialists brought a wide range of expertise to the project, including medical education, clinical psychiatry ethical pharmaceuticals sales and marketing, satellite communications, interactive videodisk production, systems analysis, artificial intelligence and sociocultural research.

We chose psychiatry because it seemed to us to be a field where we might produce something of practical value, as well as academic interest. A number of factors contributed to this judgement. Firstly, because we had a group of colleagues who were young psychiatrists in their first senior 'consultant' posts, and who were interested in research. Secondly because of the complexity of information-flow in this field of medicine, and hence the inherent technical complexity of the domain. Thirdly because of the way that new developments in diagnosis (imaging etc.) therapy (new drugs) and practice culture (multidisciplinary teamwork) were bringing about rapid changes. And finally because psychiatry was still something of a 'Cinderella' discipline, at least as far as access to computing technology was concerned, and did not have a legacy of established systems.

What Project MEDICA Set Out to Achieve

Our long-term aim was to produce an 'Intelligent Diagnostic Assistant', a future-oriented diagnostic support and case-management tool, which would support psychiatrists in both diagnosis and administration by providing them with a range of computer-based tools within a single environment. It would give rapid access to medical records, medical databases and case notes, using a variety of media including text, graphics and video. It would also provide assistance with diagnostic procedures and would support record-keeping and the communication and exchange of information between practitioners.

Project MEDICA was seen as a prototyping stage, aiming to design and develop a specification and working demonstrator of the Diagnostic Assistant, grounded in current psychiatric practice, and capable of commercial exploitation by IT industries within the European Community.

Our project objectives were:

- To develop a model of evolving psychiatric practice in order to facilitate the design process.
- To specify an appropriate hardware and software environment:
  - able to display and interact with stored data;
  - readily available to a substantial user population;
  - making use of hypertext association;
  - embodying active knowledge representation;
  - using current concepts in the design of the user interface.
- To specify methods and standards for producing and interpreting digitised material (particularly PET and CT images) for intelligent psychiatric diagnostic support.
- To produce a 'concept demonstrator'.

What did we Achieve in MEDICA?

Anything that Project MEDICA could do was inevitably going to be limited in scale. Just a look at the millions of dollars spent by Microsoft on the *launch* of Windows 95 is surely enough to convince anybody that an overall budget of 300000 ECU over two years (150000 ECU with 50% funding) is not going to result in any mega-global breakthrough. However, we covered quite a lot of ground, and, *en route*, learned enough to realise that some of our objectives were of more interest to a closed academic community, and did not address many pressing real-world issues, solutions to which were preconditions to the acceptability of any more advanced systems. The work of the project team involved:

- team-building: the establishment of a common project culture and an ethos of total quality assurance;
- detailed study of current and possible future work practices in psychological medicine by a group of clinical psychiatrists in the UK, Catalunya and Euzkady (the Basque Country);
- user requirements analysis for the design of a prototype multimedia diagnostic assistant for psychiatry;
- analysis of the dynamics of successful implementation, including attention to market acceptability and compatibility with existing systems.

This work led to:

- protocols for evaluation of such systems;
- a pragmatic assessment of the role of imaging procedures in current psychiatric practice;
- a partially validated model of current psychiatric consultation procedures in the UK, with outline models for other regions of the EC;
- a project 'vision' of the likely course of the evolution of psychiatric practice during the coming decade;
- a full IEEE functional specification[2] for a Multimedia Medical Diagnostic Assistant (The MEDICA system);
- a partial implementation of the first version prototype of the MEDICA system;
- a technology-transfer model for the marketing of the MEDICA system;
- trial video programmes and data for broadcast by means of the OLYMPUS satellite.

Where Did the 'Human-centredness' Come in?

An important factor in the success of project MEDICA was the project team's effectiveness in developing and sustaining an environment in which it was possible for all team members to evolve and communicate their ideas. This working environment was the outcome of our commitment to the ideal of 'human-centredness' and the principle of 'participatory design'. However, despite our generally favourable orientation to 'human-centredness' as an ideal, it was very difficult to find a clearly-articulated operational definition of the term. Indeed, despite the sheer volume of writing on the subject which has grown up since the publication of Mike Cooley's *Architect or Bee?*, there is still no 'Manual of Human-Centredness' to which systems developers can refer for guidance, and we found the abstract, academic and philosophical orientation of most of the domain literature of little practical help.

After some analysis and discussion, the MEDICA team took the view that human-centred (anthropocentric) design should aim to achieve an effective synergy between human skills and machine capabilities in the fulfilment of human priorities. It should therefore involve the effective design of working procedures in order to achieve the best possible division of functions between humans and machines. This focused us onto trying to achieve a high level of compatibility between human concepts and machine representations at all levels.

Fortunately, 'participatory design' is a little easier to define! It is an approach to the development of human-centred technologies which has emerged under the influence of the 'Dialogue' principle advocated by our Arbetslivscentrum partners, notably Bo Göranzon and his colleagues, and which has been further refined by workers such as Karamjit Gill. The idea of participatory design related to the dynamic processes of knowledge transfer. This transfer involves both the sharing of knowledge and its mediation by professional and social expertise. It requires an understanding of social and organisational frameworks and of the ways in which ideas are communicated with and within the target population.

The system specification which emerged was not the final repository of all knowledge necessary to meet the aims of Project MEDICA; but rather a knowledge-based tool. People interacting with this 'tool', whether as psychiatrists or computer scientists or whatever, created their own knowledge. Designing the system was in effect a process of knowledge transfer, but there was no point at which the knowledge base could be explicitly identified. Instead, it was a function of a social group, and which had properties defined partly by the professional origins and partly by the present situation of the group members.

Two distinct sorts of expertise are involved in medical informatics; namely medical expertise and computer expertise – even today, it is still rather rare to find a person who has both at the same time. The picture is further complicated by the sheer diversity of practice in almost any field of medicine: and psychiatry is even more diverse and complex than most in this respect! Using participatory design within a common working culture enabled us to produce a 'social knowledge base' which effectively integrated the diffuse strands of knowledge and expertise within the project team.

What did we Find?

The Working Culture of Psychiatry in the UK

Psychiatry is an extremely complex field of medical practice and does not readily lend itself to description in algorithmic form. The management and treatment of a patient may involve a range of agencies and actors in a variety of physical locations. Detailed job audits complemented the dialogue workshops and other techniques of participatory design. Using interviews, questionnaires and work shadowing, audits for each of the clinical members of the team revealed that the work of the psychiatrist involves:

- interaction with the patient;
- examining the patient's psychiatric history;
- studying current case notes obtained from a general practitioner and/or from the patient;
- producing an intermediate diagnosis and management plan for the patient;
- monitoring the patient's progress and modifying the diagnosis and management plan as necessary;
- interaction with other healthcare workers.

They indicated that 50% of consultants' time was spent on administration (in the broadest sense), and about 50% on direct patient care. These findings were to become important in shaping the direction of the project.

User Needs

'Requirements Capture' is far from being a trivial matter even in projects where it is possible to begin the design process with a high-level formal systems specification or clearly articulated set of user requirements. In a domain such as psychiatry, where the penetration of NICTs (New Information and Communications Technologies) is only just gaining momentum, it can be both protracted and expensive. We found that our participatory design approach allowed us to accelerate the process.

One of our problems was that potential users were generally quite unaware of what they could expect from the technology. An (unpublished) questionnaire survey carried out for MEDICA among members of a UK 'Computers in Psychiatry' SIG (Special Interest Group) revealed a very low level of 'hands-on' experience of advanced computer systems even among a self-selected group of enthusiasts. Most (82%) of the SIG members sampled had only ever used computers for word processing and as 'number-crunchers'. Their directly stated requirements and expectations tended to range from the banal to the impossible! In fact, the survey revealed near unanimity of opinion that the single device which was contributing most to practice was the word processor, with other applications far behind.

Certain user requirements could be inferred on the basis of a rather superficial analysis of the domain, For example, it was reasonable to conclude that senior clinicians worth their salt should have sufficient competence to handle the vast majority of diagnostic problems unaided by MEDICA. At this level of expertise, therefore, an 'intelligent' component of MEDICA would mainly be needed to provide support for rare case diagnosis: estimated to be no more than 10% of the workload. At lower levels of experience, it was arguable that an 'intelligent diagnostic system' could do more harm than good by obstructing the process by which tacit knowledge is acquired and expertise achieved.

The psychiatric practitioner, at whatever level, uses a lengthy interaction with the patient, together with a complex set of textual records (as well as other less formal types of information) to build up a 'picture' of underlying pathology. The patient is (by definition) often not capable of giving a clear account of his or her condition, and the psychiatrist needs substantial skills in interpreting behaviour and utterances which can be (or appear to be) totally irrational. The propositional knowledge underlying many of these skills is probably capable of formal representation in some kind of explicit knowledge base. However, tacit knowledge also plays a substantial role in diagnosis and subsequent patient management. Indeed, all of the clinicians involved with Project MEDICA attached considerable personal significance to the tacit aspects of their understanding of their various specialisms.

There are obvious limits to the extent to which this whole process could, or even should, be automated, in the sense of its substantial delegation to some kind of 'intelligent' machine. In any case, we found nothing to suggest that there was any demand from psychiatrists for 'automation' of any but the most trivial (and tedious) aspects of diagnosis and patient care. There was, however, considerable interest in functions consistent with what has been labelled 'cognitive support'.

The Principle of Cognitive Support

'Cognitive Support Systems' are systems which support techniques of AI and cognitive science and offer 'knowledgeable support' to human problem-solvers. Our partner: Mike Sharples of Sussex University (and his colleagues) identified the essential features of cognitive support systems as:

- humans and computers collaborate in the performance of a 'cognitively demanding task';

- the system provides support for 'externalising cognition' – showing the user representations of the task structure and of steps already taken towards a solution;
- the system reduces 'cognitive load' on the human user by taking over some demanding but low-level activities, such as routine arithmetic, leaving the user free to concentrate on higher level tasks;
- the system provides support for a familiar activity performed in a familiar way;
- the system supports the human's personal strategies and techniques: it does not impose 'one best way' of working.

This approach to 'intelligent' support technologies steps away from the trend towards the automation of expert domains and represents a major step towards the more 'anthropocentric' (human-centred) application of advanced informatics recently urged on the NICT community by the EC's FAST (Forecasting and Assessment in Science and Technology) Programme. Project MEDICA therefore set out to extend the cognitive support principle into the field of psychiatry. We believed that the MEDICA 'Diagnostic Assistant' system should eventually conform very closely to the criteria for CSS set out above, and this was reflected in the functional specification document.

Our preliminary consultations within the UK psychiatric profession led us to conclude that an acceptable Cognitive Support System for psychiatric practice should eventually fulfil four basic functions:

- case management;
- diagnostic support;
- training;
- clinical audit.[3]

This system should aim to help four main types of user:

- consultants and trainee psychiatrists;
- community psychiatric workers;
- general practitioners;
- patients and their families.

On this view, the MEDICA system would serve several possible groups of users by allowing them to view what was basically the same data-set in a variety of different ways. The different software modules would provide a series of 'windows' on data according to the current requirements of particular practitioners (interpreted here to include all professional medical and paramedical personnel who may have a legitimate interest in a particular patient). Access would be controlled perhaps on the basis of 'need to know'. (This presupposed some measure of agreement among the medical and allied professions concerning the function of medical records). The right of ownership/access on the part of the patient would also need to be taken into account. In view of the obvious complexity of the user requirements and the different time-scales for development of various components, we adopted a modular approach to the design of the Diagnostic Assistant. This would allow the early implementation of certain core functionalities and the phased incorporation of new modules. In order to permit adaptation to technological innovations, it was intended that the MEDICA system should support an open systems architecture, using appropriate document interchange protocols and communications standards.

The MEDICA Prototypes

In line with current software engineering practice, we adopted a 'rapid prototyping' approach to systems development. This is a design methodology which closely involves potential users in an accelerated cycle of system evaluation and refinement. Rapid prototyping creates systems which are well-attuned to practitioner requirements, as well as to users' changing perceptions of their own requirements. 'Users' have frequent opportunities to work with programmers' interpretations of their ideas, and this helps to provide an early resolution of the sorts of conflicts which frequently arise between the 'techno-centricity' (even 'gizmology') of programmers, and the users' lack of a technical language in which to articulate their needs and preferences. The prototyping process can serve a 'user education' function where user requirements cannot be defined or inferred *a priori*. As such, it is very much in the spirit of participatory design.

A 'First Version Prototype' was implemented on IBM-compatible PC machines, which appear to represent the installed user base in healthcare practice throughout much of the EC. The prototype was coded in 'C' and provided a WIMP (Windows, Icons Menus & Pointer) environment. Its data structures could support the maintenance of patient records, diagnostic support and 'hypermedia' annotations to records.

The prototype showed the 'MEDICA System' as it would appear to a psychiatrist healthcare professional (HCP) at one point in the treatment of a patient. It provided screen displays of the patient's records (including certain image data), together with 'contact cards' summarising the patient's interactions with various HCPs, and a simple graphic representation of 'patient flow': a time-line representation of the patient's progress and management to date. Limited help facilities and a glossary were implemented.

The early prototypes of the Diagnostic Assistant were delivered on a machine which was even then technologically obsolescent. Despite this, the basic IBM-compatible desktop microcomputer, with VGA and mouse was found to offer a completely satisfactory testbed for the first generation of prototypes. A certain trade-off was involved here. We were, of course, aware of the advantages of working on far more advanced systems: it is always a delight to work on top specification workstations! However, we never once met a clinician who had one (a fact which is frequently overlooked by academic developers). We gained considerably from the fact that our collaborators and evaluators were able to assess prototypes on their own computers in their own normal working environments.

The principal focus of attention at this stage of the development of MEDICA was information retrieval and management, since these represented the fundamental point of contact with existing practice. The prototypes incorporated some image-handling capabilities, but did not support any of the facilities for storing and annotating video images, for learner support or knowledge management which were planned for future implementations. However, the modular systems architecture was able to cater for these features as they were developed.

What we Learned From the MEDICA Prototypes

The Role of Imaging in Psychological Medicine

One of the first messages to emerge from the prototyping process was the irrelevance, in the short to medium terms, of imaging technology to the routine practice of psychiatry. Modern neuro-imaging techniques such as Computerised Tomographic Nuclear Magnetic Resonance (CTNMR) and PET (Positron Emission Tomography) have an evolving role in psychiatry, but the relevant hardware is both scarce and expensive, and these technologies are likely to remain relatively under-developed at least in the short to medium term. Despite this, CTNMR, PET and other imaging techniques, such as 'Brain Mapping' (computerised analysis and integration of electroencephalogram, or EEG, records) seem promising in the long run. In addition, the analysis of video and other moving images may become increasingly valuable in the diagnosis of movement disorder and non-verbal communications.

For the time being however, these techniques are likely to remain minor tools as far as routine clinical practice is concerned. The position will undoubtedly change as the technology becomes more accessible and as the quality of images improves. This latter point is very important here. One clinician informed us that he would not accept any system which gave him less image resolution than conventional X-rays (equivalent to 2000 × 2000 pixels). We have since found this to be typical of the requirements of expert practitioners.

Whilst it is true that this criterion may be relaxed in cases (such as radio-imaging) where low-resolution images provide valuable information which would not otherwise be accessible, it nonetheless establishes a minimum acceptable performance for image processing within the Diagnostic Assistant. The data storage and processing requirements associated with these informal *de facto* 'standards' are significant short-term constraints on the MEDICA system.

The Ownership of Records

Traditional roles in the healthcare professions have given rise to a situation where doctors, who carry out diagnoses, therefore 'own' the diagnosis. This ownership then defines the status and function of other healthcare professionals. Patients, and particularly psychiatric patients, tend to figure quite low in the 'information order' so created.

NICT applications can reverse this by making records more generally accessible than is possible with the traditional written notes. Indeed, the problems raised by the assertion of a personal right of ownership of one's medical records more or less defines an NICT-based solution such as 'smart-card' records. Quite how this right might be asserted by psychiatric patients will be a difficult question to answer in some cases, but it cannot be avoided and it has been given very careful consideration by the MEDICA team.

The Impact of Computers on Psychiatric Workers

The interaction between people and their tools is very complex. This is especially so in the case of tools which may radically alter the nature of professional practice in a particular domain. New technologies may influence not only the way people work, but also the ways in which they perceive themselves. This is an aspect of systems design, development and implementation to which projects can occasionally be quite insensitive: sometimes with catastrophic results. We have tried to bear in mind the simple adage that 'designing a computer system also involves designing jobs'. There is a particular responsibility in the case of systems such as the MEDICA system, where the jobs being designed are in a highly specialised professional domain. It would be quite unacceptable for lay people to assume responsibility for dictating the course and nature of psychiatric practice, and this is a major ethical issue which our working methodology has allowed us to tackle with some success.

The Pace of Systems Development

I have argued repeatedly (in *AI & Society* and elsewhere) that IT applications specialists need to break away from the mindset where the term 'system' refers only to a specific configuration of hardware and software, with real people consigned to the role of rather unreliable dependent peripherals. On the contrary, the human-centred approach demands that the developers and designers view the 'system' essentially as a human entity within the MEDICA projects, and it is an ideal we continue to maintain.

Such an approach has a number of implications. It means, for example, that a project team does not simply identify the cleverest thing it can possibly do and then try to graft it on to the outside of an existing human-activity system – at least, not with any real prospect of success. At the same time, it does not mean delivery of minimal retrograde solutions or complete inert surrender to technophobia and professional 'not invented here' attitudes. Instead, it implies a careful attention to two quite distinct rhythms: that of technological evolution and that of the evolution of practice culture.

It is the latter which has the slowest pace. People who only ever work at the leading edge of technology seem to find it difficult to appreciate just how far they are from what we have called 'the state of the desk'. I have already mentioned the lack of high-specification workstations in psychiatric practice and even of any operational experience of advanced IT applications among practitioners. This is in many ways typical of a large number of professional domains. Technology is not neutral. It may have implications (both good and bad) for the evolution of practice in a particular profession domain which are not apparent at all to technologists, and which are, indeed, only apparent to practitioners as the results of insights derived from the creative application of considerable expert analysis. These insights may then drive the evolution of practice (and hence its calls upon technology) in quite different directions from those originally anticipated. This is exactly what happened in MEDICA, and what continues to happen now.

The 'year zero' approach is seldom helpful.

After MEDICA

The exploratory phase of MEDICA must be seen as the first stage in a larger project, the fundamental objective of which was to bring the system defined in the functional specification to commercial implementation and to ensure its effective transfer into routine use in psychiatric healthcare. The principal thrust of our subsequent strategy was directed towards these targets.

This is where the real work began! MEDICA left us with a mass of experience, a set of design specifications and some early prototypes, but no product. Bringing a product to market took a further two years of intensive development and the investment of a very large sum of money. None of this came from public or quasi-public sources, except that a grant of 100000 ECU under the EC VALUE programme enabled us to investigate the feasibility of using MEDICA as a data-capture system for pharmaco-epidemiological study, as well as to develop a drug information module as an on-line facility via MEDICA.

We decided, on grounds of cost and user acceptability, to adopt a development path based on existing hardware technology and established software engineering methodologies. This, together with the proven participatory design approach, reduced the risk of developing unique, idiosyncratic or non-standard systems architectures and components. In addition, the modular approach adopted for the basic architecture of the MEDICA System reduced the difficulty of integrating new functionalities.

Now called 'AMIGOS', the first version of the fully commercial system was launched in Spring 1993. It has established itself as a tool of choice, particularly among small multidisciplinary mental-health teams. Currently, its most cost beneficial application seems to be in support of management audit and communications functions. In particular, instant access to, updating, and transmission of patient records offers significant time savings and more effective use of specialist time. It provides assistance with the procedures of medical diagnosis and record keeping and support for the communication and exchange of information between practitioners, allowing psychiatrists and other healthcare workers in the field of psychological medicine to annotate and interrogate case notes in interesting and complex ways.

However, we are still a long way from implementing the full Diagnostic Assistant as defined in Project MEDICA. We hope that the system will eventually use a variety of media, including still and moving images to create personal stores of case-related and general medical information. We believe that satellite video and data channels will eventually be used to update the medical information. But (and it is a *big* 'but') we can only do this if we generate enough sales (and, therefore, enough profits). We cannot invest losses! This means careful attention to where actual and potential users are, both financially and technologically. Getting too far ahead or behind the 'state of the desk' can be equally disastrous commercially.

This is where our original commitment to human-centredness and participatory design have proven commercially as well as intellectually productive.

The Way Ahead

Nobody can doubt that there is a substantial role for New Information and Communications Technologies (NICTs) in helping healthcare workers in organising, presenting and communicating relevant information, as well as in providing assistance with their decision-making. This is just as true in psychiatry as elsewhere in healthcare practice. However, it is important not to regard devices such as computer systems in isolation. Human factors are critically important in determining the extent to which a technical innovation can be transferred into routine practice. This is not simply a matter of attention to the ergonomic aspects of design, though these are certainly important. It is clear that the design and development of such systems must proceed in the light of some conception of the broader social, professional and cultural contexts within which they will be embedded.

Project MEDICA, and its commercial offspring AMIGOS, provides evidence that this can be achieved in a way which is commercially viable and which nevertheless remains true to the general working principles of human centredness.

Notes

1 The product in question is 'AMIGOS' (Advanced Medical Information Organisation & Guidance System), a psychiatric patient-management and professional information system, marketed by my company, AVC Multimedia Ltd.

2 A 'Functional Specification' is a software engineering tool which defines the characteristics of the proposed system in terms which can serve as a design brief for detailed programming and hardware configuration. The full functional specification for the MEDICA system comprised Project Deliverable Five (January 1990). This specification follows the ANSI/IEEE Guide to Software Requirements Specifications (ANSI/IEEE Std. 83-1984). It provides a general description of the MEDICA system, under the headings: System perspective; System functions; User characteristics; General constraints; Assumptions and dependencies. It also provides specific requirements for each of the hardware and software sub-systems.

3 'Clinical Audit' involves attention to the effectiveness of treatment. It is a concept derived from the growing trend towards 'evidence-based medicine'.

Index

Made in the USA
Monee, IL
13 September 2026

62975069R00280